Electrical and Electronic Principles 4/5

S. A. Knight B.Sc. (Hons) Lond.

Lately Lecturer in Mathematics and
Electronic Engineering, Bedford
College of Higher Education

Heinemann : London

William Heinemann Ltd
10 Upper Grosvenor Street, London W1X 9PA

LONDON   MELBOURNE   JOHANNESBURG   AUCKLAND

First published by Butterworth & Co. (Publishers) Ltd 1982
First published by William Heinemann Ltd 1986

British Library Cataloguing in Publication Data
Knight, S. A.
    Electrical and electronic principles 4/5
    1. Electric engineering
    I. Title
    621.3   TK145

ISBN 0 434 91059 7

Printed in Great Britain by
Whitstable Litho Ltd, Whitstable, Kent

Preface

This book covers the light current topics for the Higher TEC A2/A3 programme in Principles for Electrical and Electronic Engineering. The contents are a direct development from Level 3 Electrical and Electronic Principles and it is assumed that this level has been adequately reached. Because of the very wide coverage of the guide syllabus the book cannot pretend to be in any way exhaustive, but every effort has been made adequately to explain and stress the fundamentals of each section. In this way a common groundwork in Principles for both Level 4 and Level 5 is provided in a single volume.

The main difficulty that technician students have with this subject is the associated mathematics, but mathematics is unavoidable however much we may wish it otherwise, particularly at the higher levels covered by this book. So a compromise is necessary: a qualitative introduction to each section of the syllabus with sufficient mathematics to support the theory as it develops, together with many worked examples to illustrate every relevant part of the text, many self-assessment problems, graded from the easy to the difficult, at the end of each chapter, and, wherever possible, descriptive paragraphs to illuminate the mathematical concepts at those points where the mathematics alone would probably prove difficult. Such points, in the author's experience, are always found in (i) the appreciation and use of *j*-notation, (ii) transmission line theory, (iii) coupled circuit theory.

Consequently, the book opens with an introductory chapter on *j*-notation and its elementary applications. Chapters on resonant circuit theory, networks and coupled circuits follow where *j*-notation is extensively developed. Attenuators and filter systems precede transmission lines proper and provide the necessary introduction to the line as a low-pass filter network. Hyperbolic functions have been avoided here for the reason that this treatment, while precise and very elegant to the mathematically minded, is difficult for the average technician to understand, and is uninformative in that it does not describe to him (as a non-mathematician) what is actually happening as a signal passes along the line. For the same reason the dot notation associated with coupled circuits has been omitted. These aspects do not mean that the treatment is superficial; on the contrary, the treatment is extensive within the limits imposed by the length of the book and possibly enhanced by the omission of *inessential* mathematics. As a consequence of this philosophy, some small degree of mathematical licence has occasionally been taken with some of the proofs and theorems given throughout the book.

This approach clearly may not satisfy the believer in mathematical rigour, but it is hoped that it will satisfy the technician who is not equipped with mathematics beyond elementary calculus but who can put two and two together for himself when some of its higher pressures are suitably deflated.

The writing of this book has brought some not inconsiderable domestic stress and (so I am told) some rather disagreeable behaviour on my own part. I can only hope that this public admission will compensate my wife in some small way for the upheaval she has patiently endured over the many months of the book's gestation.

S. A. K.

Contents

1 Symbolic notation

The fundamental equations for the amplitudes, impedances and phase angles of linear series and parallel circuits made up from combinations of resistive, inductive and capacitive elements and supplied from sinusoidal voltage or current generators have been the subject of earlier parts of the course. There, alternating quantities have been represented in the form of vectors (phasors) and the a.c. analysis of circuits has been made by the construction of phasor diagrams. The solutions of circuit problems have then been reduced to one of two basic methods: the phasor diagrams have been drawn to an appropriate scale, and magnitudes and angles have been measured directly from the diagrams, length representing magnitude and direction representing phase; or the diagrams have been resolved into mutually perpendicular components and the solutions obtained by the application of the rules and identities of trigonometry.

While such phasor diagrams are of value in appreciating the nature and form of elementary problems, their tendency to become complicated and unwieldy as the complexity of circuit systems increase limits their field of direct application. For involved systems it becomes necessary to be able to express circuit behaviour in a formal algebraic symbolism or *symbolic notation* as it is called. Such a notation must not only be able to express both the magnitude and phase conditions of the voltages and currents under examination in an unambiguous manner, but the results obtained must be immediately decipherable in terms of these quantities. What is known as complex algebra or *j-notation* is ideally suited to this purpose.

For a large part of this section of the course, circuit theory and analysis will be carried out using complex symbolism. A brief introduction to the method may usefully be made at this point. It will be assumed that the reader is already familiar with the basic rules of complex algebra.

As is conventional, capital letters, e.g. V or I, will be used to denote any voltage or current which in the circumstances may be taken to remain constant over the interval of time considered. Where the quantities vary such capitals will denote r.m.s. values. Instantaneous values will be denoted by small letters, e.g. v or i, so that, for example, $v = f(t)$ indicates that the value of voltage v at any instant t (the instantaneous voltage) can be expressed as a function of t. In the particular case of a sinusoidal function, v might satisfy the expression

$$v = \hat{V} \cdot \sin \omega t$$

where $\hat{V}$ denotes the peak value or maximum amplitude of the voltage and ω the angular frequency of the voltage. The r.m.s. value of this voltage is, of course, given by $\hat{V}/\sqrt{2}$.

1.1 THE *j*-OPERATOR

The so-called *j*-operator and the *j*-notation in which it is used is often regarded at first sight as something of extreme difficulty

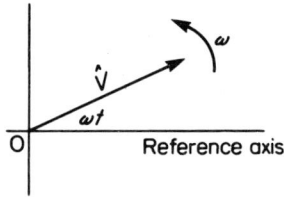

Figure 1.1

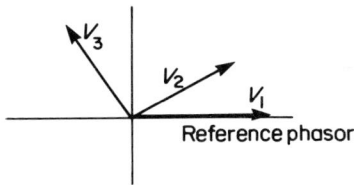

Figure 1.2

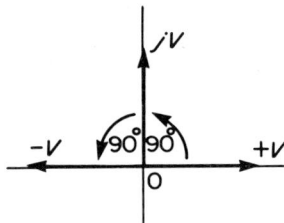

Figure 1.3

and savouring of mathematical 'trickery'. This is not so, and once the fundamental principle of the method is understood, there is no difficulty in applying its powerful technique to the solution of a great number of circuit problems.

An alternating voltage or current acting across or in a circuit system containing resistance and reactance is only completely specified when both the magnitude and the phase relationship (relative to some reference quantity) of the voltage or current is known. As illustrated in a phasor diagram, these conditions are easily identified. The phasor illustrated in *Figure 1.1*, for example, represents the voltage given by

$$v = \hat{V} \cdot \sin \omega t$$

When several phasors are involved, representing sinusoidal voltages (or currents) of the same frequency, it is usual to refer all the phasors to one of the group, this being the *reference* phasor. This reference is conventionally drawn along the positive horizontal axis and each of the other phasors is given an appropriate length and phase angle relative to those of the selected reference. Three phasors are illustrated in this way in *Figure 1.2*. The resultant phasor and its phase relative to the reference axis is then determined by the usual scaled diagram or trigonometric calculation. Similarly, if v and i represent, respectively, the sinusoidal voltage applied to a linear circuit and the current that flows in it, the ratio v/i is the circuit impedance z and the phase angle of this impedance, which is the angle by which the voltage phasor leads the current phasor, is denoted φ. For a passive network φ may have any value between $\pm 90°$ ($\pm \pi/2$ rad). If φ lies between $0°$ and $+90°$ the circuit is inductive; if φ lies between $0°$ and $-90°$ the circuit is capacitive. Any symbolic notation must clearly be able to distinguish between resistance and reactance and between inductive and capacitive reactance.

Since multiplication of a phasor by -1 reverses its direction of action or rotates it through $180°$, it follows that phasors acting along the same straight line can immediately be distinguished from each other by appropriate algebraic signs. Thus a voltage phasor of r.m.s. magnitude V can be expressed as $+V$ when acting horizontally from left to right or by $-V$ when acting horizontally from right to left (see *Figure 1.3*). [*Note. It is important to realise that a phasor quantity, unlike a vector quantity, e.g. force or velocity, does not have a direction associated with it as indicated by the arrowed end. Multiplication by -1, however, reverses it in the sense that $-V$ is antiphase to $+V$ and hence its phase has been shifted by $180°$.*]

Clearly the phasor $-V$ represents in magnitude and direction the effect of multiplying $+V$ by -1. Another phasor is seen lying along the vertical co-ordinate axis in the diagram. This is of the same length (magnitude) as $+V$ but has been turned only through an angle of $+90°$. What multiplying factor has brought about this particular rotation?

Let this multiplier be symbolised j; then $+jV$ represents the vertical phasor. But if phasor $+jV$ is in turn multiplied by j it will, by definition, rotate a further $+90°$ and become phasor $-V$. Hence, *two* successive multiplications by j are equivalent to a single multiplication by -1, whence $j^2 = -1$ and $j = \sqrt{-1}$. We cannot resolve or simplify further the square root of −1 (or of any other negative number for that matter), hence j is known as an 'imaginary' number. In many ways this is an unfortunate choice of word. Nevertheless, we accept it as a *symbol* which represents an abstract but at the same time a very practical idea, namely, that of an *operator* which acts to rotate a phasor in a positive (anticlockwise) direction through an angle of 90°. Operator j can be dealt with in calculations, as can any other algebraic quantity, manipulated in exactly the same way as 'real' numbers, and it obeys all the rules of arithmetic with the added convention that $j^2 = -1$.

The formal rules for the basic operation of j are very simple:—

$$j^2 = j \times j = -1$$
$$j^3 = j \times j \times j = j^2 \times j = -j$$
$$j^4 = j^2 \times j^2 = -1 \times -1 = +1$$
$$j^5 = j^2 \times j^3 = -1 \times -j = +j$$

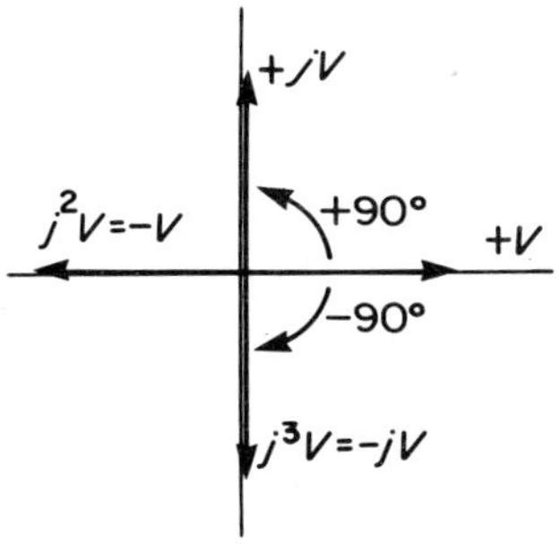

Figure 1.4

The table now repeats. Odd powers of j are therefore wholly imaginary, being equal to either $+j$ or $-j$, while even powers of j are wholly real, being equal to either $+1$ or -1. So, for any phasor quantity such as voltage V in *Figure 1.4*

jV = phasor rotated 90° ahead of phasor V
$j^2V = -V$ = phasor rotated 180° ahead of phasor V
$j^3V = -jV$ = phasor rotated 270° ahead of phasor V
$j^4V = V$ = phasor rotated back to its original position

From the third row we see that a multiplication by j^3 is identical to a multiplication by $-j$. Hence, phasor V can be considered not to have advanced 270° ahead of V by multiplication by $-j$, but to have rotated through 90° in the negative (clockwise) direction instead. It follows that successive *clockwise* rotations follow from successive multiplications by $-j$.

It is useful to keep in mind that all imaginary numbers can be built up from the basic unit j just as all real numbers are built up from the basic unit 1. For example, $\sqrt{-25} = \sqrt{25}\sqrt{-1} = j5$; $\sqrt{-7} = \sqrt{7}\sqrt{-1} = j\sqrt{7}$ and so on. Also, since $j^2 = j \times j = -1$, then $1/j = -j$.

1.2 THE COMPLEX PLANE

What is known as the complex plane is simply a convenient graphical form of representing real and imaginary numbers. Ordinary Cartesian axes can only be used for real numbers, but with a small modification, attributed to Robert Argand, a Parisian bookseller, Cartesian axes can be adapted to accommodate imaginary numbers as well. The *Argand diagram* as it is called is similar to the ordinary Cartesian plane except that the vertical axis is now used to represent imaginary numbers and is known as the imaginary axis.

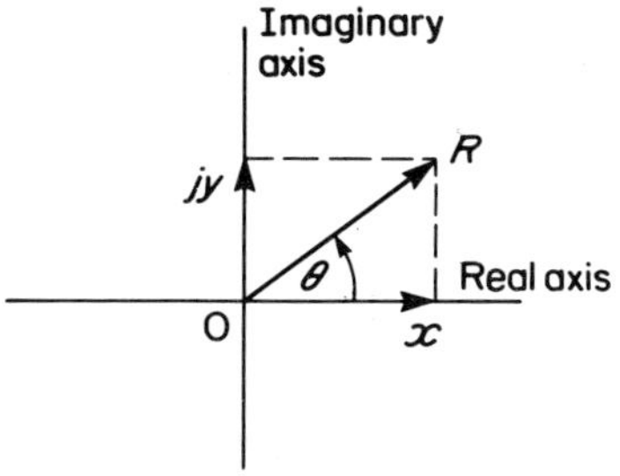

Figure 1.5

Figure 1.5 represents a phasor quantity $0R$ drawn on a complex plane. The horizontal and vertical components of $0R$ are x and jy, respectively. Thus x represents a length x along (or parallel to) the real axis, and jy a step of length y along (or parallel to) the imaginary axis. Combining these lengths, the resultant becomes, by the parallelogram rule of phasor addition,

$x + jy$ represented in the diagram by $0R$. Hence we may write
$$0R = x + jy$$
An expression of this kind, containing both real (x) and imaginary (jy) parts, is known as a *complex number*.

If now the phasor $0R$ is a voltage phasor V, then
$$V = x + jy$$
and the real and imaginary components represent, respectively, the voltages associated with the resistive and reactive components of the circuit.

As a phasor may be specified in this way in the complex Cartesian plane, so may it equally be specified in polar co-ordinates, that is, in terms of its length (magnitude) from the origin of axes and the angle which the phasor makes with the horizontal axis. The magnitude is known as the *modulus*, denoted by $|V|$, and the angle as the *argument*, denoted by θ, of the phasor. There is a simple relationship between polar co-ordinates and Cartesian co-ordinates and these may be derived from a consideration of *Figure 1.6*. Let the polar co-ordinates of phasor V be $V\underline{/\theta}$ and the Cartesian co-ordinates (x, y);

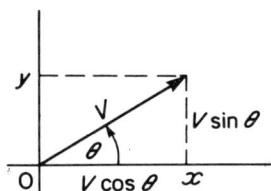

Figure 1.6

then $x = V \cos \theta, \quad y = V \sin \theta$

But $V = x + jy = V\underline{/\theta}$

whence $V\underline{/\theta} = V \cos \theta + V j \sin \theta$
$$= V(\cos \theta + j \sin \theta) \qquad (1.1)$$

The conversion from Cartesian to polar co-ordinates is equally simple for, referring again to *Figure 1.6*,
$$V = \sqrt{(x^2 + y^2)} \qquad (1.2)$$
$$\theta = \quad \tan^{-1} \frac{y}{x}$$

It is particularly important to remember that (1.1) is $V(\cos \theta + j \sin \theta)$ and *not* $V(\sin \theta + j \cos \theta$ and that in (1.2) $V = \sqrt{(x^2 + y^2)}$ and *not* $\sqrt{(x^2 + (jy)^2)}$.

Example 1. Write the phasor $4\underline{/30°}$ in Cartesian form, and the phasor $3 + j4$ in polar form.

$$4\underline{/30°} = 4(\cos 30° + j \sin 30°)$$
$$= 4\left[\frac{3}{2} + j\frac{1}{2}\right]$$
$$= \underline{6 + j^2} \qquad \text{Answer}$$

Let $x + jy = 3 + j^4 = V$, say

Then $|V| = \sqrt{(x^2 + y^2)} = \sqrt{(3^2 + 4^2)} = 5$

Also $\theta = \quad \tan^{-1} \frac{4}{3} = 53.1°$

$\therefore$ $3 + j4 = \underline{5\underline{/53.1°}} \qquad \text{Answer}$

Care is often needed when calculating the angle θ, since $\tan \theta = \tan (180° + \theta)$. The possibility of ambiguity can be avoided by considering the quadrant in which the phasor lies. As x and y are both positive in this example, the phasor lies in the first quadrant and the solution $\theta = 53.1°$ is clearly the correct one. If there is any doubt, always make a sketch diagram for

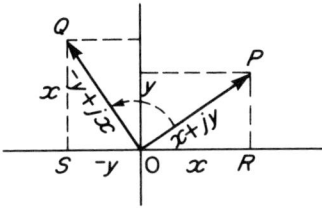

Figure 1.7

yourself. A formula such as $\tan^{-1}y/x$ is often a snare for the unwary.

Example 2. Show that any phasor of the form $x + jy$ is rotated through 90° when multiplied by j.

In *Figure 1.7* let the line $0P$ represent the phasor $(x + jy)$. Then
$$j(x + jy) = jx + j^2y = jx - y$$
replacing j^2 by -1. But phasor $-y + jx = 0Q$ and so triangles $0PR$ and $0QS$ are similar. The angle $Q0P$ is therefore 90°.

1.3 CIRCUIT APPLICATIONS

We know that current lags or leads on the voltage by 90° in purely inductive or capacitive reactance, respectively, but is in phase with the voltage in pure resistance. Hence, resistance and reactance in an a.c. circuit can be represented as mutually perpendicular axes forming a complex plane. Consider first the three basis cases of purely resistive, inductive and capacitive elements.

In a pure resistance, V and I are in phase (*Figure 1.8(a)*) and their phasors both lie along the positive horizontal axis. The circuit opposition to the flow of current is then simply $R = V/I$ and phase angle $\varphi = 0°$. In Cartesian and polar forms
$$R = R + j0$$
$$R = R\underline{/0°}$$

Both of these forms are real and independent of frequency. This is a very elementary statement and corresponds, in r.m.s. terms, exactly with that of a d.c. circuit.

In an inductance, current lags on the voltage by 90° (*Figure 1.8(b)*), so the inductive reactance X_L must be of such a

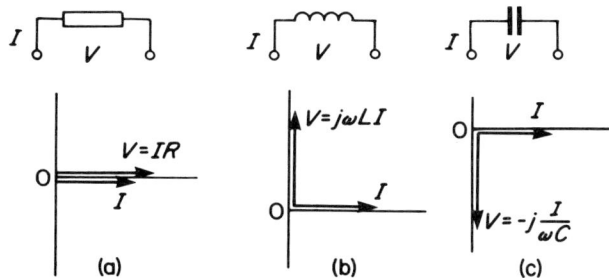

Figure 1.8

character that it produces a voltage phasor when multiplied by the current, since $V = I \cdot X_L$. This means that the inductive reactance acts as an operator which rotates the current phasor through +90°. But such an operation is equivalent to multiplication by j, hence
$$\text{inductance reactance} = jX_L = j\omega L$$
which is imaginary and *increases* with frequency. In polar form
$$X_L = \omega L\underline{/90°}$$

The conditions for V and I in a capacitive reactance are shown in *Figure 1.8(c)*. Since reactance X_c is that value by which the current must be multiplied in order to obtain voltage, it must be associated with an operator which will rotate the current phasor through $-90°$. This is equivalent to a multiplication by $-j$ and we have in this case

$$\text{capacitive reactance} = -jX_c = -\frac{j}{\omega C} = \frac{1}{jC\omega}$$

This quantity is also imaginary and *decreases* with frequency. In polar form

$$X_c = \frac{1}{\omega C} \angle{-90°} \text{ or } \frac{1}{\omega C} \angle{90°}$$

We see from these three fundamental cases that we may treat the phasor resistance of a resistive circuit as real, but the phasor reactance of either a purely inductive or capacitive circuit as wholly imaginary. This may appear strange at first sight, but since resistance will dissipate power while reactance will not, it is perfectly reasonable not to expect the dissipation of real power in an imaginary circuit 'resistance'.

Reactance cannot be isolated from resistance in real life components, particularly so in the case of large inductors; the phase difference between current and voltage is then less than $90°$, the exact value depending upon the relative magnitudes of resistance and reactance present in the circuit. Whether the angel will be positive or negative will depend upon the nature of the dominant reactance and also upon the circuit configuration. We can, in any event, confine our attention to angles lying within the first and fourth quadrants of the complex plane.

1.3.1 Series combination

It is now clear that the complex plane and the j-operator provide a means of distinguishing between resistance and reactance coupled at the same time with an algebra that is relatively simple to manipulate and interpret. Using the real positive horizontal axis for the in-phase condition of the resistive circuit, an inductive (positive) reactance of X_L ohms may be expressed as jX_L ohms and set along the imaginary positive vertical axis, while a capacitive (negative) reactance of X_c ohms may be expressed as $-jX_c$ ohms and set along the imaginary negative vertical axis. Thus the complex impedance Z of an inductive circuit, with resistance, is expressed by

$$Z_L = R + jX_L = R + j\omega L$$

and of a capacitive circuit with resistance

$$Z_c = R - jX_c = R - \frac{j}{\omega C}$$

Example 3. A coil has an inductance of $10\,\text{mH}$ and a resistance of $50\,\Omega$. If the applied frequency is $1600\,\text{Hz}$, express the circuit impedance in complex form:

$$Z = R + j\omega L = R + j2\pi fL$$
Here
$$2\pi fL = 2\pi \times 1600 \times 10 \times 10^{-3}$$
$$= 100\,\Omega$$
Hence
$$Z = 50 + j100 = \underline{50(1 + j2)\,\Omega} \qquad \text{Answer}$$

Example 4. A sinusoidal supply of frequency $1\,\text{kHz}$ and e.m.f. $200\,\text{V}$ is applied to a coil whose inductance is $10\,\text{mH}$ and resistance $25\,\Omega$. By finding the impedance in the form $R + jX$, obtain the circuit current in complex form and interpret the answer on a phasor diagram:

$$Z = R + j\omega L = 25 + j2\pi \times 10^3 \times 10 \times 10^{-3}$$
$$= 25 + j20\pi = 25 + j62.8\,\Omega$$

Now
$$I = \frac{E}{Z} = \frac{200}{25 + j62.8} \text{ A.}$$

Rationalising, by multiplying numerator and denominator by the conjugate of the denominator:

$$I = \frac{200(25 - j62.8)}{(25 + j62.8)(25 - j62.8)} = \frac{5000 - j1256}{457}$$
$$= 1.09 - j2.75 \text{ A} \qquad\qquad \text{Answer}$$

This is the required expression for the circuit current in complex form. The real part of this expression, 1.09 A, is the component of current which is in phase with the voltage; the imaginary part, $-j2.57$ A, is the component of current which is lagging by 90° on the voltage (implied by the negative j).

In the phasor diagram drawn in *Figure 1.9*, these two components of current are shown on their appropriate axes. The phase angle φ is determined from the ratio of the current component magnitudes.

The *modulus* of the impedance and the current in the circuit (as it would be read on an ammeter) are, respectively,

$$|Z| = \surd(R^2 + X_L^2) \quad = \surd(25^2 + 62.8^2)$$
$$= \quad 67.6\,\Omega$$
$$|I| = \surd(1.09^2 + 2.75^2 = \surd(1.19 + 7.56)$$
$$= \quad 2.96 \text{ A}$$

The previous examples are elementary illustrations of how, by making use of complex impedance, the various circuit components are distributed. Most problems can be solved using the same general method.

A summary of the basic complex impendances follows. It should be emphasised that knowledge of the magnitude and phase angle of an impedance gives no indication of the nature of the resistive, inductive or capacitive components forming the circuit. It merely indicates the overall effect at the frequency concerned.

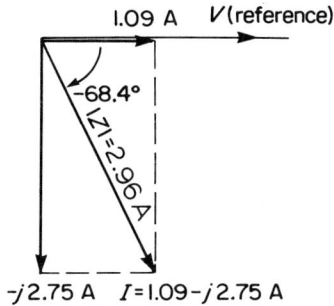

1.09 A V (reference)

$-68.4°$

$|Z| = 2.96$ A

$-j2.75$ A $I = 1.09 - j2.75$ A

Figure 1.9

1.3.2 Series combinations

(1) Inductance in series with resistance (*Figure 1.10*):

$$Z_1 = R \qquad\qquad Z_2 = +jX_L$$
$$Z = Z_1 + Z_2$$
$$\therefore \qquad Z = R + jX_L = R + j\omega L$$

(2) Capacitance in series with resistance (*Figure 1.11*):

$$Z_1 = R \qquad\qquad Z_2 = -jX_c$$
$$Z = Z_1 + Z_2$$
$$\therefore \qquad Z = R - jX_c = R - j\frac{1}{\omega C}$$

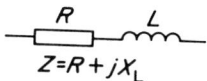

R L

$Z = R + jX_L$

Figure 1.10

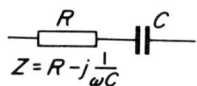

R C

$Z = R - j\frac{1}{\omega C}$

Figure 1.11

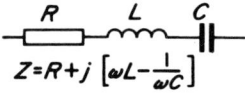

$$Z = R + j\left[\omega L - \frac{1}{\omega C}\right]$$

Figure 1.12

(3) Capacitance, inductance and resistance in series (*Figure 1.12*):

$$Z_1 = R \qquad\qquad Z_2 = +jX_L \qquad\qquad Z_3 = -jX_c$$
$$Z = Z_1 + Z_2 + Z_3$$
$$\therefore \qquad Z = R + jX_L - jX_c$$
$$= R + j\left[\omega L - \frac{1}{\omega C}\right]$$

Notice that, for series connected elements, the total circuit impedance Z is, analogous to series connected resistors, the sum of the individual impedances.

1.3.3 Parallel combinations

The analogous case for parallel connected resistors may be applied in the case of parallel impedances, that is

$$\frac{1}{Z} = \frac{1}{Z_1} + \frac{1}{Z_2} + \frac{1}{Z_3} + \quad\cdots$$

or, where only two impedances are involved, the familiar product-over-sum formula may be used, that is

$$Z = \frac{Z_1 Z_2}{Z_1 + Z_2}$$

It is often simpler, however, when dealing with parallel combinations to work in terms of admittance rather than impedance. The admittance of an a.c. circuit is the reciprocal of the impedance, denoted by Y, hence $Y = 1/Z$. For an inductive circuit, therefore, since $Z = R + jX_L$

$$Y_L = \frac{1}{Z} = \frac{1}{R + jX_L}$$

and rationalising

$$Y_L = \frac{R - jX_L}{(R + jX_L)(R - jX_L)}$$
$$= \frac{R - jX_L}{R^2 + X_L^2}$$
$$= \frac{R}{R^2 + X_L^2} - j\frac{R}{R^2 + X_L^2}$$
$$= G + jB$$

where G is the *conductance* (the resistive component of the admittance) and B is the *susceptance* (the reactive component of the admittance). Thus

$$G = \frac{R}{R^2 + X_L^2} \quad \text{siemen}$$

and

$$B = \frac{-X_L}{R^2 + X_L^2} \quad \text{siemen}$$

Notice that if the reactance X_L is zero, then $G = 1/R$, as in the case of a d.c. circuit. Also notice that B in this case is a capacitive reactance.

If the impedance is capacitive, then replacing X_L by X_c in the above calculation, we find

$$G = \frac{R}{R^2 + X_c^2} \quad \text{siemen}$$

$$B = \frac{X_c}{R^2 + X_c^2} \quad \text{siemen}$$

and B is inductive.

(1) Capacitance in parallel with resistance (*Figure 1.13*):

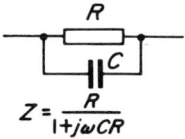

Figure 1.13

$$Z_1 = R \qquad\qquad Z_2 = -jX_c$$

Then $\qquad Y_1 = \dfrac{1}{Z_1} = \dfrac{1}{R} \qquad\qquad Y_2 = \dfrac{1}{Z_2} = -\dfrac{1}{jX_c}$

Hence, total admittance $Y = Y_1 + Y_2 = \dfrac{1}{R} - \dfrac{1}{jX_c} = \dfrac{1}{R} + j\omega C$

Hence $\qquad\qquad\qquad\qquad Z = \dfrac{1}{Y} = \dfrac{1}{\dfrac{1}{R} + j\omega C}$

$$= \frac{R}{1 + j\omega CR}$$

(2) Inductance in parallel with resistance (*Figure 1.14*):

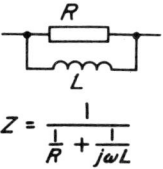

Figure 1.14

$$Z_1 = R \qquad\qquad Z_2 = jX_L$$

Then $\qquad Y_1 = \dfrac{1}{R} \qquad\qquad Y_2 = \dfrac{1}{jX_L}$

Hence, total admittance $Y = Y_1 + Y_2 = \dfrac{1}{R} + \dfrac{1}{jX_L}$

Hence $\qquad\qquad\qquad\qquad Z = \dfrac{1}{Y} = \dfrac{1}{\dfrac{1}{R} + \dfrac{1}{j\omega L}}$

We shall deal with the more complicated parallel circuits as they occur in the following pages.

Example 5. A circuit consists of an inductor of inductance 200 mH and resistance 50 Ω connected in series with a 10 μF capacitor. If a current of 0.15 A flows in the circuit at a frequency of 100 Hz, calculate the applied voltage and the circuit phase angle.

The total circuit impedance is $Z = R + j\omega L - \dfrac{j}{\omega C}$

Now $\quad R = 50\,\Omega$

$$X_L = j\omega L = j2\pi \times 100 \times 0.2 = j40\pi$$
$$= j125.6\,\Omega$$

$$X_c = \frac{-j}{\omega C} = -j\frac{10^6}{2\pi \times 100 \times 10} = -j159.2\,\Omega$$

The total circuit impedance then becomes

$$Z = 50 + j125.6 - j159.2$$
$$= 50 - j33.6\,\Omega$$

The negative j-term indicates that the capacitive reactance predominates. Since $V = IZ$ we have

$$V = 0.15(50 - j33.6)\ \text{V}$$
$$= 7.5 - j5.04\ \text{V} \qquad\qquad \text{Answer}$$

The phase angle is $\quad \varphi = \tan^{-1} -\dfrac{5.04}{7.5} = \tan^{-1} -0.672$

$$\varphi = \underline{-33.9^\circ} \qquad\qquad \text{Answer}$$

Example 6. The solenoid of an electrically driven switch, designed for operation from a 100 V, 50 Hz supply, has an inductance of 2 H and a series resistance of 250 Ω under these conditions. What power is being dissipated in the solenoid?

The impedance of the solenoid $Z = R + j\omega L$

$$= 250 + j2\pi \times 50 \times 2$$
$$= 250 + j200\pi = 100(2.5 + j2\pi)\,\Omega$$

Taking the 100 V supply as reference we can write $V = 100\underline{/0°}$ or $100 + j0$ V. Then, since $I = V/Z$

$$I = \frac{100}{100(2.5 + j2\pi)} = \frac{1}{2.5 + j2\pi} \quad \text{A}$$

The power dissipated is equal to voltage × in-phase component of the current, and this latter can be found by rationalising;

$$I = \frac{2.5 - j2\pi}{45.7} = 0.055 - j0.137 \quad \text{A}$$

The *real* term is the in-phase component, hence the power dissipated is

$$P = 0.055 \times 100 = \underline{5.5 \text{ W}} \qquad \text{Answer}$$

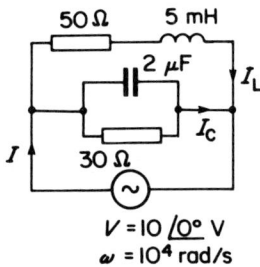

50 Ω 5 mH

2 μF

I_L

I_C

I

30 Ω

$V = 10\underline{/0°}$ V

$\omega = 10^4$ rad/s

Figure 1.15

Example 7. For the circuit shown in *Figure 1.15*, express the impedance of the inductive branch and the admittance of the capacitive branch in complex phasor form and hence calculate the current taken from the supply and its phase angle relative to that of the applied voltage.

The impedance of the inductive branch $Z_L = R + j\omega L$

$$= 50 + j10^4 \times 5 \times 10^{-3}$$
$$= 50 + j50 = 50(1 + j)\,\Omega$$

The admittance of the capacitive branch $Y_C = \dfrac{1}{Z_C}$

$$= \frac{1}{R} + j\omega C = \frac{1}{30} + j10^4 \times 2 \times 10^{-6}$$
$$= 0.033 + j0.02 \text{ S}$$

The total circuit current is the complex sum of the separate branch currents I_L and I_C. Hence

$$I = I_L + I_C = \frac{V}{Z_L} + V \cdot Y_C$$

$$= V\left[\frac{1}{50(1 + j)} + 0.033 + j0.02\right]$$

$$= V\left[\frac{1}{50}\frac{1 - j}{2} + 0.033 + j0.02\right]$$

$$= V[0.01 - j0.01 + 0.033 + j0.02]$$

$$= V[0.043 + j0.01]$$

From the diagram, $V = 10$ V. Hence

$$I = 10[0.043 + j0.01] = 0.43 + j0.1 \quad \text{A}$$

Hence $\qquad |I| = \sqrt{(0.43^2 + 0.1^2)} = \underline{0.44 \text{ A}} \qquad \text{Answer}$

$$\varphi = \tan^{-1}\frac{0.1}{0.43} = \tan^{-1} 0.2326$$

$$\varphi = \underline{13.1°} \qquad \text{Answer}$$

1.4 EQUATING REAL AND IMAGINARY PARTS

Two complex numbers are equal to each other *only* if the real parts are equal and the imaginary parts are equal, each to each. For let $x + jy$ and $a + jb$ be two equal complex numbers, where x, y, a, b are all real. Then

$$x + jy = a + jb$$

Rearranging $\qquad x - a = -j(y - b)$

Squaring $\qquad (x - a)^2 = -(y - b)^2$

This situation implies that a positive real number is equal to a negative real number, which is impossible unless both sides are zero. Hence

$$x - a = 0, \qquad y - b = 0$$
or $\qquad x = a, \qquad\qquad y = b$

It follows that, when two complex numbers are equal, we can equate their real parts and their imaginary parts.

The next example illustrates this technique.

Example 8. In the a.c. bridge circuit shown in *Figure 1.16* the inductance L and self-resistance R of a coil are being measured by adjustment of capacitor C_1 and resistor R_1, resistors P and Q being fixed and known. Assuming that the bridge is correctly balanced, find the values of L and R in terms of P, Q, R_1 and C_1.

When a bridge is balanced, the product of opposite arms of the bridge are equal, hence in this case the bridge is balanced when

$$PQ = (R + j\varphi L)\,\frac{1}{\dfrac{1}{R_1} + j\omega C}$$

Cross-multiplying $\quad \dfrac{PQ}{R_1} + j\omega C_1 PQ = R + j\omega L$

Equating the *real parts* on each side gives us $R = (PQ)/R_1$, and equating the *imaginary parts* gives us $L = PQC_1$.

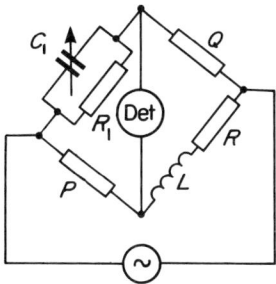

Figure 1.16

1.5 PRODUCTS AND QUOTIENTS IN POLAR FORM

Finding products and quotients of complex numbers expressed in Cartesian form is often tedious and time-consuming, particularly in the case of quotients where the necessity of rationalisation of the denominator often leads to difficulty. If complex numbers are expressed in polar form, multiplication and division are both greatly simplified.

It may be shown that if $Z_1 = r_1\underline{/\theta_1}$ and $Z_2 = r_2\underline{/\theta_2}$, then the product

$$Z_1 Z_2 = r_1 r_2 \underline{/\theta_1 + \theta_2}$$

That is, to multiply two (or more) complex numbers expressed in polar form, *multiply* the moduli and *add* the angles.

For division, the quotient

$$\frac{Z_1}{Z_2} = \frac{r_1}{r_2}\underline{/\theta_1 - \theta_2}$$

That is, to divide two complex numbers expressed in polar form, *divide* the moduli and *subtract* the angles, taking due note of the signs attached to the angles.

Example 9. If $Z_1 = 15\underline{/35°}$ and $Z_2 = 22.5\underline{/-46°}$, find Z_1Z_2 and Z_1/Z_2. Express both answers in Cartesian form.

$$Z_1Z_2 = 15\underline{/35°} \times 22.5\underline{/-46°}$$
$$= (15 \times 22.5)\underline{/35 + (-46)} = 337.5\underline{/-11°}\,\Omega$$

$$\frac{Z_1}{Z_2} = \frac{15\underline{/35}}{22.5\underline{/-46}} = \frac{15}{22.5}\underline{/35-(-46)}$$
$$= 0.67\underline{/81°}\,\Omega$$

Converting to Cartesian form:
$$337.5\underline{/-11°} = 337.5(\cos(-11°) + j\sin(-11°))$$
$$= 337.5(0.982 - j0.191) = 331.4 - j64.5\,\Omega$$
$$0.67\underline{/81°} = 0.67(\cos 81° + j\sin 81°)$$
$$= 0.67(0.156 + j0.987) = 0.105 + j0.66\,\Omega$$

Summary. This chapter has covered the basic use of the *j*-operator in the solutions of elementary a.c. problems. Throughout the remainder of the book, *j*-notation will be used as and where necessary. It is essential, then, that the contents of this introductory chapter are thoroughly understood so that the following chapters may be worked through without undue difficulty.

A number of self-assessment problems now follow.

PROBLEMS FOR CHAPTER 1

(1) Represent the following complex numbers as points on an Argand diagram, and express them in polar form: (a) $1 + j$, (b) $3 + j4$, (c) $12 + j5$, (d) $4 - j3$, (e) $-3 - j6$, (f) $(2 + j)^2$, (g) $(1 - j)(2 + j3)$.

(2) Express the following in the form $a + jb$: (a) $Z_1 = 2\underline{/30°}$, (b) $Z_2 = 5\underline{/120°}$, (c) $Z_3 = 1.75\underline{/-95°}$. Using the polar numbers calculate: (i) Z_1Z_2, (ii) $Z_1 Z_2 Z_3$, (iii) Z_1/Z_2, (iv) Z_3/Z_1.

(3) Impedances Z_1 and Z_2 are connected first in series and then in parallel. Express their combined impedance in each case in the form $x + jy$ given that $Z_1 = 4 - j3$, $Z_2 = 5 + j2$.

(4) Write down the complex impedances of the circuits shown in *Figure 1.17*, given that $\omega = 10^4$ rad/s in each case.

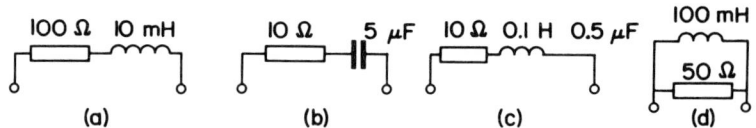

100 Ω 10 mH   10 Ω 5 μF   10 Ω 0.1 H 0.5 μF   100 mH / 50 Ω

(a)   (b)   (c)   (d)

Figure 1.17

(5) If the admittance of a circuit is given by
$$Y = \frac{1}{628} + \frac{1}{j200\pi}\ \text{S},$$
express the circuit impedance Z in polar form.

(6) The current in a circuit component is $(3.6 + j1.2)$ mA when the applied voltage is $(3 - j2)$ V. Calculate the impedance Z of the component and its modulus and phase angle.

(7) Express the impedance $Z = \dfrac{1 - j\sqrt{3}}{1 + j}$ in the form $(x + jy)$.

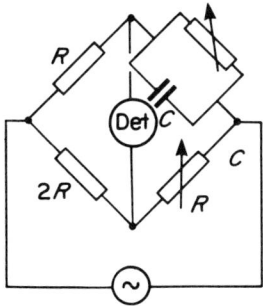

Figure 1.18

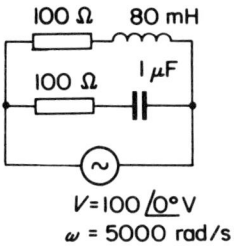

$V = 100 \angle 0° \text{V}$

$\omega = 5000 \text{ rad/s}$

Figure 1.19

Find also $|Z|$ and its phase angle.

(8) A circuit consisting of a 0.1 μF capacitor shunted by a 1000 Ω resistor is connected across a 50 V, 800 Hz supply. Show that the circuit impedance is $100(8 - j4)\,\Omega$.

(9) Given that $R = 240\,\Omega$, $L = 8\,\text{mH}$, $G = 1.8 \times 10^{-3}\,\text{S}$ and $C = 0.12\,\mu\text{F}$, calculate the value of the impedance Z in polar form, taking ω to be 10^4 rad/s, if

$$Z = \sqrt{\left(\frac{R + j\omega L}{G + j\omega C}\right)}$$

(10) A circuit has an impedance $Z = 10 + j17.3\,\Omega$. Express this as an admittance and state the conductance and susceptance of the admittance.

(11) Three impedances $8 + j10$, $6 + j5$, $8 - j6\,\Omega$ are wired in parallel and the feed current is found to be 10 A. Calculate the current in each branch of the circuit.

(12) The admittance of a circuit is $0.04 - j0.075\,\text{S}$. Find the values of the resistance and the reactance in the circuit if they are (a) wired in series, (b) wired in parallel.

(13) For the circuit shown in *Figure 1.18*, find the current taken from the supply and its phase angle relative to the applied voltage.

(14) Show that the bridge of *Figure 1.19* will balance when the applied frequency is $f = 1/(2\pi CR)$.

2 Network theory

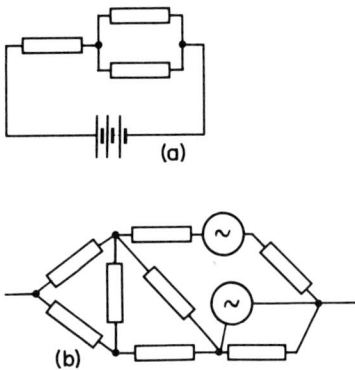

(a)

(b)

Figure 2.1

A network may be defined as any general electrical circuit made up of generator and impedances. The generators may consist simply of batteries or other sources of direct current, or they may consist of alternators, electronic oscillators or the outputs of amplifiers. The impedances may consist simply of ordinary resistors or they may consist of inductive or capacitive reactances in combination with resistive components. Whatever the make-up of the network it will, in general, consist of either a simple network such as that sketched in *Figure 2.1(a)* or it will be a complex network such as that shown in *Figure 2.1(b)*, where a number of generators and impedances are interconnected each with the others.

The basic approach to the solution of network problems lies in the application of Kirchhoff's voltage and current laws, although in the case of a complex network the processes of calculation, involving as they do a number of simultaneous equations, can become lengthy and tedious. Other network theorems have therefore been devised which enable a degree of simplification to be made in the analysis involved, and we shall be mainly concerned with these theorems in this chapter.

2.1 KIRCHHOFF'S LAWS

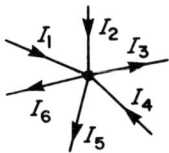

(a)

(b)

Figure 2.2

Kirchhoff's first law (which is known as the current law) states that the *algebraic* sum of the currents meeting at any point (or node) in a network is zero. A *node* is defined as a region in a network where a number of conductors meet. Hence the current law simply means that at any meeting point of conductors, such as is shown in *Figure 2.2.(a)*, the total current flowing into the point must equal the total current flowing out of the point, that is,

$$I_1 + I_2 + I_4 = I_3 + I_5 + I_6$$
or
$$I_1 + I_2 + I_4 - I_3 - I_5 - I_6 = 0$$

Here a positive sign has been assigned to currents flowing into the node and a negative sign to currents flowing out of the node. This convention is quite arbitrary; the signs can be reversed without in any way affecting the validity of the law.

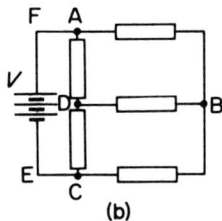

Kirchhoff's second law (the voltage law) is as follows: in moving around any closed loop of a network, the sum of the e.m.f.s is equal to the sum of the products of the resistances of, and the currents in, the various parts of the loop. *Figure 2.2(b)* shows a completely closed circuit. In this circuit there are a number of closed loops, or meshes. ABDA and BCDB are loops. Route ABCEFA is also a loop for that matter. The first two loops do not include the battery, so the total e.m.f. acting in either of them is zero. The other loop includes the battery, so the total e.m.f. acting in it is E volts. Kirchhoff's voltage law simply tells us that we can take any such meshes in a network, add up the *IR* voltage drops as we proceed around it, and equate the sum of these potential drops to the e.m.f. acting in the loop. When applying this law we assume a direction of current flow round the loop, which may be clockwise or anticlockwise, to be positive. On going around the loop in this positive direction, any current found to be flowing, or any voltage found to be acting, contrary to this assumption, must be given a negative sign.

To summarise, then, we apply Kirchhoff's law to a network problem as follows:

(1) Insert current symbols, I_1, I_2, I_3, etc., to represent the unknown currents whose values are required, assigning to them the *probable* directions in which they are flowing and keeping their numbers as small as possible.

(2) With these current distributions marked in, apply the voltage law to as many *closed* loops as there are unknown currents. A number of simultaneous equations can then be set up from which the unknown currents may be calculated.

These procedures are perhaps best illustrated by considering some worked examples.

Example 1. Using Kirchhoff's law, calculate the currents in each branch of the network shown in *Figure 2.3*.

The first step is to insert the unknown currents. The assigned directions are quite arbitrary, but Kirchhoff's current law must be obeyed at the junctions. We assign currents I_1 and I_2, therefore, to the $2\,\Omega$ and $3\,\Omega$ branches, respectively, as shown in the figure; the current in the $5\,\Omega$ branch is then clearly $I_1 + I_2$. Notice that the number of unknown currents has been kept to two, the third current being simply the sum of these.

Taking the loop ABCDA and working clockwise,

$$5(I_1 + I_2) + 2I_1 = 4 \qquad \text{(i)}$$

and for the loop ABFEA including the $3\,\Omega$, the $2\,\Omega$ and the batteries,

$$-3I_2 + 2I_1 = -1 + 4$$

or $$2I_1 - 3I_2 = 3 \qquad \text{(ii)}$$

Make a special point of noticing that the batteries are taken to act in the direction negative-to-positive *inside* the battery, that is, conventional current flow is assumed.

Multiplying (i) by 2 and (iii) by 7 and subtracting,

$$31I_2 = -13$$

$$I_2 = -\frac{13}{31} = -0.419\ \text{A}$$

The negative sign merely indicates that the actual direction of current flow is opposite to that assumed in the diagram. Now substituting for I_2 in equation (i) gives

$$7I_1 + 5\left[-\frac{13}{31}\right] = 4$$

whence $$I_1 = \frac{27}{31} = 0.871\ \text{A}$$

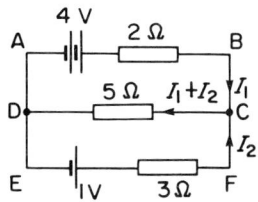

Figure 2.3

Finally
$$I_3 = I_1 + I_2 = 0.871 - 0.419$$
$$= 0.452 \text{ A}$$

The currents are therefore $I_1 = 0.871$ a, $I_2 = 0.419$ A, $I_3 = 0.452$ A with the direction of I_2 actually opposite to that assumed.

2.1.1 Maxwell's cyclic currents

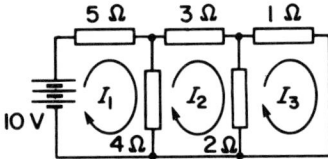

Figure 2.4

This is a method of applying Kirchhoff's laws which often helps in the solution of more complicated networks. A current is assumed to flow in a clockwise direction in *every* closed loop of the network; *Figure 2.4* shows a network with circulating currents I_1, I_2 and I_3 added in this manner. Where a component belongs only to a particular loop, the current in that component is equal to the loop current. In the diagram, the $5\,\Omega$ resistor, for example, carries only current I_1. If, however, a component is common to two loops, the current in that component is equal to the difference between the two loop currents. In the diagram, the $2\,\Omega$ resistor, for example, carries a current $(I_2 - I_3)$, flowing downwards. Equations may be set up for the three loops as follows:

For the I_1 loop, we have $10 = 5I_1 - 4(I_1 - I_2)$ and for the I_2 and I_3 loops, respectively,
$$0 = 3I_2 + 2(I_2 - I_3) + 4(I_1 - I_2)$$
$$0 = I_3 + 2(I_3 - I_2)$$
Gathering up the terms and rearranging
$$10 = 9I_1 - 4I_2$$
$$0 = 4I_1 - 10I_2 - 2I_3$$
$$0 = 2I_2 - 3I_3$$
The solutions of these equations then give the required loop currents.

2.2 OTHER NETWORK THEOREMS

While the laws of Ohm and Kirchhoff may be considered the fundamental working tools in the solution of network problems, much time can often be saved by the application of certain other theorems which have been specially developed from these laws. A few of the more important of these extended theorems will now be discussed and illustrated with worked examples.

2.2.1 Superposition theorem

This theorem states that in any network made up of linear impedances, the current flowing at any point is the sum of the currents that would flow if each generator was considered separately, all other generators being replaced at the time by impedances equal to their internal impedances.

The use of this theorem permits the solution of network problems without setting up a large number of simultaneous equations as required by the direct application of Kirchhoff's laws, since only one generator at a time need be considered. This does not necessarily mean that the working will be shorter but it will in general be simpler, as each generator can be considered to act independently and its own contribution of current through a particular part of the circuit may be separately calculated.

Returning to the simple network of Example 1 above, we may verify the superposition theorem by considering it to operate in the circuit of *Figure 2.5(a)*, where the earlier network is repeated, this time with Maxwell's currents assumed to be flowing in a clockwise direction through the loops. We require to find, as in the illustration, the current through the 5 Ω branch of the network.

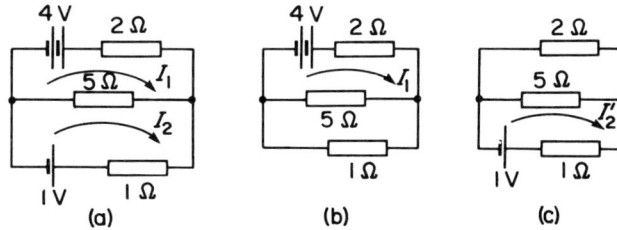

Figure 2.5

Let the current due to the 4 V source acting *alone* be I_1' and let the current due to the 1 V source acting *alone* be I_2'. Then for the 4 V source alone (see *Figure 2.5(b)*), we have from a direct application of Ohm's law

$$I_2' = \frac{4}{2 + \dfrac{3 \times 5}{3 + 5}} = \frac{32}{31} \text{ A}$$

Of this current, a proportion $3/(3 + 5) = \frac{3}{8}$ flows through the 5 Ω branch.

Hence, the current through the 5 Ω branch due to the 4 V source acting alone is

$$\frac{32}{31} \times \frac{3}{8} = \frac{12}{31} = 0.387 \text{ A}$$

For the 1 V source alone (see *Figure 2.5(c)*), we have

$$I_1' = \frac{-1}{3 + \dfrac{2 \times 5}{2 + 5}} = -\frac{7}{31} \text{ A}$$

and of this current a proportion $2/(2 + 5) = \frac{2}{7}$ flows through the 5 Ω branch.

Hence, the current through the 5 Ω branch due to the 1 V source alone is

$$-\frac{7}{31} \times \frac{2}{7} = -\frac{2}{31} = -0.065 \text{ A}$$

The *algebraic* sum of I_1' and I_2' gives the required total current through the 5 Ω branch, that is, $0.387 - (-0.065) = 0.452 \text{ A}$, as calculated by Kirchhoff earlier. The other branch currents may be similarly found.

Notice particularly the signs which appear in the above working; the 4 V source is positive with respect to the assumed flow of current I_1^+; the 1 V source is negative with respect to the assumed flow of I_1^+. The current in the 5 Ω branch is then $I_2' - (-I_2')$.

In a particularly elementary example such as this there is not a lot to choose between either method of working and, if anything, the direct use of Kirchhoff may well be easier for the beginner. It is in the solution of more complex networks that the superposition theorem often proves of greatest benefit.

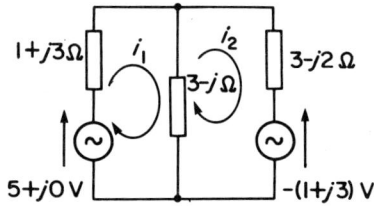

Figure 2.6

Example 2. Calculate the current supplied by the 5 V generator in the circuit of *Figure 2.6*.

Assign currents i_1 and i_2 acting in a clockwise direction as shown in the figure. Then, working round the two loops and taking due note of the generator polarities, we have

$$5 + j0 = i_1(1 + j3 + 3 - j) - i_2(3 - j) \qquad \text{(i)}$$
$$-(1 + j3) = i_2(3 - j + 3 - j2) - i_1(3 - j) \qquad \text{(ii)}$$

Since we require current i_1 (which is supplied by the 5 V generator) we eliminate i_2 by multiplying (i) by $(3 - j + 3 - j2) = (6 - j3)$ and (ii) by $(3 - j)$. This gives

$$5(6 - j3) = i_1(4 + j2)(6 - j3) - i_2(3 - j)(6 - j3)$$
$$-(3 - j)(1 + j3) = -i_1(3 - j)(3 - j) + i_2(3 - j)(6 - j3)$$

and by adding:

$$5(6 - j3) - (1 + j3)(3 - j) = i_1[(4 + j2)(6 - j3) - (3 - j)^2]$$
$$30 - j15 - 3 + j - j9 - 3 = i_1[30 - 8 + j6]$$
$$\therefore \qquad 24 - j23 = i_1[22 + j6]$$
$$\therefore \qquad i_1 = \frac{24 - j23}{22 + j6} \text{ A}$$

Converting to polar form:

$$i_1 = \frac{\sqrt{(24^2 + 23^2)}.\underline{/\tan^{-1} - 23/24}}{\sqrt{(22^2 + 6^2)} \ \underline{/\tan^{-1} 6/22}}$$
$$= \frac{33.24}{22.8} \underline{/-43.8° - 15.3°} = 1.46\underline{/-59.1°} \text{ A}$$

2.3 THÉVENIN'S THEOREM

This theorem states that any two-terminal network consisting of linear impedances and generators may be replaced, as far as the output terminals are concerned, by a single constant voltage generator in series with a single impedance. The theorem is illustrated in *Figure 2.7*. Let a network of impedances and generators be contained inside the box, then the equivalent circuit as viewed at the terminals is as shown on the right. E_{oc} is the voltage measured at the terminals when these are open-circuited, that is, when no load or other circuit is connected. Z_i is the impedance measured between the terminals when all internal generators have been suppressed or replaced by their internal impedances.

Thévenin's theorem is basically an application of the superposition theorem. For if the open-circuit terminal voltage is E_{oc} and the impedance looking back into the network is Z_i when the internal generators have stopped generating, we may, by connecting an e.m.f. equal to $-E_{oc}$ in series with a load impedance Z_L, stop any flow of current into the load from the network terminals. By the superposition principle this zero current may be regarded as the sum of two equal and opposite currents circulated by the internal generators and by $-E_{oc}$. The current circulated by $-E_{oc}$ is plainly $-E_{oc}/(Z_i + Z_L)$ and so the current which could flow by the connection of Z_L would be $E_{oc}/(Z_i + Z_L)$. Hence, the network behaves as though it was the simple series circuit of e.m.f. E_{oc} and internal impedance Z_1.

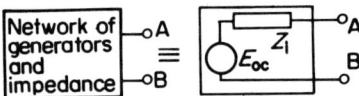

Figure 2.7

Example 3. Deduce the Thévenin equivalent circuit for the terminals AB in the circuit of *Figure 2.8*. Hence, calculate the

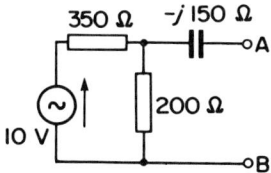

Figure 2.8

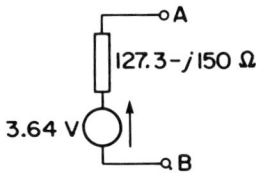

Figure 2.9

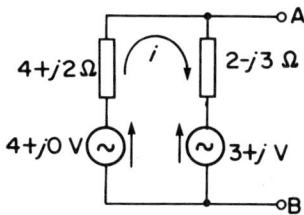

Figure 2.10

current that would flow in an impedance $(100 - j50)\,\Omega$ connected across AB.

When terminals AB are open-circuited, no current flows through the capacitor, hence the voltage across the terminals will be

$$10 \times \frac{200}{200 + 350} = 3.64\,\text{V}$$

The impedance measured across the terminals when the generator is replaced by a short-circuit is

$$\frac{200 \times 350}{200 + 350} - j150 = 127.3 - j150\,\Omega$$

Hence, the Thévenin equivalent circuit for the terminals AB of the given network comprises a generator having an e.m.f. of 3.64 V and an internal impedance of $(127.3 - j150)\,\Omega$ (see *Figure 2.9*).

When an impedance of $100 - j50\,\Omega$ is connected across AB, the current drawn from the network is

$$\frac{3.64}{127.3 - j150 + 100 - j50} = \frac{3.64}{227.3 - j200}\,\text{A}$$

$$\therefore \quad |I| = \frac{3.64}{302.7} = 1.2\,\text{mA}$$

Example 4. Find the Thévenin equivalent circuit for the network illustrated in *Figure 2.10*.

To find E_{oc} at terminals AB we require the voltage across *either* of the parallel branches. For a circulating current i we have

$$(4 + j0) - (3 + j) = i(4 + j2 + 2 - j3)$$
$$1 - j = i(6 - j)$$
$$i = \frac{1 - j}{6 - j}$$

Hence, $E_{oc} = (3 + j) + i(2 - j3)$ *or* $(4 + j0) - i(4 + j2)$. Make certain you understand why there are two possible expressions for E_{oc} like this and that both lead to the same value for E_{oc}.

Taking the second of these

$$E_{oc} = 4 - i(4 + j2) = 4 - \frac{(4 + j2)(1 - j)}{6 - j}$$

$$= 4 - \frac{4 + j2 - j4 + 2}{6 - j} = 4 - \frac{6 - j2}{6 - j}$$

Rationalising, we get

$$E_{oc} = 4 - \frac{(6 - j2)(6 + j)}{37}.$$

$$= 4 - \frac{36 - j12 + j6 + 2}{37}$$

$$= 4 - \frac{38 - j6}{37} = \frac{148 - 38 + j6}{37}$$

$$= \frac{110 + j6}{37} = 2.97 + j0.16\,\text{V}$$

Now use the alternative expression for E_{oc} and show that it leads to the same result.

We require now the impedance looking back into the terminals AB, with the generators short-circuited. This impedance will be $(4 + j2)$ in parallel with $(2 - j3)\,\Omega$. Using the product over sum formula, we have

$$Z_i = \frac{(4 + j2)(2 - j3)}{6 - j} = \frac{8 + j4 - j12 + 6}{6 - j}$$

$$= \frac{(14 - j8)(6 + j)}{37} = \frac{84 - j48 + j14 + 8}{37}$$

$$= \frac{92 - j34}{37} = 2.48 - j0.92\,\Omega$$

2.3.1 Norton's theorem

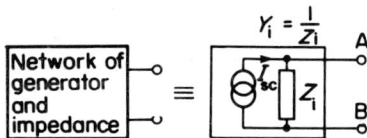

Figure 2.11

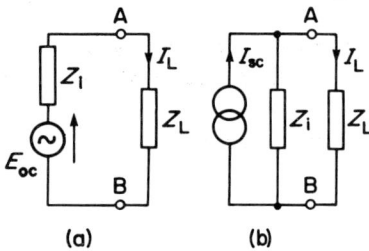

(a)   (b)

Figure 2.12

This theorem is simply the dual of Thévenin's theorem in that it enables a network to be replaced by a single generator and impedance, but in this instance the generator is of the constant-current type and the impedance is in parallel with it. In *Figure 2.11*, the network is shown as a box and Norton's equivalent circuit is shown beside it. The components of this equivalent circuit are found as follows: the generator delivers a *constant* current equal to the current that would flow in a short-circuit connected across terminals AB, I_{sc}. Z_i is the impedance seen between the terminals with all internal generators suppressed and replaced by their internal impedances. As we are concerned with parallel elements here, it is often best to work in terms of admittances. Y_i is the admittance measured between the terminals under the above conditions, hence Y_i is the reciprocal of the Thévenin equivalent Z_i.

The equivalent circuits as obtained by Norton's and Thévenin's theorems give the same current in, and voltage across, a load impedance, and are therefore effectively identical to one another. For, referring to *Figure 2.12(a)*, which is Thévenin's equivalent circuit, we have

$$I_L = \frac{E_{oc}}{Z_i + Z_L}$$

and the current on short-circuit would be

$$I_{sc} = \frac{E_{oc}}{Z_i} \qquad \text{(i)}$$

In *Figure 2.12(b)*, which is Norton's equivalent circuit, the load current

$$I_L = \frac{I_{sc} Z_i}{Z_i + Z_L} \qquad \text{(ii)}$$

Combining (i) and (ii), Norton's load current is

$$I_L = \frac{E_{oc}}{Z_i + Z_L}$$

which is identical with Thévenin's load current.

In any particular problem, either theorem may therefore be used, and the choice is usually one of convenience. A multi-source network is best reduced to a single source equivalent by the repeated application of Thévenin's or Norton's theorems as the following example shows.

Example 5. Calculate the current in the $6\,\Omega$ resistor shown in the circuit of *Figure 2.13*.

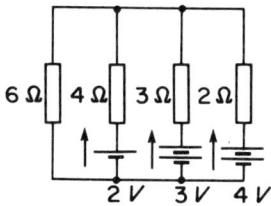

Figure 2.13

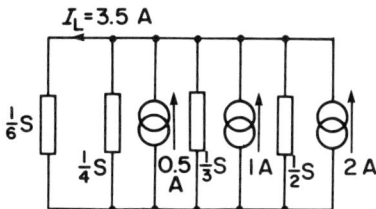

Figure 2.14

As the elements of the network are in parallel branches, the approach to the problem is to convert each branch into its Norton's equivalent. The current sources can then be directly added.

For the $2\,\Omega$ branch, $I_{sc} = \frac{4}{2} = 2\,\text{A}$, and $Y_i = \frac{1}{2}\,\text{S}$

For the $3\,\Omega$ branch, $I_{sc} = \frac{3}{3} = 1\,\text{A}$, and $Y_i = \frac{1}{3}\,\text{S}$

For the $4\,\Omega$ branch, $I_{sc} = \frac{2}{4} = 0.5\,\text{A}$, and $Y_i = \frac{1}{4}\,\text{S}$

For the $6\,\Omega$ load, $\quad Y_L = \frac{1}{6}\,\text{S}$

Figure 2.14 shows the complete Norton's equivalent of the given circuit, From this, the total load current $I_L = 2 + 1 + 0.5 = 3.5\,\text{A}$, and the total admittance $Y = 1/2 + 1/3 + 1/4 + 1/6 = 15/12\,\text{S}$. Hence, the load voltage

$$V = \frac{I_L}{Y} = \frac{3.5 \times 12}{15} = 2.80\,\text{V}$$

Problems concerning a number of current sources in *series* can be dealt with similarly by conversion to Thévenin equivalents. The resulting voltage sources can then be directly added as can the impedances.

A common pitfall in such problems is failure to make a note of the direction of action of currents or voltages. In *Example 5*, for instance, the direction in which the batteries of *Figure 2.13* were acting was upwards; the current generators which replaced them also acted upwards. If one of the batteries had been reversed in polarity, its equivalent current generator would have been shown as acting downwards and its current output would have been subtracted from the other two.

Example 6. By successive conversion between Thévenin and Norton circuits for the branches of the circuit shown in *Figure 2.15*, find a Thévenin equivalent for the terminals AB.

The approach here is to convert first of all the series Norton circuit into a Thévenin equivalent. A current of $1\,\text{mA}$ flowing through $1.5\,\text{k}\Omega$ develops a voltage of $1.5\,\text{V}$; hence, the Thévenin equivalent is as drawn in *Figure 2.16(a)*, a $1.5\,\text{k}\Omega$ resistor in series with a $1.5\,\text{V}$ constant voltage generator. Notice the direction in which this voltage acts, in conformity with the direction that *either* circuit would cause a current to flow in an external load.

To find the Thévenin equivalent of the two parallel branches, the circulating current in this loop alone is clearly

$$\frac{5 - 3}{2500} = \frac{2}{2500}\,\text{A} = 0.8\,\text{mA}$$

Hence, the voltage across the parallel branches (which is equal to the open-circuit voltage of the Thévenin equivalent) is

$$5 - \frac{0.8}{1000} \times 2000 \quad \text{or} \quad 3 + \frac{0.8}{1000} \times 500$$

either of which lead to $3.4\,\text{V}$.

The resistance looking into the parallel branches with the

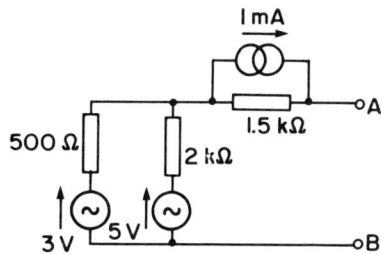

Figure 2.15

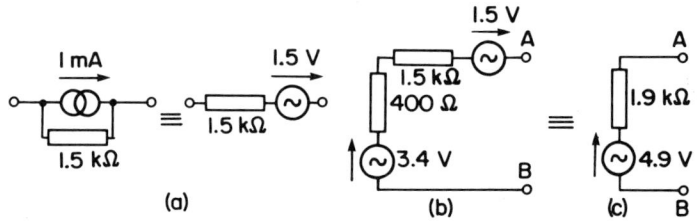

Figure 2.16

generators short-circuited is $(500 \times 2000)/(500 + 2000) = 400\,\Omega$. Thus, the Thévenin equivalent of the whole circuit is as shown in *Figure 2.16(b)*. The e.m.f.s and resistances can now be directly added to give the final equivalent circuit of *Figure 2.16(c)*.

A final word: the equivalence afforded by Thévenin's or Norton's theorems holds for the *current in the load* which is connected to the network and *not* for conditions within the network itself. It is the failure to recognise this limitation that wrong and even ridiculous answers are sometimes obtained to problems when these theorems are applied to their solutions.

2.4 T- AND π-NETWORK TRANSFORMATIONS

When a network has three impedances, these may be arranged either in the form of a T-section or a π-section, each of which is illustrated in *Figure 2.17(a)* and *2.17(b)*, respectively. It is possible to replace a T-section by an equivalent π-section, and

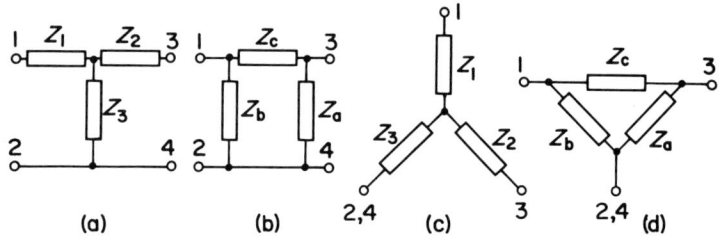

Figure 2.17

conversely, so that as far as the terminals A, B and C are concerned, no external measurements can distinguish between the circuits. Since the T-section may be drawn as a *star*, (*Figure 2.17(c)*), and the π-section as a mesh (or *delta*), the conversion is sometimes known as a star-delta transformation.

2.4.1 π to T transformation

If a T (or star) network is to be equivalent to a π (or delta) network, the impedance measured between any pair of the terminals of *Figure 2.17(c)* must be the same as that measured between the same pair of terminals of *Figure 2.17(d)*. Thus, if we consider terminals 1 and 3 for the two circuits and equate them, we get

$$Z_1 + Z_2 = \frac{Z_c(Z_a + Z_b)}{Z_a + Z_b + Z_c}$$

Similarly, equating impedances for terminals 3 and 4

$$Z_2 + Z_3 = \frac{Z_a(Z_b + Z_c)}{Z_a + Z_b + Z_c}$$

and for terminals 1 and 2

$$Z_3 + Z_1 = \frac{Z_b(Z_b + Z_c)}{Z_a + Z_b + Z_c}$$

From these three equations Z_1, Z_2 and Z_3 may be evaluated in terms of Z_a, Z_b and Z_c. Doing this, we obtain

$$Z_1 = \frac{Z_b Z_c}{Z_a + Z_b + Z_c} \tag{2.1}$$

$$Z_2 = \frac{Z_a Z_c}{Z_a + Z_b + Z_c} \tag{2.2}$$

$$Z_3 = \frac{Z_a Z_b}{Z_a + Z_b + Z_c} \tag{2.3}$$

There is a symmetry about these three expressions which should be noticed: the letter subscripts in the numerators on the right-hand side are those of the impedances connected to the terminals corresponding to the numerical subscripts on the left-hand side. The equations may therefore be memorised thus: the equivalent T (or star) impedance connected to a given terminal is equal to the product of the two π (or delta) impedances *connected to the same terminal* divided by the sum of the π impedances. These equations are used to convert from *delta* into *star*.

2.4.2 T to π transformations

By transposition, the equivalent equations for the star to delta transformations are

$$Z_a = Z_2 + Z_3 + \frac{Z_2 Z_3}{Z_1} \tag{2.4}$$

$$Z_b = Z_1 + Z_3 + \frac{Z_1 Z_3}{Z_2} \tag{2.5}$$

$$Z_c = Z_1 + Z_2 + \frac{Z_1 Z_2}{Z_3} \tag{2.6}$$

Once again, the symmetry of these equations enable us to memorise them easily: the equivalent π (or delta) impedance connected *between* any two terminals is the sum of the T (or star) impedances *connected between those same two terminals*, plus the product of the same two T impedances divided by the remaining T impedance.

It may be shown that any network of impedances, however complex, can be reduced to a T-network of three impedances only. This theorem follows from the fact that the number of meshes in a complex network is reduced by one each time that a π network is replaced by an equivalent T.

Example 7. Derive the star equivalent of the delta network shown in *Figure 2.18*.

The terminals of the delta network are numbered 1, 2 and 3. The actual number positions is not important provided that the numbering on the equivalent star conforms to the requirements of the transforming equations. We now use equations (2.1) to (2.3) above:

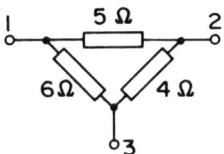

Figure 2.18

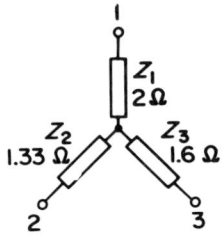

Figure 2.19

For terminal 1　$Z_1 = \dfrac{5 \times 6}{15} = 2\,\Omega$

For terminal 2　$Z_2 = \dfrac{5 \times 4}{15} = 1.33\,\Omega$

For terminal 3　$Z_3 = \dfrac{4 \times 6}{15} = 1.60\,\Omega$

The equivalent star network is then as shown in *Figure 2.19*.
　You might like to convert this star back into the equivalent delta to verify that you finish up with the original impedances.

Example 8. What current will the bridge circuit shown in *Figure 2.20* draw from the 2 V battery?
　This kind of problem can be solved by Kirchhoff's law quite easily, but the star–delta transformation reduces the working considerably. Let the delta mesh ABD be transformed into a star; this transformation is illustrated in *Figure 2.21*, where the calculations are identical with those of the previous example.

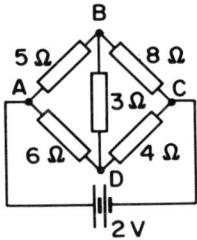

Figure 2.20

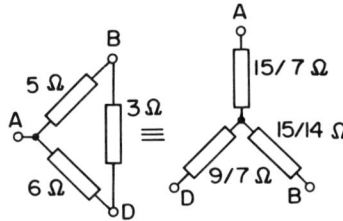

Figure 2.21

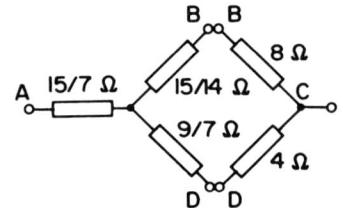

Figure 2.22

Combining the resulting star with the remaining two resistances of the bridge circuit leads to *Figure 2.22*. The circuit can now be reduced to a single resistance immediately:

$$R_{\text{total}} = \frac{15}{7} + \frac{\left[8 + \dfrac{15}{14}\right] \times \left[4 + \dfrac{9}{7}\right]}{8 + \dfrac{15}{14} + 4 + \dfrac{9}{7}}$$

$$= 5.4\,\Omega$$

The current drawn from the battery is consequently
$2/5.48 = 0.365$ A

2.5 REPETITIVE SECTIONS

Let a network be made up of an infinite series of repetitive sections, as shown in *Figure 2.23(a)*. The sections may take one of two forms: the *symmetrical* form in which the electrical

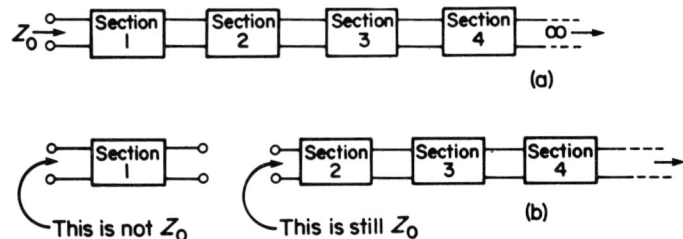

(a)

(b)

Figure 2.23

properties are unaffected by interchanging the input and output terminals, or the *asymmetrical* form in which the input and output terminals cannot be so interchanged without affecting the electrical properties of the section. Symmetrical sections have two important and fundamental electrical characteristics: (i) the *characteristic impedance*, designated Z_0; (ii) the *propagation constant*, designated γ (gamma).

The impedance measured at the input terminals of the infinite network will have a value which will depend only upon the nature and circuit configuration of the individual sections. This impedance is the characteristic impedance of the network Z_0. In due course we will find that Z_0 can be derived from a knowledge of the component values making up the sections of the network.

Suppose now that the first section of the infinite network is disconnected, as shown in *Figure 2.23(b)*. The input impedance of the main network will be unaffected, for the removal of this one section (or for that matter, any finite number of sections) from the infinite series will not affect its infinite nature; the input impedance, in other words, will still be Z_0. The input impedance of the section removed, however, will not, in general, be Z_0. It can be made so either (a) by replacing the rest of the infinite line, or (b) by connecting across its output terminals 3 and 4 an impedance equal in value to Z_0. This second alternative will clearly be true no matter how many (finite) sections are removed from the infinite line, since these sections are effectively terminated by Z_0 when reconnected to the rest of the infinite line. Hence, the properties of an infinite network of sections will be displayed by a finite number of sections terminated in the characteristic impedance Z_0 of the infinite network.

2.5.1 Propagation constant

The propagation constant of a network is the quantity (usually complex) representing the relationship between the input and output currents in the network. In the repetitive system shown in *Figure 2.24*, let the input current be I_i and the output current I_0. Then, for a number of identical sections, the ratios I_i/I_1, I_1/I_2, etc., will be equal, so that each section attenuates at the

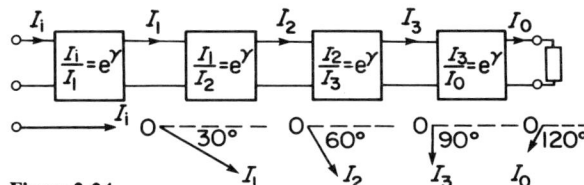

Figure 2.24

same *rate* but not by the same *amount*. Suppose, for example, that we lose half of the input each time we go through a section. Then if we start off with unit input, the output from section 1 is $\frac{1}{2}$ and this is the input to section 2. Here again half the input is lost, so making the output of section 2 equal to $\frac{1}{2}$ of $\frac{1}{2} = \frac{1}{4}$, and so on; the *amount* of loss in each section being different because the magnitude remaining at any point becomes progressively smaller. The attenuation suffered is therefore in the form of a logarithmic decay and for the successive current ratios

$$\frac{I_i}{I_1} = \frac{I_1}{I_2} = \frac{I_2}{I_3} = \ldots = e^\gamma$$

Hence

$$\gamma = \ln \frac{I_i}{I_1}$$

In general, there will be a phase change in each section; hence, the ratio of the current entering a section to that leaving it will be a phasor quantity having both modulus and angle; let

$$\frac{I_i}{I_1} = \frac{I_1}{I_2} = \ldots = e^\gamma = e^{\alpha + j\beta}$$

Then

$$e^\gamma = e^\alpha \cdot e^{j\beta} = e^\alpha(\cos\beta + j\sin\beta)$$
$$= e^\alpha \sqrt{(\cos^2\beta + \sin^2\beta)} \Big/ \tan^{-1}\frac{\sin\beta}{\cos\beta}$$
$$= e^\alpha \underline{/\beta}$$

Then, in general

$$\frac{I_i}{I_0} = e^{n\gamma} = e^{n\alpha}\underline{/n\beta}$$

where there are n sections in a network terminated in Z_0.

The real part of the propagation constant is the *attenuation constant* measured in népers* or decibels. [*Note. It is because it is derived from logarithms to base e instead of to base 10 that the néper is particularly convenient for expressing attenuation along transmission lines where the decay takes an exponential form.*] Hence, for

$$e^\alpha = \left| \frac{I_i}{I_1} \right|$$

$$\alpha = \ln \left| \frac{I_i}{I_1} \right| \text{ népers}$$

$$= 20 \lg \left| \frac{I_i}{I_1} \right| \text{ dB}$$

So for n sections

$$e^{n\alpha} = \left| \frac{I_i}{I_0} \right|$$

$$n\alpha = \ln \left| \frac{I_i}{I_0} \right| \text{ népers}$$

The imaginary part of the propagation constant (β) is the *phase constant* and is the angle (in radians) by which the current leaving a section lags behind the current entering it. Obviously for a phase shift of β per section, the total phase shift for n sections is $n\beta$, as indicated above.

The phasor diagrams shown below the sections of *Figure 2.24* illustrate the meaning of the propagation constant for the case where $\alpha = 1.1$, i.e. $|I_i/I_1| = 3$ and $\beta = \pi/6$ rad, or 30°. The phasor I_i representing the input current rotates by 30° for each section and shortens its length by one-third. The propagation constant here would be expressed

$$\gamma = 1.1 + j.\frac{\pi}{6} \text{ per section}$$

Example 9. Show that a power ratio may be expressed as $\frac{1}{2}\ln[P_1/P_2]$ népers, where P_1 and P_2 are the input and output power levels of a network, respectively. Hence show that 1 dB = 0.115 néper.

Assuming that the input and output resistances of the network are equal, then

$$P_1 = I_1^2 R \text{ and } P_2 = I_2^2 R$$

Hence

$$\frac{P_1}{P_2} = \left[\frac{I_1}{I_2}\right]^2 \text{ and } \sqrt{\frac{P_1}{P_2}} = \frac{I_1}{I_2}$$

$$\therefore \quad \alpha = \ln\left[\frac{P_1}{P_2}\right]^{\frac{1}{2}} = \tfrac{1}{2}\ln\left[\frac{P_1}{P_2}\right] \text{ népers}$$

Decibel power ratios are expressed as $10\lg\left[\dfrac{P_1}{P_2}\right]$ dB

Now

$$\tfrac{1}{2}\ln\left[\frac{P_1}{P_2}\right] = \tfrac{1}{2}\lg\left[\frac{P_1}{P_2}\right] \times \ln 10$$

$$= 2.3026 \times \tfrac{1}{2}\lg\left[\frac{P_1}{P_2}\right]$$

$$= 1.15\,\lg\left[\frac{P_1}{P_2}\right]$$

But

$$\lg\left[\frac{P_1}{P_2}\right] = \frac{\text{decibels}}{10}$$

Hence $\quad$ 1 dB $= 0.115$ néper

From this $\quad$ 1 néper $= 8.686$ dB

Both these relationships should be kept in mind.

2.5.2 Characteristic impedance

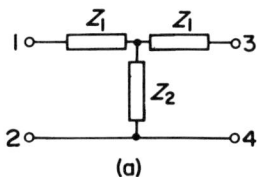

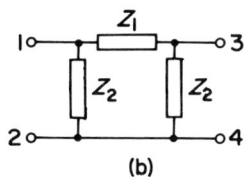

(a)

(b)

Figure 2.25

The forms taken by the individual sections in repetitive networks are, or may be reduced to, T or π circuits, as illustrated respectively in *Figure 2.25(a)* and *2.25(b)*. Used between equal impedances both these sections are symmetrical. For an analysis of these sections, we consider primarily the T-section. Suppose the output terminals of a T-section to be open-circuited, then the input impedance is clearly

$$Z_{oc} = Z_1 + Z_2$$

Let the output terminals now be short-circuited, then the input impedance becomes

$$Z_{sc} = Z_1 + \frac{Z_1 Z_2}{Z_1 + Z_2}$$

These are the extreme cases of termination. If the output is terminated in some finite impedance Z_R, then the input impedance will be

(2.7)
$$Z_i = Z_1 = \frac{Z_2(Z_1 + Z_R)}{Z_1 + Z_2 + Z_R}$$

It is clear from this that if Z_R is altered, Z_i will alter also. It is reasonable, therefore, to ask if there is some value of Z_R which will make Z_i also equal to Z_R. Assume then that when Z_R has some particular value, say Z_k, Z_i has this value also. Substituting Z_k for Z_i and Z_R in equation (2.7) then gives

$$Z_k = Z_1 = \frac{Z_2(Z_1 + Z_k)}{Z_1 + Z_2 + Z_k}$$

whence $\quad Z_k^2 = Z_1^2 + 2Z1Z_2$

But the input impedance Z_k of this section must be equal to the characteristic impedance of the section since, from our earlier definition of characteristic impedance, the section is correctly terminated by Z_k:

$$\therefore \qquad Z_0 = \surd(Z_1^2 + 2Z_1Z_2) \qquad (2.8)$$

Now the product $Z_{oc}Z_{sc} = (Z_1 + Z_2) \cdot \left(Z_1 + \dfrac{Z_1Z_2}{Z_1 + Z_2}\right)$

$$= Z_1^2 + 2Z_1Z_2$$

which, from formula (2.8) is equal to Z_0^2. Hence

$$Z_0 = \surd(Z_{oc}Z_{sc}) \qquad (2.9)$$

These are two most important formulae for they enable the characteristic impedance of a section to be determined from a knowledge either of the section impedances themselves or from the nature of the input impedance when the output terminals are open- and short-circuited in turn. The characteristic impedance of the π-section is also given by formula (2.8), and you should be able to verify this by an analysis similar to that used for the T-section.

Example 10. A T-network is shown in *Figure 2.26*. What resistance R_0 must be placed across terminals 3–4 in order that the resistance measured across terminals 1–2 is also R_0?

The required value of R_0 is clearly the characteristic resistance of the T-section. Here $R_1 = 10\,\Omega$ and $R_2 = 50\,\Omega$, so using formula (2.8) we have

$$\begin{aligned} R_0 &= \surd(R_1^2 + 2R_1R_2) \\ &= \surd(10^2 + 2 \times 10 \times 50) \\ &= 33.16\,\Omega \end{aligned}$$

Try using equation (2.9) for yourself to verify that it leads to the same solution.

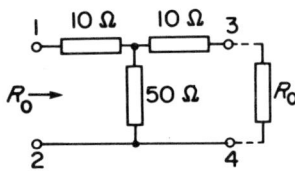

Figure 2.26

2.5.3 Ladder networks

Any repetitive network, or *ladder network*, may be considered as a series of T- or π-sections in cascade. Such a ladder network is shown in *Figure 2.27*. Now either a T- or a π-section can be cut from this network by cutting along the lines AA or BB, respectively. A T-section is obtained by cutting along the centre of any of the successive series arms; a π-section is obtained by cutting along the centre of any of the successive shunt arms. For

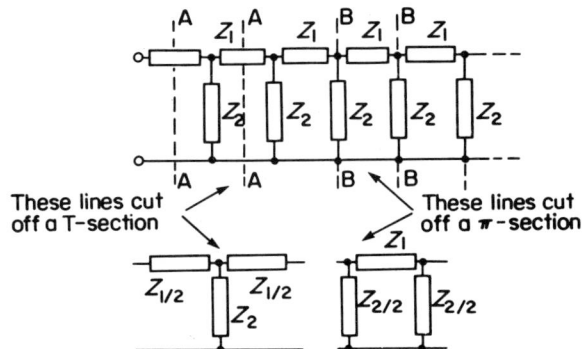

These lines cut off a T-section

These lines cut off a π-section

Figure 2.27

the removed sections, therefore, the T-section elements are $Z_{1/2}$ as the series elements and Z_2 as the shunt element; for the π-section Z_1 is the series element and $Z_{2/2}$ are the shunt elements. Formulae (2.8) and (2.9) are unaffected provided that the appropriate impedance values are inserted.

2.6 MAXIMUM POWER TRANSFER

Any network containing one or more current or voltage sources and linear impedances can be reduced to a Thévenin equivalent circuit. When a load is connected to the terminals of this equivalent circuit, power is transferred from the circuit to the load. We require to find the conditions existing between the effective generator circuit and the load for the maximum power to be transferred to the load.

From our earlier work, when the load and the generator internal impedances are *both* purely resistive, maximum power is transferred when these resistances are equal. When the load is reactive, that is, of the form $Z_L = R_L + jX_L$, the maximum transfer of power depends upon the relationship between R_L and X_L, and there are several cases which have to be separately considered.

Figure 2.28 illustrates the general circuit of a Thévenin equivalent generator and a complex load. We consider three possible load conditions:

(1) R_L *and* X_L *are both independently adjustable.* Since power will be dissipated only in the resistive part of the load impedance, this power will be

$$P_L = |I| R_L$$

where

$$|I| = \frac{E}{\sqrt{[(R + R_L)^2 + (X + X_L)^2]}}$$

If X_L is adjusted to be equal to $-X$, i.e. a reactance of opposite form to that of the source reactance,

$$|I| = \frac{E}{R + R_L}$$

and

$$P_L = E^2 \cdot \frac{R_L}{(R + R_L)^2}$$

Differentiating this

$$\frac{dP_L}{dR_L} = E^2 \cdot \frac{(R + R_L)^2 - 2R_L(R + R_L)}{(R + R_L)^4}$$

and equating the numerator to zero for maximum power in R_L

$$(R + R_L)^2 = 2R_L(R + R_L)$$

whence

$$R = R_L$$

Hence, maximum power transfer is obtained when $X_L = -X$ and $R_L = R$. What we have here is a series resonant circuit, the reactances being equal but of opposite sign. The circuit is consequently purely resistive and the maximum power is transferred when the internal and external resistances are equal.

(2) *Load purely resistive and adjustable.* Here $X_L = 0$. Then

$$|I| = \frac{E}{\sqrt{[(R + R_L)^2 + X^2]}}$$

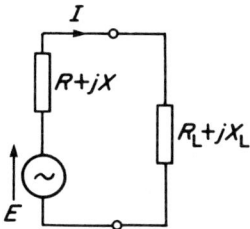

I

$R + jX$

$R_L + jX_L$

E

Figure 2.28

and

$$P_{\mathrm{L}} = E^2 \cdot \frac{R_{\mathrm{L}}}{(R + R_{\mathrm{L}})^2 + X^2}$$

Differentiating

$$\frac{\mathrm{d}P_{\mathrm{L}}}{\mathrm{d}R_{\mathrm{L}}} = E^2 \cdot \frac{[(R + R_{\mathrm{L}})^2 + X^2] - R_{\mathrm{L}}(R + R_{\mathrm{L}})^2}{[(R + R_{\mathrm{L}}) + X^2]}$$

and equating to zero

$$(R + R_{\mathrm{L}})^2 + X^2 = R_{\mathrm{L}}(R + R_{\mathrm{L}})^2$$

from which

$$R_{\mathrm{L}}^2 = R^2 + X^2$$

and

$$R_{\mathrm{L}} = \sqrt{(R^2 + X^2)}$$
$$= |Z_{\mathrm{s}}|$$

where $|Z_{\mathrm{s}}|$ is the modulus of the source impedance.

(3) *Load resistive and adjustable but X_{L} is fixed.* This case is left as an exercise. You should be able to show that maximum power is transferred to the load when

$$R_{\mathrm{L}} = \sqrt{[R^2 + (X + X_{\mathrm{L}})^2]}$$

Example 11. Referring to *Figure 2.29*, calculate the load required for maximum power transfer and the value of that maximum power for the case of a purely resistive load.

This is an example of maximum power being transferred to a load when R_{L} is equal to the source impedance $|Z_{\mathrm{s}}|$. Here the source impedance is $200\,\Omega$ in parallel with a $0.75\,\mu\mathrm{F}$ capacitor. Hence

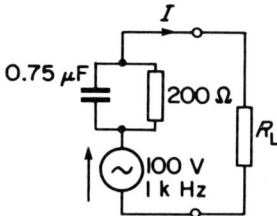

Figure 2.29

$$Z_{\mathrm{s}} = \frac{\dfrac{R}{j\omega C}}{R + \dfrac{1}{j\omega C}} = \frac{R}{1 + j\omega C R}$$

$$= \frac{200}{1 + j2\pi \times 1000 \times 0.75 \times 10^{-6} \times 200}$$

$$= \frac{200}{1 + j0.942} = \frac{200(1 - j0.942)}{1.887}$$

$$= 106 - j100\,\Omega$$
$$= 146\underline{/-43.3°}\,\Omega$$

Hence

$$|Z_{\mathrm{s}}| = 146\,\Omega$$

∴   Maximum power is transferred when $R_{\mathrm{L}} = 146\,\Omega$

Now the total circuit impedance $Z = 146 + 106 - j100$

$$= 252 - j100\,\Omega$$
$$= 271\underline{/-21.6°}\,\Omega$$

∴

$$I = \frac{100\underline{/0°}}{271\underline{/-21.6°}} = 0.37\underline{/21.6°}\ \mathrm{A}$$

∴

$$P = |I|^2 R_{\mathrm{L}} = 0.37^2 \times 146\ \mathrm{W}$$
$$= 19.9\ \mathrm{W}$$

Summary. This chapter has covered a number of network theorems and their application to the analysis and solution of various network problems. Only rarely does a network need to be solved 'completely', and where conditions in only one or two of the network branches are required, the application of Kirchhoff's law or the superposition theorem will usually provide a solution without difficulty.

For complicated networks, the 'reduction' theorems of the star–delta transformations can be used prior to the application of Kirchhoff, and for currents in the load of a network, Thévenin's and Norton's theorems may be used. The analysis of ladder networks with concepts of characteristic impedance and propagation constant will turn up again in later chapters dealing with filters and transmission line theory.

In the following assessment problems, methods of solution are suggested in some cases but not in others. It is useful to work such problems out by alternative methods where possible; in this way not only is the working checked, but experience in choosing the best method is gained.

PROBLEMS FOR CHAPTER 2

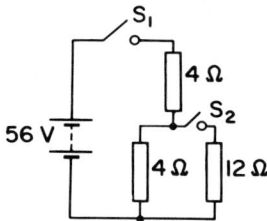

Figure 2.30

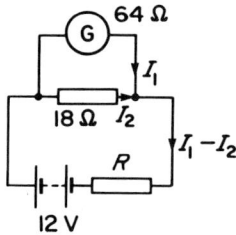

Figure 2.31

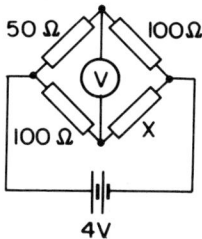

Figure 2.32

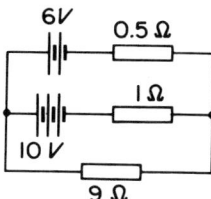

Figure 2.33

(1) In the circuit of *Figure 2.30*, switch S_1 is closed, then after a short while switch S_2 is closed. Show in separate diagrams, the distribution of current (and current values) in the circuit before and after closing S_2.

(2) In *Figure 2.31* the meter G is used to measure a resistance R. Using Kirchhoff's laws, the currents I_1 and I_2 are related by the equations

$$64I_1 - 18I_2 = 0$$
$$18I_2 + R(I_1 + I_2) = 12$$

Verify that these equations satisfy Kirchhoff's laws.

(a) If $R = 84\,\Omega$, calculate the current through the meter; (b) If the full-scale deflection of the meter is $50\,\text{mA}$, what is the smallest value of resistance that can be measured with this circuit system?

(3) In a network the mesh currents I_1, I_2 and I_3 are related by the equation

$$8I_2 - 3I_3 = 10$$
$$5(I_1 - I_2) - 10I_3 = 0$$
$$2I_1 - 3I_2 + 15I_3 = 0$$

Calculate the current I_3 to the nearest milliampere.

(4) In the bridge circuit of *Figure 2.32*, the voltmeter V has an internal resistance of $200\,\Omega$ and when connected as shown reads $1\,\text{v}$. What is the value of resistor X?

(5) State and explain Kirchhoff's laws. A resistance of $22\,\Omega$ is supplied by 16 cells each of e.m.f. $1.5\,\text{V}$ and internal resistance $0.5\,\Omega$. The cells are used in two parallel rows, each of eight cells in series. Find the currents in each branch of the circuit.

If, due to an error, two of the cells in one of the rows were reversed, what will the branch currents be now?

(6) Find the current flowing in each battery of the circuit of *Figure 2.33*, using (a) Kirchhoff, (b) the superposition theorem.

When a certain resistor is connnected in parallel with the $9\,\Omega$ load the current in the 6 V battery branch falls to zero. What is the value of this resistor?

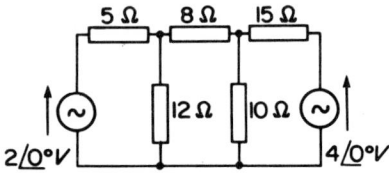

Figure 2.34

(7) Each branch of a delta-connected system has a resistance of 120 Ω. What will be the branch resistances of an equivalent star-connected system?

(8) Using Maxwell's circulating currents, or otherwise, calculate the current in the 10 Ω resistor in the circuit of *Figure 2.34*.

(9) By conversion to Norton equivalent circuits, calculate the voltage across the 5 Ω load resistor in the circuit of *Figure 2.35*.

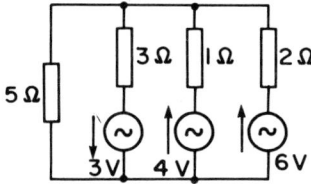

Figure 2.35

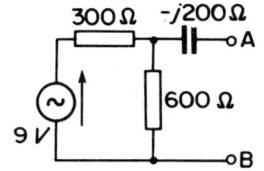

Figure 2.36

(10) State Thévenin's theorem. Deduce the Thévenin equivalent circuit for terminals AB in the circuit of *Figure 2.36*. Hence, calculate the magnitude of the current flowing in an impedance $100(2 - j)\,\Omega$ when it is connected across AB.

What impedance will take maximum power from the source and what will be the value of this power?

(11) A network consisting of generators and linear impedances has two output terminals A and B. The following measurements are made at these terminals.

(a) Open-circuit voltage = 100 mV.
(b) Short-circuit current = 7.7 mA.
(c) A current of 6.4 mA flows in a 5 Ω resistor connected across the terminals.

Deduce the Thévenin equivalent circuit of the network. Hence, find the current that would flow in an impedance $5.39\underline{/68°}$ wired across AB.

(12) Calculate the current flowing in the load resistor of *Figure 2.37* and the power dissipated in this resistor.

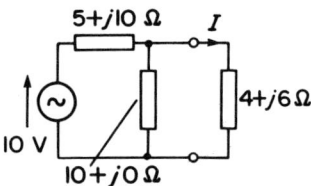

Figure 2.37

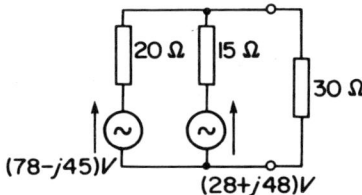

Figure 2.38

(13) Referring to Figure 2.38, what must be the value of the source resistance R if the maximum power is to be dissipated in the 25 Ω load resistor? What will this maximum power be?

(14) Find the Thévenin equivalent circuit for the generator shown in *Figure 2.39*. Hence, calculate the current flowing in the load impedance $(4 + j6)\,\Omega$ connected across the generator terminals.

(15) Explain why it is not possible to find a load value that will dissipate maximum power from the source generator shown in *Figure 2.40*.

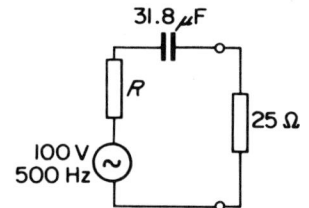

Figure 2.39

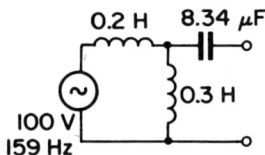

Figure 2.40

(16) Deduce the Thévenin equivalent circuit for the network AB terminals shown in *Figure 2.41*.

(17) In the T- and π-sections shown in *Figure 2.42*, $Z_1 = 10\,\Omega$ and $Z_2 = 40\,\Omega$. Using the formula $Z_o = \sqrt{(Z_{oc}Z_{sc})}$ calculate the characteristic impedance for each network.

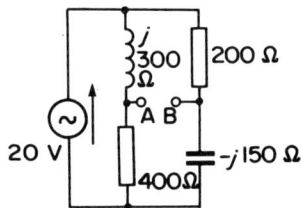

Figure 2.41

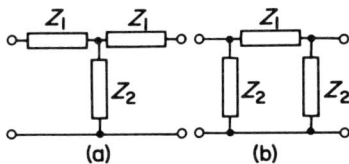

Figure 2.42

Verify that $Z_{0T}Z_{0\pi} = Z_1Z_2$ where Z_{0T} and $Z_{0\pi}$ are the characteristic impedances of the T- and π-sections, respectively.

(18) Now prove in general that $Z_{oT}Z_{o\pi} = Z_1Z_2$, where Z_1 and Z_2 are the same for both a T- and a π-section. (Hint: work out Z_0 for a π-section, then compare the result with equation (2.8) for a T-section.)

(19) In *Figure 2.43*, the angular frequency of the generators is 5000 rad/s and all resistors are $100\,\Omega$. Calculate the currents supplied by each generator in polar form.

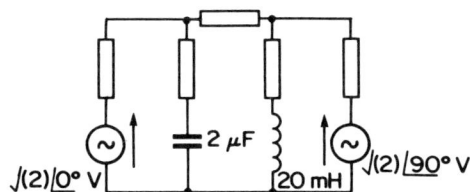

Figure 2.43

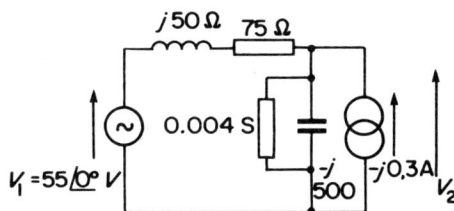

Figure 2.44

(20) Solve the circuit of *Fig. 2.44* for V_2 using Thévenin's theorem and confirm your result by the use of Norton's theorem.

(21) Determine the magnitude and phase of the currents supplied by the two generators in the circuit of *Figure 2.45*, given $\omega = 5000$ rad/s.

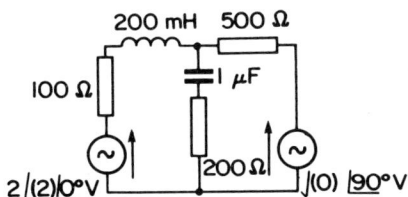

Figure 2.45

3 General circuit theory

We shall be mainly concerned in this chapter with the frequency selective characteristics of two-terminal a.c. steady-state resonant circuits. Resonance occurs in a circuit containing reactances when the input impedance is purely resistive, i.e. when the phase angle between the applied voltage and the input current is zero. Two types of resonance occur: one when the input impedance is a minimum at, or near to, the natural resonant frequency, usually called series resonance; and the other when the input impedance is a maximum at, or near to, the resonant frequency, usually called parallel resonance. The terms 'resonance' for the series case and 'anti-resonance' for the parallel case are used in general since they may be applied to the resonances occurring in feeders, lines and cavities where the actual series and parallel circuits are not identifiable as lumped components. In complex networks such as filters, where a number of resonances may be present, it can be shown that resonant and anti-resonant frequencies will occur alternately.

3.1 FREQUENCY CHARACTERISTICS

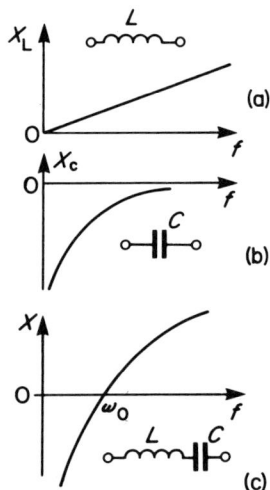

Figure 3.1

The frequency characteristics of simple two-terminal networks can be readily deduced to a first approximation if the presence of resistive elements is neglected. The impedance of any of the remaining elements, or combinations of these elements, is then always purely reactive or, in the case of resonance conditions, zero or infinite.

The reactance X_L of a loss-free coil is ωL; this is directly proportional to frequency and hence the graph of X_L against frequency is a straight line, as shown in *Figure 3.1(a)*. The reactance X_c of a loss-free capacitor is $-1/\omega C$; this is inversely proportional to frequency, so that the graph of *Figure 3.1(b)* showing the variation of X_c against frequency is a rectangular hyperbola. The variation in the resultant reactance of these two elements connected in series is found by adding the ordinates of the individual curves, and this summation is shown in *Figure 3.1(c)*. For a certain value of the frequency denoted by ω_0, the effective circuit reactance is zero. This is the series resonant frequency. For frequencies below ω_0 the circuit is purely capacitive, and for frequencies above ω_0 the circuit is purely inductive.

For parallel connection it is more convenient for purposes of calculation to use the reciprocal of reactance, or the *susceptance*, instead of reactance. The susceptance B_L of a loss-free coil is $-1/X_L$ and the graph of B_L against frequency is shown in *Figure 3.2(a)*. Similarly, the susceptance of a loss-free capacitor, B_c, is given by $1/X_c = \omega C$, and the graph of B_c against frequency is shown in *Figure 3.2(b)*. The resultant susceptance of the two elements connected in parallel is found by adding the ordinates of the individual curves, and this summation is illustrated in *Figure 3.2(c)*. As for the reactance curve of the series circuit,

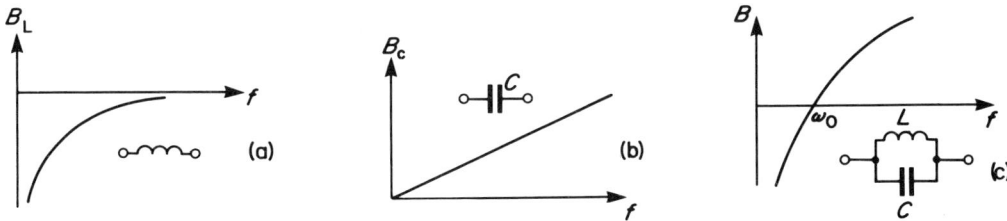

Figure 3.2

this curve has a frequency point ω_0 at which the total susceptance is zero. This is the parallel anti-resonant frequency. For frequencies below ω_0 the circuit this time is purely inductive, and for higher frequencies, purely capacitive.

It is clear that whenever the reactance curve of a series circuit is zero, the susceptance curve must be infinite; and whenever the susceptance curve of a parallel circuit is zero, the reactance curve must be infinite. Hence for series resonance

$$X(\omega_0) = 0 \text{ or } B(\omega_0) \to \infty$$

and for parallel anti-resonance

$$B(\omega_0) = 0 \text{ or } X(\omega_0) \to \infty$$

Points at which the curves cross the frequency axis are known as *zeros*, and points at which the curves recede to infinity are known as *poles*. A reactance pole is therefore a susceptance zero, and vice versa.

3.1.1 Multiple resonant circuits

Circuits containing a number of L and C elements may exhibit both series and anti-resonant conditions at a number of different frequencies. The sketching of reactance and susceptance curves for such circuits can be considerably simplified if a few basic facts are kept in mind. First, it is not very often that the *exact* form of the variation of reactance or susceptance with frequency is needed. It is usually sufficient to locate approximately the poles and zeros of the curves and then to take note of the fact that the gradient of the curve is always positive. The reason for this is not difficult to understand. For the simple expressions $X = \omega L \text{ or } -1/\omega C$, $B = \omega C \text{ or } -1/\omega L$, the differential coefficients with respect to ω yield in each case a positive answer. In general terms, if X_1 is the reactance of *any* combination of L and C, and either L_1 or C_1 is added in series with X_1, then if $dX_1/d\omega$ is positive, so also is $d(X_1 + \omega L_1)/d\omega$ or $d(X_1 - 1/\omega C_1)/d\omega$. Similarly, if a susceptance is added in parallel with B_1. Hence, since $dX/d\omega$ and $dB/d\omega$ are always positive for the basic networks, they are always positive for any series or parallel arrangements of such networks.

It follows from this that the poles and zeros of a particular diagram must occur alternately along the frequency axis. For if the gradient of the curves is always positive, the reactance or susceptance must change from $-$ to $+$ as the frequency is increased through a zero, and jump discontinuously from $+\infty$ to $-\infty$ as the frequency is increased through a pole.

Some reactance and susceptance sketches for simple networks are shown in *Figure 3.3*. The first two of these illustrate the cases of series and parallel connection of a single L and C element, as already covered. *Figure 3.3(c)* is constructed, for the reactance graph, by adding the reactance ωL_1 and the reactance

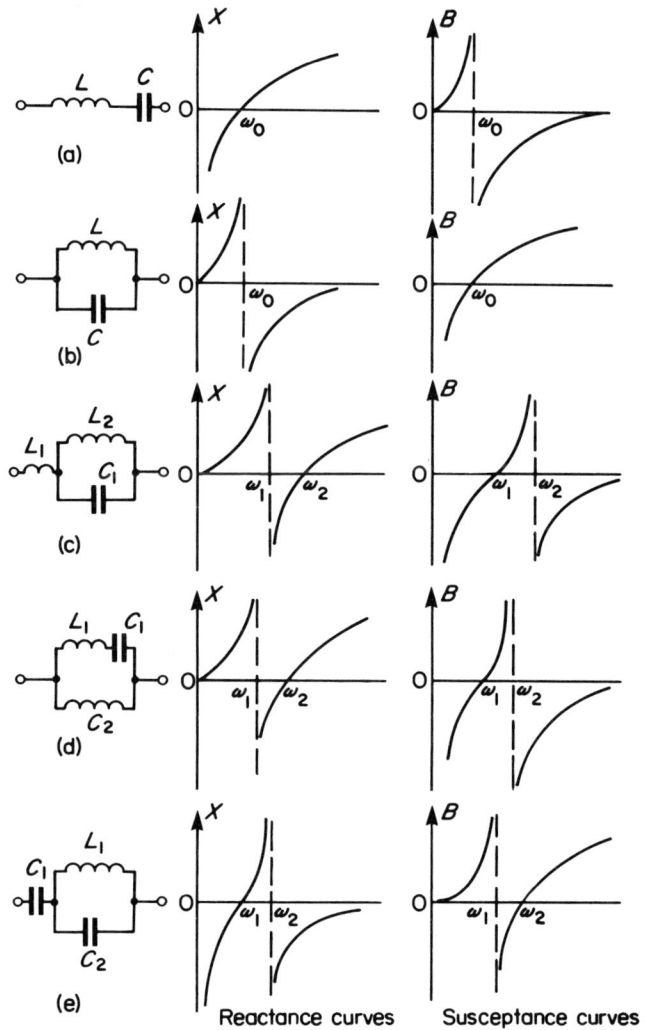

Figure 3.3

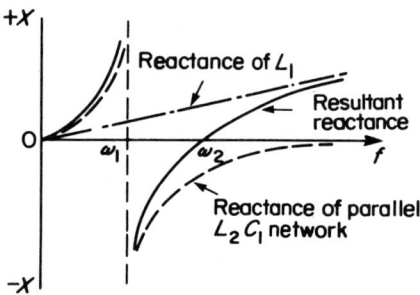

Figure 3.4

of the parallel L_2C_1 network. *Figure 3.4* shows the process in detail. The broken line construction follows from the reactance curve for a parallel circuit shown in *Figure 3.3(b)*, and the chain-dotted line represents the reactance of L_1 alone. The resultant curves are then shown in full line. There is a pole at frequency ω_1 and a zero at frequency ω_2. The susceptance curve for this case is shown on the right in *Figure 3.3(c)*. The subsequent examples of the figure should be deduced as an exercise.

3.1.2 Effect of resistance

The components so far considered have been assumed to be loss-free, that is, they have been purely reactive. In practice all components have resistance, however small, and for accurate calculation and representation the diagrams so far sketched require some modification.

Consider for a moment the diagrams of *Figure 3.5*. The result of converting the reactance sketch of *Figure 3.1(c)* for the series resonant circuit, to a diagram representing the magnitude of the

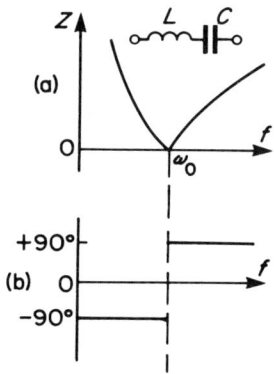

Figure 3.5

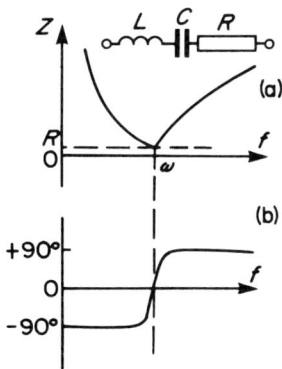

Figure 3.6

circuit impedance, is shown in *Figure 3.5(a)*. Since R was assumed zero, clearly $Z = |X|$; hence, the impedance curve is obtained by changing the sign of X in *Figure 3.1(c)* where this is negative. *Figure 3.5(b)* shows the variation in the phase angle of this same circuit. This is clearly $-90°$ below the resonant frequency ω_0 and $+90°$ above ω_0, the transition being abrupt at ω_0 itself. These curves are purely theoretical.

Consider now the effect of a *small* resistance inserted in series with L and C, small enough, that is, for the reactance X_L at resonance (which then equals X_c) to be much greater than R. The effect of R is then of importance only close to the resonant frequency, since the reactances are cancelling each other at this point and the circuit impedance reduces to R. For only small deviations from resonance the resulting reactance is large compared with R and the effect of R is negligible. *Figure 3.6(a)* shows the variation of circuit impedance when resistance is present.

The diagram showing the variation in phase angle also is modified by the presence of R. There can now be no abrupt change at resonance, as there was in the loss-free case shown in *Figure 3.5(b)*, because the circuit impedance is no longer purely reactive on either side of resonance. All angles from $0°$ to $\pm 90°$ now have to be passed through as the frequency deviates from resonance, and *Figure 3.6(b)* shows the practical effect of this.

Corresponding modifications are needed to the susceptance diagrams since the addition of resistance converts them to admittance sketches. Where the magnitude of the circuit impedance is small, but not zero as in *Figure 3.6(a)*, that of the admittance ($= 1/Z$) is large but not infinite. Conversely, zeros in the susceptance curves are replaced by finite admittances of small magnitude; these correspond to impedances of large, but not infinite, magnitudes.

Example 1. Sketch the ideal and the practical forms of the impedance and phase angle curves for the parallel combination of L and C.

Looking at the reactance curves of the ideal LC parallel circuit of *Figure 3.3(b)*, the impedance curve can be drawn by

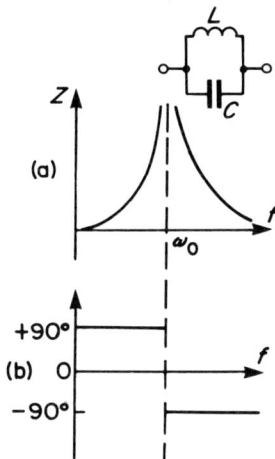

Figure 3.7

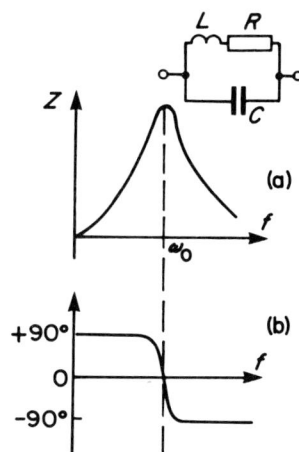

Figure 3.8

changing the sign of X where this is negative, since for $R = 0$, $Z = X$. The resulting curve is shown in *Figure 3.7(a)*, where the resonance impedance is seen to be infinite. Correspondingly, since the circuit is purely reactive at all times, purely inductive below resonance and purely capacitive above, the phase variation is as shown in *Figure 3.7(b)*, an abrupt transition from $+90°$ to $-90°$ occurring at resonance.

When the circuit has resistance, the susceptance cannot be zero at ω_0, hence the impedance cannot be infinite. Further, on either side of resonance the circuit is no longer purely reactive, hence all angles from $0°$ to $\pm 90°$ have to be passed through as the frequency deviates from resonance. *Figure 3.8(a)* and *3.8(b)* show, respectively, the practical curves for parallel circuit impedance and phase angle.

3.2 Q-FACTOR

The phenomenon of resonance occurs in series and parallel LC circuits because both reactive elements store energy during one quarter-cycle of the applied alternating input and return it to the circuit source during the following quarter-cycle. Since the inductor is storing energy in its magnetic field when the capacitor is returning energy from its electric field, and conversely, it is possible for the inductive and capacitive elements to transfer energy from one to the other successively, with the source of supply ideally providing no additional energy at all.

Practical reactors, however, both store and dissipate energy. The 'goodness' or *quality-factor* of a reactive component may therefore be expressed as the ratio of the energy stored to the energy dissipated as heat in equal intervals of time, that is

$$Q = \frac{I^2 \omega L t}{I^2 R t} = \frac{\omega L}{R} \quad \text{for an inductor}$$

$$Q = \frac{I^2 t}{I^2 \omega C R t} = \frac{1}{\omega C R} \quad \text{for a capacitor}$$

where Q is known as the quality-factor of the component. Clearly

$$Q = \frac{\text{Reactance}}{\text{Resistance}}$$

Q-factor is a very important parameter in the analysis of resonant conditions. It is used to measure both the 'goodness' of individual reactors and to describe the frequency selective properties of resonant circuits.

A practical coil can be closely represented as an ideal inductor in series with the resistance R of the coil. We dealt with such a case under *Figure 1.10* earlier. In this case the Q of the coil is $\omega L/R$ and it follows from elementary phasor trigonometry that $Q = \tan \varphi$, where φ is the circuit phase angle. Practical Q values for a coil at frequencies of the order of a few megahertz range from 50 to 300.

A practical capacitor, unlike a coil, has a very low effective series resistance and so on the whole exhibits a much higher Q-factor than does the average coil. Capacitor losses can occur either in the plates or, for dielectrics other than air, almost entirely within the dielectric. However, capacitor Q is still usually very much greater than coil Q at the same frequency, and since $Q = \tan \varphi$ and is large, $\varphi \approx 90°$ and $\tan \varphi \approx 1/\cos \varphi$. Hence the Q-factor of a capacitor with a solid dielectric is approximately equal to 1/(power factor) of the dielectric, and this may have a value of several thousands.

3.2.1 Variations in Q-factor

Although the expressions for Q suggest that the Q of either a coil or a capacitor is proportrional to frequency, in practice this is not the case. First of all, the representation of the coil as being an ideal inductor in series with a resistance is reasonably true only at low frequencies. At very high frequencies the shunt reactance of the self-capacitance of the coil reduces the effective inductance and eventually the coil resonates with its own self-capacity as a parallel resonant circuit. Similarly, at very high frequencies the plates and lead connections of a capacitor exhibit appreciable inductive reactance so that self-resonance is again possible.

Additionally, in the case of coils, the resistance is not constant but increases with frequency, due to the fact that the increased inductance of the central region of the conductor making up the coil confines the current to the outer layers (the 'skin effect'), thus effectively reducing the cross-sectional area of the conductor. The problem is eased somewhat by the fact that a coil is usually designed for a particular frequency range, often a restricted one, and, by the use of stranded wire and special techniques of winding, the ratio $\omega L/R$ can be made to remain substantially constant over the frequency range for which the coil is intended. It is quite valid in problems concerned with resonant conditions to assume that the Q of a circuit remains constant over a small range of frequency on either side of resonance, and the Q value is stated at the frequency of resonance. It follows that, at resonance,

$$Q = \frac{\omega_0 L}{R} = \frac{1}{\omega_0 CR} = \frac{1}{R}\sqrt{\frac{L}{C}} \qquad (3.1)$$

since $\omega_0 L = \dfrac{1}{\omega_0 C}$.

Hence Q-factor may also be defined at resonance as

$$\frac{\text{Reactance of one kind}}{\text{Total circuit resistance}}$$

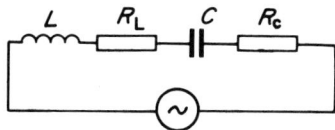

Figure 3.9

Example 2. Figure 3.9 shows a series LC circuit which is assumed to be at resonance. Deduce a value for the Q-factor of the complete circuit in terms of the Q-factors of the individual components.

When L and C are brought together in a single circuit, the circuit Q-factor will obviously be a function of the Q-factors of the coil and capacitor. There is no question, of course, of simply adding the individual Q-factors together for the total circuit resistance, which is the degrading quantity in Q-factor expressions, is greater than either component resistance on its own.

So, since $\qquad Q = \dfrac{\text{Reactance of one kind}}{\text{Total circuit resistance}}$

we have
$$Q_{circuit} = \frac{\omega_0 L}{R_L + R_c}$$

But for the coil alone $Q = \omega_0 L / R_L$ and for the capacitor alone $Q = 1/\omega_0 C R_c$. Now

$$\frac{1}{Q_{coil}} + \frac{1}{Q_{cap}} = \frac{R_L}{\omega_0 L} + \omega_0 C R_c$$

$$= \frac{R_L + R_c}{\omega_0 L}$$

since at resonance $\omega_0 L = \dfrac{1}{\omega_0 C}$

But
$$\frac{R_L + R_c}{\omega_0 L} = \frac{1}{Q_{circuit}}$$

Hence
$$\frac{1}{Q_{circuit}} = \frac{1}{Q_{coil}} + \frac{1}{Q_{cap}} \qquad (3.2)$$

In practice, since $Q_{cap} \gg Q_{coil}$, to a good approximation
$$Q_{circuit} = Q_{coil}$$

3.3 SERIES RESONANCE

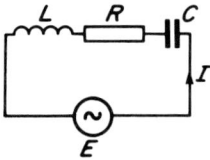

Figure 3.10

Figure 3.10 shows the general series combination of L, C and R. The circuit losses are considered to be entirely concentrated in R, that is, R does not only represent the resistance of the inductor but includes the minor losses associated with the capacitor and the connecting leads. The impedance of the circuit is given by

$$Z = R + j\omega L - \frac{j}{\omega C}$$

$$= R + j\left[\omega L - \frac{1}{\omega C}\right]$$

The circuit is resonant when the reactances cancel each other, or when $\omega_0 L = 1/\omega_0 C$. Hence

$$\omega_0^2 = \frac{1}{LC}$$

and the resonant frequency is

$$f_0 = \frac{1}{2\pi \sqrt{(LC)}}$$

In this condition the power factor of the circuit is unity, the impedance has reduced to its minimum value R, and the circuit current is a maximum. For a constant source voltage E, at any frequency

$$I = \frac{E}{Z} = \frac{E}{R + j\left[\omega L - \frac{1}{\omega C}\right]}$$

$$|Z| = \sqrt{\left\{ R^2 + \left[\omega L - \frac{1}{\omega C}\right]^2 \right\}}$$

and at resonance

$$|Z| = R$$
$$I = I_0 = \frac{E}{R}$$

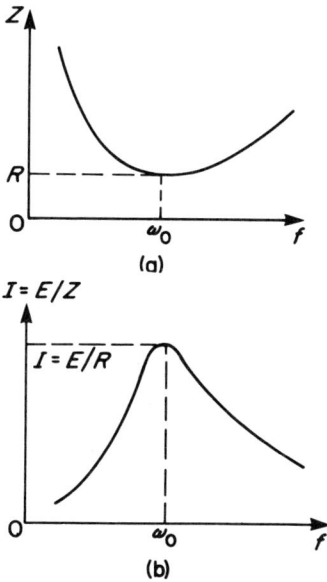

(a)

(b)

Figure 3.11

Curves showing the general variation in circuit impedance and circuit current are given in *Figure 3.11(a)* and *3.11(b)*, respectively. Notice that these curves are symmetrical only in the vicinity of resonance.

The voltages developed across the inductor and the capacitor at resonance are usually very much greater than that of the supply. For a circuit current I_0, the voltage across the inductor, $I_0\omega_0L$, and the capacitor I_0/ω_0C, are equal since the reactances are equal, and so

$$V_L = I_0\omega_0L = E\frac{\omega_0L}{R}$$

$$V_c = \frac{I_0}{\omega_0C} = E\frac{1}{\omega_0CR}$$

But the factors $(\omega_0L)/R$ and $1/(\omega_0CR)$ are the component Q-factors, hence

$$V_L = V_c = QE$$

For this reason, Q is often known as the *circuit magnification* factor.

3.3.1 Conditions at resonance

Resonance in the series circuit may be obtained either by varying the input frequency, or by keeping the frequency constant and varying one of the reactive components, usually the capacitor.

Let the frequency be variable, then the voltage developed across the capacitor may be used to indicate resonance. However, maximum capacitor voltage does not coincide with resonance unless the circuit Q is large, for with applied voltage E

$$\frac{V_c}{E} = \frac{\text{Capacitor reactance}}{\text{Circuit impedance}}$$

$$\therefore \quad \left|\frac{V_c}{E}\right| = \frac{1}{\omega C\sqrt{\left[R^2 + \left(\omega L - \frac{1}{\omega C}\right)^2\right]}}$$

and

$$\left|\frac{V_c}{E}\right|^2 = \frac{1}{\omega^2C^2\left[R^2 + \left(\omega L - \frac{1}{\omega C}\right)^2\right]}$$

Let this ratio be N. Then to find the value of ω which makes the capacitor voltage a maximum, we differentiate with respect to ω^2 and equate to zero in the usual way:

$$\frac{d(N^2)}{d(\omega^2)} = \frac{2\omega^2L^2C^2 - \left[C^2\left(R^2 - \frac{2L}{C}\right)\right]}{\left[\omega^2C^2\left(R^2 - \frac{2L}{C}\right) + \omega^4L^2C^2 + 1\right]} = 0$$

Hence the numerator is equal to zero or

$$2\omega^2L^2C^2 = C^2\left(R^2 - \frac{2L}{C}\right)$$

or

$$\omega^2 = \frac{1}{LC} - \frac{R^2}{2L^2}$$

But $\dfrac{1}{LC} = \omega_0^2$ $\quad \therefore \quad$ $\omega^2 = \omega_0^2 - \dfrac{R^2}{2L^2}$

$$= \omega_0^2\left[1 - \frac{R^2}{2\omega_0^2 L^2}\right]$$

This brings us to a form where we have $R/(\omega_0 L) = 1/Q$ in the right-hand term

$\therefore \qquad\qquad \omega^2 = \omega_0^2\left[1 - \dfrac{1}{2Q^2}\right]$

Hence $\qquad\qquad f = f_0\sqrt{\left(1 - \dfrac{1}{2Q^2}\right)}$

(3.3)

So the indicated resonance when V_c is a maximum differs from the true resonance by the term under the root sign. f is always *less* than f_0, but for Q greater than 10, the error is less than 0.05%.

If the circuit Q is very low, the magnification at resonance is relatively small, the frequency of maximum current deviates from the resonant frequency, and the circuit does not discriminate sharply between the resonant frequency and frequencies on either side of the resonance. *Figure 3.12* shows the variation in the voltages developed across the reactive components for both high and low circuit Q values. V_c and V_L are equal at resonance but unequal at all other frequencies. For the high Q condition, the maximum voltages occur coincidentally at f_0 but in the low Q circuit there is no such coincidence and the frequency difference is large.

It is left as an exercise for you to show that the frequency at which the voltage across L is a maximum is given by

$$f = \frac{f_0}{\sqrt{\left(1 - \dfrac{1}{2Q^2}\right)}}$$

This is always *greater* than f_0.

Let the *frequency* now be held constant and the capacitor varied. Then

$$I = \frac{E}{R + j\left[\omega L - \dfrac{1}{\omega C}\right]} \quad \text{and } V_c = -j\frac{I}{\omega C}$$

$\therefore \qquad V_c = \dfrac{E}{j\omega C\left[R + j\left(\omega L - \dfrac{1}{\omega C}\right)\right]} = \dfrac{E}{1 - \omega^2 LC + j\omega CR}$

and $\qquad |V_c| = \dfrac{|E|}{\sqrt{[(1 - \omega^2 LC)^2 + (\omega CR)^2]}}$

The maximum value of $|V_c|$ will occur now when the denominator of this last expression is a minimum. To find this, we differentiate the denominator and equate to zero:

$$\frac{d(\text{denom})}{dC} = 2(1 - \omega^2 LC)(-\omega^2 L) + 2(\omega CR) = 0$$

$\therefore \qquad\qquad -L + \omega^2 L^2 C + R^2 C = 0$

$$C = \frac{L}{R^2 + \omega^2 L^2}$$

Let C_m be the value of C when $|V_c|$ is a maximum. Then

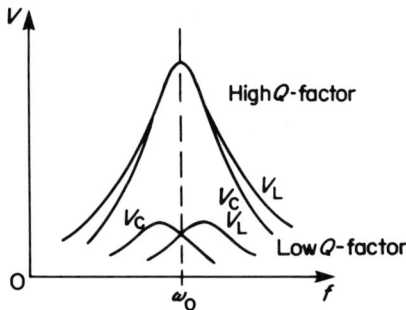

Figure 3.12

(graph axes labelled V, f, ω_0, with curves: High Q-factor, Low Q-factor, V_C, V_L)

$$C_{\mathrm{m}} = \frac{L}{R_2 + \omega^2 L^2} = \frac{L}{\omega^2 L^2}\left[\frac{1}{1 + R^2/\omega^2 L^2}\right]$$

$$= \frac{1}{\omega^2 L}\left[\frac{1}{1 + 1/Q^2}\right] = \frac{1}{\omega^2 L}\left[1 + \frac{1}{Q^2}\right]^{-1}$$

Using the binomial approximation on the right-hand side

$$C_{\mathrm{m}} \simeq \frac{1}{\omega^2 L}\left[1 - \frac{1}{Q^2} + \dots\right]$$

For $R = 0$, $C_0 = 1/(\omega^2 L)$, where C_0 is the capacitance *producing resonance* for fixed values of frequency and inductance. Hence

$$C_{\mathrm{m}} = C_0\left[1 - \frac{1}{Q^2}\right]$$

For $Q \geqslant 10$, clearly we may take $C_{\mathrm{m}} = C_0$.

Example 3. A coil of unknown inductance was connected in series with a variable capacitor. When the capacitor setting was 800 pF the circuit was resonant at 110 kHz, and when the capacitor was adjusted to 300 pF, the resonant frequency was found to be 170 kHz. The circuit was known to possess stray capacitance which was assumed to be constant and effectively in parallel with the capacitor. Calculate the values of the coil inductance and stray circuit capacitance.

$$\text{Resonant frequency} = f_0 = \frac{1}{2\pi\sqrt{(LC)}}$$

Here the capacitance C includes the parallel stray capacitance C_{s}, i.e. $C = C_1 + C_{\mathrm{s}}$.

At 110 kHz, $C_1 = 800$ pF, therefore

$$110 \times 10^3 = \frac{1}{2\pi\sqrt{[L(800 + C_{\mathrm{s}})]}} \qquad \text{(i)}$$

At 170 kHz, $C_1 = 300$ pF, therefore

$$170 \times 10^3 = \frac{1}{2\pi\sqrt{[L(300 + C_{\mathrm{s}})]}} \qquad \text{(ii)}$$

Dividing (ii) by (i)

$$\frac{17}{11} = \sqrt{\left(\frac{800 + C_{\mathrm{s}}}{300 + C_{\mathrm{s}}}\right)}$$

or
$$17^2(300 + C_{\mathrm{s}}) = 11^2(800 + Cs)$$

$$168C_{\mathrm{s}} = 10\,100$$

$$\therefore \qquad C_{\mathrm{s}} = 60\,\mathrm{pF}$$

Substituting this value for C_{s} into equation (i), (although (ii) may be used instead), we have

$$110 \times 10^3 = \frac{1}{2\pi\sqrt{[(860 \times 10^{-12})L]}}$$

$$\therefore \qquad 860L = \frac{10^{12}}{4\pi^2 \times 110^2 \times 10^6} \quad \mathrm{H}$$

$$\therefore \qquad L = \frac{10^{12}}{4\pi^2 \times 110^2 \times 860} \quad \mathrm{\mu H}$$

$$= 2434\,\mathrm{\mu H}$$

3.3.2 Universal resonance curve

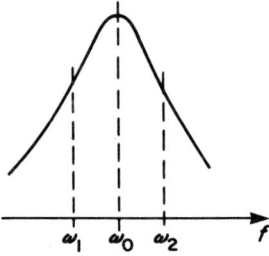

Figure 3.13

A more precise knowledge of the behaviour of the series circuit in the neighbourhood of the resonant frequency can be obtained from what is known as the *universal resonance curve* and equation. This equation simplifies the treatment of many problems because it is derived in terms of the fractional deviation from resonance and the circuit Q-factor.

Consider any series resonant circuit, and let the response curve be cut at frequencies ω_1 and ω_2 on either side of resonance, as shown in *Figure 3.13*. Provided that we are close enough to resonance for the curve to be considered symmetrical about ω_0, which will be true for a large Q-factor, then $\omega_0 - \omega_1 = \omega_2 - \omega_0$.

Let
$$\delta = \frac{\omega_2 - \omega_0}{\omega_0} = \frac{\text{Fractional deviation from}}{\text{resonance}}$$

that is,
$$\omega_2 = \omega_0(1 - \delta), \text{where } \delta \text{ is small.}$$

Now at resonance $I_0 = \dfrac{E}{R}$

and at any *other* frequency $I = \dfrac{E}{Z}$

Hence
$$\frac{I}{I_0} = \frac{R}{Z} = \frac{R}{R + j\left[\omega L - \dfrac{1}{\omega C}\right]}$$

So, since ω_2 above is some frequency off-resonance, by substitution for ω in the last expression:

$$\frac{I}{I_0} = \frac{R}{R + j\left[\omega_0 L(1 + \delta) - \dfrac{1}{\omega_0 C(1 + \delta)}\right]}$$

At resonance $\omega_0 L = \dfrac{1}{\omega_0 C}$, hence

$$\frac{I}{I_0} = \frac{1}{1 + j\dfrac{\omega_0 L}{R}\left[\dfrac{2\delta + \delta^2}{1 + \delta}\right]} = \frac{1}{1 + jQ\delta\left[\dfrac{2 + \delta}{1 + \delta}\right]}$$

For $\delta \ll 1$
$$\frac{I}{I_0} \approx \frac{1}{1 + j2Q\delta}$$

$$\left|\frac{I}{I_0}\right| \approx \frac{1}{\sqrt{[1 + (2Q\delta)^2]}} \tag{3.4}$$

The approximations involved in the derivation of this equation depend, as already mentioned, on δ being small. The curve of (3.4) has been plotted in *Figure 3.14* to give the universal resonance curve. It should be noted that $Q\delta$ is taken as the independent variable rather than δ alone, for the curve is then made independent of the specific value of Q at resonance. It is instructive to draw the curve accurately to a large scale, taking values of Q from 0 up to 1.6 in steps of 0.1.

From equation (3.4), the phase angle is given at once by

$$\varphi = \tan^{-1} 2Q\delta$$

and this is plotted for convenience on the same axes as the universal curve.

Example 4. The frequency of the generator shown in *Figure 3.15*

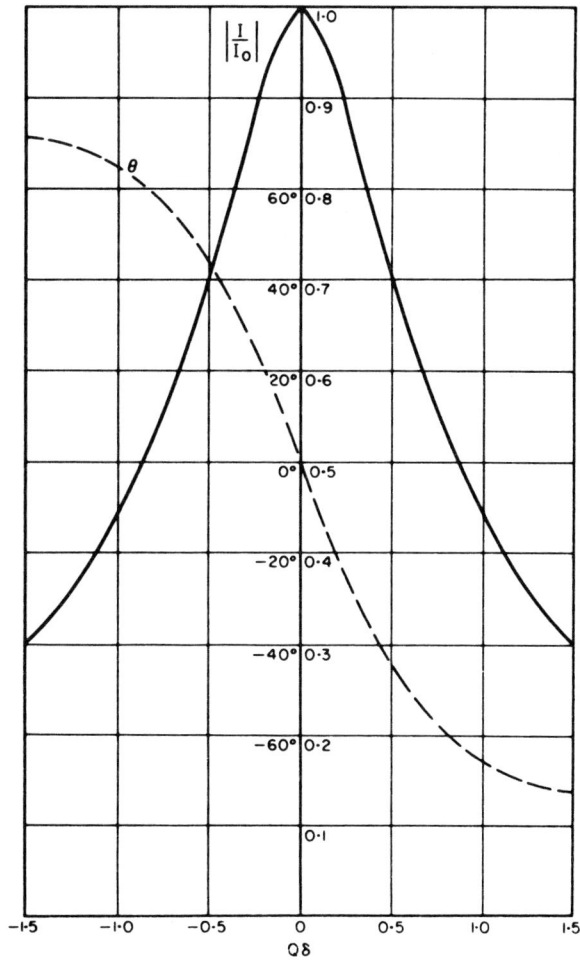

Figure 3.14

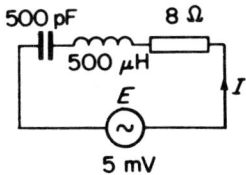

Figure 3.15

can be varied. Calculate for this circuit (a) the resonant frequency, (b) the resonant current, (c) the capacitor voltage at resonance, and (d) the current at frequencies 0.2% above and below the resonant frequency:

(a) At resonance

$$\omega_0^2 = \frac{1}{LC} = \frac{1}{5 \times 10^{-4} \times 500 \times 10^{-12}}$$

$$= \frac{10^{14}}{25}$$

$$\therefore \qquad \omega_0 = \frac{10^7}{5} \quad \text{and} \quad f_0 = \frac{1}{2\pi} \times \frac{10^7}{5} \text{ Hz}$$

$$= 318.3 \text{ kHz}$$

(b) $$I_0 = \frac{E}{R} = \frac{5 \times 10^{-3}}{8} \text{ A}$$

$$= 0.625 \text{ mA}$$

(c) $$V_c(\text{max}) = \frac{I_0}{\omega_0 C} = \frac{0.625 \times 10^3 \times 5}{10^7 \times 500 \times 10^{-12}} \text{ V}$$

$$= 0.625 \text{ V}$$

(d) $\left|\dfrac{I}{I_0}\right| = \dfrac{1}{\sqrt{[1 + (2Q\delta)^2]}}$

Now $Q = \dfrac{\omega_0 L}{R} = \dfrac{10^7 \times 5 \times 10^{-4}}{5 \times 8} = 125$

Therefore, since $\delta = 0.2/100 = 0.002$

$$\left|\dfrac{I}{I_0}\right| = \dfrac{1}{\sqrt{[1 + (2 \times 125 \times 0.002)^2]}} = \dfrac{1}{\sqrt{1.25}}$$

$\therefore \qquad |I| = \dfrac{1}{\sqrt{1.25}}|I_0| = 0.89 \times 0.625 = 0.556\,\text{mA}$

Example 5. The circuit shown in *Figure 3.16* is tuned to a frequency of 2 MHz. When the frequency of an e.m.f. injected in series with the circuit deviates by 10 kHz from the resonant frequency, the current falls to one-half its resonant peak. What is the circuit Q-factor?

$$\delta = \dfrac{f_2 - f_0}{f_0} = \dfrac{10\,\text{kHz}}{2\,\text{MHz}} = \dfrac{10^4}{2 \times 10^6}$$
$$= 0.005$$

$$\left|\dfrac{I}{I_0}\right| = \dfrac{1}{\sqrt{[1 + (2Q\delta)^2]}} = 0.5$$

$\therefore \qquad (2Q\delta)^2 = 3 \quad \therefore \quad Q = \dfrac{\sqrt{3}}{2\delta} = 173$

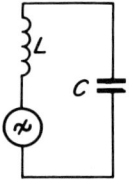

Figure 3.16

3.3.3 Selectivity

From the universal resonance equation (3.4), when $2Q\delta = 1$, $|I| = I_0/\sqrt{2}$ and then $Z = \sqrt{(2)}.R$, since R is the resonant impedance. The resistance is then equal to the reactance and

$$\delta = \pm \dfrac{1}{2Q}$$

Hence $\qquad \dfrac{\omega_1 - \omega_0}{\omega_0} = -\dfrac{1}{2Q} \; ; \; \dfrac{\omega_2 - \delta_0}{\delta_0} = +\dfrac{1}{2Q}$

Subtracting $\qquad \omega_2 - \omega_1 = \dfrac{\omega_0}{Q}$

$\therefore \qquad Q = \dfrac{\omega_0}{\omega_2 - \omega_1} = \dfrac{f_0}{f_2 - f_1} = \dfrac{f_0}{B_{3\,\text{dB}}}$

where $B_{3\,\text{dB}}$ is the bandwidth at the half-power points. Thus Q is a measure of the circuit selectivity in terms of the points on each side of resonance where the circuit current has fallen to $1/\sqrt{2}$ of its resonant peak value.

Example 6. In the circuit of *Figure 3.17*, resonance is obtained by adjustment of capacitor C and detected by the maximum voltage developed across C. What is the value of C and the voltage across it when this occurs? What is the 3 dB bandwidth of this circuit?

$$Q = \dfrac{\omega L}{R} = \dfrac{10^5 \times 3000 \times 10^{-6}}{100} = 3$$

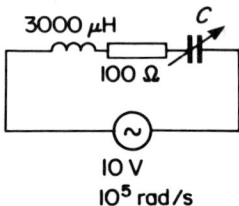

3000 μH

C

100 Ω

10 V

10^5 rad/s

Figure 3.17

This value of Q is very low and the resonant condition in terms of the values of L and C will not coincide with the applied frequency of 10^5 rad/s.

The capacitor voltage will be a maximum when

$$\omega = 10^5 = \omega_0 \left[1 - \frac{1}{2Q^2} \right] = \omega_0 \left[1 - \frac{1}{18} \right]$$

$\therefore \qquad \omega_0 = \dfrac{10^5}{0.945} = 1.059 \times 10 \text{ rad/s}$

or $\qquad f_0 = 16.85 \text{ kHz}$

In this condition $\qquad C = \dfrac{1}{\omega_0^2 L}$

$$= \frac{10^6}{(1.059 \times 10^5)^2 \times 3000 \times 10^{-6}} \, \mu F$$

$$= \frac{10^2}{1.059^2 \times 3000} = 0.03 \, \mu F$$

Now the voltage across C will be Q times the applied voltage or $3 \times 10 = 30$ V, but as an exercise in the use of the universal equation we find the circuit current and hence V_c.

If $\omega = \omega_0$, $\quad I_0 = \dfrac{10}{100} = 0.1 \text{ A}$

and at $\omega = 10^5$ rad/s $\quad I = \dfrac{I_0}{1 + j2Q\delta}$

where $\qquad \delta = \dfrac{1.059 - 1}{1.059} = 0.056$

$\therefore \qquad (2Q\delta)^2 = (6 \times 0.056)^2 = 0.113$

$\therefore \qquad |I| = \dfrac{0.1}{\surd(1 + (0.113^2))} = 0.099 \text{ A}$

Then $\qquad V_c = \dfrac{|I|}{\omega C} = \dfrac{0.099}{10^5 \times 0.03 \times 10^{-6}}$

$$= 31.6 \text{ V}$$

For the bandwidth $\quad B_{3\,dB} = \dfrac{F_0}{Q} = \dfrac{16.85}{3}$

$$= 5.17 \text{ kHz}$$

Clearly, this is not a very selective circuit!

Example 7. A coil having an inductance of $100 \, \mu H$ and a Q-factor of 30 is tuned to resonance with a variable series capacitor at a frequency of 250 kHz. Calculate the value of the capacitor.

The circuit is now detuned by alteration of C. Calculate the variation required to reduce the circuit current to one-half of its maximum value.

The value of C can be obtained from the basic resonance formula

$$f_0 = \frac{1}{2\pi\surd(LC)} \text{ or } C = \frac{1}{4\pi^2 f_0^2 L}$$

$\therefore \qquad C = \dfrac{10^{12}}{4\pi^2 \times (250 \times 10^3)^2 \times 10^{-4}} = 405 \text{ pF}$

We have not, so far, examined the effect of constant

frequency and variable capacitance as it affects the universal response equation, but the same approach can be used. As before, in the derivation of equation (3.4)

$$\frac{I}{I_0} = \frac{R}{R + j\left[\omega L - \dfrac{1}{\omega C}\right]}$$

which we may write as $\dfrac{I}{I_0} = \dfrac{R}{R + j\left[\omega L - \dfrac{C_0}{\omega C . C_0}\right]}$

where C_0 is the *value of capacitance producing resonance*.

Now, since ω is constant, $\omega L = \dfrac{1}{\omega C_0}$

$$\frac{I}{I_0} = \frac{1}{1 + j\dfrac{1}{\omega C_0 R}\left[1 - \dfrac{C_0}{C}\right]}$$

$$= \frac{1}{1 + jQ\left[1 - \dfrac{C_0}{C}\right]}$$

noticing that $\dfrac{1}{\omega C_0 R} = Q$

$$\therefore \qquad \left|\frac{I}{I_0}\right| = \frac{1}{\sqrt{\left(1 + Q^2\left[1 - \dfrac{C_0}{C}\right]^2\right)}} \qquad (3.5)$$

Rearranging this we get

$$\frac{C_0}{C} = 1 \pm \frac{1}{Q}\sqrt{\left(\left|\frac{I_0}{I}\right|^2 - 1\right)}$$

and substituting the values given in the problem

$$\frac{C_0}{C} = 1 \pm \frac{1}{30}\sqrt{[(2)^2 - 1]} \text{ since } \frac{I}{I_0} = 0.5$$

$$= 1 \pm 0.058 = 1.058 \text{ or } 0.942$$

$\therefore \qquad C = 1.058C_0$ or $0.942C_0$, where $C_0 = 405\,\text{pF}$

$\therefore \qquad C = 428\,\text{pF}$ or $381\,\text{pF}$

The capacitor has to be varied, therefore, about $23\,\text{pF}$ either side of resonance.

Example 8. Derive an expression for the value of C in terms of C_0 and Q if a series circuit is detuned by variation of C to the $3\,\text{dB}$ points.

$$\text{For } 3\,\text{dB } \left|\frac{I}{I_0}\right| = \frac{1}{\sqrt{2}}$$

$\therefore \qquad$ from equation (3.5) above $1 + Q^2\left[1 - \dfrac{C_0}{C}\right]^2 = 2$

$$\therefore \qquad Q^2\left[1 - \frac{C_0}{C}\right]^2 = 1 \text{ or } Q\left[1 - \frac{C_0}{C}\right] = \pm 1$$

$$\therefore \qquad \frac{C_0}{C} = 1 \pm \frac{1}{Q}$$

$$C = \frac{C_0}{1 \pm \frac{1}{Q}} = \frac{C_0 Q}{Q \pm 1}$$

The alternative signs, of course, indicate that the capacitor may be tuned above or below the resonance point.

3.4 PARALLEL RESONANCE

From the definition of resonance, the impedance of the circuit must be resistive at one or more resonant frequencies. It is more convenient to work in terms of admittance for a parallel circuit, so that the circuit admittance becomes conductive at resonance. Since

$$Y = G + jB$$

where G is the conductance and B the susceptance of the circuit, the condition of resonance is satisfied when the total susceptance is zero. If therefore the variation in susceptance for each branch of the parallel circuit is examined as the frequency is changed, the sum of the susceptances will be zero at resonance.

The reactance and susceptance curves of *Figure 3.3(b)* are brought forward as *Figure 3.18*. These curves are drawn for ideal components L and C; the impedance at resonance is infinite and the two susceptances ωC and $-1/\omega L$ are equal, making the total susceptance zero at ω_0. In practical circuits, both capacitors and inductors have losses and these must be considered. Since the bulk of the loss occurs in the inductor, it is usual to include a resistive component in series with L, so that the parallel circuit becomes as shown in *Figure 3.19*.

Let B_1 be the susceptance of the capacitive branch and B_2 the susceptance of the inductive branch. Then

$$B_1 = \omega C \text{ which is positive and proportional to frequency}$$
so that its graph will be a straight line

and the admittance of the inductive branch

$$G + jB_2 = \frac{1}{Z} = \frac{1}{R + j\omega L}$$

Rationalizing this we have

$$G + jB_2 = \frac{R - j\omega L}{R^2 + \omega^2 L^2}$$

so that

$$G = \frac{R}{R^2 + \omega^2 L^2} \text{ and } B_2 = -\frac{\omega L}{R^2 + \omega^2 L^2}$$

The curves of B and B_2 are shown on the graph of *Figure 3.20*. B_1, as noted above, is a straight line. B_2 would be a rectangular hyperbola if resistance was absent, but it approximates to such a hyperbola as ωL becomes very much greater than R for then $B \to -1/\omega L$. Also, as $\omega \to 0$, $B \to 0$, hence the curve exhibits a minimum point as shown. This minimum point can be located by means of the calculus. Writing $\omega L = X$ for convenience, we have

$$B = -\frac{X}{R^2 + X^2}$$

$$\frac{dB}{dX} = \frac{-(R^2 + X^2) + X(2X)}{(R^2 + X^2)^2}$$

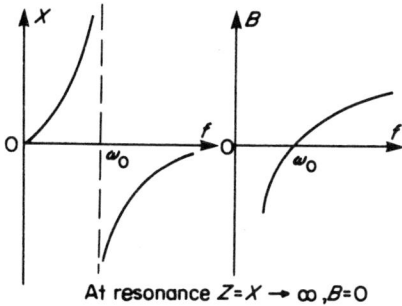

At resonance $Z = X \to \infty, B = 0$

Figure 3.18

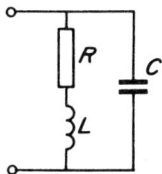

Figure 3.19

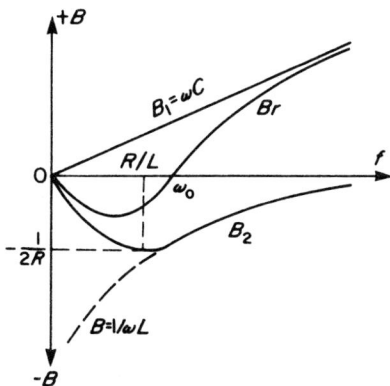

Figure 3.20

and equating the numerator to zero

$$2X^2 = R^2 + X^2$$

$$\therefore \qquad R = X$$

Hence, since $X = \omega L$, $R = \omega L$ and $\omega = R/L$ rad/s. At this frequency

$$B_2 = -\frac{R}{2R^2} = -\frac{1}{2R}$$

The resonant frequency ω_0 can now be obtained by considering the total susceptance $B_1 = B_1 + B_2$ and equating this to zero:

$$B_1 = \omega_0 C - \frac{\omega_0 L}{R^2 + \omega_0^2 L_2} = 0$$

$$\therefore \qquad R^2 + \omega_0^2 L^2 = \frac{L}{C}$$

$$\omega_0^2 = \frac{1}{LC} - \frac{R^2}{L^2}$$

$$\omega_0 = \sqrt{\left(\frac{1}{LC} - \frac{R^2}{L^2}\right)} \qquad (3.6)$$

Writing this in the form

$$\omega_0^2 = \frac{1}{LC}\left[1 - \frac{CR^2}{L}\right]$$

$$= \frac{1}{LC}\left[1 - \frac{1}{Q^2}\right]$$

since $Q = 1/R\sqrt{(L/C)}$, shows that if Q is large, the resonant frequency reduces to that of the series circuit. For this reason, the result is sometimes written

$$\omega_{op} = \omega_{os}\left[1 - \frac{1}{Q^2}\right]$$

Notice that unless $CR^2/L < 1$, there is no frequency at which the susceptance is zero and the circuit is capacitive at *all* frequencies.

3.4.1 Resonant impedance

At resonance the susceptance is zero and the remaining conductance is that of the inductive branch. Now

$$G = \frac{R}{R^2 + \omega^2 L^2}$$

and so

$$Z = \frac{1}{G} = \frac{R^2 + \omega_0^2 L^2}{R}$$

But

$$R^2 + \omega_0^2 L^2 = \frac{L}{C}$$

$\therefore$ the resonant impedance $= \dfrac{L}{CR}\Omega$ and is resistive.

This is usually referred to as the *dynamic resistance* R_d. Since $Q = L/CR^2$, $R_D = Q^2 R$.

3.4.2 Universal resonance

Provided Q is large the universal resonance curve can be used to give accurately the variation of impedance (or admittance) of a parallel resonant circuit in the neighbourhood of resonance.

In general, the impedance of a parallel circuit is expressed as

$$Z = \frac{(R + j\omega L) \cdot 1/j\omega C}{R + j\omega L + 1/j\omega C}$$

For Q large, $R \ll L$, and the expression reduces to

$$Z = \frac{L/C}{R + j\left(\omega L - \dfrac{1}{\omega C}\right)} = \frac{L/CR}{1 + j2Q\delta}$$

since the denominator is the same as that for the coil and capacitor connected in series. Hence, we obtain the impedance form of the universal resonance equation for parallel circuits:

$$Z = \frac{R_D}{1 + j2Q\delta}$$

or

$$\left|\frac{Z}{R_D}\right| = \frac{1}{\sqrt{[1 + (2Q\delta)^2]}} \tag{3.7}$$

A curve of impedance against frequency may be drawn from this equation in the same manner as was used for *Figure 3.14*. The bandwidth of the circuit is defined as the range of frequency between which the circuit impedance falls to $R_D/\sqrt{2}$ or $Z = R_D/(1 + j)$. Here the resistance is equal to the reactance and the phase shift is $\pm 45°$. As for the series circuit

$$B_{3\text{dB}} = \frac{f_0}{Q}$$

Example 9. The resonant frequency of a parallel tuned circuit is 5 MHz and the effective shunt capacitance is 200 pF. When the signal frequency differs by 15 kHz from resonance, the circuit impedance falls to 0.707 of the impedance at resonance. Calculate the dynamic resistance and the Q-factor of this circuit, and the 3 dB bandwidth.

Knowing the deviation and the ratio $|Z/R_D|$ we find first the circuit Q-factor:

$$\left|\frac{Z}{R_D}\right| = \frac{1}{\sqrt{2}} = \frac{1}{\sqrt{[1 + (2Q\delta)^2]}}$$

$\therefore$

$$1 + (2Q\delta)^2 = 2$$

and

$$Q = \frac{1}{2\delta}$$

Now the fractional detuning $\delta = \dfrac{15\,\text{kHz}}{2\,\text{MHz}} = \dfrac{15 \times 10^3}{2 \times 10^6}$

$$= 0.003$$

$\therefore$

$$Q = \frac{1}{2 \times 0.003} = 166.7$$

For the dynamic resistance R_D we need an expression for R_D in terms of Q, C and ω_0, since R and L are not known. We can obtain such an expression by writing

$$R = \frac{L}{CR} = \frac{\omega_0 L}{\omega_0 CR} = \frac{Q}{\omega_0 C}$$

Hence, using the known values of Q, C and ω_0, we have

$$R_D = \frac{166.7}{2\pi \times 5 \times 10^6 \times 200 \times 10^{-12}}$$

$$= \frac{166.7 \times 10^6}{2\pi \times 5 \times 200} = 26.53\,\text{k}\Omega$$

The 3 dB bandwidth is clearly $2 \times 15\,\text{kHz}$, since the impedance has fallen to $1/\sqrt{2}$ of its maximum (R) value at 15 kHz deviation on either side of resonance. But we could use

$$B_{3\,\text{dB}} = \frac{f_0}{Q} = \frac{5 \times 10^6}{166.7} = 30\,\text{kHz}$$

3.4.3 Bandwidth and loading

The 3 dB bandwidth for a parallel tuned circuit is, as for the series case, given by f_0/Q when the Q-factor is that derived for the inductive branch. Loading, or damping, is often necessary to modify the bandwidth as many electronic applications require a broad response. Since such loading is nearly always placed in parallel with the circuit, this being the most convenient way of adding extra resistance, it is convenient to consider the Q-factor as resulting from the capacitor in parallel with the total *shunt* loss resistance. This shunt resistance must include the coil conductance and any external loading, and *Figure 3.21* shows the effective circuit.

It is necessary, then, to convert the series loss of the inductor, R_s, into an equivalent parallel loss. Now for this branch we have

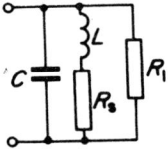

Figure 3.21

$$Z = R_s + j\omega L \quad \text{and} \quad Y = \frac{1}{Z} = \frac{1}{R_s + j\omega L}$$

The coil conductance $G_c = \dfrac{R_s}{R_s^2 + \omega^2 L^2}$

$$= \frac{1/R_s}{1 + \dfrac{\omega^2 L^2}{R_s^2}} = \frac{1/R_s}{1 + \left(\dfrac{\omega}{\omega_0}\right)^2 \dfrac{\omega_0^2 L^2}{R_s^2}}$$

$$= \frac{1}{R_s\left(1 + \left(\dfrac{\omega}{\omega_0}\right)^2 Q^2\right)}$$

For Q very much greater than 1 this reduces to

$$G_c = \frac{1}{R_s Q^2 \left(\dfrac{\omega}{\omega_0}\right)^2}$$

and so the equivalent parallel resistance of the coil is

$$R_p = \frac{1}{G_c} = R_s Q^2 \left(\frac{\omega}{\omega_0}\right)^2$$

For frequencies close to resonance $\omega \simeq \omega_0$, hence

$$R_p \simeq Q^2 R_s$$

Hence a lossy inductor can be represented as an inductance L connected in parallel with a loss resistance of $Q^2 R_s$ (*Figure 3.22*).

Now

$$R_s = \frac{\omega_0 L}{Q}$$

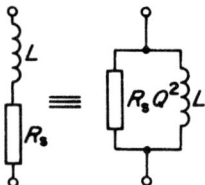

Figure 3.22

$$\therefore \qquad R_p = Q^2 \frac{\omega_0 L}{Q} = \omega_0 L Q$$

Then, when the circuit is externally loaded with R_1 (see *Figure 3.21*), the total shunt resistance

$$R_T = \frac{R_p R_1}{R_p + R_1} = \frac{\omega_0 L Q R}{\omega_0 L Q + R_1}$$

Then the equivalent Q-factor becomes

$$Q_e = \frac{R_T}{\omega_0 L}, \text{ the loaded } Q\text{-factor}$$

Example 10. A coil of inductance $150\,\mu H$ has a Q-factor of 50 at a frequency of 1 MHz. Calculate the required value of parallel capacitor to obtain resonance at this frequency and find the dynamic resistance of this resonant circuit.

The circuit is to be used as the tuned load of an amplifier stage which has an output impedance of $50\,k\Omega$. Calculate the bandwidth of this amplifier stage.

To find C we can use the formula

$$f_0 = \frac{1}{2\pi \sqrt{(LC)}}$$

from which

$$C = \frac{10^{12}}{4\pi^2 \times 10^{12} \times 150 \times 10^{-6}} \text{ pF}$$

$$= 169 \text{ pF}$$

The dynamic resistance

$$R_D = \omega_0 L Q = 2\pi 10^6 \times 150 \times 10^{-6} \times 50$$

$$= 47.1 \text{ k}\Omega$$

You can check that this value is equivalent to the parallel loss resistance R_p by finding R_s for the coil and calculating $Q^2 R_s$. Now, from *Figure 3.23*, the circuit will have an additional shunting resistance in parallel with it when it is connected to the amplifier stage, that is, the output resistance of the stage. The total effective load is therefore

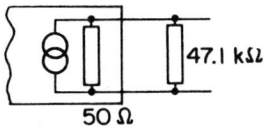

47.1 kΩ

50 Ω

Figure 3.23

$$R_T = \frac{50 \times 47.1}{50 + 47.1} = 24.3 \text{ k}\Omega$$

and the effective Q-factor is

$$Q_e = \frac{R_T}{\omega_0 L} = \frac{24.3 \times 10^3}{2\pi \times 10^6 \times 150 \times 10^{-6}}$$

$$= 25.8$$

The amplifier bandwidth $= \dfrac{f_0}{Q}$

$$= \frac{10^6}{25.8} = 38.8 \text{ kHz}$$

3.5 APPLICATIONS TO TUNED AMPLIFIERS

The previous example has illustrated a very common use for parallel resonant circuits; the loads of tuned amplifier stages. The function of a tuned amplifier is to provide a specified gain

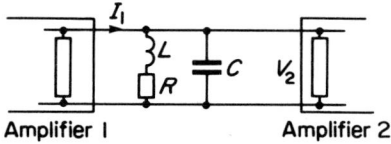

Amplifier 1   Amplifier 2

Figure 3.24

from a source to a load over a specified frequency range. Let amplifier 1 in *Figure 3.24* have an output conductance Y_0 and deliver a current I_1 to the tuned circuit consisting of C, L and R. Let amplifier 2 have an input conductance of Y_i, and let the input voltage to its terminals be V_2.

Then the admittance Y of the tuned circuit is

$$Y \simeq j\omega C + \frac{1}{j\omega L} + \frac{1}{RQ_2} \qquad \text{(i)}$$

(remember how this last term arrived!).

Let $\quad Y_0 = G_0 + j\omega C_0 \quad$ and $\quad Y_i = G_i + j\omega C_i$

Then the signal transfer from amplifier 1 to amplifier 2 is

$$\frac{I_1}{V_2} \simeq \frac{1}{R_T} + j\omega C_T + \frac{1}{j\omega L} \qquad \text{(ii)}$$

where $\dfrac{1}{R_T} = G_0 + G_i + \dfrac{1}{RQ^2}$ and $C_T = C_0 + C_i + C$

Now $\qquad \omega_0 = \dfrac{1}{\sqrt{(LC_T)}}$ and $Q_e = \dfrac{R_T}{\omega_0 L}$

Rewriting (ii) in terms of ω_0 and Q_e we have

$$\frac{I_1}{V_2} \simeq \frac{1}{R_T}\left[1 + jQ_e\left(\frac{\omega}{\omega_0} - \frac{\omega_0}{\omega}\right)\right]$$

$$\simeq \frac{1}{R_T}\left[1 + jQ_e\left(\frac{f}{f_0} - \frac{f_0}{f}\right)\right]$$

As before, let the fractional detuning $\delta = \dfrac{\omega - \omega_0}{\omega_0} = \dfrac{\omega}{\omega_0} - 1$

Then $\qquad \dfrac{\omega}{\omega_0} - \dfrac{\omega_0}{\omega} = (1 + \delta) - \dfrac{1}{1 + \delta} = \dfrac{\delta(2 + \delta)}{1 + \delta}$

and for $\delta \ll 1$, this approximates to 2δ.

$$\therefore \qquad \frac{I_1}{V_2} \simeq \frac{1}{R_T}\left[1 + j2Q_e\delta\right]$$

or $\qquad \left|\dfrac{V_2}{I_1}\right| \simeq \dfrac{R_t}{\sqrt{[1 + (2Q_e\delta)^2]}} \qquad \text{(3.8)}$

As *Figure 3.25* shows, the circuit bandwidth is defined as the points separation between which V_2/I falls to $R/\sqrt{2}$. Hence

$$\frac{R_T}{\sqrt{2}} = \frac{1}{\sqrt{[1 + (2Q_e\delta)^2]}}$$

whence $\qquad \delta = \pm\dfrac{1}{2Q_e}$

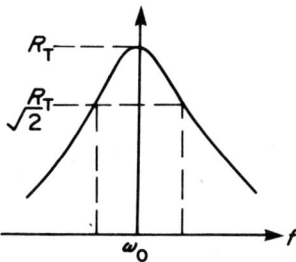

Figure 3.25

It should not be difficult for you to show that $B_{3\,dB} = \omega_0/Q_e$.

We shall have occasion to return to equation (3.8) in the following chapter.

Example 11. At $\omega_0 = 5 \times 10^5$ rad/s, two amplifying devices have $G_i = 800\,\mu S$, $C_i = 500\,pF$, $G_0 = 120\,\mu S$ and $C_0 = 150\,pF$. A tuned parallel circuit is connected between the devices having $L = 1\,mH$ and $Q = 100$. Calculated (a) the required tuning capacitance for resonance to occur, (b) the loaded Q-factor, (c) the 3 dB bandwidth. Comment on the results.

(a) The *total* capacitance required $C_T = \dfrac{1}{\omega_0^2 L}$

$\therefore$ $\qquad C_T = \dfrac{1}{(5 \times 10^5)^2 \times 10^{-3}}$ F $= 4000 \, \text{pF}$

Therefore the actual tuning capacitance $C = C_T - (500 + 120)$

$\qquad$ d $4000 - (500 + 120) = 3380 \, \text{pF}$

(b) The coil resistance $R = \dfrac{\omega_0 L}{Q} = \dfrac{5 \times 10^5 \times 10^{-3}}{100}$

$\qquad\qquad\qquad = 5 \, \Omega$

Then $\dfrac{1}{R_T} = (800 \times 10^{-6}) + (120 + 10^{-6}) + \dfrac{1}{5 \times 100^2}$

$\qquad = 940 \times 10^{-6} \, \text{S}$

$\therefore \quad R_T = 1.06 \, \text{k}\Omega$

$\therefore \quad Q_e = \dfrac{1.06 \times 10^3}{5 \times 10^5 \times 10^{-3}} = \dfrac{10.6}{5} = 2.1$

Notice how the effective Q is extremely low.

(c) $\qquad\qquad B_{3\,\text{dB}} = \dfrac{F_0}{Q_e} = \dfrac{5 \times 10^5}{2\pi \times 2.1} = 37.9 \, \text{kHz}$

This very broad (unselective) response follows from the low Q factor.

PROBLEMS FOR CHAPTER 3

(1) An e.m.f. of 1 V is applied to a series circuit of an inductor of 0.1 H and resistance 10 Ω and a capacitor of 0.25 μF. Find the resonant frequency of the circuit, the circuit resistance at resonance and the voltage across the capacitor at resonance.

(2) If the voltage at the extremes of the bandwidth of a tuned circuit is 3 dB down on the voltage at resonance, show that these two frequencies represent the 'half-power' points.

A circuit consists of a 180 μH coil of resistance 10 Ω in series with a 300 pF capacitor of negligible loss. Calculate the 3 dB bandwidth of the circuit and the circuit Q.

(3) A circuit consists of a 200 μH coil of resistance 40 Ω in parallel with a 200 pF capacitor. Find the impedance of the circuit at resonance and the impedance at the 3 dB points.

(4) A series circuit consisting of a 200 pF capacitor, an inductor of 800 μH and resistance 10 Ω, is connected to a constant voltage generator of e.m.f. 10 mV. Calculate (a) the frequency of resonance, (b) the current at resonance, (c) the capacitor voltage at resonance.

Find also the current at frequencies 0.5% above and below resonance and sketch the current–frequency response curve.

(5) A coil of inductance 0.1 H and Q-factor 10 is connected in series with a capacitance of 0.1 μF. Calculate (a) the frequency at which the current is a maximum, (b) the circuit resistance at resonance, (c) the impedance of the circuit at a frequency 5% lower than the resonant frequency, assuming the resistance to be constant, (d) the resistance of the circuit at resonance assuming that the capacitor had a Q-factor of 100.

(6) What is meant by the Q-factor of a tuned circuit? Derive an expression for the resonant frequency and impedance at

resonance of a circuit consisting of an inductor L having a resistance R connected in parallel with a capacitance C.

(7) A coil of inductance $0.02\,\text{H}$ and resistance $10\,\Omega$ is connected in parallel with a capacitor of $100\,\mu\text{F}$. Calculate (a) the total current taken by the circuit, (b) the additional capacitance which must be wired in parallel with the circuit to produce resonance, (c) the impedance at resonance.

(8) A circuit comprises an inductor of $100\,\mu\text{H}$ and a loss-free capacitor of $100\,\text{pF}$ connected in parallel. Assuming that the effective resistance of the inductor is independent of frequency and that the resonant Q-factor is 100, calculate (a) the resonant frequency, (b) the reactance of the inductor at the resonant frequency, (c) the circuit impedance at resonance, (d) the 3 dB bandwidth.

If the components are now connected in series, what is the impedance at resonance?

(9) In the circuit of *Figure 3.26*, show that the Q-factor at resonance is given by

$$\frac{\omega_0 L}{R_1} \cdot \frac{R_2}{R_2 + R_D}$$

and the resonant impedance by

$$\frac{R_2 R_D}{R_2 + R}$$

where in each case

$$R_D = \frac{L}{CR_1}$$

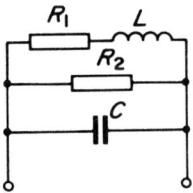

Figure 3.26

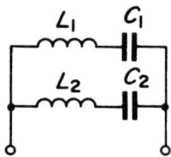

Figure 3.27

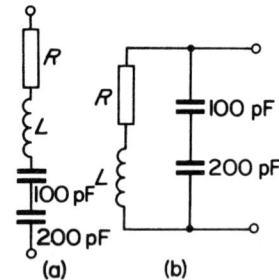

Figure 3.28

(10) Sketch the reactance and susceptance curves for the circuit shown in *Figure 3.27*.

(11) A series circuit has a Q-factor of 150 and, when connected as shown in *Figure 3.28(a)*, resonates at a frequency of 10 MHz. If the circuit is rearranged as a parallel circuit, *Figure 3.28(b)*, calculate, at resonance, the impedance as seen (i) across the coil, (ii) across the 200 pF capacitor.

(12) A series tuned circuit resonant at frequency f_0, has two frequencies f_1 and f_2, respectively, below and above f_0 at which the current in the circuit is one-half its value measured at f_0. Show that

$$f_0^2 = f_1 \times f_2 \quad \text{and that } f_2 - f_1 = \frac{\sqrt{(3)}R}{2\pi L}$$

(13) A sinusoidal waveform of 100 V r.m.s. at a frequency of 15 kHz is maintained across a circuit, the circuit current being $0.1\underline{/-60°}\,\text{A}$. Calculate (a) the circuit admittance in the form $G + jB$, (b) the shunt capacitance required to make the circuit non-reactive, (c) the value of the impedance when this value of capacitance is introduced.

Derive curves to show the variation of (a) reactance, (b) phase angle, (c) impedance, when the capacitance is varied so that the susceptance of circuit (b) above changes from $+1.5\,\text{mS}$ to $-1.5\,\text{mS}$.

(14) Two impedances $R_1 + jX_1$ and $R_2 + jX_2$ are connected in parallel. Obtain expressions for the conductance, susceptance and impedance of the combination.

A circuit consists of an inductor of inductance 0.05 H and resistance 100 Ω in parallel with a non-inductive resistor of 200 Ω. Find the conductance, susceptance and impedance of the combination at an angular frequency of 2×10^4 rad/s. Hence, find the value of capacitance needed in parallel with the circuit to produce resonance.

(15) A tuned circuit consists of a coil of 16 mH inductance and 10 Ω resistance in parallel with a loss-free 0.016 μF capacitance. Sketch curves showing how the conductance and susceptance vary with frequency over the range 7 kHz to 14 kHz.

(16) Find the relationship between the components of the circuits shown in *Figure 3.29* so that the impedances of the circuits are equal at all frequencies.

(17) The series circuit shown in *Figure 3.30* is tuned so that the reading on the thermal milliammeter is a maximum. It is then detuned by adjustment of the variable capacitor C. It is found that when C is set to readings of 130 pF and 145 pF, the current falls to $1/\sqrt{2}$ of the maximum value observed. If maximum current occurred at an angular frequency of 10^6 rad/s and the resistance of the ammeter is 100 Ω, calculate the inductance and the resistance of the coil.

(18) Prove, for the circuit system of *Figure 3.31*, that if $R = \sqrt{(Z_1 Z_2)}$, the circuit is resistive at all frequencies.

(19) A constant-voltage generator of e.m.f. 10 V .r.m.s. and negligible internal resistance is connected to a circuit whose conductance is 200 μS and susceptance −1 mS. Calculate (a) the magnitude and phase of the current, (b) the power dissipated, (c) the resistance and reactance of the circuit.

(20) The conductance of a circuit is 30 μS and the susceptance is −50 μS. Calculate (a) the power dissipated when 30 V r.m.s. is maintained across the circuit, (b) the impedance in magnitude and phase, (c) the series reactance required to increase the susceptance to +50 μS.

(21) The networks A and B of *Figure 3.32* present an equal impedance across their terminals at an angular frequency ω if

$$R_2 - \frac{j}{\omega C_2} = \frac{1}{\dfrac{1}{R_1} + j\omega C_1}$$

Obtain expressions for R_2 and C_2 in terms of ω, R_1 and C_1. Prove also that ω is the geometric mean of the frequencies ω_1 and ω_2 given by

$$\omega_1 = \frac{1}{R_1 C_1} \; ; \; \omega_2 = \frac{1}{R_2 C_2}$$

i.e. that $\omega = \sqrt{(\omega_1 \omega_2)}$.

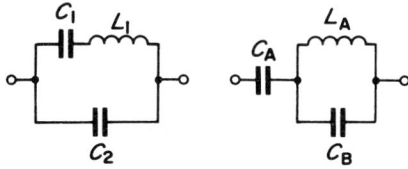

Figure 3.29

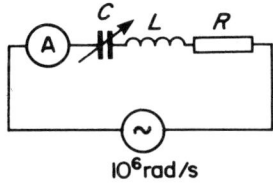

Figure 3.30

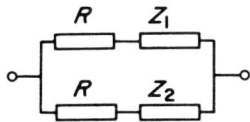

Figure 3.31

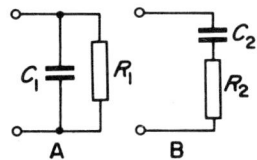

Figure 3.32

4 Coupled circuits

Coupled circuits are usually employed to provide (i) an impedance transformation, and (ii) a controlled bandwidth, when used as the coupling elements between active devices such as bipolar transistors amplifiers operating at frequencies above some 100 kHz. Such coupling systems consist of two windings that are coupled together in some way so that energy is transferred from one winding to the other, but the windings are tuned and the degree of coupling is 'loose', that is, the energy transfer is small. In this respect such couplings differ from the tight-coupling, large inductance requirements of untuned transformer couplings often used between active devices at audio frequencies. A different approach must therefore be made in the analysis of high frequency coupling transformers to that employed at power and speech frequencies.

Although we shall particularly investigate the performance of circuits coupled together solely by mutual inductance, a brief mention is necessary of the other forms of coupling which may be made between two circuits.

4.1 INDUCTIVE AND CAPACITIVE COUPLING

Figure 4.1(a) and 4.1(b) show examples of coupling through an impedance that is not common to either circuit, whereas Figure 4.2(a) and 4.2(b) show examples of coupling through an impedance that is common to both circuits. In all the diagrams, L_m or C_m are the impedances coupling together the two main circuits $L_1 C_1$ and $L_2 C_2$.

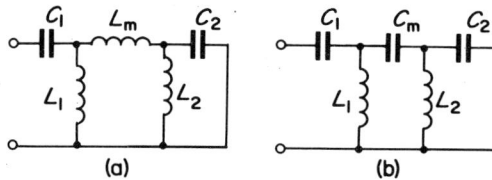

Figure 4.1

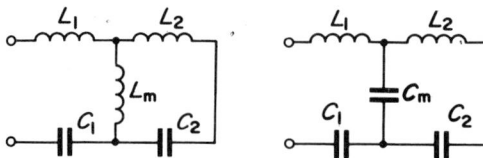

Figure 4.2

When circuits are coupled in this way, the resonant frequency of each of them is usually set to be the same, that is, the product $L_1 C_1$ is made to equal $L_2 C_2$. When the circuits are coupled, however, it is found that resonance of the complete system is not confined to a single frequency but occurs at two frequencies. This happens because there are two *effective* circuits in which resonance can occur: one is the basic tuned circuit

$L_1 C_1$ which on its own would resonate at a frequency given by $\frac{1}{2}\pi\sqrt{(L_1 C_1)}$, and the other is the second circuit $L_2 C_2$ which on its own would resonate at a frequency given by $\frac{1}{2}\pi\sqrt{(L_2 C_2)}$. The actual frequencies of resonance, however, depend upon the degree of coupling existing between the circuits, so that each circuit influences the other and the two resonant frequencies are found to be shifted, to a greater or lesser extent, one on each side of the common resonant frequency.

The degree of coupling between the tuned circuits is known as the *coupling factor* or the *coupling coefficient*, and is denoted by the symbol k. It is a ratio always less than unity and expresses the amount of coupling *actually present* to the maximum coupling *possible*. Any two circuits that are coupled together by some common impedance have a coupling factor that is equal to the ratio of the common impedance to the square root of the product of the total impedances of the same kind (i.e. inductive or capacitive) as the coupling impedance that is common to both circuits.

This is rather a mouthful but it simply expresses formulae giving k in terms of the basic circuit components L and C, and the coupling impedances L_m or C_m. These are, referring to *Figure 4.2(a)* and *4.2(b)*, respectively:

$$k = \frac{\omega L_m}{\sqrt{[\omega(L_1 + L_m).\omega(L_2 + L_m)]}} = \frac{L_m}{\sqrt{[(L_1 + L_m)(L_2 + L_m)]}}$$

where there is common inductive coupling, and

$$k = \frac{1/\omega C_m}{\sqrt{\left(\dfrac{C_1 + C_m}{\omega C_1 C_m} \cdot \dfrac{C_2 + C_m}{\omega C_2 C_m}\right)}} = \sqrt{\left(\dfrac{C_1 C_2}{(C_1 + C_m)(C_2 + C_m)}\right)}$$

where there is common capacitive coupling.

Fortunately, we shall not be so much interested in these methods of coupling as in that of mutually coupled coils, in which two circuits are coupled electromagnetically by virtue of their mutual inductance M, i.e. the coils are placed side by side and the coupling factor depends upon the amount of physical spacing between the coils.

4.1.1 Mutual inductive coupling

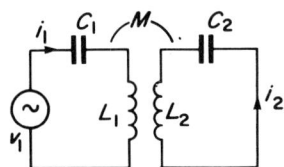

Figure 4.3

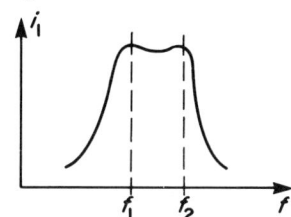

Figure 4.4

Let two resonance circuits be coupled together as shown in *Figure 4.3*. With this arrangement two resonance conditions are again, in general, found to exist as a signal v_1 is applied to the primary coil and adjusted in frequency . The primary current i_1 reaches two distinct maxima at two distinct frequencies. So the response curve of primary current plotted against frequency has, for such mutually coupled circuits, a shape rather of the form shown in *Figure 4.4*. This curve is distinctly different from the response curves of single tuned resonant circuits such as were discussed in the previous chapter. The height of the peaks, the depth of the trough between the peaks, and the frequency separation of the peaks themselves, are all functions of the mutual inductance M and so of the coupling coefficient k.

Circuits are said to be either closely or loosely coupled according to whether the proportion of energy transferred from primary to secondary is large or small. By moving the coils closer together an increased proportion of the primary flux links with the turns of the secondary and so both M and k increase. By moving the coils further apart the reverse happens and both M and k decrease. Now the coupling impedance in this case is ωM, and the circuit impedances of the same kind are ωL_1 and ωL_2. From our wordy definition above, therefore, we express the coefficient of coupling this time as

$$k = \frac{\omega M}{\sqrt{(\omega L_1 . \omega L_2)}} = \frac{M}{\sqrt{(L_1 L_2)}} \tag{4.1}$$

and for $k = 1$, $M = \sqrt{(L_1 L_2)}$. This condition would represent the maximum possible mutual inductance between the circuits and would assume that all the primary flux linked with the secondary turns. This is impossible to achieve in practice of course, and so M is always less than $\sqrt{(L_1 L_2)}$ and k is always less than 1.

4.2 REFLECTED IMPEDANCE

In *Figure 4.5*, let Z_1 be the impedance of the primary circuit alone and let Z_2 be the impedance of the secondary circuit alone. When a current i_1 flows in the primary coil, it will induce a voltage e_2 in the secondary winding where $e_2 = M . di_1/dt$. This, of course, follows from the definition of mutual inductance: two coils have a mutual inductance M of 1 H when a current changing at the rate of 1 A/s in one of them induces an e.m.f. of 1 V in the other.

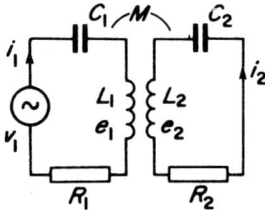

Figure 4.5

Let $i_1 = \hat{I}_1 \sin \omega t$, then $\dfrac{di_1}{dt} = \omega \hat{I}_1 \cos \omega t$

$$= j\omega i_1$$

$\therefore$ $$e_2 = j\omega M i_1$$

This voltage causes a current i_2 to flow in the secondary circuit given by

$$i_2 = \frac{e_2}{Z_2} = \frac{j\omega M i_1}{Z_2}$$

This current in turn will induce an e.m.f. e_1 in the primary winding where

$$e_1 = M . \frac{di_2}{dt} = \frac{j\omega M}{Z_2} . \frac{di_1}{dt}$$

$$= j^2 \frac{\omega^2 M^2}{Z_2} i_1 = - \frac{\omega^2 M^2}{Z_2} i_1$$

The net primary voltage is therefore $v_1 + e_1$ so that

$$v_1 - \frac{\omega^2 M^2}{Z_1} i_1 = i_1 Z_1$$

Hence, the impedance presented by the primary to the source generator is

$$Z_p = \frac{v_1}{i_1} = Z_1 + \frac{\omega^2 M^2}{Z_2} \tag{4.2}$$

Hence, the presence of the secondary current has increased the effective input impedance by the quantity $\omega^2 M^2 / Z_2$. This term is known as the *reflected impedance* from the secondary.

4.2.1 Nature of the reflected impedance

The reflected impedance $\omega^2 M^2/Z_2$ will be complex, made up of a resistive and a reactive component. Let R_2 and X_2 be the resistance and reactance of the secondary circuit, respectively. Then

$$\frac{\omega^2 M^2}{Z_2} = \frac{\omega^2 M^2}{R_2 + jX_2}$$

The right-hand term of this equation can be resolved into its real and imaginary components by rationalisation of the denominator in the usual way. Then

$$\frac{\omega^2 M^2}{R_2 + jX_2} = \frac{(\omega^2 M^2) \cdot (R_2 - jX_2)}{R_2^2 + X_2^2}$$

$$= \frac{\omega^2 M^2 R_2}{|Z_2|^2} - j\frac{\omega^2 M^2 X_2}{|Z_2|^2}$$

Hence, the impedance added to the primary consists of

(1) A resistive component $\dfrac{\omega^2 M^2 R_2}{|Z_2|^2}$

(2) A reactive component $-j\dfrac{\omega^2 M^2 X_2}{|Z_2|^2}$ which is opposite in sign to the reactance of the secondary circuit, i.e. if X_2 is capacitive, the reflected reactive component will be inductive, and conversely.

The equivalent primary resistance is then

$$R_\mathrm{p} = R_1 + R_2\frac{\omega^2 M^2}{|Z_2|^2} \tag{4.3}$$

and the equivalent primary reactance is

$$X_\mathrm{p} = X_1 - X_2\frac{\omega^2 M^2}{|Z_2|^2} \tag{4.4}$$

Example 1. In the coupled circuits of *Figure 4.6*, the mutual inductance between the coils is 0.05 H. Calculate the impedance seen by the generator at the input terminals at an angular frequency of 5000 rad/s.

The effective impedance at the input terminals is the phasor sum of the effective resistance and reactance; we therefore find separately these two component parts of the impedance.

For the primary circuit

$$R_1 = 10\,\Omega$$

$$X_1 = \omega L_1 - \frac{1}{\omega C_1} = (5000 \times 0.4) - \frac{10^6}{5000}$$

$$= 1800\,\Omega$$

For the secondary circuit $R_2 = 20\,\Omega$

$$X_2 = \omega L_2 - \frac{1}{\omega C_2} = (5000 \times 0.1) - \frac{10^6}{5000 \times 0.5}$$

$$= 500 - 400 = 100\,\Omega$$

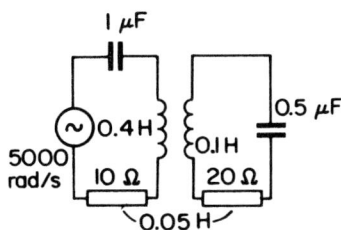

Figure 4.6

The effective primary resistance is made up of the primary self-resistance of $10\,\Omega$ plus the resistive component reflected from the secondary, that is

$$R_p = R_1 + R_2 \frac{\omega^2 M^2}{|Z_2|^2}$$

$$= 10 + 20 \left[\frac{5000 \times 0.05}{102}\right]^2$$

since $|Z_2| = \sqrt{(20^2 + 100^2)}$. Hence

$$R_p = 10 + (20 \times 6) = 130\,\Omega$$

The effective primary reactance is found similarly:

$$X_p = X_1 - X_2 \frac{\omega^2 M^2}{|Z_2|^2}$$

$$= 1800 - (100 \times 6) = 1200\,\Omega$$

The primary impedance is now

$$Z_p = \sqrt{(R^2 + X^2)}\bigg/\!\tan^{-1}\frac{X}{R}$$

$$= \sqrt{(130^2 + 1200^2)}\big/\!\tan^{-1}9.2$$

$$= 1207\big/83.8°\,\Omega$$

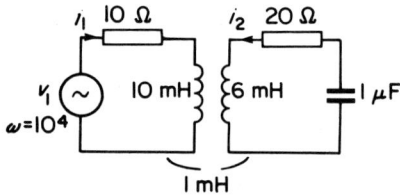

Figure 4.7

Example 2. By applying Kirchhoff's laws to the circuit shown in *Figure 4.7*, derive expressions for i_1 and i_2 in terms of the *general* circuit parameters. Hence, using the component values given, find the voltage developed across the secondary capacitor in polar form, given that $v_1 = 38$.

In general terms, we allocate the primary and secondary components and impedances the subscripts $_1$ and $_2$, respectively. Then, for the primary and secondary loops, we have

$$Z_1 = R_1 + j\omega L_1$$

$$Z_2 = R_2 + j\omega L_2 + \frac{1}{j\omega C_2}$$

and by Kirchhoff:

$$v_1 = i_1 Z_1 + j\omega M i_2 \qquad \text{(i)}$$

$$0 = j\omega M i_1 + i_2 Z_2 \qquad \text{(ii)}$$

Multiplying (i) by Z_2 and (ii) by $j\omega M$ to eliminate i_2 we get

$$v_1 Z_2 = i_1 Z_1 Z_2 + j\omega M i_2 Z_2$$

$$0 = j^2 \omega^2 M^2 i_1 + j\omega M i_2 Z_2$$

and subtracting

$$v_1 Z_2 = i_1 (Z_1 Z_2 + \omega 2 M^2)$$

whence

$$i_1 = \frac{v_1 Z_2}{Z_1 Z_2 + \omega^2 M^2}$$

From this, as an aside, we can verify our expression for the input impedance obtained in equation (4.2) earlier:

$$Z_p = \frac{v_1}{i_1} = \frac{Z_1 Z_2 + \omega^2 M^2}{Z_2}$$

$$= Z_1 + \frac{\omega^2 M^2}{Z_2} \quad \text{as before}$$

Solving for i_2 from the above simultaneous equations we get

$$i_2 = -j\frac{\omega M v_1}{Z_1 Z_2 + \omega^2 M^2}$$

$$\therefore \qquad |i_2| = \frac{\omega M v_1}{Z_1 Z_2 + \omega^2 M^2}$$

The voltage across the capacitor will be $V_c = |i_2| X_c$.

Now $Z_1 = R_1 + j\omega L_1 = 10 + j10^4 \cdot 10 \cdot 10^{-3} = 10(1 + j10)\,\Omega$

$$Z_2 = R_2 + j\omega L_2 - \frac{j}{\omega C_2}$$

$$= 20 + j\left[10^4 \cdot 6 \cdot 10^{-3} - \frac{10^0}{10^4}\right] = 20(1 - j2)\,\Omega$$

Also $\qquad \omega M = 10^4 \times 10^{-3} = 10\,\Omega$

$$X_c = \frac{10^6}{10^4} = 100\,\Omega$$

$$\therefore \qquad V_c = |i_2| X_c = \frac{10 \times 3 \times 100}{200(1 + j10)(1 - j2) + 100}$$

$$= \frac{30}{(1 + j10)(1 - j2) + 1}$$

$$= \frac{30}{43 + j16} = \frac{30\underline{/0°}}{45.8\underline{/20°}} = 0.65\underline{/-20°}\,\text{V}$$

4.3 RESPONSE CURVES

We will now examine the current and voltage conditions in a coupled circuit of the form shown in *Figure 4.5* when, first, the coupling is very loose and, secondly, when the coupling is very tight. The primary and secondary circuits will be assumed to be identical in every respect and tuned separately to the same frequency.

4.3.1 Loose coupling

The coils are well separated and the energy passed from the primary to the secondary is small. In this condition the secondary winding will have little influence on the primary circuit, for the secondary current will be very small and the reflected impedance into the primary will also be very small. Hence, if v_1 is constant in amplitude, i_1 will vary with frequency very much as it does in a single series resonant circuit; *Figure 4.8(a)* shows the response curve to be expected.

The secondary voltage $e_2 = \omega M i_1$, and since ω does not change very much over the small range of frequencies in the vicinity of resonance ω_0, the curve of e_2 against frequency will be proportional to the curve of primary current and so of similar shape to that shown in *Figure 4.8(a)*. The change in secondary current with frequency, however, will be different from either of the foregoing curves, for

$$i_2 = \frac{\omega M i_1}{Z_2} = \frac{e_2}{Z_2}$$

and so as resonance is approached, e_2 increases but Z_2 decreases. The effect on i_2 is that there is a rapid increase in the current as resonance is approached from below, and an equally rapid fall as the frequency passes resonance. The curve of i_2

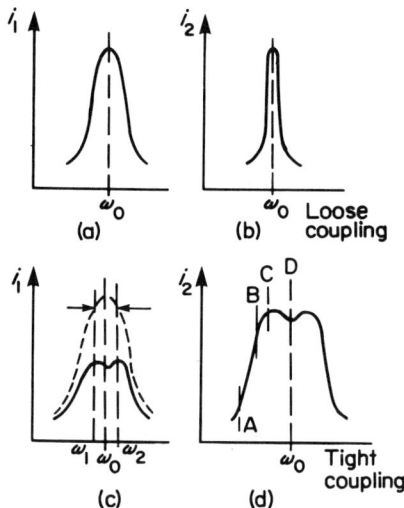

Figure 4.8

against frequency is therefore very selective, and is shown in *Figure 4.8(b)*.

This is a characteristic of loosely coupled circuits: the output at the secondary is small but the circuit is very selective.

4.3.2 Very tight coupling

The coils are placed in close proximity to each other and the energy transferred from primary to secondary is relatively large. The primary current i_1 is now very much affected by the secondary reflected impedance. As resonance is approached, the primary impedance Z_1 falls but the reflected impedance $(\omega M)^2/Z_2$ increases. There will be a point somewhere along the frequency scale where this reflected impedance will be increasing at exactly the *same rate* as Z_1 is decreasing. At this point i_1 will be momentarily steady. Nearer resonance, the reflected impedance increases more rapidly still, and there is a fall in the primary current which drops to a minimum value at the resonant frequency itself. *Figure 4.8(c)* shows the curve of primary current against frequency for this case. Compare the peak frequencies ω_1 and ω_2 with the positions they occupy on the curve of a single tuned circuit which is shown in broken lines.

Now as $e_2 = \omega M i_1$, the e_2 curve will have the same shape as that of primary current and need not be separately drawn. It is in the variation of secondary current i_2 that we are now particularly interested. Again consider $i_2 = e_2/Z_2$ and refer now to *Figure 4.8(d)*. Let the frequency increase from point A to point B. Over this range the numerator e_2 is increasing but the denominator Z_2 is decreasing, since resonance is being slowly approached; therefore i_2 increases rapidly. Let the frequency increase over the range from B to C. e_2 is now approximately constant, for although Z_2 is still decreasing, the ratios of the change of e_2 and Z_2 are roughly equal. Therefore i_2 still increases but much less rapidly, and at point C becomes stationary.

As the frequency now increases to resonance at point D, e_2 decreases at a faster rate than Z_2 so that i_2 falls. As resonance is passed, the whole cycle of events repeats in the opposite direction, and the complete response curve of i_2 against frequency is as shown in the diagram. We notice that the secondary current curve has steeper sides, closer peaks and a much shallower centre trough than that of the primary current curve. This is the kind of resonance curve we were looking for in the previous chapter, for it approximates to the ideal rectangular shape with vertical sides and a flat top. So an over-coupled pair of tuned circuits will provide us, if we choose the degree of coupling properly, with an increased bandwidth over which the impedance is low when compared with the response of a single tuned circuit.

4.4 DEGREES OF COUPLING

What is the best degree of coupling? Too loose gives us a highly selective curve but a small output; too tight gives us two peaks with a deep centre trough. Clearly the best situation would appear to be that in which the peaks are relatively close together so that the trough is levelled out to give a good approximation to a flat top. The degree of coupling which gives this desired effect is known as *critical coupling*. At this point,

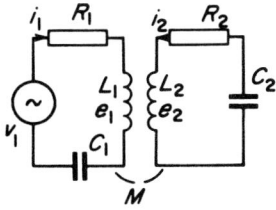

Figure 4.9

the degree of coupling is such that the resistance reflected back into the primary circuit at resonance is equal to the primary self-resistance. Any increase in this value just causes the response curve to double-hump.

Let both primary and secondary circuits of the system shown in *Figure 4.9* be previously tuned to the same resonant frequency ω_0. Now let the coupling be slowly increased. When the coupling is still loose, the curve of i_2 against frequency will show a high degree of overall selectivity, as we have recently seen. If the signal source is now set to ω_0, i_1 and i_2 are both at a maximum for that particular value of coupling. As the coupling is now increased, the secondary induced e.m.f. e_2 and the resulting current i_2 will rise, since

$$e_2 = -j\omega_0 M i_1 \quad \text{and} \quad \hat{i}_2 = -j\frac{\omega_0 M i_1}{R_2}$$

(at resonance, remember). As M increases so does the coupled resistance $(\omega_0 M)^2/R_2$ but its effect upon the primary current depends upon its magnitude relative to the total value of R_1. As M increases further, so does the coupled resistance, and this increase continues until the coupled resistance is approximately equal to R_1. In this condition i_1 tends to fall and limit any further rise in i_2. A further increase in M then causes a rapid fall in i_1 with a consequent fall in i_2 also. This we have already noted. Thus there is an *optimum* value of i_2 which is governed by the value of M.

Now
$$i_1 = \frac{v_1}{Z_1 + \dfrac{\omega^2 M^2}{Z_2}}$$

and
$$i_2 = -j\frac{M v_1}{Z_1 Z_2 + \omega^2 M^2}$$

This has a maximum value at resonance of

$$|\hat{i}_2| = \frac{M v_1}{R_1 R_2 + \omega_0^2 M^2} \tag{4.5}$$

since both X_1 and X_2 are zero. To find the optimum value of M we differentiate (4.5):

$$\frac{d|\hat{i}_2|}{dM} = \frac{(R_1 R_2 + \omega_0^2 M^2).\omega_0 v_1 - 2\omega_0 v_1.\omega_0^2 M^2}{(R_1 R_2 + \omega_0^2 M^2)^2}$$

and equating the numerator to zero

$$\omega_0 v_1 (R_1 R_2 + \omega_0^2 M^2) = 2\omega_0 v_1.\omega_0^2 M^2 = 0$$

from which
$$\omega_0 M = \sqrt{(R_1 R_2)} \tag{4.6}$$

Hence
$$M_{\text{opt}} = \frac{\sqrt{(R_1 R_2)}}{\omega_0}$$

Substituting this into equation (4.5) gives us

$$|i_2|_{\text{opt}} = \frac{v_1}{2\sqrt{(R_1 R_2)}}$$

and then
$$R_1 = \frac{\omega_0^2 M^2}{R_2}$$

i.e. the coupled resistance equals the primary resistance (which of course includes the source resistance). These are the conditions of critical coupling. We can now obtain an important relationship between the coupling coefficient and the circuit Q-factor.

Since $k = M/\sqrt{(L_1 L_2)}$, by substitution into equation (4.6) we have

$$\omega_0 k \sqrt{(L_1 L_2)} = \sqrt{(R_1 R_2)}$$

$$\therefore \qquad \omega_0 k = \sqrt{\left(\frac{R_1 R_2}{L_1 L_2}\right)}$$

or critical $k = k_c = \sqrt{\left(\frac{R_1}{\omega_0 L_1} \cdot \frac{R_2}{\omega_0 L_2}\right)} = \frac{1}{\sqrt{(Q_1 Q)}}$

If $Q_1 = Q_2 = Q$, $k_c = \dfrac{1}{Q}$

It is usual in practice to make k rather larger than critical (about $1.5 k_c$) to get a bandpass characteristic to the response curve. This is known as *optimum coupling*.

Figure 4.10(a) and *4.10(b)* show, respectively, curves for primary and secondary currents against frequency for various degrees of coupling. The tuned circuits in question consisted of coils of inductance $150\,\mu\text{H}$ and resistance $9\,\Omega$, series tuned with $170\,\text{pF}$ capacitors, the input being $1\,\text{V}$. Critical k was 0.01, optimum k was 0.015. Notice that the condition of critical coupling gave the maximum secondary current out of *all* the possible response curves.

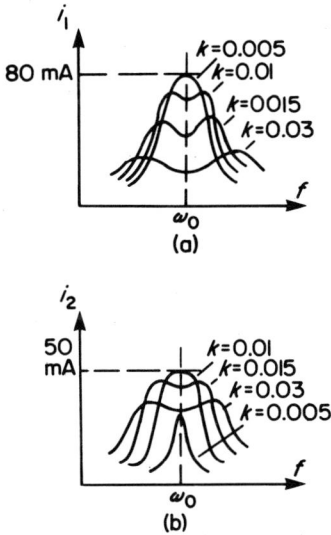

Figure 4.10

4.5 PEAK SEPARATION

To calculate the precise peak separation of coupled circuits for given conditions is very difficult, but a sufficiently accurate estimation may be made by making the simplification that the peaks of primary and secondary current occur at the same frequencies. This is very nearly true if the Q-factors are large. With this assumption, a peak of current occurs when the primary is in a state of induced resonance, i.e. when the total effective reactance is zero. For this to happen, the reflected impedance must equal the self-reactance of the primary, that is

$$X_1 = \frac{\omega^2 M^2 X_2}{|Z_2|^2}$$

or $\qquad \omega L_1 - \dfrac{1}{\omega C_1} = \dfrac{\left[\omega L_2 - \dfrac{1}{\omega C_2}\right] \omega^2 M^2}{R_2^2 + \left[\omega L_2 - \dfrac{1}{\omega C_2}\right]^2}$

$$\therefore \quad \omega L_1 \left[1 - \frac{1}{\omega L_1 C_1}\right]\left\{R_2^2 + \omega^2 L_2^2\left[1 - \frac{1}{\omega^2 L_2 C_2}\right]^2\right\}$$

$$= \omega L_2 \left[1 - \frac{1}{\omega^2 L_2 C_2}\right]\omega^2 M^2$$

At resonance $\qquad \omega_0^2 = \dfrac{1}{L_1 C_1} = \dfrac{1}{L_2 C_2}$

Also $\qquad Q = \dfrac{\omega L_1}{R_1} = \dfrac{\omega L_2}{R_2}$

$$k^2 = \frac{M^2}{L_1 L_2}$$

Substituting:

$$\omega L_1 \left[1 - \frac{\omega_0^2}{\omega^2}\right] \left\{ R_2^2 + \omega^2 L_2^2 \left[1 - \frac{\omega_0^2}{\omega^2}\right]^2 \right\} = \omega L_2 \left[1 - \frac{\omega_0^2}{\omega^2}\right] \omega^2 M^2$$

$$\therefore \quad \omega L_1 \left[R_2^2 + \omega^2 L_2^2 \left\{ 1 - \frac{\omega_0^2}{\omega^2} \right\}^2 \right] = \omega L_2 \omega^2 M^2$$

$$\omega L_1 \left[1 + Q^2 \left\{ 1 - \frac{\omega_0^2}{\omega^2} \right\}^2 \right] = \frac{Q\omega^2 L_1 L_2 k^2}{R_2} = Q^2 \omega L_1 k^2$$

$$\therefore \quad 1 + Q^2 \left[1 - \frac{\omega_0^2}{\omega^2} \right]^2 = Q^2 k^2$$

and

$$\frac{1}{Q^2} + \left[1 - \frac{\omega_0^2}{\omega^2} \right] = k^2$$

For $Q > 10$, $1/Q^2$ is negligible, hence

$$k = 1 - \frac{\omega_0^2}{\omega^2} \text{ or } \omega = \frac{\omega_0}{\sqrt{(1 \pm k)}}$$

Hence, if k is very small with respect to unity, the two current peaks are symmetrically placed about the natural resonant frequency ω_0.

The actual peak separation $(\omega_2 - \omega_1)$ can be found by squaring the last equation:

$$\omega_2^2 - \omega_1^2 = \omega_0^2 \left[\frac{1}{1 - k} - \frac{1}{1 + k} \right]$$

$$= \frac{2\omega_0^2 k}{1 - k^2}$$

k is normally about 0.01, so neglecting k^2 gives

$$\omega_2^2 - \omega_1^2 = 2\omega_0^2 k$$

or

$$(\omega_2 + \omega_1)(\omega_2 - \omega_1) = 2\omega_0^2 k$$

Since $\omega_0 = \frac{1}{2}(\omega_2 + \omega_1)$, $\quad 2\omega_0(\omega_2 - \omega_1) = 2\omega_0^2 k$

$$\therefore \quad \omega_2 - \omega_1 = k\omega_0$$

If k is now adjusted so that $k = k_c = 1/Q$, the two peaks merge to give a flat-topped curve of bandwidth

$$\omega_2 - \omega_1 = \frac{\omega_0}{Q}$$

which compares with the expression for the bandwidth of a single tuned resonant circuit.

When the coupling is optimum, the response curve exhibits definite double peaking and the bandwidth is then given not in terms of the peak separation but in terms of the points at which the response is 3 dB down on the peak amplitudes. It may be shown that this bandwidth is $\sqrt{2}$ times the peak separation bandwidth. *Figure 4.11* shows the distinction. Hence

$$B_{3\,dB} = \sqrt{(2)} . B_p$$

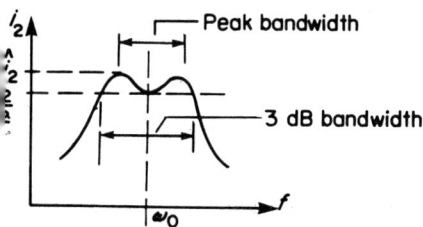

Figure 4.11

The 3 dB bandwidth level coincides with the trough level as the diagram shows. This is often a useful fact when problems on the shape of the response curve are encountered.

Example 3. An intermediate-frequency transformer having identical primary and secondary circuits has these circuits tuned with 150 pF capacitors. If the transformer has a peak separation of 9 kHz centred on 470 kHz and the coupling is critical, calculate (a) the inductance of each coil, (b) the coupling coefficient, (c) the mutual inductance, and (d) the Q-factor of each winding.

(a) This is simply a case of finding L when the capacitance and frequency are known. So

$$L = \frac{1}{\omega_0^2 C} = \frac{10^{12}}{(2\pi \times 470 \times 10^3)^2 \times 150}$$

$$= \frac{10^3}{(2\pi \times 47)^2 \times 15} \text{ H}$$

$$= 765 \mu\text{H}$$

(b) For peak separation, the bandwidth $B_p = k_c f_0$

$$\therefore \qquad k_c = \frac{B_p}{f_0} = \frac{9}{470}$$

$$= 0.019$$

(c) As $L_1 = L_2$, then $M = k_c L$

$$= 0.019 \times 765 = 14.5 \text{ }\mu\text{H}$$

(d) $$Q = \frac{f_0}{B_p} = \frac{1}{k_c}$$

$$= 52.2$$

Example 4. For the circuit shown in *Figure 4.12*, calculate the mutual inductance between the windings and the capacitance of C required if the maximum power is to be transferred from the source generator to the 200 Ω load resistor.

Gathering up the given information we have

$$f = 1 \text{ MHz, hence } \omega = 2\pi \cdot 10^6 \text{ rad/s}$$

$$\omega L_1 = 2\pi \times 10^6 \times 100 \times 10^{-6} = 200\pi = 628 \text{ }\Omega$$

$$\omega L_2 = 2\pi \times 10^6 \times 31.8 \times 10^{-6} = 63.6\pi = 200\Omega$$

The secondary impedance $Z_2 = 200 + j200 \text{ }\Omega$

$$\therefore \qquad |Z_2| = 200\sqrt{2}\,\Omega$$

The reflected impedance into the primary is

$$\frac{\omega^2 M^2}{Z_2} = \frac{\omega^2 M^2}{200 + j200} = \frac{\omega^2 M^2 (200 - j200)}{8 \times 10^4}$$

$$= \frac{\omega^2 M^2}{400} - j\frac{\omega^2 M^2}{400}$$

Hence, the reflected resistance $R_R = \dfrac{\omega^2 M^2}{400}$

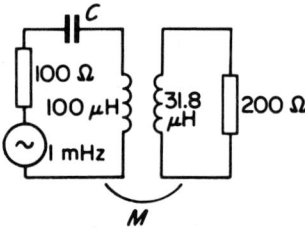

Figure 4.12

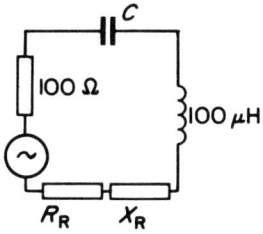

Figure 4.13

and the reflected reactance $\quad X_R = \dfrac{-\omega^2 M^2}{400}$

The circuit consequently reduces to the equivalent shown in *Figure 4.13*. The maximum power will be delivered to the secondary load when maximum power is dissipated in the transferred resistance R_R. This will happen when R_R is equal to the source resistance and the total primary reactance is zero, that is, the primary is resonant.

Hence $R_R = 100\,\Omega$ and $\omega L_1 - \dfrac{1}{\omega C} + X_R = 0$

$\therefore \qquad \dfrac{\omega^2 M^2}{400} = 100$ and $628 - \dfrac{1}{\omega C} - \dfrac{\omega^2 M^2}{400} = 0$

From the first of these equations

$$\omega^2 M^2 = 4 \times 10^4 \quad \therefore \quad \omega M = 100\sqrt{4} = 200$$

$$M = \dfrac{200}{2\pi} \times 10^{-6}\,\text{H}$$

$$= \dfrac{200}{2\pi}\,\mu\text{H} = 31.8\,\mu\text{H}$$

For the second equation, using the value for ωM above,

$$\dfrac{1}{\omega C} = 628 - \dfrac{200^2}{400} = 528\,\Omega$$

$$\therefore \qquad C = \dfrac{10^{12}}{2\pi \times 10^6 \times 528}\,\text{pF}$$

$$= 301\,\text{pF}$$

Example 5. For the coupled circuits shown in *Figure 4.14* find (a) whether the response curve is single- or double-humped, (b) the 3 dB bandwidth, and (c) the secondary current at resonance.

Both circuits will be separately resonant at the same frequency which is

$$\omega_0^2 = \dfrac{1}{250 \times 10^{-6} \times 10^3 \times 10^{-12}}$$

$$= 4 \times 10^{12}$$

$$\therefore \qquad \omega_0 = 2 \times 10^6$$

$$f_0 = \dfrac{2 \times 10^6}{2\pi} = 318\,\text{kHz}$$

Now the circuit Q-factors are

$$Q = \dfrac{\omega_0 L_1}{R_1} = \dfrac{2 \times 10^6 \times 250 \times 10^{-6}}{20} = 25$$

$$Q = \dfrac{\omega_0 L_2}{R_2} = \dfrac{2 \times 10^6 \times 400 \times 10^{-6}}{30} = 26.7$$

Hence $\quad k_c = \dfrac{1}{\sqrt{(Q_1 Q_2)}} = \dfrac{1}{\sqrt{(25 \times 26.7)}} = 0.039$

The *actual* coupling coefficient is different from this, i.e.

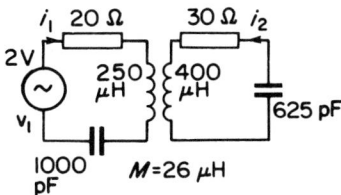

Figure 4.14

$$k = \frac{M}{\sqrt{(L_1 L_2)}} = \frac{26 \times 10^{-6}}{\sqrt{(250 \times 10^{-6} \times 400 \times 10^{-6}}}$$
$$= 0.082$$

(a) From these results, $k > k_c$, hence the curve will be double-humped.

(b) The positions of the peaks need to be found in order to establish the peak bandwidth. Now

$$\omega_1 = \frac{\omega_0}{\sqrt{(1+k)}} \text{ and } \omega_2 = \frac{\omega_0}{\sqrt{(1-k)}}$$

Here $\omega_1 = \dfrac{2 \times 10^6}{1.082} = 1.923 \times 10^6 \text{ rad/s}$

$\therefore$ $f_1 = \dfrac{1.932 \times 10^6}{2\pi} \text{ Hz} = 306 \text{ kHz}$

$\omega_2 = \dfrac{2 \times 10^6}{0.918} = 2.087 \times 10^6 \text{ rad/s}$

$\therefore$ $f_2 = \dfrac{2.086 \times 10^6}{2\pi} \text{ Hz} = 332 \text{ kHz}$

The *peak* bandwidth is therefore $332 - 306 = 26$ kHz

$\therefore$ the 3 dB bandwidth $= \sqrt{(2)} \times 26 = 36.8$ kHz

(c) Secondary current $i_2 = -j\dfrac{\omega_0 M v_1}{Z_1 Z_2 + \omega_0^2 M^2}$

At resonance $Z_1 = R_1$ and $Z_2 = R_2$

$$|i_2| = \frac{\omega_0 M v}{R_1 R_2 + \omega^2 M^2}$$

Now $\omega_0 M = 2 \times 10 \times 26 \times 10^{-6} = 52\,\Omega$

$\therefore$ $(\omega_0 M)^2 = 2704\Omega$

Also $v_1 = 2$ V

$\therefore$ $|i_2| = \dfrac{52 \times 2}{(20 \times 30) + 2704}$ A

$= \dfrac{104}{3304} \times 10^3 = 31.4$ mA

A sketch of the response curve is given in *Figure 4.15*.

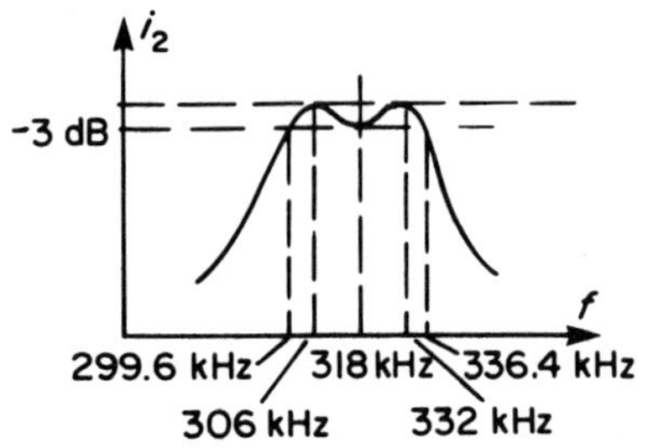

i_2
-3 dB
f
299.6 kHz
318 kHz
336.4 kHz
306 kHz
332 kHz

Figure 4.15

A summary of the foregoing. The impedance coupled into the primary circuit from the secondary $= (\omega^2 M^2)/Z_2$. The equivalent primary impedance is then $Z_1 + (\omega^2 M^2)/Z_2$. The voltage induced into the secondary circuit $= -j\omega M i_1$, and the secondary current $= -j.(\omega M i_1)/Z_2$.

These formulae are true *whether the circuits are tuned or untuned*, and hold at all frequencies.

When both primary and secondary circuits are tuned to the same frequency (ω_0), then the coupling coefficient

$k = M/\surd(L_1L_2)$ and for critical coupling $k = 1/Q$, where $Q = Q_1 = Q_2$. In this condition, the maximum energy is transferred from primary to secondary circuits. With tight coupling there are two resonant frequencies in both primary and secondary, given closely by $\omega_0/\surd(1 \pm k)$.

4.6 TRANSFORMER COUPLINGS

If we refer back to Section 3.5 of the previous chapter, it was shown there that the use of a single parallel tuned circuit as the load and frequency discriminating element between amplifier stages led to a very serious reduction in Q-factor with the consequent unselective response. Although the technique of using a series of such stages with the frequency of each tuned circuit staggered about the wanted centre frequency is often employed where a broad response is needed, the use of coupled circuits as inter-stage tuned transformers has certain advantages in that the input and output impedances of the amplifying devices can be matched and a relatively broad response can be obtained without the serious reduction in Q-factor and loss of gain that a single tuned circuit introduces.

4.6.1 Tuned primary coupling

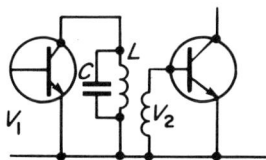

Figure 4.16

Consider *Figure 4.16* and let n such identical stages be tuned to a centre frequency ω_0. Such a system is referred to as being *synchronously tuned*. Let the voltage gain of the stage be A_v and let $A_{v\,max}$ be the gain at resonance when the tuned primary impedance is greatest.

From equation (3.8) previously, we have

$$\left|\frac{V_2}{I_1}\right| \simeq \frac{R_T}{\surd[1 + (2Q_e\delta)^2]}$$

But $I_1 = \dfrac{V_1}{R_T}$, hence $\left|\dfrac{V_2}{I_1}\right| = \left|\dfrac{V_2}{V_1}\right|R_T = A_vR_T$

$$= \frac{1}{\surd[1 + (2Q_e\delta)^2]} \times A_v$$

Hence for n stages $|A_v| = \dfrac{A_{v\,max}}{[1 + (2Q_e\delta)^2]^{n/2}}$

When $\delta = 0$, that is, at resonance, clearly $A_v = A_{v\,max}$.

Now when $A_v = \dfrac{A_{v\,max}}{\surd2}$, that is, at the half-power points on the gain characteristic response curve,

$$\surd2 = [1 + (2Q_e\delta)^2]^{n/2}$$

$$\therefore \qquad \delta = \pm\frac{1}{2Q}[2^{1/n} - 1]^{\frac{1}{2}}$$

From this it follows that the bandwidth of n stages is

$$B_n = \frac{\omega_0}{Q_e}[2^{1/n} - 1]^{\frac{1}{2}}$$

and that $\qquad \dfrac{B_n}{B} = [2^{1/n} - 1]^{\frac{1}{2}}$ where β_1 is the bandwidth of a single stage

Clearly, as n increases, this ratio decreases, i.e. the overall bandwidth decreases with an increasing number of stages. For $n = 3$, the bandwidth is halved.

4.6.2 Bandpass coupling

In this system, both primary and secondary windings of the coupling transformer is tuned and δ is no longer small. It may be shown that this leads to a modification in the gain equation in so far as a further term is necessary in the denominator:

$$A_v = \frac{A_{v\,max}}{1 + j2Q_e\delta - 2Q_e^2\delta^2}$$

At resonance $\delta = 0$ and $A_v = A_{v\,max}$.

$$\therefore \quad \left|\frac{A_v}{A_{v\,max}}\right| = \left[\frac{1}{(1 - 2Q_e^2\delta^2)^2 - (2Q_e\delta)^2}\right]^{\frac{1}{2}}$$

$$= \left[\frac{1}{1 + 4\delta^4 Q_e^4}\right]^{\frac{1}{2}}$$

and for n stages

$$\left|\frac{A_v}{A_{v\,max}}\right|^n = \frac{1}{[1 + 4\delta^4 Q_e^4]^{n/2}} = \frac{1}{\sqrt{2}} \text{ for 3 dB down}$$

$$\therefore \quad \delta = \frac{[2^{1/n} - 1]^{\frac{1}{4}}}{\sqrt{(2)} \cdot Q_e}$$

But the bandwidth $B_n = 2\omega_0\delta = \dfrac{\sqrt{(2)}\omega_0[2^{1/n} - 1]^{\frac{1}{4}}}{Q_e}$

$$\therefore \quad B_n = \sqrt{(2)}B_1[2^{1/n} - 1]^{\frac{1}{4}}$$

and

$$\frac{B_n}{B_1} = \sqrt{2}[2^{1/n} - 1]^{\frac{1}{4}}$$

When $n = 1$, $B_n = \sqrt{(2)}B_1$, i.e. a bandpass coupling has a bandwidth $\sqrt{2}$ times that of a single tuned circuit.

PROBLEMS FOR CHAPTER 4

(1) Define mutual inductance. If an e.m.f. of 1 V is induced into the secondary coil of a coupled pair of windings by an alternating current of 5 mA at a frequency of 100 kHz in the primary, calculate the mutual inductance between the coils.

(2) The mutual inductance M between a pair of coupled circuits in 25 μH. If a current of 1 mA flows in the primary at a frequency of 0.5 MHz, find the voltage induced in the secondary.

(3) The coils of a coupled circuit have inductances of 150 μH and 180 μH, respectively. If the mutual inductance between the coils is 20 μH, what is the coupling coefficient?

(4) If two resonant circuits are tuned to the same frequency their resulting behaviour, when coupled together, depends very largely upon the degree of coupling. Explain, by reference to curves, how this property may be used to obtain a choice of bandpass characteristics.

(5) The primary and secondary coils of an intermediate-frequency transformer are each tuned to a frequency of 465 kHz by parallel 220 pF capacitors. The coupling is then adjusted to give a peak bandwidth of 12 kHz. Assuming that the coefficient of coupling is small, find the mutual inductance between the coils.

(6) An intermediate-frequency transformer has identical primary and secondary windings inductive coupled. Each winding has an inductance of 500 μH and is tuned by a capacitor. If the 3 dB bandwidth of the circuit is 16 kHz centred on 450 kHz, calculate for critical coupling conditions: (a) the coupling coefficient, (b) the mutual inductance, (c) the Q-factor of each winding, (d) the value of the tuning capacitor. (C. & G.)

(7) The primary winding of a coupled circuit has an inductance of 300 μH and resistance 20 Ω, and is tuned with a capacitor of 0.005 μF capacitance. This winding is inductively coupled to a second coil which has an inductance of 100 μH, resistance 10 Ω and tuning capacitor of 0.01 μF capacitance. If the mutual inductance is 20 μH and a signal of frequency 0.1 mHz is applied to the primary terminals, calculate the effective primary impedance.

(8) Two resonant circuits having high Q-factors are tuned to the same frequency and coupled together by a mutual inductance M henries. Prove that the ratio of the voltage across the secondary capacitor when the coupling is critical to that obtained across the capacitor of either circuit alone is one-half. Assume that the applied e.m.f. is unchanged.

(9) The secondary winding of a coupled circuit has a resistance of 100 Ω and a self-inductance of 40 mH. The mutual inductance between the primary and secondary windings is 200 mH. When the primary is connected to a signal source, a current of 10 mA flows in a non-reactive resistance of 100 Ω connected across the secondary. If the signal frequency is 5000 rad/s, determine the magnitude of the primary current.

(10) Two coils L_1 and L_2 are closely coupled together with a coupling coefficient of 0.9. L_1 has an inductance of 8 mH and a resistance of 6 Ω; L_2 has an inductance of 640 μH and a resistance of 2 Ω. The primary coil L_1, is connected to a generator of e.m.f. 10 V and internal resistance 600 Ω, and a load resistor of 50 Ω is connected across the secondary coil L_2. Calculate (a) the effective input impedance at the primary coil terminals, (b) the power supplied by the generator into the primary terminals, (c) the power dissipated in the 50 Ω load. The input frequency is 1 kHz.

(11) Two identical inductors are coupled together as shown in *Figure 4.17*. With the secondary terminals short-circuited, the capacitor C is set to 500 pF and resonance is obtained. When the secondary terminals are open-circuited, the circuit resonates at the same frequency as before when C is adjusted to 450 pF. Prove that the coupling coefficient is about 0.32.

(12) The bandwidth of a synchronously tuned four-stage amplifier is 4.35 kHz. Show that the bandwidth of each stage considered individually is about 10 kHz.

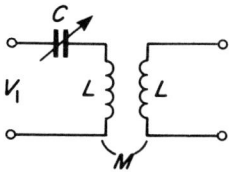

Figure 4.17

5 Attenuators

An ideal attenuator is a device which enables us to obtain as output some desired fraction of the input, this fraction being constant irrespective of frequency. It follows that ideal attenuators should be made up of pure resistances, since reactive elements will give frequency discrimination. A fixed attenuator network is generally known as a 'pad'.

Such a pad may, by the choice of suitable resistive elements, introduce any required degree of attenuation, but the input and output impedances of the pad must be such that the impedance conditions existing in the circuit into which it is connected are not disturbed. The attenuator must consequently fulfil three conditions: it must give the correct input and output impedances and it must provide the specified attenuation. In general, attenuator networks are made up of repetitive sections, the most common being the symmetrical T- and π-sections.

5.1 THE T-SECTION ATTENUATOR

In the T-section attenuator of *Figure 5.1(a)*, made up of a series arm of two equal resistances, each R_1, and a centre shunt arm R_2, let a terminating resistance R_0 be connected as shown. From our earlier work on sections, the input resistance of the section will be R_0 also if

$$R_0 = \surd(R_1^2 + 2R_1R_2) = \surd(R_{oc}R_{sc})$$

Similarly, if the terminating resistance R_0 is transferred to the opposite end of the pad, then the input resistance looking this time from the right-hand terminals will again be R_0 (see *Figure 5.1(b)*). The pad is therefore symmetrical in terms of its input and output resistance and may as a consequence be inserted into a circuit whose resistance is also equal to R_0. This value of R_0 which gives an input resistance equal to the terminating resistance is, of course, the characteristic resistance of the section.

Consider now a T-section (A), as shown in *Figure 5.2*, correctly terminated by R_0. Its input resistance is also R_0, so when a similar section (B) is connected to it in place of R_0, as shopwn in the figure, B in turn correctly terminates A. Section A is unable to distinguish between the presence of section B or the presence of a single terminating resistance equal in value to

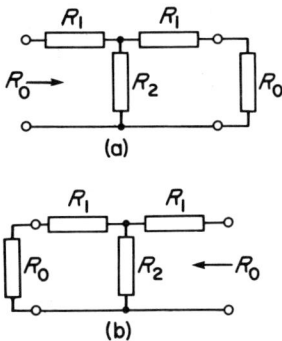

(a)

(b)

Figure 5.1

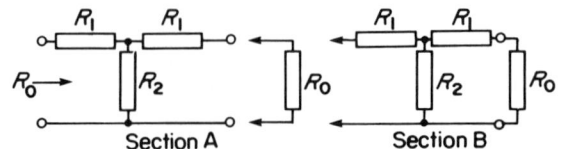

Section A

Section B

This section cannot distinguish between this resistance R_0 and section B

This section can be substituted for R_0 without affecting the input resistance of section A

Figure 5.2

R_0. The input resistance of the combined sections will consequently also be R_0.

Evidently, *any* number of similar sections can be wired in tandem in this manner, each section correctly terminating the one before it and each introducing a fixed amount of attenuation. Any desired amount of attenuation may therefore be obtained by the use of a sufficient number of repetitive sections, the characteristic resistance of the total system being equal to that of a single section. As an attenuator is made up of purely resistive elements, there will be zero phase shift in a section or any number of sections and only the attenuation constant is of immediate concern.

Example 1. Obtain the characteristic resistances of the three attenuating networks shown in *Figure 5.3*.

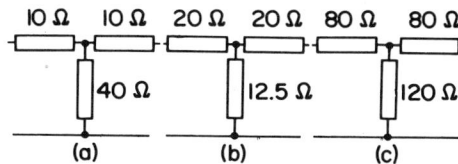

Figure 5.3

Using the formula $R_0 = \sqrt{(R_1^2 + 2R_1R_2)}$ for each of the networks, we have

(a) $R_0 = \sqrt{[10^2 + (2 \times 10 \times 40)]}$
$\qquad = \sqrt{900} = 30\,\Omega$

(b) $R_0 = \sqrt{[20^2 + (2 \times 20 \times 12.5)]}$
$\qquad = \sqrt{900} = 30\,\Omega$

(c) $R_0 = \sqrt{[80^2 + (2 \times 80 \times 120)]}$
$\qquad = \sqrt{25\,600} = 160\,\Omega$

5.1.1 Attenuation factor

For a single section pad, a desired characteristic resistance R_0 may be obtained with numerous combinations of R_1 and R_2; the examples (a) and (b) above, for instance, both have a characteristic resistance of $30\,\Omega$ but the section at (b) will be found to provide a greater degree of attenuation than that at (a). The problem in the design of an attenuator pad is, given R_0 and the desired attenuation, to find suitable values for R_1 and R_2.

The attenuation may be expressed as a voltage ratio V_i/V_o, where V_i is the input and V_o the output voltage, or quoted in decibels where

$$\text{Attenuation} = 20\lg\left[\frac{V_i}{V_o}\right]\text{dB}$$

or in terms of power
$$= 10\lg\left[\frac{P_i}{P_o}\right]\text{dB}$$

$$= 20\lg\left[\frac{P_i}{P_o}\right]^{\frac{1}{2}}\text{dB}$$

whence $V_i/V_o = \sqrt{(P_i/P_o)} = N$, which we will call the *attenuation factor*. This relationship follows from the fact that if the section is symmetrical and is matched to *equal* resistances, then

$$\frac{P_i}{P_o} = \left[\frac{V_i}{V_o}\right]^2 = \left[\frac{I_i}{I_o}\right]^2$$

Hence, for a pad in a symmetrical system, the attenuation factor $N = V_i/V_o = I_i/I_o$. If the circuit is asymmetrical the value $N = \sqrt{(P_i/P_o)}$ must be used.

5.1.2 The symmetrical T-section

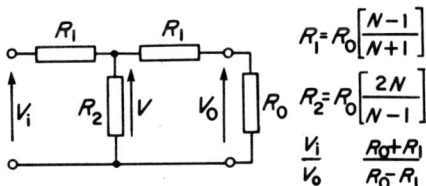

$$R_1 = R_0\left[\frac{N-1}{N+1}\right]$$

$$R_2 = R_0\left[\frac{2N}{N-1}\right]$$

$$\frac{V_i}{V_o} \quad \frac{R_0 + R_1}{R_0 - R_1}$$

Figure 5.4

For the symmetrical T-section shown in Figure 5.4 let the attenuation factor $N = V_i/V_o$. Then from the circuit

$$I_i = \frac{V_o}{R_0} \quad \text{and} \quad V = V_o - R_1 I_i$$

Hence $V = V_i - R_1\frac{V_i}{R_0} = V_i\left[1 - \frac{R_1}{R_0}\right]$

$\therefore$ $V_o = V\frac{R_0}{R_1 + R_0} = V_i\left[1 - \frac{R_1}{R_0}\right]\left[\frac{R_0}{R_1} + R_0\right]$

$$= V_i\left[\frac{R_0 - R_1}{R_0}\right]\left[\frac{R_0}{R_1 + R_0}\right]$$

$$\frac{V_i}{V_o} = N = \frac{R_0 + R_1}{R_0 - R_1} \tag{5.1}$$

This formula gives us the attenuation factor in terms of R_0 and R_1. We require now to derive expressions for R_1 and R_2 in terms of N and R_0. From (5.1) above we have by cross-multiplication

$$NR_0 - NR_1 = R_0 + R_1$$
$$R_0(N - 1) = R_1(N + 1)$$

from which $$R_1 = R_0\left[\frac{N - 1}{N + 1}\right] \tag{5.2}$$

Now $$R_0^2 = R_1^2 + 2R_1 R_2$$

$\therefore$ $$R_2 = \frac{R_0^2 - R_1^2}{2R_1}$$

Substituting the value of R_1 obtained in (5.2), we have

$$R_2 = \frac{R_0^2 - \left[R_0\left\{\frac{N-1}{N+1}\right\}\right]^2}{2R_0\left[\frac{N-1}{N+1}\right]}$$

$$= \frac{R_0[(N+1)^2 - (N-1)^2]}{2(N+1)(N-1)}$$

$$= R_0\left[\frac{4N}{2(N^2-1)}\right] = R_0\left[\frac{2N}{N^2-1}\right] \tag{5.3}$$

Using these formulae, an attenuator can be designed to provide the specified attenuation factor and characteristic resistance.

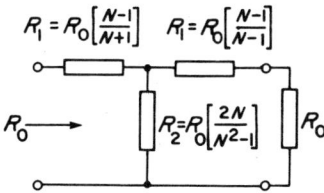

Figure 5.5

The calculated values for R_1 and R_2 have been inserted in the T-section of *Figure 5.5*, where the input and output resistance is R_0.

Example 2. Design a T-type symmetrical pad to provide 20 dB of voltage attenuation with a characteristic resistance of 600 Ω.

Here
$$N = \text{antilog } \frac{20}{20} = 10$$

Then
$$R_1 = R_0\left[\frac{N-1}{N+1}\right] = 600 \times \frac{9}{11}$$
$$= \underline{491\ \Omega}$$

$$R_2 = R_0\left[\frac{2N}{N^2-1}\right] = 600 \times \frac{20}{99}$$
$$= \underline{121\ \Omega}$$

Notice at all times that the attenuation factor N is expressed as a number greater than unity. The attenuator output voltage is, of course, only $1/N$ of the input voltage, in the present example one-tenth.

Example 3. A generator having an internal resistance of 600 Ω supplies current to a 600 Ω load. Design a resistance network of T formation having a characteristic resistance of 600 Ω which, when connected between the generator and the load, will reduce the load current to one-fifth of its original value. In what proportion would the load current be reduced if two such networks were connected in series between the generator and the load?

We note that the network will be symmetrical as it has to match between equal resistances. Also, as the load current is to be reduced to one-fifth, that $N = 5$. The appropriate circuit system is drawn in *Figure 5.6*. Then

$$R_1 = R_0\left[\frac{N-1}{N+1}\right] = 600 \times \frac{4}{6}$$
$$= \underline{400\ \Omega}$$

and
$$R_2 = R_0\left[\frac{2N}{N^2-1}\right] = 600 \times \frac{10}{24}$$
$$= \underline{250\ \Omega}$$

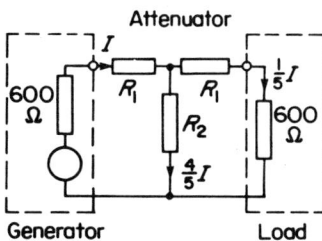

Figure 5.6

It should be particularly noted that the current I supplied by the generator will be the same *whether the attenuator network is in circuit or not*, as the generator sees a load resistance of 600 Ω in either case and cannot distinguish between them. The current in the load with the attenuator in circuit, however, is only one-fifth I as required; the remaining four-fifths I passes along the shunt arm, the distribution being indicated in the diagram.

As the present network reduces the load current to one-fifth of its original value, the insertion of a second similar network will reduce the load current to $1/5 \times 1/5 = 1/25$ of the original current. The 600 Ω characteristic resistance of the system will be unaffected.

5.1.3 The asymmetrical T-section

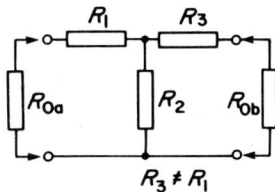

Figure 5.7

An asymmetrical T-pad is shown in *Figure 5.7*. Here the network has *different* characteristic reisistances as viewed from the input and output terminals. Viewed from the input, R_{0b} provides the proper termination; viewed from the output R_{0a} provides the proper termination. To distinguish this case from the symmetric network, two new terms are introduced: iterative resistance and image resistance.

(1) Iterative resistance is the resistance measured at one pair of terminals when the other pair is terminated with a resistance of the same value. Referring to *Figure 5.8*, the resistance seen

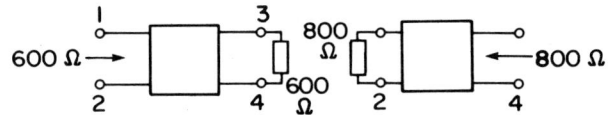

Iterative resistances of an asymmetrical network

Figure 5.8

looking into terminals 1 and 2 is, say, $600\,\Omega$ when terminals 3 and 4 are terminated in $600\,\Omega$; and the resistance seen looking into terminals 3 and 4 is, say, $800\,\Omega$ when terminals 1 and 2 are terminated in $800\,\Omega$. In symmetrical networks the two iterative resistances are of course equal, their common value being the characteristic resistance of the section.

(2) Image resistances are deflned as those resistances that are of such values that when *one* of them is connected across the appropriate pair of terminals of the network, the *other* is presented at the opposite terminals. An example is illustrated in Figure 5.9. Here the resistance seen looking into terminals 1 and

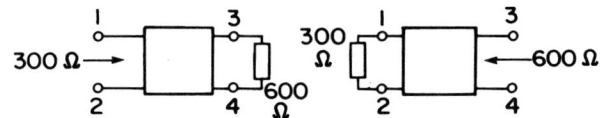

Image resistance of an asymmetrical network

Figure 5.9

2 is, say, $300\,\Omega$ when terminals 3 and 4 are terminated in, say, $600\,\Omega$; and the resistance seen looking into terminals 3 and 4 is $600\,\Omega$ when terminals 1 and 2 are terminated in $300\,\Omega$. An asymmetrical network is correctly terminated when it is terminated in its image resistances. If the image resistances are equal, their common value is the characteristic resistance.

Example 4. An asymmetrical T-section attenuator has series arms of $100\,\Omega$ and $200\,\Omega$ with a shunt arm of $50\,\Omega$. Calculate the image and iterative resistances of this section.

The section is drawn in *Figure 5.10(a)*. Dealing first with the image resistances, the image resistance R_{0a} seen at terminals 1 and 2 is, using the open- and short-circuit formula,

$$R_{0a} = \sqrt{(150 \times 140)} = \underline{145\,\Omega}$$

Likewise, the image resistance seen at terminals 3 and 4 is

$$R_{0b} = \sqrt{(250 \times 233.3)} = \underline{241.5\Omega}$$

Now, before going any further, verify for yourself that the

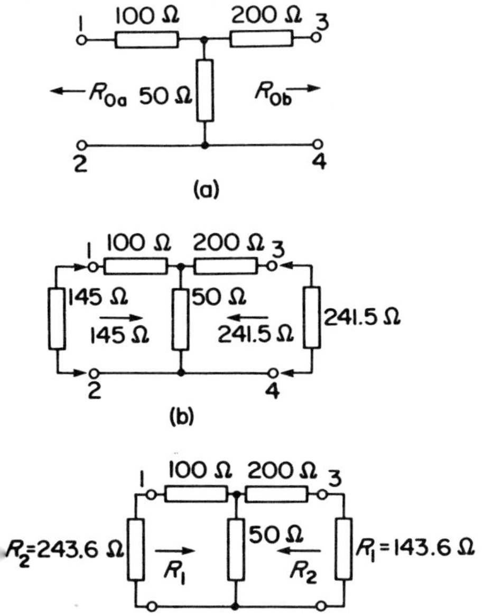

Figure 5.10

Figure 5.11

resistance seen at terminals 1 and 2 when R_{0b} (241.5 Ω) is connected to terminals 3 and 4 is R_{0a} (145 Ω); and that the resistance seen at terminals 3 and 4 when R_{0a} is connected to terminals 1 and 2 is R_{0b} (241.5 Ω). This will explain the meaning of image resistances to you—see *Figure 5.10(b)*.

For the iterative resistances, referring to *Figure 5.10(c)*, seen from terminals 1 and 2

$$R_1 = 100 + \frac{50(200 + R_1)}{250 + R_1}$$

$$\therefore \quad R_1^2 + 150R_1 - 25\,000 = 10\,000 + 50R_1$$

$$R_1^2 + 100R_1 - 35\,000 = 0$$

This is a quadratic equation and the solution is found by the usual formula:

$$R_1 = \frac{-100 \pm \sqrt{(10\,000 + 140\,000)}}{2}$$

$$= -50 \pm 193.6$$

Taking only the positive solution $R_1 = \underline{143.6\,\Omega}$

Seen from terminals 3 and 4

$$R_2 = 200 + \frac{50(100 + R_2)}{150 + R_2}$$

$$\therefore \quad R_2^2 - 50R_2 - 30\,000 = 5000 + 50R_2$$

$$R_2^2 - 100R_2 - 35\,000 = 0$$

Solving as before, we find $R_2 = \underline{243.6\,\Omega}$

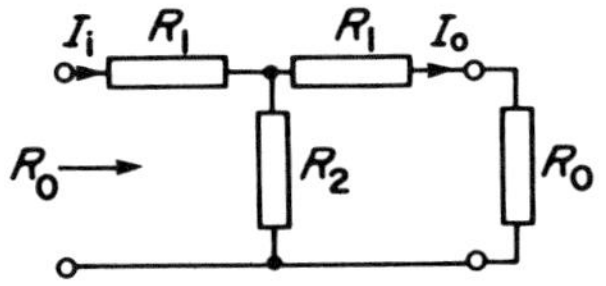

Example 5. Derive an expression for the attenuation of the network shown in *Figure 5.11* in terms of R_1 and R_2 only. Hence, given $R_1 = 75\,\Omega$ and $R_2 = 150\,\Omega$, find the dB attenuation introduced by the network.

If I_i and I_o are the currents entering and leaving the section, then

$$I_o = \frac{R_2}{R_1 + R_2 + R_0}\,I_i$$

since the input and output resistances are each equal to R_0.

$$\therefore \quad \frac{I_i}{I_o} = \frac{R_1 + R_2 + R_0}{R_2} = 1 + \frac{R_1}{R_2} + \frac{R_0}{R_2}$$

Substituting $R_0 = \sqrt{(R_1^2 + 2R_1R_2)}$ into this gives

$$N = \frac{I_i}{I_o} = 1 + \frac{R_1}{R_2} + \frac{1}{R_2}\sqrt{(R_1^2 + 2R_1R_2)}$$

Hence, the anttenuation introduced by the section is

$$20\lg\left[1 + \frac{R_1}{R_2} + \frac{1}{R_2}\sqrt{(R_1^2 + 2R_1R_2)}\right]\text{ dB}$$

In this example, $R_1 = 75\,\Omega$ and $R_2 = 150\,\Omega$, therefore $R_1/R_2 = \frac{1}{2}$ and also $R_0 = \sqrt{[75^2 + (2 \times 75 \times 150)]} = 167.7\,\Omega$

Hence. the attenuation is

$$20 \ \lg\left[1 + \frac{1}{2} + \frac{167.7}{150}\right] = 20\lg 2.618$$

$$= \underline{8.36\,\text{dB}}$$

5.2 THE π-SECTION ATTENUATOR

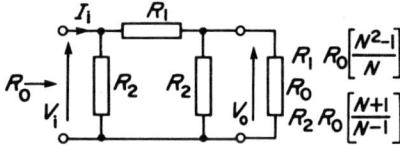

This form of attenuator pad consists of two shunt arms and one series arm and is illustrated in *Figure 5.12*. When used as a symmetrical pad between equal resistances, the two shunt arms are equal in value.

The calculations for the evaluation of R_1 and R_2, given the characteristic resistance R_0 and the desired attenuation factor N are similar to those used for the T-section, and part of the procedure is illustrated in the following worked example.

$$R_1 \ R_0\left[\frac{N^2-1}{N}\right]$$

$$R_2 \ R_0\left[\frac{N+1}{N-1}\right]$$

Figure 5.12

Example 6. A resistive network of π configuration is to have a characteristic resistance R_0 and an attenuation factor N. Derive expressions for the series and shunt resistances and evaluate these in the case when $R_0 = 600\,\Omega$ and the attenuation is 30 dB.

Referring to *Figure 5.12*, let the attenuation factor $N = V_i/V_o$ as before. Then

$$V_o = \frac{V_i}{N}$$

and

$$I_i = \frac{V_i}{R_0} = \frac{V_i}{R_2} + \frac{V_o}{R_2} + \frac{V_o}{R_0}$$

$$= \frac{V_i}{R_2} + \frac{V_i}{NR_2} + \frac{V_i}{NR_0}$$

$$= V_i\left[\frac{1}{R_2} + \frac{1}{NR_2} + \frac{1}{NR_0}\right]$$

$$\therefore \quad \frac{1}{R_0} = \frac{1}{R_2} + \frac{1}{NR_2} + \frac{1}{NR_0}$$

$$\frac{1}{R_0}\left[1 - \frac{1}{N}\right] = \frac{1}{R_2}\left[1 + \frac{1}{N}\right]$$

$$R_2 = R_0\left[\frac{N+1}{N-1}\right] \tag{5.4}$$

It is left as an exercise for the reader to show that

$$R_1 = R_0\left[\frac{N^2 - 1}{2N}\right] \tag{5.5}$$

noting as a clue that the voltage across $R_1 = V_i - V_o$.

Compare these last equations with those derived for the T-section, (5.2) and (5.3) earlier.

In the problem, $R_0 = 600\,\Omega$ and the attenuation is 30 dB. Since the attenuation is given by $20\lg N$, $\lg N = 1.5$ and so $N = 31.6$. Substituting these values into the appropriate formulae, we have

$$R_1 = 600\left[\frac{31.6 - 1}{2 \times 31.6}\right] = 600 \times \frac{30.6}{63.2}$$

$$= 290\,\Omega$$

$$R_2 = 600\left[\frac{31.6 + 1}{31.6 - 1}\right] = 639\,\Omega$$

5.3 THE L-SECTION ATTENUATOR

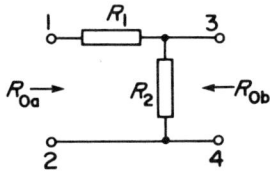

Figure 5.13

The basic L-section attenuator pad is illustrated in *Figure 5.13*. It is essentially a T-section with the right-hand series member removed or a π-section with the left-hand shunt member removed. An L-section is fundamentally used for matching purposes only, the design being such that the attenuation introduced is a minimum. To derive values for R_1 and R_2 we consider the resistances seen from either end of the section.

Seen from terminals 1 and 2

$$R_{0a} = R_1 + \frac{R_2 R_{0b}}{R_2 + R_{0b}}$$

whence $R_{0a}R_2 + R_{0a}R_{0b} = R_1R_2 + R_1R_{0b} + R_2R_{0b}$ (5.6)

and from terminals 3 and 4

$$R_{0b} = \frac{R_2(R_1 + R_{0a})}{R_1 + R_2 + R_{0a}}$$

whence $R_{0b}R_2 + R_{0b}R_1 + R_{0a}R_{0b} = R_1R_2 + R_2R_{0a}$ (5.7)

Adding (5.6) and (5.7) we have

$$2R_{0a}R_{0b} = 2R_1R_2$$

$$\therefore \quad R_1 = \frac{R_{0a}R_{0b}}{R_2}$$

Substituting this value for R_1 back into equation (5.6) gives

$$R_2 = \sqrt{\left(\frac{R_{0a}R_{0b}^2}{R_{0a} - R_{0b}}\right)}$$ (5.8)

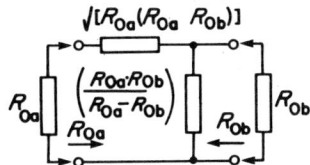

Figure 5.14

Hence $R_1 = \sqrt{[R_{0a}(R_{0a} - R_{0b})]}$ (5.9)

We see from *Figure 5.14* that the L-section has image resistances R_{0a} and R_{0b}. Clearly from a study of the formulae for R_1 and R_2, R_{0a} must be greater than R_{0b}.

5.4 MULTI-SECTION ATTENUATOR SYSTEMS

It is not always possible to obtain the required attenuation for a given value of R_0 in a single T- or π-section attenuator, nor indeed is it desirable. It may be necessary to cover a range of attenuations, either in discrete steps or a continuous variation, and a single section is then not a practicable method.

A switched type attenuator consisting of a number of fixed pads of equal characteristic resistance but having different attenuation factors is shown in *Figure 5.15*. Each pad may be switched in or out of circuit as required and, with the attenuations illustrated, any attenuation up to a total of 31 dB may be obtained in steps of 1 dB. As shown, there is a total attenuation of 10 dB in circuit.

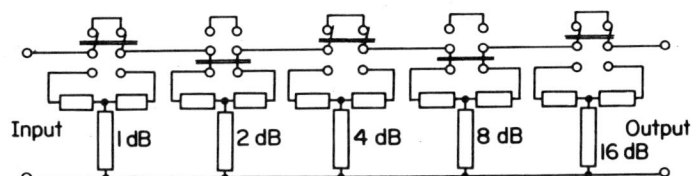

Figure 5.15

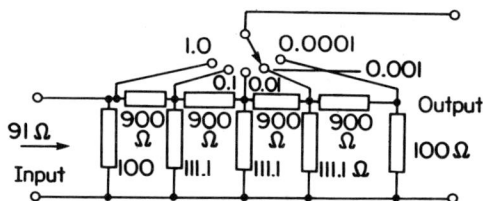

Figure 5.16

Figure 5.16 shows an alternative arrangement. Here a number of sections are cascaded and switched to provide a decimal attenuator, the characteristic resistance with the values shown being 91 Ω. In some cases this form of attenuator is made with the series arm (or a part of it) in the form of a continuous resistance, the output being taken from a sliding contact moving along it. The output attenuation is then continuously variable or variable between steps.

Care must be exercised in the design of such variable multi-section attenuators to ensure that they are as purely resistive as possible. This means that stray capacitance across the sections must be reduced to a minimum and that the interconnecting wires are kept as short as possible to eliminate the effects of inductance. The sections are usually screened from each other to avoid capacitive transfer, and the series arms are fed through the screening walls. The power ratings of the resistors making up the sections must also be adequate enough to ensure that no appreciable temperature rise can occur in normal use.

5.4.1 Balanced attenuators

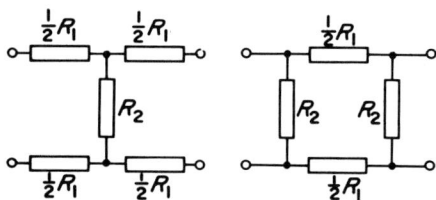

Figure 5.17

It is sometimes necessary for an attenuator to be balanced with respect to the line into which it is inserted. A balanced attenuator is made from the basic T- or π-secion by dividing the series arm into two equal parts and inserting one part into the opposite series lead. *Figure 5.17* shows the arrangement for T- and π-sections. The calculation of the component parts is straightforward, being simply that of the basic (unbalanced) sections formulae already derived, with the series arm division being carried out afterwards. The characteristics of the balanced sections are completely identical with those of the unbalanced sections from which they are derived. Great care should be taken not to confuse the word 'balanced' with 'symmetrical' when dealing with attenuator sections.

Summary. This chapter has described resistive attenuator sections for insertion between resistive sources and loads. Such attenuators are required to yield without distortion at their output terminals a predetermined fraction of the input voltage, current or power. For an ideal attenuator, the propagation constant γ is real so that the phase constant β is zero; also, the characteristic impedance is purely resistive.

Attenuators may be either of balanced or unbalanced form. The properties of balanced networks are readily derived from those of unbalanced networks.

PROBLEMS FOR CHAPTER 5

(1) Construct a T-section attenuator to provide 14 dB attenuation per section and to match to a 2000 Ω line.

(2) Compute the values of the shunt and series resistances for

T-pad attenuator sections having a characteristic resistance of 600 Ω to provide the following attenuations: (a) 10 dB, (b) 15 dB, (c) 25 dB, (d) 40 dB. Give your answers to the nearest whole number.

(3) A π-section attenuator has series resistor 1000 Ω and shunt resistors each 4000 Ω. Find the characteristic resistance of the section and the attenuation produced.

(4) A television receiver has an input resistance of 75 Ω and is fed through a coaxial cable whose characteristic resistance is also 75 Ω. If the signal input has to be attenuated by 10 dB before entering the receiver, design a suitable T-section to give this reduction. Would this attenuator section be balanced or unbalanced? What nearest 'preferred' values of resistance would you choose?

(5) In the switched attenuator shown in *Figure 5.15*, calculate the values of the resistors used in the five sections to provide the attenuations shown; the input and output resistances are 600 Ω. What special precuations would you exercise in building this attenuator into a small metal box?

(6) Compute the values of the shunt and series resistances for π-pad attenuator sections of characteristic resistance 300 Ω to give the following attenuations: (a) 6 dB, (b) 12 dB, (c) 20 dB, (d) 25 dB.

(7) The T-section shown in *Figure 5.18* is connected to a generator of e.m.f. 1 V and internal resistance 40 Ω. What value of load resistor connected across the output terminals will absorb the maximum power? What will be the magnitude of this power? Explain what is meant by insertion loss and calculate the insertion loss of the network under these conditions.

(8) A battery of negligible internal resistance is connected to the input terminals of the T-network shown in *Figure 5.19*. Calculate the current drawn from the battery when (a) the network is open-circuited, (b) the network is short-circuited, (c) the network is correctly terminated. In this last case, calculate the voltage developed across the termination and so state the attenuation of the network in dB.

(9) Deduce that the symmetrical lattice of *Figure 5.20(a)* is electrically equivalent to the bridge network shown in *Figure 5.20(b)*. Prove that the characteristic resistance of the lattice is $\sqrt{(R_1R_2)}$. If the bridge in *Figure 5.20(b)* has $R_1 = 10\,\Omega$ and $R_2 = 40\,\Omega$, and a resistance of 20 Ω is connected across terminals 3 and 4, what current will be drawn from a 2 V battery connected to terminals 1 and 2?

(10) Let the characteristic resistance of the lattice of *Figure 5.20(a)* be R_0. Prove that

$$R_1 = R_0\left[\frac{N-1}{N+1}\right]$$

$$R_2 = R_0\left[\frac{N+1}{N-1}\right]$$

where N is the attenuation ratio.

(11) A generator of internal resistance 600 Ω is connected to a 150 Ω load resistor via an L-section matching pad, as shown in *Figure 5.21*. Calculate the values of R_1 and R_2, the attenuation and the insertion loss of the pad.

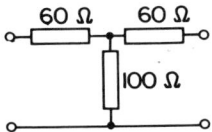
Figure 5.18

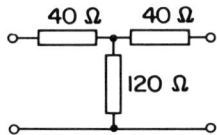
Figure 5.19

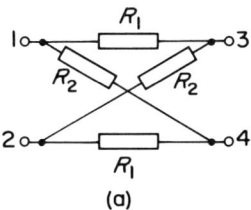
(a)

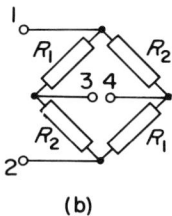
(b)
Figure 5.20

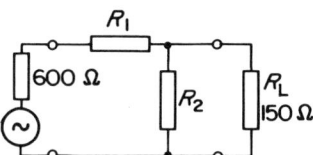
Figure 5.21

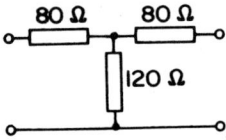

80 Ω   80 Ω

120 Ω

Figure 5.22

(12) Calıculate the characteristic resistance of the network shown in *Figure 5.22*. A source, having an e.m.f. of 1 V and internal resistance equal to R_0 is connected to the input terminals of the network. Calculate the open-circuit voltage across the output terminals and the resistance measured across them.

From this result, or otherwise, show that the insertion loss of the network between this source and a load is independent of the resistance of the load. For a given value of load resistance, will the insertion loss of the network depend upon the internal resistance of the source?

(13) A symmetrical attenuator section having a characteristic resistance R_0 is divided into two electrically equal halves. Make diagrams showing how this can be done for both T- and π-sections. Deduce that the image resistance of one of the asymmetrical sections so formed, at the terminals which correspond to the terminals of the original section, is R_0.

(14) A T-section and a π-section have identical resistance values for their R_1 and R_2 resistors, respectively. Show that the product of their respective characteristic resistances, R_{0T} and $R_{0\pi}$, is equal to $R_1 R_2$.

6 Filters

A filter is a network, which may be passive or active, designed to attenuate certain ranges of frequency and to pass others without loss. It is therefore unlike an attenuator network composed only of resistive elements in which the attenuation is constant and independent of frequency. As a filter is frequency sensitive, we shall expect it to be composed of reactive elements; further, since certain frequencies are to be passed without loss, it follows that these reactors must be ideal, for the presence of resistance will result in some attenuation at all frequencies.

A filter will possess at least one band of frequencies in which the attenuation is zero (the *pass-band*) and at least one band of frequencies in which the attenuation is finite (the *attenuation band*). The frequency at which the attenuation changes from zero to some finite value is called the cut-off frequency, and for some filter designs there may be several of these. Ideally, any unavoidable phase shift in the pass-band should be proportional to frequency, and since the load is normally resistive, the characteristic impedance of the section in the pass-band should be a resistance of constant value. Practically it is not possible to make the phase shift proportional to frequency over the whole of the pass-band, and the transition from pass-band to attenuation band at the cut-off frequency involves a gradual increase in the attenuation rather than an abrupt change as the frequency recedes from its cut-off value.

We shall consider here only those cases of filter sections composed of ideal reactances. In practice, the results obtained agree closely with those derived from this simplification provided high-Q elements are used. The filters considered will, in general, be symmetrical unbalanced T- or π-sections, and examples are illustrated in *Figure 6.1(a)* and *6.1(b)*, respectively. The case of balanced sections will not be dealt with since their performance can be deduced from the theory of unbalanced section in exactly the same way as balanced attenuators were derived from unbalanced forms at the end of the previous chapter.

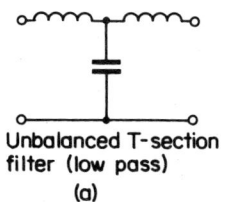

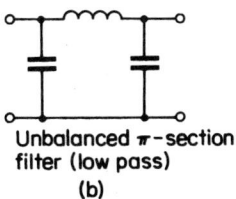

Unbalanced T-section filter (low pass)
(a)

Unbalanced π-section filter (low pass)
(b)

Figure 6.1

6.1 THE T-SECTION

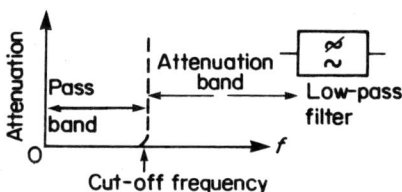

The simple *prototype* T-section filter shown in *Figure 6.1(a)* will, if connected into a circuit and fed with a constantly increasing frequency, be found to exhibit a frequency-attenuation characteristic in form basically like the one ideally presented in *Figure 6.2*. It will be found to pass all frequencies up to a certain value (the *cut-off frequency*) without attenuation, and then to attenuate all frequencies above this limit. For this reason it is known as a *low-pass filter* and its symbol is shown at the side of the figure.

If, on the other hand, capacitors are placed in the series arm of *Figure 6.1(a)* instead of the inductors and an inductor is

Figure 6.2

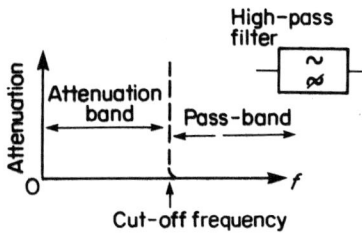

Figure 6.3

substituted for the shunt capacitor, the filter section now derived will be found to exhibit a frequency-attenuation characteristic in form basically like that ideally presented in *Figure 6.3*. All frequencies *below* the cut-off frequency are now attenuated and all frequencies above this limit are passed without loss. Such a filter is known as a *high-pass filter* and its circuit symbol is again illustrated at the side of the figure.

Such filters, then, are kinds of electronic sieves, permitting frequencies below (or above) a certain limit to pass through their reactive mesh, while blocking all others. Both of the characteristic curves have of course been ideal. They have assumed that there is zero attenuation in the pass-bands and infinite attenuation in the attenuation bands, conditions which cannot be realised in practice. For just as the mesh grading of an actual sieve is bound to have slight irregularities which allow a few larger pieces of material to pass through with the smaller, so in a practical filter section the attenuation in the attenuation band is finite; and, because of the presence of resistance, mainly in the inductive elements, the attenuation in the pass-band is not zero. Further, the filter has to be inserted into a circuit and ideally it must be matched to the impedance of that circuit. But the characteristic impedance of a filter will vary with frequency, and the termination itself may well be an impedance that does not vary with frequency in the same way as the impedance of the filter. So to resistive loss there must be added loss due to mismatching. Filters are used under image impedance conditions as far as possible to minimise such losses.

6.1.1 Nature of the impedance

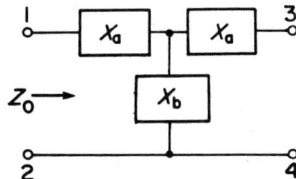

Figure 6.4

Begin with the T-section shown in *Figure 6.4*. Let X_a and X_b be reactances of opposite kind, then the input impedance of the section Z_0, when the output terminals 3 and 4 are open- and short-circuited can be either inductive or capacitive, depending upon the frequency of input signal. Let the magnitude of the reactance on open-circuit be X_1 and the magnitude of the reactance on short-circuit be X_2. As the filter elements are all purely reactive they may be expressed in the form jX, where X is real, but may be positive or negative in sign. Four combinations of jX for the input impedance are now possible, namely:

(a) $Z_{oc} = jX_1$ and $Z_{sc} = jX_2$

(b) $Z_{oc} = jX_1$ and $Z_{sc} = -jX_2$

(c) $Z_{oc} = -jX_1$ and $Z_{sc} = -jX_2$

(d) $Z_{oc} = -jX_1$ and $Z_{sc} = jX_2$

Now, from general section theory, we know that

$$Z_0 = \surd(Z_{oc}Z_{sc})$$

Taking cases (b) and (d) above, we have

$$Z_0 = \surd(-j^2X_1X_2) = \surd(X_1X_2)$$

which is real. The input impedance will therefore be purely *resistive*.

Taking cases (a) and (c), we have

$$Z_0 = \surd(j^2X_1X_2) = \pm j\surd(X_1X_2)$$

which is imaginary. The input impedance will therefore be purely *reactive*.

Hence, as already remarked, since both the magnitudes and the nature of Z_{oc} and Z_{sc} depend upon frequency, so also will the magnitude and nature of Z_0 depend upon frequency.

6.2 ATTENUATION AND CHARACTERISTIC IMPEDANCE

The characteristic impedance of a section can be derived from a knowledge of its open- and short-circuited impedance, i.e.

$$Z_0 = \sqrt{(Z_{oc}Z_{sc})}$$

Let the section be terminated in an impedance Z_0; the input impedance is then also Z_0. We may now state an important theorem:

'Throughout the frequency range for which the characteristic impedance is purely resistive, the attenuation is zero. Throughout the frequency range for which the characteristic impedance is purely reactive, the attenuation is finite.'

The validity of this theorem is readily demonstrated by elementary considerations. Let the low-pass filter shown in *Figure 6.5* be working over a range of frequencies such that Z_0 is purely resistive. Let also the input voltage and current be V_i and I_i, respectively, and the output voltage and current V_o and I_o, respectively. Then clearly

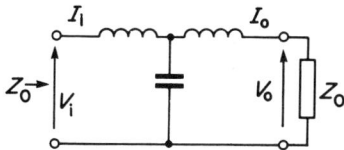

Figure 6.5

$$Z_0 = \frac{V_i}{I_i} = \frac{V_o}{I_o}$$

and the power dissipated in the termination is I_oV_o. This will be the same as the power delivered at the input of the section, I_iV_i, that is $V_i = V_o$ and $I_i = I_o$. No power can be absorbed by the filter members as they are purely reactive. Hence, if the filter is properly terminated and working in such a frequency range that Z_0 is resistive, all the power delivered to the input must be passed to the output *and there is no attenuation*.

If now the frequency range is such that Z_0 is purely reactive, then

$$\frac{V_i}{I_i} = jZ_0 = \frac{V_o}{I_o}$$

but this time the voltage and current are everywhere in quadrature, hence the circuit can neither accept nor deliver any power from the source to the load. Theoretically, then, there is infinite attenuation. The attenuation in practice, however, is finite, for the condition $V_i/I_i = V_o/I_o$ can hold for V_o less than V_i and I_o less than I_i, since V and I are 90° out of phase. This fact is illustrated in *Figure 6.6*.

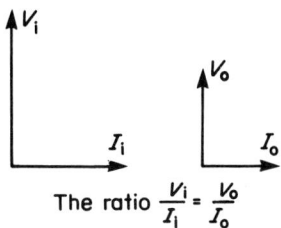

The ratio $\dfrac{V_i}{I_i} = \dfrac{V_o}{I_o}$

Figure 6.6

The theorem given above is important because it can be applied to determine the cut-off frequency point of any filter section from a knowledge of the nature of Z_0. For Z_0 is *real* in the pass-band and *imaginary* in the attenuation band; and so the cut-off frequency is that point along the frequency scale at which Z_0 changes from a real to an imaginary quantity. This point is readily determined.

6.2.1 Ladder analysis

At this stage we go back for a moment to our earlier consideration of a ladder network. A repetitive network may be

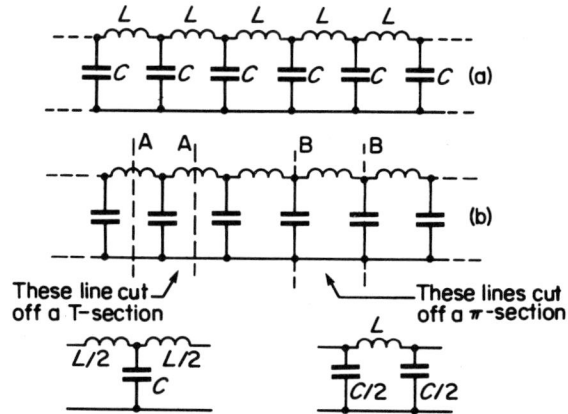

Figure 6.7

considered as a series of T- or π-sections in cascade. A low-pass network is shown as a ladder in *Figure 6.7(a)*. Now either a T-section or a π-section can be taken from this ladder by cutting along the lines AA or BB, respectively, as shown in detail in *Figure 6.7(b)*. The T-section is clearly composed of reactive elements $L/2$ and C as shown, and the ladder may be considered as being built up of a large number of such sections in cascade. Notice that the section has been cut in the centre of each of its *series* elements. The π-section, cut off by the lines BB, is also shown in detail in *Figure 6.7(b)*. Again, the ladder may be considered as being built up of a large number of such sections in cascade. Notice that the reactive elements are this time L and $C/2$, and that the section has been cut along the centre of each of its *shunt* elements.

6.2.2 The cut-off frequency

From general theory

$$Z_0 = \sqrt{(X_a^2 + 2X_a X_b)}$$

In the low-pass T-section of *Figure 6.7*

$$X_a = \frac{j\omega L}{2} \text{ and } X_b = \frac{1}{j\omega C}$$

$$\therefore \quad Z_0 = \sqrt{\left(\frac{j^2\omega^2 L^2}{4} + \frac{j\omega L}{j\omega C}\right)}$$

$$= \sqrt{\left(\frac{L}{C} - \frac{\omega^2 L^2}{4}\right)} \tag{6.1}$$

Z_0 will be real if $L/C - (\omega^2 L^2)/4$ is positive, that is, if L/C is greater than $(\omega^2 L^2)/4$. Hence, attenuation will commence when

$$\frac{L}{C} = \frac{\omega^2 L^2}{4} \text{ or when } \omega_c^2 = \frac{4}{LC} \tag{6.2}$$

where $\omega_c = 2\pi f_c$ and f_c is the cut-off frequency.

$$\therefore \quad f_c = \frac{1}{\pi\sqrt{(LC)}} \tag{6.3}$$

When the frequency is very low, ω is small and the term $\omega^2 L^2/4$ in equation (6.1) may be neglected. The characteristic impedance then reduces to

$$Z_0 = \sqrt{\frac{L}{C}} = R_0 \tag{6.4}$$

which is purely resistive. This particular value of Z_0 is known as the *design* or *nominal impedance* of the section.

It follows from a consideration of equation (6.1) that Z_o is a function of frequency and if the design impedance is made to equal the load or circuit impedance into which the filter feeds (as it is in practice), then the matching deteriorates as the frequency increases from zero towards f_c. It is a convention to make the terminating impedance equal to the value of Z_0 *well within the pass-band*, that is, to take as the characteristic impedance the *limiting* value of Z_0 as the frequency approaches zero. This limit is plainly $\sqrt{(L/C)}$. The filter is then properly terminated at very low frequencies (relative to f_c) and the effect of mismatching is not pronounced until the cut-off frequency is approached, particularly if the source impedance has itself a resistance equal to $\sqrt{(L/C)}$. *Figure 6.8* illustrates the effect of this. Here the loss-frequency characteristic is plotted for a single T-section terminated in $\sqrt{(L/C)}$. The input impedance is resistive only at zero frequency and some loss is introduced into the pass-band. If several sections are cascaded and terminated in $\sqrt{(L/C)}$ or if a matching section is inserted between the T-network and the load, the loss-frequency characteristic for each section will, in general, lie between the two curves of *Figure 6.8*.

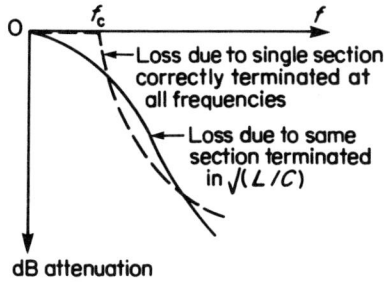

Figure 6.8

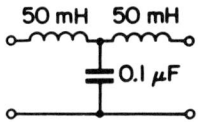

Figure 6.9

Example 1. Calculate the cut-off frequency and the nominal impedance of the low-pass filter section shown in *Figure 6.9*. Here

$$L = 100\,\text{mH} = 0.1\,\text{H}$$
$$C = 0.1\,\mu\text{F}$$

$$f_c = \frac{1}{\pi\sqrt{(LC)}} = \frac{1}{\pi\sqrt{(0.1 \times 0.1 \times 10^{-6})}}$$

$$= \frac{10^3}{0.1\pi} = 3.183\,\text{kHz}$$

$$R_0 = \sqrt{\frac{L}{C}} = \sqrt{\left(\frac{0.1}{0.1 \times 10^{-6}}\right)} = 1\,\text{k}\Omega$$

Example 2. By considering the T- and π-sections shown in *Figure 6.10*, and assuming that the input frequency is well below their cut-off frequencies, show that $R_0^2 = Z_{0T}Z_{0\pi}$.

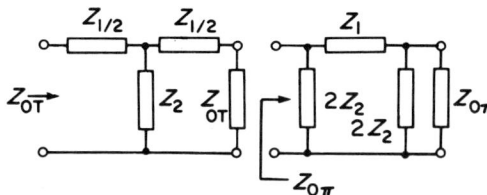

Figure 6.10

From general theory $\quad Z_{0T}\,Z_{0\pi} = Z_1 Z_2$

So in the filter sections shown where $Z_1 = j\omega L$ and $Z_2 = 1/j\omega C$, we have

$$Z_1 Z_2 = \frac{j\omega L}{j\omega C} = \frac{L}{C} = R_0^2$$

Hence
$$Z_{0T} Z_{0\pi} = R_0^2$$

Any T- or π-section having series and shunt impedances Z_1 and Z_2, respectively, connected by the relationship given in the solution of this last example is known as a *constant-k* section. Clearly Z_{0T} and $Z_{0\pi}$ will both be real or both be imaginary together; and when Z_{0T} changes from real to imaginary at the cut-off frequency, so also will $Z_{0\pi}$. The two sections will therefore have identical cut-off frequencies and so identical pass-bands. Constant-k sections of any kind of filter are known as *prototypes* and from these prototypes more complex sections may be derived.

Example 3. Design a low-pass T-section filter having a characteristic impedance at zero frequency of $600\,\Omega$ and a cut-off frequency of $4\,\text{MHz}$.

This problem requires equations connecting R_0 and f_c with L and C. Since

$$f_c = \frac{1}{\pi\sqrt{(LC)}} \quad \text{and} \quad R_0 = \sqrt{\frac{L}{C}}$$

by substitution and elimination we obtain

$$L = \frac{R_0}{\pi f_c} \quad \text{and} \quad C = \frac{1}{\pi R_0 f_c}$$

In the problem
$$R_0 = 600\,\Omega$$
$$f_c = 4 \times 10^6\,\text{Hz}$$

Then
$$L = \frac{600}{4\pi \times 10^6} = 47.75 \times 10^{-6}\,\text{H}$$
$$= 47.75\,\mu\text{H}$$

Also
$$C = \frac{1}{4\pi \times 10^6} \times \frac{1}{600}$$
$$= \frac{1}{24\pi} \times 10^{-8}\,\text{F} = 132.6\,\text{pF}$$

The section will have two series inductors each of about $23.8\,\mu\text{H}$ and a shunt capacitance of $132.6\,\text{pF}$.

6.3 THE LOW-PASS PROTOTYPE

The characteristic impedance of a low-pass T-section
$$Z_{0T} = \sqrt{\left(\frac{L}{C} - \frac{\omega^2 L^2}{4}\right)}$$

may be written
$$= \sqrt{\frac{L}{C}}\sqrt{\left(1 - \frac{\omega^2 LC}{4}\right)} = R_0\sqrt{\left(1 - \frac{\omega^2 LC}{4}\right)}$$

But from equation (6.2)
$$\frac{4}{LC} = \omega_c^2$$

Hence
$$Z_{0T} = R_0\sqrt{\left(1 - \left[\frac{\omega}{\omega_c}\right]^2\right)} \tag{6.5}$$

Further, since $Z_{0\pi} Z_{0T} = R_0^2$ we may write

$$Z_{0\pi} = \frac{R_0}{\sqrt{\left(1 - \left[\dfrac{\omega}{\omega_c}\right]^2\right)}} \qquad (6.6)$$

When $\omega = \omega_c$, which is at the cut-off frequency, expression (6.5) reduces to zero, that is, Z_{0T} falls from the nominal impedance R_0 at zero frequency to zero ohms at f_c. Conversely, expression (6.6) becomes infinite at the cut-off frequency, that is, $Z_{0\pi}$ is infinite at f_c.

The variation of the input impedance for both T- and π-sections with frequency can be shown graphically by plotting these last two expressions, and the results are seen in *Figure 6.11 (a)*. Z_0 is not depicted for frequencies above cut-off as it becomes purely reactive in the attenuation band. The departure of Z_0 from the nominal impedance R_0 as the frequency changes from zero up to f_c should be noted, particularly in the case of the π-section.

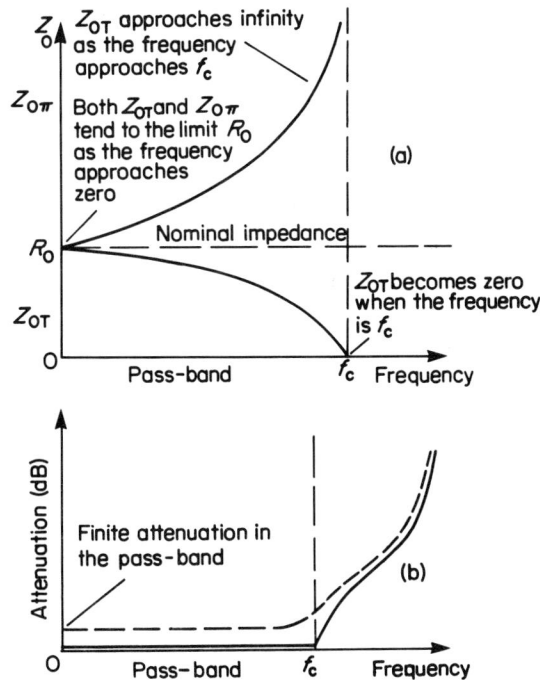

Figure 6.11

Figure 6.11(b) shows the frequency–attenuation characteristic of a low-pass prototype section, the general form of the curve obtained in practice being shown dotted. A comparison of this curve with the ideal characteristic shown earlier in *Figure 6.2* demonstrates at once that the practical prototype falls a long way short of perfection in this respect; in fact, the attenuation provided by a prototype section is only 10 dB at a frequency $f = 1.2f_c$. The situation can be improved to a certain extent by connecting two or more identical sections in cascade. Although the attenuation over the pass-band is then increased, it is still sufficiently small to ignore, but the increased attenuation in the attenuation band provides a much sharper cut-off characteristic.

Example 4. Show that $Z_{0\pi}$ is equal to $2.3 R_0$ at a frequency $f = 0.9 f_c$. Verify this by finding the impedance for the π-section equivalent to the T-section used in Example 3 earlier at a frequency of 3.6 MHz. Compare this impedance with that of the T-section at the same frequency.

From equation (6.6) $Z_{0\pi} = \dfrac{R_0}{\sqrt{\left(1 - \left[\dfrac{\omega}{\omega_c}\right]^2\right)}} = \dfrac{R_0}{\sqrt{[1 - (0.9)^2]}}$

$$= 2.3 R_0$$

For the T-section of Example 3, $R_0 = 600\,\Omega$ and $f_c = 4\,\text{MHz}$. For the equivalent π-section, these values will be the same. At $f = 0.9 f_c$, the frequency is 3.6 MHz, hence

$$Z_0 = 600 \times 2.3 = 1380\,\Omega$$

Also at 3.6 MHz $Z_{0T} = R_0 \sqrt{\left(1 - \left[\dfrac{\omega}{\omega_c}\right]^2\right)}$

$$= 600\sqrt{[1 - (0.9)^2]} = 600 \times 0.436$$
$$= 262\,\Omega$$

The departure from the nominal impedance of $600\,\Omega$ is much greater in the π-section than it is in the T-section, as the curves of *Figure 6.11(a)* have already indicated.

6.3.1   The high-pass prototype

Figure 6.12(a) and *6.12(b)* show the prototype high-pass T-section and π-section, respectively. As for the low-pass cases, the sections are constant-k sections with $R_0 = \sqrt{(L/C)}$.

The cut-off frequency is determined as before by considering the basic equation

$$Z_0 = \sqrt{(X_a^2 + 2X_a X_b)}$$

where this time $X_a = j\omega 2C$ and $X_b = j\omega L$

$\therefore$
$$Z_{0T} = \sqrt{\left(\frac{4}{j^2\omega^2 C^2} + \frac{2j\omega L}{2j\omega C}\right)}$$

$$= \sqrt{\frac{L}{C}}\sqrt{\left(1 - \frac{1}{4\omega^2 LC}\right)} = R_0 \sqrt{\left(1 - \frac{1}{4\omega^2 LC}\right)}$$

$$(6.7)$$

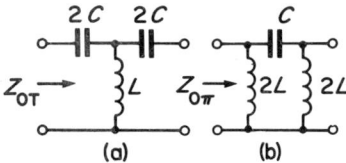

2C 2C   C

$Z_{0T} \rightarrow$ L $Z_{0\pi} \rightarrow$ 2L   2L

(a)   (b)

Figure 6.12

This will be real for $1/(4\omega^2 LC)$ less than 1, hence the filter will pass all frequencies above the point where $4\omega^2 LC = 1$, that is, where

$$\omega_c = \frac{1}{2\sqrt{(LC)}}$$

or
$$f_c = \frac{1}{4\pi\sqrt{(LC)}} \qquad (6.8)$$

This is the cut-off frequency of a high-pass prototype. We note that equation (6.7) may be written

$$Z_{0T} = R_0 \sqrt{\left(1 - \left[\frac{\omega_c}{\omega}\right]^2\right)} \qquad (6.9)$$

and that since $Z_{0T} Z_{0\pi} = R_0^2$

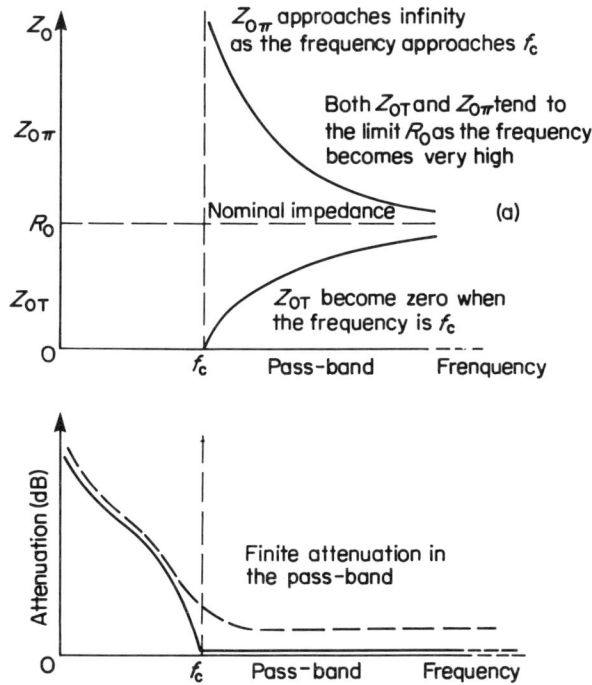

Figure 6.13

$$Z_{0\pi} = \frac{R_0}{\sqrt{\left(1 - \left[\dfrac{\omega_c}{\omega}\right]^2\right)}} \qquad (6.10)$$

These last two equations again enable the variation of impedance with frequency to be plotted graphically, and this has been done in *Figure 6.13(a)*. Z_0 is not shown for frequencies below f_c as it is purely reactive in the attenuation band. The departure of the impedance from the nominal R_0 as the frequency approaches f_c should again be noted. *Figure 6.13(b)* shows the frequency–attenuation characteristic of a high-pass prototype filter section, the practical curve again being shown in dotted line.

To determine values for the circuit elements L and C, we make use of equation (6.8) and that for R_0:

$$f_c = \frac{1}{4\pi\sqrt{(LC)}} \text{ and } R_0 = \sqrt{\frac{L}{C}}$$

From these we obtain

$$L = \frac{R_0}{4\pi f_c}, \quad C = \frac{1}{4\pi R_0 f_c}$$

Care must be taken in the way these values are actually distributed in the filter sections.

Example 5. A filter is required to pass all frequencies above $5\,\text{kHz}$ and to have a nominal impedance of $600\,\Omega$. Calculate the circuit values for both T- and π-section prototypes and illustrate the filters.

For the high-pass filter

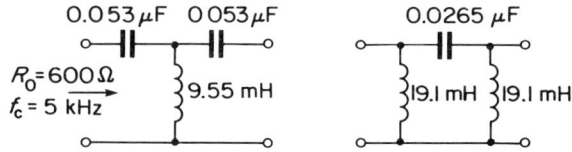

Figure 6.14

$$L = \frac{R_0}{4\pi f_c} = \frac{600}{4\pi \times 5000} = 9.55 \times 10^{-3}\,\text{H}$$

$$= 9.55\,\text{mH}$$

$$C = \frac{1}{4\pi R_0 f_c} = \frac{10^6}{4\pi \times 600 \times 5000}$$

$$= 0.0265\,\mu\text{F}$$

The sections are illustrated in *Figure 6.14*.

6.4 PHASE RELATIONSHIPS

As far as attenuation is concerned, the filter sections so far discussed have worked satisfactorily in that they have provided a pass-band throughout which the attenuation has been small, and an attenuation band throughout which the loss has been reasonably large. But while the attenuation throughout the pass-band has been for all practical purposes constant, the phase shift between the input and output voltages will be found to vary considerably over the range of frequencies comprising the pass-band. We have now to consider not only the attenuation constant α (which was all that was necessary for resistive attenuator sections), but the phase shift β which makes up the full propagation coefficient $\alpha + j\beta$ of the network.

In the prototype T-section of *Figure 6.15(a)* let the termination be the nominal impedance R_0. The input impedance for frequencies well removed from the cut-off frequency is then also R_0 and resistive. Further, let the distribution of current and voltage throughout the network be as shown.

To draw the phasor diagram of the section, we start with I_i in phase with V_i (*Figure 6.15(b)*), recalling that the input impedance is resistive. The voltage across the inductance $V_L = \omega L I_i/2$ will lead I_i by 90° and as the phasor sum of V_L and V_c must be V_i, phasor V_c ($= 0B$) may be drawn as shown to complete the parallelogram 0DAB of which V_i is the resultant diagonal 0A. If we ignore the small attenuation introduced in the pass-band, the termination R_0 will receive the whole of the transmitted power, hence V_0 will equal V_i in magnitude. Further, the voltage developed across the second part of the series inductance is $\omega L I_0/2 = \omega L I_i/2 = V_L$. The triangle 0BC may now be drawn, since 0B is already completed and V_0 and V_L are known. 0BC is of course right-angled at C. Having thus obtained V_0, I_0 ($= I_i$) may be drawn in phase with V_0.

The phase lag experienced over the section is then β where

(a)

(b)

Figure 6.15

$$\tan\frac{\beta}{2} = \frac{V_L}{V_i} = \frac{\omega L I_i}{2} \times \frac{1}{I_i R_0}$$

$$= \frac{\omega L}{2R_0} \tag{6.11}$$

Since $R_0 = \sqrt{(L/C)}$ we may write

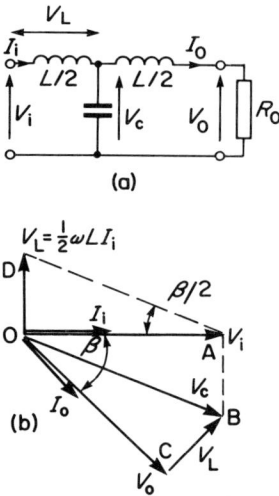

$$\tan \frac{\beta}{2} = \frac{\omega L}{2} \times \sqrt{\frac{C}{L}} = \frac{\omega \sqrt{(LC)}}{2}$$

For small angles (up to about 20°) $\tan \beta/2 \simeq \beta/2$ rad, hence

$$\beta = \omega \sqrt{(LC)} \qquad (6.12)$$

Thus the angle of phase shift over a section is proportional to the frequency which, ideally, is what is required. For frequencies within the attenuation band the phase shift is unimportant since all voltages having such frequencies are suppressed.

The ideal graph showing phase shift against frequency would be a straight line of gradient $2\pi \sqrt{(LC)}$, from equation (6.12), passing through the origin. This is not the case in a practical filter and the assumption of a straight-line relationship is only valid up to a frequency of about $0.7f_c$ for a low-pass filter.

It can be shown that

$$\beta = \cos^{-1}\left[1 - 2\left\{\frac{f}{f_c}\right\}^2\right] \text{ rad}$$

for the low-pass filter, and

$$\beta = \cos^{-1}\left[1 - 2\left\{\frac{f_c}{f}\right\}^2\right] = -2\sin - \left[\frac{f_c}{f}\right] \text{ rad}$$

for the high-pass filter.

The first of these equations is illustrated in the curve of *Figure 6.16* which shows phase shift β plotted against frequency (in terms of the ratio f/f_c), for a constant-k low-pass section. The graph can be interpreted for the high-pass section by setting the frequency axis to read the ratio f_c/f.

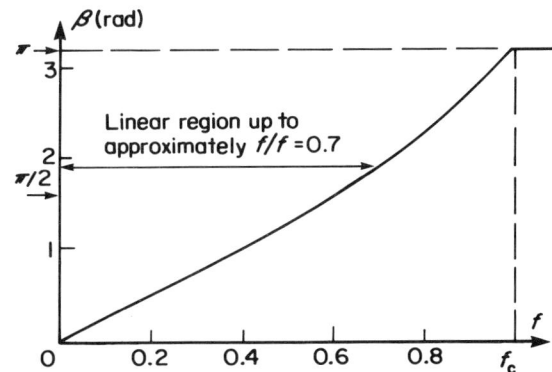

Figure 6.16

In the low-pass section, the phase angle increases from zero at $f = 0$ to π rad at $f = f_c$, remaining at this value within the attenuation band. For the high-pass section, the phase angle falls from $-\pi$ rad to zero as the frequency increases from f_c to infinity.

6.4.1 Time delay

The change of phase occurring in a section depends upon the time that the signal takes to pass through the section. Expressing phase shift as a time delay, if the frequency is f, the period (time of one cycle) is $1/f$, hence the time delay T will be

$$T = \frac{\beta}{2\pi} \times \frac{1}{f} = \frac{\beta}{\omega}$$

where $\beta \simeq \omega\sqrt{(LC)}$. Hence $T = \sqrt{(LC)}$. This result is independent of the frequency. So, by having a phase shift which is *proportional* to frequency, we have a time delay which is *independent* of frequency. This means that if the input to the section or sections consists of a complex wave made up of many harmonic components of differing frequencies, the output will consist of a complex wave made up of the sum of corresponding components all delayed by the *same* amount. There will consequently be no distortion due to varying time delays for the separate frequency components, or phase distortion as it is called.

In practice, β is not constant and the increase in time delay with rising frequency causes distortion of non-sinusoidal inputs, this distortion being superimposed on that due to the attenuation of components whose frequency is higher than f_c. This problem will be further discussed in relation to transmission along lines.

Example 6. Derive an expression for the time of transit of a signal through a filter section at the cut-off frequency. Calculate this delay for the case of the filter section shown in *Figure 6.17*.

At the cut-off frequency $\beta = \pi$ for a low-pass prototype. Hence the transit time

$$T = \frac{\pi}{2\pi f_c} = \frac{1}{2f_c} = \frac{\pi}{2}\sqrt{(LC)}$$

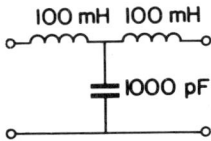

100 mH 100 mH

1000 pF

Figure 6.17

since $f_c = \dfrac{1}{\pi\sqrt{(LC)}}$.

In the filter section shown, $L = 200\,\text{mH}$ and $C = 1000\,\text{pF}$ (recall that 100 mH here is actually $L/2$). Then

$$T = \frac{\pi}{2}[200 \times 10^{-3} \times 1000 \times 10^{-12}]^{\frac{1}{2}}$$

$$= 2.22 \times 10^{-5}\,\text{sec} = 22.2\,\mu\text{s}$$

Example 7. A series of cascaded sections is to form a delay line which will pass signals of all frequencies up to 3 MHz and provide a total delay of 5 μs. If the characteristic impedance of the circuit into which the delay line is inserted is 2 kΩ, calculate the section element values and the number of such sections required.

$$f_c = \frac{1}{\pi\sqrt{(LC)}} = 3 \times 10^6; \quad R_0 = \sqrt{\frac{L}{C}} = 2000\,\Omega$$

As for Example 3, we obtain

$$L = \frac{R_0}{\pi f_c} = \frac{2000}{\pi \times 3 \times 10^6} = 2.12 \times 10^{-4}\,\text{H}$$

$$= 212\,\mu\text{H}$$

$$C = \frac{10^{12}}{\pi \times 2000 \times 3 \times 10^6}\,\text{pF} = 53\,\text{pF}$$

The time delay for one section is
$$\sqrt(LC) = \sqrt(212 \times 10^{-6} \times 53 \times 10^{-12}) = 106 \times 10^{-9}\,\text{s or } 0.106\,\mu\text{s}.$$
For a total delay of $5\,\mu\text{s}$, therefore, the number of cascaded sections required is

$$\frac{5}{0.106} \simeq 47$$

6.5 BAND-PASS AND BAND-STOP FILTERS

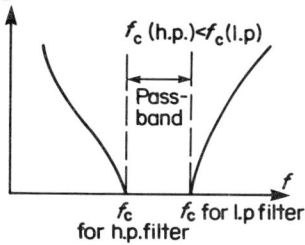

Figure 6.18

The characteristics of a band-pass filter are shown in *Figure 6.18*. This kind of filter passes a band of frequencies which does not extend to zero frequency in one direction nor to infinite frequency in the other. Ideally, the characteristic should conform to the broken lines indicated on the diagram, but the practical curve more generally conforms to the indicated full lines.

A band-pass filter may be formed of a high-pass and a low-pass filter in cascade. Obviously, f_c (high-pass) must be smaller than f_c (low-pass), the pass-band being given approximately by the difference between these values. Problems arise from the impossibility of terminating both filters correctly at all frequencies within the pass-band.

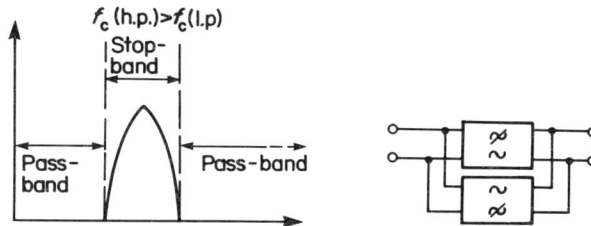

Figure 6.19

Similarly a band-stop filter may be formed by connecting a high-pass and a low-pass filter in parallel, as shown in *Figure 6.19*, where the characteristic is also shown. Here a particular band of frequencies is effectively removed from the frequency spectrum. Clearly this time f_c (high-pass) must be greater than f_c (low-pass), the stop-band being given approximately by the difference between these values.

6.6 m-DERIVED SECTIONS

It is not practicable to wire prototype constant-k sections in cascade in order to obtain a composite filter having a clearly defined cut-off frequency followed by high attenuation. What is required is a section having the same f_c as the prototype but with a rapidly rising characteristic beyond cut-off. Such a section must also have the same R_0 as the prototype at all frequencies since otherwise the two forms could not be connected together without mismatch. This requirement appears formidable, but it can be accomplished quite simply in spite of the apparent complexity. Make a note of the fact that if the two sections have the same R_0, they will have identical pass-bands.

Consider the prototype T-section. Let a new section be constructed from it having a series arm of the same type but of different value, that is, for convenience make the new $Z_1 = mZ_1$, where m is some constant. The new shunt arm must now be an adjustment of the original arm in order for Z_0 to be maintained. Hence, it will not be Z_2 but Z_2', say. Now it is necessary to find Z_2' such that the two sections, prototype and *m-derived* (as the new section is called), have the same impedance.

For the prototype $\quad Z_{0T} = \sqrt{\left(\dfrac{Z_1^2}{4} + Z_1 Z_2\right)}$ (from *Figure 6.20*)

For the new section $\quad Z_{0T}' = \sqrt{\left(\dfrac{m^2 Z_1^2}{4} + m Z_1 Z_2'\right)}$

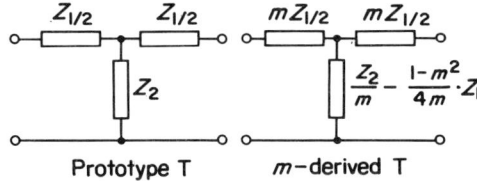

Figure 6.20

These will be identical if

$$\frac{Z_1^2}{4} + Z_1 Z_2 = \frac{m^2 Z_1^2}{4} + m Z_1 Z_2'$$

or if

$$Z_2' = \frac{Z_2}{m} + \frac{1 - m^2}{4m} \cdot Z_1$$

This means that Z_2' must have an impedance of Z_2/m in series with $Z_1[(1 - m^2)/4m]$. Hence, an additional component has to be introduced into the shunt arm of the section. This is possible provided $0 < m < 1$. Notice that if $m = 1$, the m-derived section reduces to the prototype. *Figure 6.20* illustrates the formation of the m-derived section from the prototype.

6.6.1 The low-pass sections

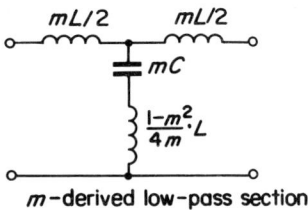

m-derived low-pass section

Figure 6.21

Consider now the low-pass T-section filter shown in *Figure 6.21*. At some frequency the shunt arm will resonate as a series circuit, so short-circuiting the transmission path. In the prototype, infinite attenuation is obtained only at infinite frequency. In the m-derived section, let the frequency of infinite attenuation be f_∞, then

$$\omega_\infty^2 = \frac{1}{\dfrac{1 - m^2}{4m} L \cdot mC} = \frac{4}{LC(1 - m^2)}$$

$$= \frac{\omega_c^2}{1 - m^2}$$

where ω_c is the cut-off frequency of the *prototype*. Therefore

$$\omega_\infty = \frac{\omega_c}{\sqrt{(1 - m^2)}}$$

If ω_∞ and ω_c are known, the value of m can be calculated, i.e.

$$\frac{\omega_\infty}{\omega_c} = \frac{1}{\sqrt{(1 - m^2)}} \quad \therefore \quad m = \sqrt{\left[1 - \left(\frac{f_c}{f_\infty}\right)^2\right]}$$

Hence f_∞ must be *greater* than f_c.

Clearly it seems reasonable to make f_∞ as close to f_c as

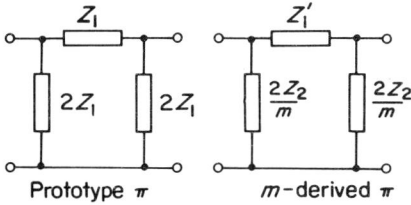

Figure 6.22

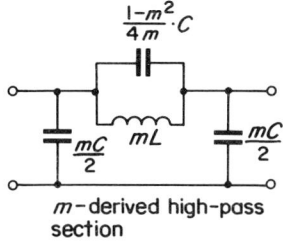

m-derived high-pass
section

Figure 6.23

possible, so making m very small, but there are other considerations and these will be discussed shortly.

For the low-pass π-section, both prototype and m-derived section must have the same Z_0, hence, referring to *Figure 6.22*, the series arm will be Z_1' and the shunt arms each $2Z_2/m$ such that

$$Z_{0\pi}' = \frac{Z_1 Z_2}{Z_{0T}} = \frac{Z_1 Z_2}{\sqrt{\left(\frac{Z_1^2}{4} + Z_1 Z_2\right)}}$$

which becomes

$$= \frac{Z_1' \cdot \frac{Z_2}{m}}{\sqrt{\left(\frac{Z_1^2}{4} + \frac{Z_1' Z_2}{m}\right)}}$$

from which $\quad 4Z_1 Z_2 m + m^2 Z_1 Z_1' = 4Z_2 Z_1' + Z_1 Z_1'$

$$\therefore \qquad Z_1' = \frac{4m Z_1 Z_2}{4Z_2 + (1 - m^2)Z_1}$$

This amounts to an inductive impedance mZ_1 in parallel with a capacitive impedance $4mZ_2/(1 - m^2)$. This leads to a similar value for f_∞ as for the T-section already discussed—see *Figure 6.23*.

6.6.2   The high-pass sections   Similar considerations apply in the case of T- and π-section high-pass filters as those for the low-pass cases. The corresponding configurations for the m-derived sections in

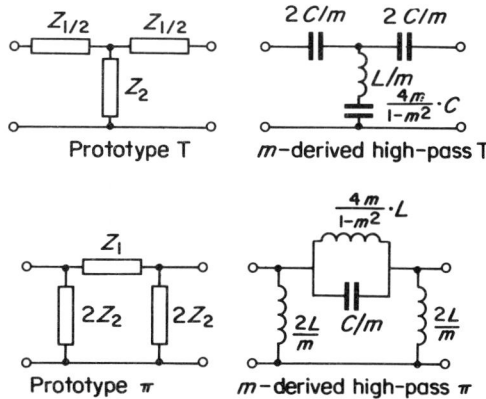

Prototype T m-derived high-pass T

Prototype π m-derived high-pass π

Figure 6.24

relation to the prototypes are shown in *Figure 6.24*. It is left as an exercise for you to verify the component values shown, and to prove that

$$m = \sqrt{\left[1 - \left(\frac{f_\infty}{f_c}\right)^2\right]}$$

where this time $f_\infty < f_c$.

Example 8. Design an m-derived T- and π-low-pass section having $f_c = 1\,\text{kHz}$, $R_0 = 600\,\Omega$ and a frequency of infinite attenuation 1050 Hz.

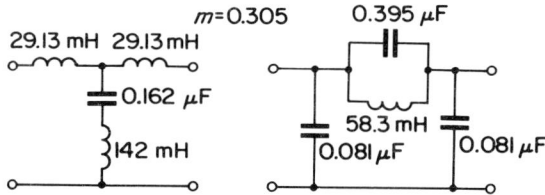

Figure 6.25

From the appropriate formula

$$m = \sqrt{\left[1 - \left(\frac{f_c}{f_\infty}\right)^2\right]} = \sqrt{\left[1 - \left(\frac{1000}{1050}\right)^2\right]}$$

$$= 0.305$$

For the prototype

$$L = \frac{R_0}{\pi f_c} = \frac{600}{\pi \times 1000} \text{ H} = 191 \text{ mH}$$

$$C = \frac{1}{\pi R_0 f_c} = \frac{10^6}{\pi \times 600 \times 1000} \text{ μF}$$

$$= 0.5304 \text{ μF}$$

The m-derived filters are consequently arranged as shown in *Figure 6.25*.

6.7 COMPOSITE FILTERS

A practical arrangement consists of the prototype ($m = 1.0$) with an m-derived section in series. The object of this arrangement is that the m-derived section improves the attenuation *immediately after* cut-off, while the prototype improves it *well after* cut-off. Further, terminating half-sections can be used at each end to obtain a constant match over the pass-band. These latter are derived, for a T-section, from the complete section, as

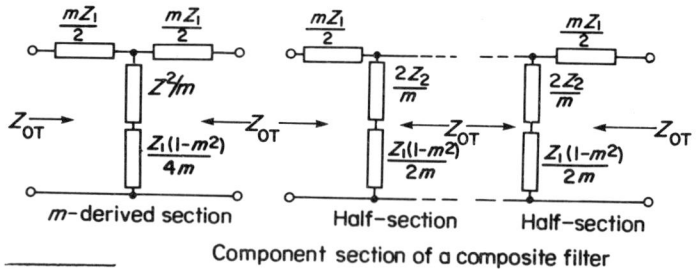

Component section of a composite filter

Figure 6.26

Figure 6.26 shows. It is necessary to find suitable values of m for such composite filters so that the proper characteristics are obtained, together with a constancy in impedance.

6.7.1 Variations in Z_0 with m

As we have seen, the series derived m-type section is based on a prototype which presents its own Z_0 at its terminals, that is, the prototype of a T-section leads to an m-derived T-section. Hence for equivalence

$$Z_{0T} = Z_{0T}'$$

Because Z_{0T} has a non-linear characteristic against frequency, so too will Z_{0T}'.

Since

$$Z_0\pi = \frac{Z_1 Z_2}{Z_{0T}}, \quad Z'_{0\pi} = \frac{Z'_1 Z'_2}{Z'_{0T}} = \frac{Z'_1 Z'_2}{Z_{0T}}$$

Substituting for Z'_1 and Z'_2 shows that

$$Z'_{0\pi} = \frac{mZ_1}{Z_{0T}}\left[\frac{Z_2}{m} + \frac{(1-m^2)}{4m}.Z_1\right]$$

$$= \frac{Z_1 Z_2}{Z_{0T}}\left[1 + \frac{(1-m^2)}{4Z_2}.Z_1\right]$$

or

$$Z'_{0\pi} = Z_{0\pi}\left[1 + \frac{(1-m^2)}{4Z_2}.Z_1\right]$$

This shows that the impedance of the m-derived section is related to the impedance of the prototype by a factor

$$1 + \frac{(1-m^2)}{4Z_2}.Z_1$$

and varies as m varies.

When $m = 1$, $Z'_{0\pi} = Z_{0\pi}$ and when $m = 0$, $Z'_{0\pi} = Z_{0T}$; hence, a family of curves drawn for $Z'_{0\pi}$ for values of m between 0 and 1 will show forms that are a compromise between the two extremes of prototype variations.

Two of the intermediate curves, for $m = 0.3$ and $m = 0.6$, are shown in *Figure 6.27*, together with the bounding curves for the prototypes the form of which has been discussed earlier. It is seen that at the value $m = 0.6$ the impedance is practically constant at $\sqrt{(L/C)}$ for most of the pass-band. The diagram is for a low-pass filter, but similar curves may be drawn for the high-pass case.

The closeness of f_∞ to f_c calls for m to be small, but practical limitations make very small values of m difficult to achieve. Anything below $m = 0.3$ is very unusual. The effect of m on the frequency of infinite attenuation is shown in *Figure 6.28*. As always, resistive losses modify the idealised curves, introducing a certain amount of attenuation in the pass-band, and generally decreasing the sharpness of the distinction between pass and attenuation bands. High-Q circuits and components are needed if the ideal curves are to be obtained closely.

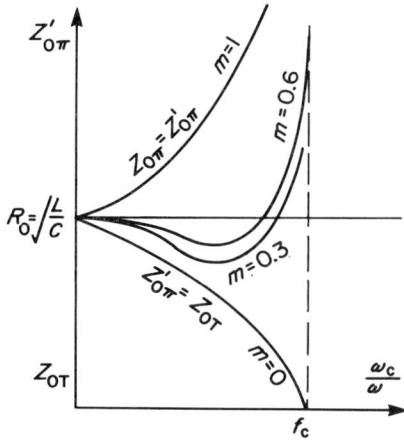

Figure 6.27

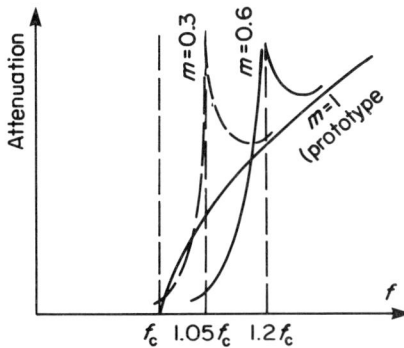

Figure 6.28

Example 9. Design a composite filter with $f_c = 1$ kHz, $f_\infty = 1.2$ kHz, and $R_0 = 600\,\Omega$. It must consist of a prototype T-section, one m-derived section and two terminating half-sections.

Since $f_c < f_\infty$ the filter is low-pass.

For the prototype $L = \dfrac{R_0}{\pi f_c} = \dfrac{600}{1000\pi} = 0.191\,\text{H}$

$$C = \frac{10^6}{\pi f_c R_0} = \frac{10^6}{\pi \times 600 \times 1000}\,\mu\text{F}$$
$$= 0.53\,\mu\text{F}$$

For the m-derived section:

$$m = \sqrt{\left[1 - \left(\frac{1000}{1200}\right)^2\right]} = 0.553$$

Then $\qquad mL = 0.553 \times 0.191 = 0.106\,\text{H}$

$$mC = 0.553 \times 0.53 \ = 0.293 \, \mu F$$

$$L\frac{(1 - m^2)}{4m} = \frac{1 - (0.553)^2}{4 \times 0.553} \, 0.191 = 0.06 \, H$$

For the half-sections:

$$\frac{mL}{2} = \frac{0.6 \times 0.191}{2} = 0.057 \, H$$

$$\frac{mC}{2} = \frac{0.6 \times 0.53}{2} \ = 0.16 \, \mu F$$

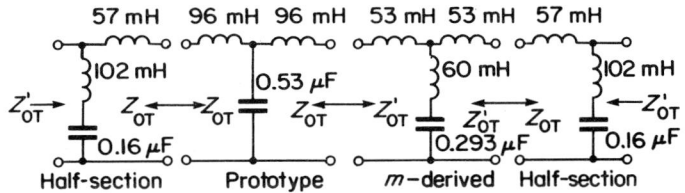

Figure 6.29

We have taken $m = 0.6$ here for the best constancy of Z_0.

$$L\frac{(1 - m^2)}{2m} = 0.191 \cdot \frac{1 - (0.6)^2}{2 \times 0.6} = 0.1018 \, H$$

The complete filter is illustrated in *Figure 6.29*.

PROBLEMS FOR CHAPTER 6

(1) Calculate the circuit component values required in a prototype T-section filter to operate with a terminating load of $600 \, \Omega$ and passing all frequencies up to 3 kHz.

(2) A π-section low-pass filter has a characteristic resistance of 600Ω and a cut-off frequency of 1.5 kHz. Calculate the values of the series and shunt arm components of this filter.

(3) A low-pass filter is in the form of a π-section having a series inductance L and shunt capacitances each $C/2$. Derive an expression for the cut-off frequency of such a filter. What will be the cut-off frequency if $L = 25 \, mH$ and $C = 0.04 \, \mu F$?

(4) A filter is to be designed to pass all frequencies above 3.5 kHz and to have a nominal impedance of $600 \, \Omega$. Calculate circuit values for both T- and π-section prototypes.

(5) Calculate the pass-band and the cut-off frequency of the filter network drawn in *Figure 6.30*, assuming the termination to be ideal. Make a sketch of what you think the attenuation–frequency curve of this network would look like compared with the characteristic of one section only.

(6) A signal of frequency 1 kHz is applied to the filter of Figure 6.31. Calculate the total time delay experienced by the signal in passing through the filter.

(7) Construct a delay line having an impedance of 2 kΩ that will pass all frequencies up to 3 MHz with a total delay of 10 μs. How many sections will be required in the line?

(8) A network consists of six sections in cascade having an impedance of $800 \, \Omega$. Find the component values for each section if the total delay time is 0.5 μs.

(9) Show that an infinite uniform ladder network composed of pure reactances exhibits pass-bands and attenuation bands.

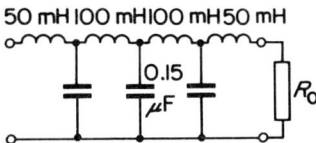

Figure 6.30

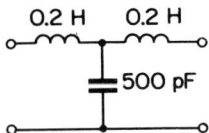

Figure 6.31

Employ the relation derived to determine the cut-off frequency of a low-pass filter made up of π-sections each having a series arm of 100 mH inductance and shunt arms of 0.05 µF capacitance.

(10) What effect does resistance have in modifying the performance of filter networks? The coil used in a prototype low-pass π-section filter has a resistance of r ohms and Q-factor Q_0. The shunt arms may be assumed to be loss-free. Prove that $r = 2R_0/Q_0$.

(11) A prototype T-section low-pass filter has a nominal impedance R_0. Prove that $Z_{0\pi}$ at a frequency $f = 0.9f_c$ is $2.3R_0$. Hence, find the impedance $Z_{0\pi}$ for the equivalent π-section to the T-section of problem 1 at a frequency of 2.5 kHz.

(12) Prove that if the shunt and series reactances of a T-section filter have the same sign at a frequency f_1, then f_1 lies in the attenuation band.

(13) Make a sketch of the sort of attenuation–frequency graph you would expect to get if a low-pass T-section filter with $f_c = 6$ kHz was connected in series with a high-pass T-section filter with $f_c = 3.5$ kHz, assuming ideal matching throughout. Calculate the values you would use in such a composite filter for a terminating impedance of 600 Ω. What kind of filter is this?

(14) Derive an expression for the frequency of infinite attenuation of a T-section m-derived filter in terms of the prototype cut-off frequency and m. Hence, design a low-pass T-section m-derived filter having a cut-off frequency of 3.5 kHz and working into a 600 Ω line, being given $F_\infty = 3.6$ kHz.

(15) In m-derived filter design, why must m lie between 0 and 1? Why do you think it is not practicable to make m very small?

7 Transmission lines — 1

Line transmission is the propagation of electrical energy along conducting wires or (as they are more generally known) transmission lines. The physical construction of such lines will be discussed in due course, but for the moment it is sufficient to visualise them as a pair of parallel conductors which are uniformly constructed and spaced throughout their entire length. Provided that this uniformity is observed, it is immaterial for general line theory whether the two wires are separated by air or by some other dielectric medium. Moreover, not only do such lines simply convey electrical energy from one point to some other point, but short sections of such lines (short relative to the wavelength of the transmission they convey) exhibit at very high frequencies the properties associated with series and parallel tuned circuits, and such line sections are often employed as resonant elements. In this first of two chapters on line transmission we shall develop the foregoing work on attenuator sections and filter networks to take in the basic theory of transmission systems, considering only steady-state voltages and currents.

7.1 TRAVELLING WAVES

Suppose the near end of a straight length of light uniform rope lying along the ground is moved rhythmically up and down, as drawn in *Figure 7.1*. Then a wave motion will be seen to travel along the rope. We notice three things about this wave motion:

(1) At any point along the length of the rope, such as at P, the rope undergoes up and down movements identical with those being made at the near end of the rope, but not necessarily in time-step with them.

(2) The height or amplitude of the wave diminishes as it travels along the rope, eventually becoming so small that the far end of the rope is for all practical purposes undisturbed.

(3) Each wave takes a finite time to travel along the length of the rope.

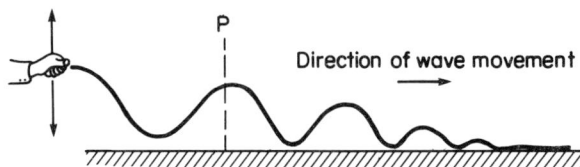

Figure 7.1

Such a wave motion travelling along the rope may be compared in a number of ways with an electrical wave travelling along a length of uniform conducting line, an a.c. generator providing the input (sinusoidal) variations. The transmission of electrical energy along any such uniform line may be visualised best by imagining the line to be of infinite length. Let the a.c. generator of *Figure 7.2* be connected to the parallel wire line at some instant $t = 0$, the generator voltage at that instant passing through zero. Then in a quarter of the periodic time T of one cycle of the generated frequency, the voltage at point A (the generator terminals) will have risen to its maximum value $\hat{V}$, and the zero voltage condition which previously existed at A will have moved on to point B, one-quarter wavelength distant along the line. The

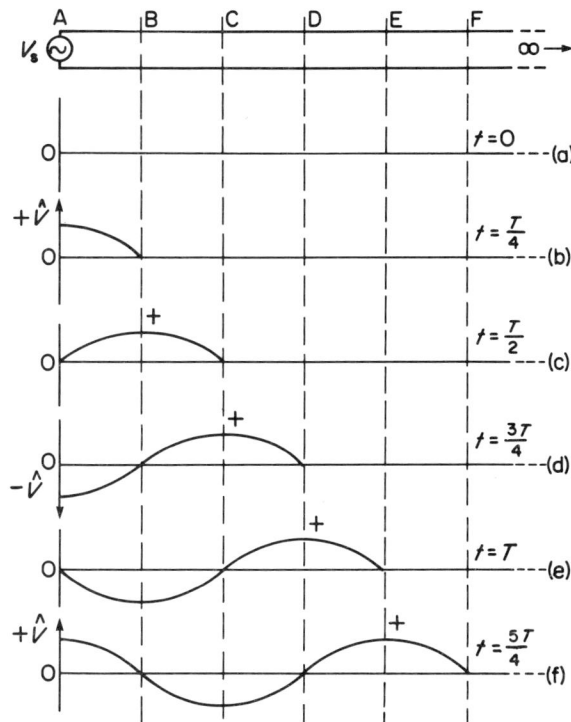

Figure 7.2

voltage distribution along the line at this instant will then be as illustrated in *Figure 7.2(b)*.

After a further quarter of the periodic time of the transmission, the voltage condition at B will have advanced to C; that at A will have advanced to B; and the voltage at the generator terminals will have returned to zero. This process continues and the figure shows from diagram (c) through (f) how the wave motion is progressing along the line. This *forward travelling wave* as it is called is seen to correspond with the progressive wave motion travelling along the length of rope. If now the generator is assumed to have been connected to the line for a considerable time (relative to a period of the transmission) and we consider the voltage variation at any given point on the line, such as B, we observe that during one period the point undergoes one complete sinusoidal cycle of the voltage. And any other point undergoes a similar variation, differing only from that at the first point (except when the points are spaced by one wavelength) in the matter of phase. This is analogous to a point on the rope, such as P in *Figure 7.1.*

Now just as the amplitude of the wave crests diminish along the length of the rope, so does the amplitude of the voltage wave moving along a conducting line decay away as it progresses further and further from the generator terminals. In the electrical case, this fall in amplitude is the result of the attenuation introduced by the ohmic resistance of the wires making up the line; and also the leakage conductance present between the wires. This decay is logarithmic in nature and is measured either in decibels or (more usually) in népers per unit length of line.

Lastly in the comparison, there is the matter of the time taken by the energy to travel along a given length of rope or line. Returning to the experiment with the rope, if we care to perform it, we shall find that the time taken by any particular wave to travel a given distance will depend upon the frequency of the up and down movement imparted by our hand to the near end of the rope; a leisurely movement will generate a leisurely moving wave, a fast movement will generate a fast-moving wave. So, although in this particular instance the comparison of the rope and the transmission line is far from being scientifically ideal as a demonstration of what is actually taking place, we might expect the time taken by an electrical signal to travel along a particular length of line to depend upon its frequency; in other words, the velocity of propagation of the signal will be a function of its frequency.

7.2 THE GENERAL LINE

When an a.c. generator is connected to the input terminals of an infinite line, a voltage travelling wave will move along the line and a finite current will flow into the line. The instantaneous conditions present at the generator terminals will be progressively passed on into the line and will process along it *ad infinitum*. Given sufficient time, a suitable measuring instrument placed at *any* point along the line will eventually record the arrival and passage of the travelling wave. Thus the line does not behave to the generator as an open circuit but presents a finite load Z_0, where Z_0 is the ratio of the applied voltage V_s to the current I_s passing into the line. This means that the whole of the energy is absorbed by the line, and the line behaves to the generator exactly as would a single lumped impedance equal in value to Z_0 connected directly across its terminals. What meaning can be attached to Z_0 in the case of an infinite line? We can look at the line as a perfectly uniform network made up of elemental sections, each section having resistance, leakance, inductance and capacitance, yet so uniformly distributed throughout the length of the line are these that at no point is there a 'discontinuity' in the form of a lumped component of any of these quantities.

This is the true picture of a perfectly uniform line; for our purposes of analysis, however, we may represent each section of the line as the network shown in *Figure 7.3(a)*, or in a simplified form as in *Figure 7.3(b)*. The finite current which flows into the line when the generator is connected is then made up of two parts: a part flowing through the leakage conductance G between the wires of the line (which is lost), and a part which progressively charges each elemental capacitance C and which constitutes the voltage travelling wave moving along the line. The attenuation loss is contributed by both G and the series-arm resistance R.

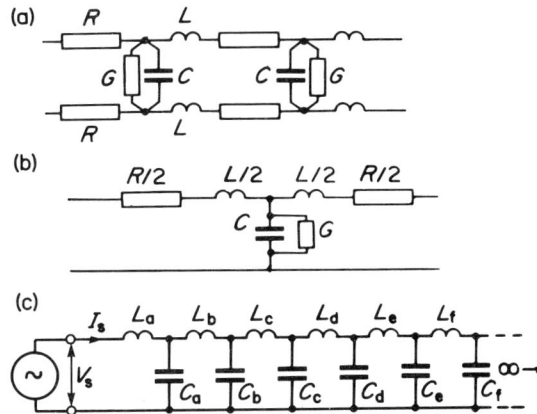

Figure 7.3

Now *Figure 7.3(b)* is clearly a low-pass filter section with losses. If we neglect these losses and remove R and G, the circuit simplifies further and the infinite line reduces to a repetitive network of T-section low-pass filters as shown in *Figure 7.3(c)*. Suppose now that the generator voltage is rising to a positive peak just at the moment that the line represented at *Figure 7.3(c)* is connected to it. The following sequence of events will then occur:

(1) A current will begin to flow through L_a into C_a.
(2) C_a will charge and a voltage will develop across it.
(3) This voltage will send a current through L_b into C_b.
(4) C_b will charge in turn and the voltage developed across it will send a current through L_c into C_c.
(5) This process will continue indefinitely and all capacitors will charge up in turn to the instantaneous peak input voltage.

The peak voltage at the input is thus progressively passed on along the line as a maximum instantaneous voltage on each capacitor element. When the line input voltage falls, each capacitor is progressively charged in opposite polarity and again the input charge is handed on, as it were, from each capacitor element to the next. In this way voltage and current waves travel along the line together and are mutually dependent. This process takes time, for by the time capacitor C_d, for example, has reached its peak voltage, the generator input may be at zero or moving towards its negative peak voltage. There is therefore a time (and so a phase) difference between the generator input and any point along the line, or between any two points on the line. We may use as an analogy here a row of billiard balls placed in contact, as shown in *Figure 7.4(a)*. If the end ball is struck smartly with the cue in the direction of the line, the ball at the far end receives the impulse practically instantaneously, that is, the propagation of the energy is extremely fast. If, however, every alternate ball is replaced by one made of a soft sponge rubber, as shown in *Figure 7.4(b)*, an impulse to the end ball will first have to compress the following ball a certain amount before the energy is fully conveyed to the ball following that, and so on along the line. This time the propagation of energy will be considerably slower than the previous case. The sponges have to be 'charged' as it were, like the elemental capacitors of the transmission line, before they pass on their stored energy to the following component.

Solid balls pass on an impulse instantaneously

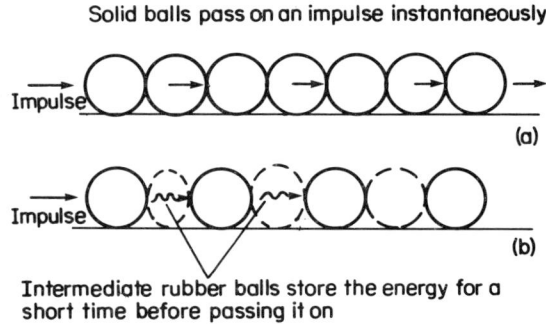

Intermediate rubber balls store the energy for a short time before passing it on

Figure 7.4

Line inductance, too, has its influence on retarding the transmission, since the current in an inductance cannot rise or fall instantaneously, and the opposition to change is a function of the frequency. And this is what we should expect, because the line being a ladder network of low-pass T-section filters, has a phase delay given by equation (6.12) earlier:

$$\beta = \omega \sqrt{(LC)}$$

and frequency, inductance and capacitance all influence the value of β. In a practical line, capacitance usually predominates, and the bulk of the delay is brought about by the time constant CR rather than L/R.

Let the phase shift along the line be β rad/km. Then

$$\lambda = \frac{2\pi}{\beta} \text{ km}$$

where λ is the wavelength *on the line*, that is, the distance between any given point and the next point further along the line at which the voltage is in the same phase. Looking back at *Figure 7.2(f)* for a moment, A and E are two such points, as are also B and F. Although the voltage at A (or B) reaches its maximum at the same instant as the voltage at E (or F), the voltage at A (or B) is leading the voltage at E (or F) by 2π rad or 360°. Further, the velocity of propagation $v = \text{frequency} \times \text{wavelength}$, hence

$$v = \frac{2\pi f}{\beta} = \frac{\omega}{\beta}$$

Now the velocity of electrical energy along a line (or for that matter, in any media other than free space) is *less* than the velocity of propagation in free space. The free space velocity is the same as that of light, approximately 300×10^6 m/s, and the wavelength of radiation in space is c/f, where c is the velocity of light. As the velocity along a line is always less than c (this is known as the *phase velocity*), the wavelength corresponding to any particular applied frequency is always *shorter* on the line than it would be in free space. This is an important point.

Example 1. A parallel-wire air-spaced line has a phase shift β of 0.04 rad/km. Find the wavelength on the line and the speed of transmission of a signal of frequency 1592 Hz.

For $f = 1592$ Hz, $\omega = 10^4$ rad/s

$$\lambda = \frac{2\pi}{\beta} = \frac{6.28}{0.04} = 157 \text{ km}$$

Then
$$v = \frac{\omega}{\beta} = \frac{10^4}{0.04} = 250 \times 10^3 \text{ km/s}$$
$$= 250 \times 10^6 \text{ m/s}$$

This is five-sixths the velocity of light.

Notice from this example that the velocity of propagation along wires separated by air is not far short of that for free space, so that a wavelength on the line is closely equal to its value in free space. This is generally true of air-spaced open lines and follows from the fact that the permittivity of air is virtually the same as that of space. If the conductors of the line are embedded in a dielectric other than air, the capacitance C is increased to $\varepsilon_r C$, where ε_r is the relative permittivity of the dielectric material. Since $\beta = \omega\sqrt{(LC)}$, the phase change per unit length is ε_r time its air-space value, and for a shift of 2π rad we require a length of line $1/\varepsilon_r$ of what it is for air-spacing. This is always less than the original length. The phase shift on a loaded underground cable may be as much as 1 rad/km, which leads to a corresponding line velocity as low as 10 000 km/s at a frequency of 1600 Hz.

7.3 CHARACTERISTIC IMPEDANCE

Let the infinite line be again connected to the generator so that a current I_s flows into the line and the generator experiences a load Z_0, where $Z_0 = V_s/I_s$. Now cut from the line a finite portion of length l; the load that the generator now sees will not, in general, be Z_0. It can be restored to Z_0, however, either by replacing the rest of the infinite line or by terminating the finite portion with an impedance equal to Z_0.

Exactly as it was in the case of repetitive attenuator networks, this second alternative will be true whatever the length of line in the finite portion, since this portion is effectively a ladder of low-pass filter sections and Z_0 is that impedance which properly terminates each section. The current and voltage at all points along the finite portion will therefore be exactly the same as if it was part of an infinite line, for the generator cannot distinguish between an infinite line and a finite line terminated in Z_0. Thus the properties of an infinite line are manifested by a properly terminated finite line. We conclude that the finite section of a line, if terminated in an impedance Z_0, has an input impedance also equal to Z_0.

It must be completely understood that Z_0, which is known as the characteristic impedance of a line, is *absolutely independent of the physical length of the line*; it simply describes a property of a line that is a function of the physical construction of the line.

Because a short length of line may be considered as a ladder of identical low-pass filter sections, its characteristic impedance may be conveniently determined from the formula

$$Z_0 = \sqrt{(Z_{oc} Z_{sc})}$$

since Z_{oc} and Z_{sc} may be readily measured on a suitable a.c. impedance bridge.

Z_0 will, of course, be a vector quantity, having a modulus $|Z_0|$ and argument ϕ. The generator is looking into a network which is reactive and the current will lead or lag on the voltage by an angle that is characteristic of the particular piece of line being considered. As a line is, in general, capacitive, the angle is negative and the current will lead. Both the modulus and angle vary with frequency and it is usual to state the frequency at which the impedance is measured although, as we shall see later, Z_0 tends to become constant and resistive as the frequency becomes high.

Example 2. Measurements were made at a frequency of 1600 Hz on a length of line with the following results:

$$Z_{oc} = 1000\underline{/-45°} \qquad Z_{sc} = 250\underline{/-15°}$$

What is the characteristic impedance of the line at 1600 Hz?

$$
\begin{aligned}
Z_0 &= \sqrt{(Z_{oc} Z_{sc})} \\
&= \sqrt{[(1000\underline{/-45°})(250\underline{/-15°})]} \\
&= \sqrt{(250\,000\underline{/-60°})} \\
&= 500\underline{/-30°} \ \Omega
\end{aligned}
$$

7.4 CURRENT AND VOLTAGE RELATIONSHIPS

Let a voltage V_s be applied at the input (sending end) terminals of an infinite line (or a line terminated in Z_0) so that a current I_s flows into the line. At a point B, one unit distance down the line, let the current be I_1. I_1 will not have the same magnitude as I_s because of line attenuation, further, I_1 will lag on I_s by some angle β because of phase shift. The ratio I_s/I_1 will therefore be a phasor quantity.

Now let the current at further unit intervals along the line be represented by I_2, I_3, etc., as shown in *Figure 7.5*. These currents will be in a definite proportion to I_s and to each other and since all unit portions of the line are identical, the phasor quantity representing the ratio of each input current to each output current per section will be identical for all sections. This relationship between the input and output currents for each unit section is the *propagation constant* of the line denoted by the symbol γ. As γ is a phasor quantity, we may write it in the Cartesian form $a + jb$. The real part a of this gives us the attenuation constant of the line, symbolised α, and the imaginary part jb expresses the phase constant so that $b = \beta$. Hence

$$\gamma = \alpha + j\beta$$

a relationship we have already considered in an earlier chapter. Since also

$$\frac{I_s}{I_1} = \frac{I_1}{I_2} = \frac{I_2}{I_3} = \cdots$$

and we define attenuation α in népers as

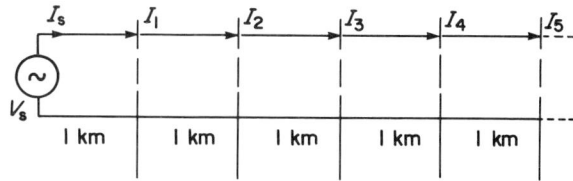

Each km length introduces an attenuation of α nepers and a phase shift of β rad

Figure 7.5

$$\ln \left[\frac{\text{Input current}}{\text{Output current}} \right]$$

then clearly

$$= \ln \left[\frac{I_s}{I_1} \right] = \ln \left[\frac{I_1}{I_2} \right] = \ldots$$

and the fall of current along each unit length is expressed by

$$\left| \frac{I_s}{I_1} \right| = e^{\alpha}$$

Thus over a distance of n units the relationship between the input current I_s and the received current I_R is

$$\left| \frac{I_s}{I_R} \right| = e^{n\alpha}$$

It must be noted that e^{α} gives the ratio of the *absolute* value of the current sent to that received; the phase angle between the two currents is β, measured in radians per unit length of line. Hence, the ratio of the two currents is completely expressed as a complex quantity, i.e.

$$\frac{I_s}{I_1} = e^{\alpha} \underline{/\beta} = e^{\alpha + j\beta} = e^{\gamma}$$

So for n units of line

$$\frac{I_s}{I_R} = e^{n\gamma} = e^{n\alpha} \underline{/n\beta}$$

whence

$$I_R = I_s \cdot e^{-n\gamma} \qquad (7.1)$$

The attenuation on the line is therefore $n\alpha$ nepers and the phase shift is $n\beta$ rad.

What we have done here, in fact, is to treat each unit length of line as a section of repetitive network; the general theory outlined in Chapter 2 then applies to the general line.

An equation of similar form to that given in (7.1) can be derived for the received voltage V_s, since at all points along an infinite (or properly terminated) line, the ratio of voltage to current is Z_0. So for n units of line

$$V_R = V_s \cdot e^{-n\gamma} \qquad (7.2)$$

Z_0, α, β and γ are referred to as the *secondary line constants*, or coefficients. The primary line constants will turn up in due course.

Example 3. A cable has an attenuation of 2.5 dB per km and a phase constant of 0.236 rad per km when working with an input frequency of 1600 Hz. If 1 V r.m.s. is applied at the sending end of the line, calculate the voltage at a point 12 km down the line when the termination is equal to the characteristic impedance of the line.

The voltage at a point n km distant from the sending end is

$$V_R = V_s . e^{-n\gamma}$$
$$= V_s . e^{-n\alpha}\underline{/-n\beta}$$

where α is in népers/km and β is in rad/km.

As the attenuation is stated as 2.5 dB/km, this equals 2.5×0.115 népers/km, so that

$$\alpha = 0.2875 \text{ népers/km}$$

The voltage at a point 12 km from the sending end is then

$$V_R = 1 \times e^{-0.2875 \times 12}\underline{/-0.236 \times 12}$$
$$= e^{-3.45}\underline{/-2.83}$$
$$= 0.0317\underline{/-2.83} \quad \text{V}$$

Thus the voltage at a point 12 km down the line is 0.0317 V r.m.s. lagging 2.83 rad (162°) behind the sending end phase.

We are now in a position to obtain a graphical representation of the current and voltage variations along an infinitely long line.

7.4.1 Current variation The equation

$$I_R = I_s . e^{-n\alpha}\underline{/-n\beta}$$

cannot be represented as a two-dimensional graph because I_s is itself an alternating quantity and so I_R varies both with the distance n and the time t.

We may, however, represent I_s in the usual manner of a rotating phasor of length I_s turning anticlockwise with an angular velocity ω radians per second. The projection of this phasor on to the vertical axis will then give the instantaneous value of I_s at any time t.

Let l be the general length of the line. The current at all points along the line will have the same frequency as that of the generator but will be progressively attenuated from its initial value of I_s to $I_s . e^{-l\alpha}$ at a distance l from the generator. At any point P along the line, therefore, the current will be found to vary sinusoidally between the limits $I_s . e^{-l\alpha}$ at the peak of the positive half-cycle and $-I_s . e^{-l\alpha}$ at the peak of the negative half-cycle, as *Figure 7.6(a)* illustrates. So *any* point on the line may be

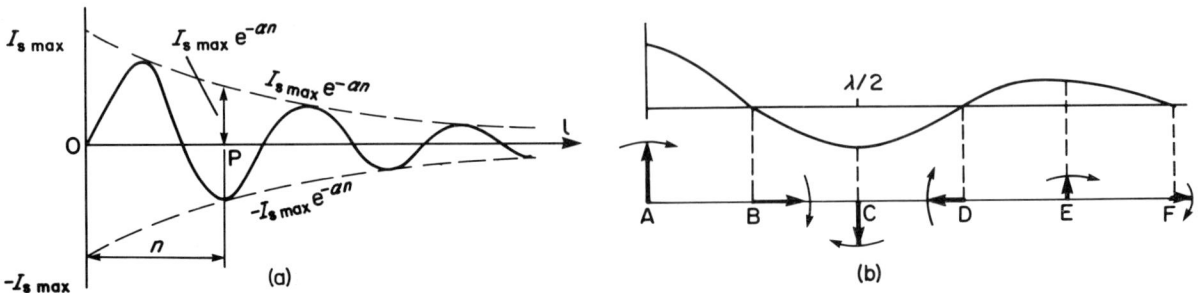

Figure 7.6

considered to have attached to it a rotating phasor in the manner shown at points A, B, C, etc., in *Figure 7.6(b)*, the length of a particular phasor representing the peak (or the r.m.s.) value of the current at that particular point. All phasors are rotating *simultaneously* at the same angular velocity ω; thus at a time $\pi/2\omega$ sec (a quarter period) after the instant illustrated in the figure, phasor C will have rotated a further 90° and the instantaneous value at C will be zero. And at points such as D and E these phasors, too, will have rotated by 90° and the instantaneous values of the current at these points will be a positive maximum and zero, respectively.

In this way, as the simultaneous rotation of diminishing phasors, the travelling current wave may be pictured in its progress along the line.

7.4.2 Voltage variation

The voltage equation has been given in equation (7.2) above and may be expressed as

$$V_R = V_s \cdot e^{-n\alpha} \underline{/-n\beta}$$

For a line of impedance $Z_0\underline{/-\phi}$ the voltage will lag on the current by angle ϕ at all points and, with this difference of phase in mind, we may now depict the voltage variations along a line in exactly the same manner as that used to display the variations in current.

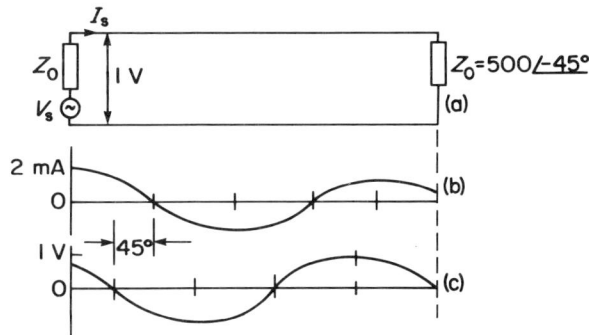

Figure 7.7

Figure 7.7(a) shows a line terminated in its characteristic impedance, assumed here to be $500\underline{/-45°}$ Ω. The instantaneous distributions of current and voltage are shown, respectively, in *Figures 7.7(b)* and *7.7(c)*.

7.5 SPIRAL DIAGRAMS

The tip of a rotating phasor of constant length r and angular velocity ω rad/s describes a circle, and its projection on to the vertical axis represents a sine wave of amplitude r and frequency $= \omega/2\pi$ Hz. If this phasor-generator is connected to the sending end of a transmission line, the effect of attenuation is to decrease the amplitude of the phasor as its rotation proceeds; the tip of such a phasor will now clearly trace out some kind of spiral path such as that shown in *Figure 7.8*, the rate at which this spiral 'winds up' depending upon the rate at which the wave it represents is attenuated along the line.

Consider again the instantaneous picture of a current wave on a

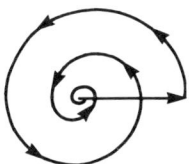

Figure 7.8

long line such as that of *Figure 7.6(a)*. The broken curves showing the current peak distribution along the line may be represented at any point by a peak phasor of length $I_s \cdot e^{-n\alpha}$ and angle $-n\beta$. As n increases the phasor rotates in a clockwise direction, but at the same time its amplitude decreases logarithmically. The locus of the tip of the generating phasor is consequently a *logarithmic* (or equi-angular) spiral. The complete spiral now rotates in an anticlockwise direction as *Figure 7.8* depicts.

The following example will illustrate this aspect of spiral graph work as well as a revision of some of the foregoing theory.

Example 4. A uniform transmission line of infinite length has a propagation coefficient per kilometre of $0.526\underline{/84.5°}$ and a characteristic impedance $Z_0 = 1000\underline{/-\pi/4}\ \Omega$.

A sinusoidal alternating voltage having a peak amplitude of 10 V is applied to the line input terminals. Draw spiral and rectangular graphs to show the instantaneous values of current and voltage along the first 18 km of the line at the moment when the sending end voltage is a maximum.

We find first the attenuation coefficient:

$$\gamma = 0.526\underline{/84.5°} = \alpha + j\beta$$
$$= 0.526\ (\cos 84.5° + j \sin 84.5°)$$
$$= 0.526(0.096 + j0.995)$$
$$= 0.05 + j0.523$$

whence $\alpha = 0.05$ népers/km and $\beta = 0.523$ rad/km ($= \pi/6$ rad/km)

Now
$$\alpha = 0.05 = \ln \frac{V_1}{V_2} = \ln \frac{I_1}{I_2}$$

where V_1 and I_1 are the voltage and current at *any* point on the line and V_2 and I_2 are the voltage and current at a point 1 km further down the line.

$\therefore$
$$\left| \frac{V_1}{V_2} \right| = \left| \frac{I_1}{I_2} \right| = e^\alpha = e^{0.05} = 1.051$$

whence
$$e^{-\alpha} = \frac{1}{1.051} = 0.951$$

The voltage V_R at a point n km along the line is given by

$$V_R = V_s \cdot e^{-n\alpha}\underline{/-n\beta}$$

where V_s is the sending end voltage, in this example 10 V peak. The peak voltage at a point n km along the line is therefore

$$|V_R| = 10e^{-n\alpha} = 10(0.951)^n\ \text{V}$$

Further, the phase angle at a distance n km along the line relative to the sending end voltage phase is

$$-n\beta = -n \cdot \frac{\pi}{6}\ \text{rad}$$

Thus V_R lags behind V_s by $n\pi/6$ rad; this becomes 2π rad when $n = 12$, so that a wavelength on the line is 12 km.

Now the current at a distance n km from the sending end is

$$I_R = \frac{V_R}{Z_0} = \frac{V_R\underline{/\pi/4}}{|Z_0|}$$

that is, I_R leads V_R by $\pi/4$ rad.

We can now substitute values for n in the appropriate equations and so build up a plotting table (*Table 7.1*) for the amplitude and phase values of the voltage and current at all points along the line when the sending end voltage is at its maximum of 10 V. For convenience, we take steps of 3 km units along the required 18 km of line.

Table 7.1

Distance (n km)	$e^{-n\alpha}$ $= (0.951)^n$	$V_R = 10\,e^{-n\alpha}$ Amplitude (V)	Phase	$I_R = V_R/1000$ Amplitude (mA)	Phase
0	1.0	10.0	0	10.0	$+\pi/4$
3	0.86	8.6	$-\pi/2$	8.6	$-\pi/4$
6	0.74	7.4	$-\pi$	7.4	$-3\pi/4$
9	0.64	6.4	$-3\pi/2$	6.4	$-5\pi/4$
12	0.55	5.5	-2π	5.5	$-7\pi/4$
15	0.47	4.7	$-5\pi/2$	4.7	$-9\pi/4$
18	0.41	4.1	-3π	4.1	$-11\pi/4$

The polar spirals for voltage and current are plotted from these figures, as shown in *Figure 7.9(a)* and *7.9(b)*, respectively, the phasors being shown in broken lines 'second time round' to avoid confusion. From these spirals the instantaneous values of voltage and current can be found from the vertical projections of the phasors in the usual way. When this is done a table of instantaneous values may be constructed (*Table 7.2*), from which graphs of the voltage and current variations along the line may be directly plotted on to rectangular axes. Alternately, direct projections from the spirals may be made on to the rectangular axes. The graphs are shown in *Figure 7.9(c)*.

Table 7.2

Distance (km)	Voltage (V)	Current (I mA)
0	10	7.1
3	0	6.1
6	−7.4	−5.2
9	0	−4.5
12	5.5	−3.9
15	0	3.3
18	4.1	2.9

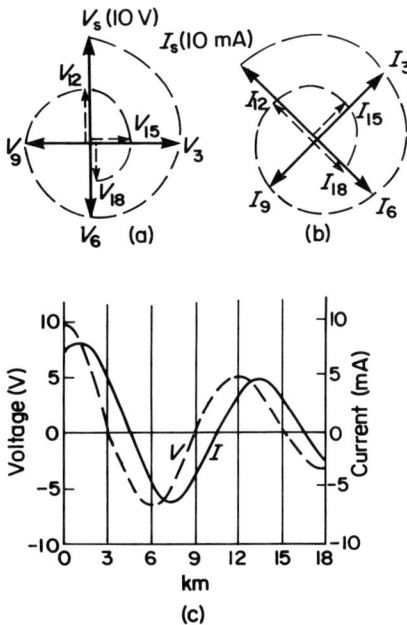

Figure 7.9

7.6 THE PRIMARY LINE COEFFICIENTS

Important formulae for characteristic impedance Z_0 and propagation constant γ of a line can be determined from a knowledge of what are called the *primary constants* or coefficients of the line, that is, the resistance R, leakage conductance G, inductance L and capacitance C, all measured per unit length of line. These primary coefficients are assumed to be independent of frequency and to a reasonable degree they are so. The word 'constant' is consequently more appropriate here than it used to be when it was used to describe Z_0, α, β and γ. R, G, L and C may be obtained by measurements on a sample of the line, and manufacturers usually state them for a standard length.

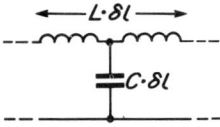

Figure 7.10

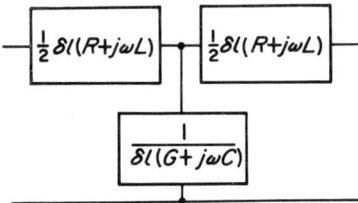

Figure 7.11

Let a line have an inductance L henries and capacitance C farads per unit length. We neglect for a moment the losses due to R and G. Then, for a very short length of time δl (*Figure 7.10*), the inductance will be approximately equal to $L \cdot \delta l$ and the capacitance will be approximately equal to $C \cdot \delta l$. Hence, as the input impedance for a section is (from Chapter 6) given by

$$\sqrt{\frac{L}{C} - \frac{\omega^2 L^2}{4}}$$

provided that the frequency is less than the cut-off value $f_c = 1/\pi\sqrt{(LC)}$, the input impedance for the small length being considered will be

$$\sqrt{\frac{L \cdot \delta l}{C \cdot \delta l} - \frac{\omega^2 L^2 (\delta l)^2}{4}}$$

provided that the frequency is less than

$$\frac{1}{\pi\sqrt{(L \cdot \delta l \cdot C \cdot \delta l)}}$$

The accuracy of these expressions increase as δl decreases, that is, as δl approaches zero. As $\delta l \to 0$, the term containing $(\delta l)^2$ may be neglected in both, so giving

$$Z_0 = \sqrt{\frac{L}{C}} \quad \text{and} \quad f_c = \infty$$

as limiting values for an element of the line. Thus a line without loss will pass all frequencies without attenuation and will have an input resistance equal to $\sqrt{(L/C)}$.

However, a practical line will have resistance R and leakage conductance G in addition to inductance and capacitance. A short section may therefore be represented as the T-network shown in *Figure 7.11*, where series impedance Z_a is given by

$$\tfrac{1}{2}(R \cdot \delta l + j\omega L \cdot \delta l)$$

and shunt impedance Z_b is

$$\frac{1}{(G \cdot \delta l + j\omega C \cdot \delta l)}$$

Substituting these values into the general equation for a section $Z_0 = \sqrt{(Z_a{}^2 + 2Z_a Z_b)}$ gives us

$$Z_0 = \sqrt{\frac{(R + j\omega L)^2 \delta l^2}{4} + \frac{2(R + j\omega L)\delta l}{2(G + j\omega C)\delta l}}$$

$$= \sqrt{\frac{R + j\omega L}{G + j\omega C} + \frac{(R + j\omega L)^2 \delta l^2}{4}}$$

As $\delta l \to 0$, the term containing δl^2 may be neglected, so giving

$$Z_0 = \sqrt{\frac{R + j\omega L}{G + j\omega C}} \tag{7.3}$$

It may not appear at first that this equation will reduce to the simple form $Z_0 = \sqrt{(L/C)}$ when the frequency becomes high, that is, when ω is very large. It does so, however, because in spite of the fact that the resistance of the line increases as frequency increases due to skin effect, this increase is negligible in

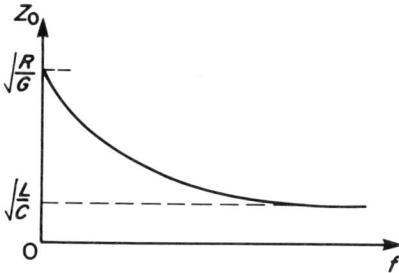

comparison with the increase in the inductive reactance, $j\omega L$. R increases only as the square root of the frequency, whereas reactance increases directly with the frequency.

Similarly, the leakance G is negligible in comparison with the capacitive admittance, ωC. R and G consequently become small relative to ωL and ωC, and so the expression of (7.3) approximates more and more to $\sqrt{(L/C)}$ (a purely resistive term) as ω increases. At very low frequencies, both ωL and ωC approach zero and so the expression approaches the limit $\sqrt{(R/G)}$. Since R/G is in all cases greater than L/C, the variation in Z_0 with frequency will be somewhat as shown in *Figure 7.12*, being equal to $\sqrt{(R/G)}$ at $f = 0$ and constantly approaching $\sqrt{(L/C)}$ as f increases.

Figure 7.12

This variation of characteristic impedance with frequency is not advantageous for the distortionless transmission of a range of frequencies over a transmission line, and in practice a means has to be found of making Z_0 substantially independent of frequency.

The following example illustrates the use of the formula just mentioned and its working should be carefully followed.

Example 5. An open-wire line has the following primary coefficients per kilometre run: $R = 10\ \Omega$, $G = 2 \times 10^{-6}\ \text{S}$, $L = 3.6\ \text{mH}$, $C = 0.012\ \mu\text{F}$. Calculate the characteristic impedance of this line when the frequency of transmission is 1 kHz.

We calculate $R + j\omega L$ and $G + j\omega C$ separately, bringing both results to polar form:

$$R + j\omega L = 10 + j2\pi \times 1000 \times 3.6 \times 10^{-3}$$
$$= 10 + j7.2\pi = 10 + j22.6$$
$$= 24.7\underline{/66.1°}\ \Omega$$
$$G + j\omega C = 10^{-6}(2 + j2\pi \times 1000 \times 0.012)$$
$$= 10^{-6}(2 + j75.4)$$
$$= 75.4 \times 10^{-6}\underline{/88.5°}\ \Omega$$

Then
$$Z_0 = \sqrt{\frac{24.7\underline{/66.1°}}{75.4 \times 10^{-6}\underline{/88.5°}}}$$
$$= 572\underline{/-11.2°}\ \Omega$$

If the primary coefficients are known, the propagation constant can be calculated by using the formula

$$\gamma = \sqrt{[(R + j\omega L)(G + j\omega C)]}$$

which is quoted without proof. Since $\gamma = \alpha + j\beta$

$$\alpha + j\beta = \sqrt{[(R + j\omega L)(G + j\omega C)]}$$

As before, α is the attenuation constant and β the phase constant of the line.

The following worked examples will illustrate further the working and application of the various formulae.

Example 6. A transmission line of negligible loss has the following primary coefficients at a frequency of 0.5 MHz: $L = 0.4\ \text{mH/km}$, $C = 0.08\ \mu\text{F/km}$. Calculate the characteristic impedance, the propagation constant, the line wavelength and the velocity of propagation at the given frequency.

Since the line is loss free

$$Z_0 = \sqrt{\frac{L}{C}}$$

$$Z_0 = \sqrt{\left(\frac{0.4 \times 10^{-3}}{0.08 \times 10^{-6}}\right)} = \sqrt{\left(\frac{4 \times 10^4}{8}\right)}$$

$$= \sqrt{5000} = 70.7 \ \Omega$$

$$\gamma = \sqrt{(j\omega L \times j\omega C)} = j\omega\sqrt{(LC)}$$

$$= j(\pi \cdot 10^6) \cdot \sqrt{(4 \times 10^{-4} \times 8 \times 10^{-8})}$$

since $\omega = 10^6 \pi$.

$$\therefore \qquad \gamma = j\pi\sqrt{32} = j17.8$$

As $\gamma = \alpha + j\beta$ and $\alpha = 0$ (the line is loss-free), we see from the last answer that $\beta = 17.8$ rad/km. Hence

$$\text{Wavelength} = \frac{2\pi}{\beta} = \frac{2\pi}{17.8}$$

$$= 0.353 \text{ km}$$

$$\text{Velocity } v = \frac{\omega}{\beta} = \frac{\pi \times 10^6}{17.8}$$

$$= 176\,500 \text{ km/s}$$

Example 7. A line has the following primary coefficients at $\omega = 10\,000$ rad/s; calculate its characteristic impedance, attenuation and phase constants at this frequency: $L = 4$ mH/km, $R = 20 \ \Omega$/km, $C = 0.025 \ \mu$F/km, $G = 100 \ \mu$S/km.

In this example the line is not loss-free and the full equations for Z_0 and γ must be used. At $10\,000$ rad/s, we have

$$\omega L = 4 \times 10^{-3} \times 10^4 = 40 \ \Omega$$
$$\omega C = 0.025 \times 10^{-6} \times 10^4 = 2.5 \times 10^{-4} \ \Omega$$
$$G = 100 \times 10^{-6} = 10^{-4} \ \text{S}$$

Calculating $R + j\omega L$ and $G + j\omega C$ separately

$$R + j\omega L = 20 + j40 = 44.72 \underline{/63.4°}$$
$$G + j\omega C = 10^{-4}(1 + j2.5) = 2.69 \times 10^{-4} \underline{/68.2°}$$

Always convert to polar form as this makes the division easier.

$$Z_0 = \sqrt{\left(\frac{R + j\omega L}{G + j\omega C}\right)} = \sqrt{\frac{44.72 \underline{/63.4°}}{2.96 \times 10^{-4} \underline{/68.2°}}}$$

$$= 408 \underline{/-2.34°} \ \Omega$$

Propagation constant $\gamma = \sqrt{[(R + j\omega L)(G + j\omega C)]}$

$$= \sqrt{[(44.72 \underline{/63.4°})(2.69 \times 10^{-4} \underline{/68.2°})]}$$

$$= \sqrt{(0.012 \underline{/131.6°})}$$

$$= 0.11 \underline{/65.8°}$$

$$\therefore \qquad \gamma = \alpha + j\beta = 0.11(\cos 65.8° + j \sin 65.8°)$$

$$= 0.045 + j0.1$$

$$\therefore \qquad \alpha = 0.045 \text{ néper/km}, \qquad \beta = 0.1 \text{ rad/km}$$

Example 8. The voltage on a transmission line, which is properly terminated in its characteristic impedance, falls by 8% over a distance of 2 km. The phase change over the same distance, when the angular frequency is 5000 rad/s, is 15°. Calculate the line attenuation in dB/km and the velocity of propagation.

Let the sending end voltage be V_s; then at a distance of 2 km the voltage will be $0.92V_s$. The line impedance will be the same at all points on the line and equal to Z_0, hence the attenuation over the 2 km will be

$$20 \lg \left| \frac{V_s}{0.92V_s} \right| \, dB = 20 \times 0.0362$$

$$= 0.724 \, dB$$

Over 1 km, the attenuation is clearly 0.362 dB.

On this line the phase change per kilometre is constant and equal to 7.5°.

∴

$$\beta = \frac{7.5\pi}{180} \, rad = 0.131 \, rad$$

But

$$\text{Velocity } v = \frac{\omega}{\beta} = \frac{5000}{0.131}$$

$$= 38\,168 \, km/s$$

Example 9. The input impedance of a line is $2000 \underline{/20°}$ with the termination open-circuited and $270 \underline{/-26°}$ with the termination short-circuited, the frequency being 796 Hz. Given that the propagation constant for the line is $0.025 + j0.3$, obtain values for the primary line coefficients.

$$Z_0 = \sqrt{(Z_{oc}Z_{sc})} = \sqrt{[(2000 \underline{/20°})(270 \underline{/-26°})]}$$

$$= 735 \underline{/-3°}$$

We now need a relationship between Z_0 and γ in terms of the primary coefficients. Since

$$Z_0 = \sqrt{\left(\frac{R + j\omega L}{G + j\omega C} \right)} \quad \text{and} \quad \gamma = \sqrt{[(R + j\omega L)(G + j\omega C)]}$$

we may write $\qquad \gamma Z_0 = R + j\omega L$

and $\qquad \dfrac{\gamma}{Z_0} = G + j\omega C$

But $\qquad \gamma = 0.025 + j0.3 = 0.3 \underline{/85.2°}$

Then $\qquad R + j\omega L = (0.3 \underline{/85.2°})(735 \underline{/-3°})$

$$= 220 \underline{/82.2°}$$

$$= 29.8 + j218 \, \Omega$$

Equating real and imaginary parts:

$$R = 29.8 \, \Omega, \quad \omega L = 218 \, \Omega$$

∴

$$L = \frac{218 \times 10^{-3}}{2\pi \times 796} \, mH = 43.6 \, mH$$

In the same way

$$G + j\omega C = \frac{0.3 \underline{/85.2°}}{735 \underline{/-3°}}$$
$$= 4.1 \times 10^{-4} \underline{/88.2°}$$
$$= 1.28 \times 10^{-5} + j4.1 \times 10^{-4}$$

Hence $\qquad G = 12.8 \times 10^{-6} \text{ S} = 12.8 \text{ μS}$

$$\omega C = 4.1 \times 10^{-4}$$

∴ $\qquad C = \dfrac{4.1 \times 10^{-4}}{2\pi \times 796} = 0.082 \text{ μF}$

PROBLEMS FOR CHAPTER 7

(1) The voltages at the input and at the output of a line, properly terminated in its characteristic impedance, are respectively 5 V and 1 V r.m.s. What would the output voltage be if the length of the line was doubled?

(2) A high-frequency transmission line consists of a pair of open wires having a capacitance of 0.01 μF/km and an inductance of 3.2 mH/km. What is the characteristic impedance of the line?

(3) An underground cable has a phase shift β of 0.168 rad/km. Find the wavelength on the line and the speed of transmission if a signal of frequency of 1.6 kHz is sent along it.

(4) What do you understand by the primary coefficients of a line? A transmission line of negligible loss has the following primary coefficients at a frequency of 796 Hz: $L = 1.2$ mH/km, $C = 0.075$ μF/km. Calculate the characteristic impedance, propagation constant, wavelength and velocity of transmission at this frequency.

(5) Measurements were made on a length of D.8 twisted cable at a frequency of 1592 Hz with the following results:
$Z_{oc} = 730 \underline{/-40°}$, $Z_{sc} = 310 \underline{/-20°}$. Calculate Z_0 for this line at this frequency.

If the attenuation of this line is 2.4 dB/km and the phase constant is 0.34 rad/km, what will be the voltage at a point 10 km down a properly terminated run of the line if 1 V r.m.s. is applied at the sending end?

(6) Write down general expressions for the characteristic impedance and the propagation constant of a uniform transmission line. What are the approximate expressions for these parameters when the frequency is high?

A line has the following primary coefficients at $\omega = 10^4$ rad/s: $L = 1.75$ mH, $C = 0.095$ μF, $R = 78$ Ω and $G = 62$ μS, all per kilometre run. Obtain values for the secondary line coefficients, at this frequency.

(7) Describe qualitatively how alternating currents are propagated along an infinite uniform line of negligible loss.

Show in the form of a sketch, a graphical representation of the instantaneous distribution of current and voltage along an infinite line having a characteristic impedance $1000 \underline{/-30°}$ Ω. Show at least two wavelengths of the signal on the line, assuming that the instantaneous current at the input is at its positive maximum value of 1 mA.

(8) Explain what is meant by a 'travelling wave' on a

transmission line, and explain the terms 'propagation constant' and 'phase constant' as applied to such a line.

A transmission line has an attenuation coefficient of 1.5 dB/km and a phase constant of $\pi/4$ rad/km. If Z_0 for this line is $2000\underline{/-30°}$, construct polar diagrams to show how the current and voltage vary along the first 8 km of an infinite length of such a line when 1 V r.m.s. is applied at its input.

(9) A line with $Z_0 = 836 - j43.8$ Ω has a propagation coefficient of $\gamma = 0.026 + j0.3$. Calculate the primary constants of this line.

(10) A transmission line of infinite length is fed at the input with an alternating voltage of 1 V r.m.s. If Z_0 for this line is $2000\underline{/-20°}$, calculate the input current to the line and, assuming the line is loss-free, show on a single phasor diagram the voltage and current relationship at the input and at a point three-quarters of a wavelength along the line.

(11) A voltage $V_s . \sin \omega t$ is applied across the sending end of a uniform transmission line which may be assumed to be infinite in length. Explain carefully why the instantaneous current at the sending end and the voltage at a point l km distant from the sending end can have non-zero values at an instant when the voltage across the sending end is not zero. Illustrate your answer with sketch graphs.

(12) The velocity of propagation of a signal along a certain line is $0.8c$, where c is the velocity of light. If the signal frequency is 10 MHz and the line attenuation is 1.06 népers/km, what is the propagation coefficient of the line at this frequency?

(13) A 4-metre length of transmission line is terminated in its characteristic impedance of $100\underline{/0°}$ Ω and fed at the input with an alternating signal given by the expression $\sqrt{(2)} . \sin 2\pi . 10^8 t$ volts. Assuming the line to be loss-free, and taking the free-space velocity of light as 3×10^8 m/s, obtain: (a) the wavelength on the line, (b) the power dissipated in the termination, (c) an expression for the instantaneous current in the termination, (d) the phase shift along the line, (e) the total energy (joules) transmitted along the line in 60 s.

(14) When a voltage $\sqrt{(2)} . \cos \pi . 10^8 t$ volts is applied at the sending terminals of a certain line, the input current is given by $12.5\sqrt{(2)} \cos \pi . 10^8 t$ mA. Assuming that the line is loss-free and properly terminated and that the wave velocity is two-thirds that of light, calculate: (a) the characteristic impedance, (b) the wavelength, (c) the input power, (d) an expression for the instantaneous voltage three-quarters of a wavelength down the line.

(15) A transmission line 3 km long has a characteristic impedance $Z_0 = 1200\underline{/-30°}$ Ω. At a certain frequency the attenuation coefficient of the line is 3.0 dB/km and the phase constant is 0.2 rad/km.

If the sending end voltage is 1 V r.m.s. what will be the received current amplitude and its phase relative to the sending end voltage?

8 Transmission lines — 2

8.1   LINE DISTORTION

If a signal transmitted along a line is to be faithfully reproduced at the receiving end, all possible causes of line distortion must be reduced as far as possible. There are the three main causes of distortion on transmission lines:

(1) The characteristic impedance of the line is a function of frequency and the terminating impedance does not vary with frequency in a like manner.

(2) The attenuation of the line varies with frequency so that waves of differing frequencies, and component frequencies of complex waves, are attenuated by different amounts.

(3) The propagation time varies with frequency so that waves of different frequencies arrive at the termination with differing time delays.

These three cases will be examined in turn to find out whether there is a condition of relationship between the primary constants of the line for which the three forms of distortion may be eliminated or reduced.

8.1.1   The distortionless line

First, distortion arising from variations in the characteristic impedance of the line as the frequency changes. Referring back for a moment to *Figure 7.12*, we recall that the characteristic impedance of a line varied from a value approximating to $\sqrt{(R/G)}$ at very low frequencies to a value constantly approaching $\sqrt{(L/C)}$ at very high frequencies. If Z_0 is to be constant throughout the entire usable frequency range, then we require that

$$\sqrt{\frac{L}{C}} = \sqrt{\frac{R}{G}}$$

for then

$$Z_0 = \sqrt{\frac{L}{C}} \underline{/0°} = \sqrt{\frac{R}{G}} \underline{/0°}$$

So, for a line in which $LG = CR$, it will be possible to provide a termination equal to the characteristic impedance at *all* frequencies, so eliminating this first form of distortion.

Consider now the case of attenuation varying with frequency, and suppose the line has the above relationship between the primary constants. We know that

$$\gamma = \sqrt{[(R + j\omega L)(G + j\omega C)]}$$

whence

$$\gamma^2 = RG + j\omega(CR + LG) - \omega^2 LC$$

When $LG = CR$ we may write $(CR + LG) = 2\sqrt{(CR . LG)}$

and so

$$\gamma^2 = RG + 2j\omega\sqrt{(CR . LG)} - \omega^2 LC$$
$$= [\sqrt{(RG)} + j\omega\sqrt{(LC)}]^2$$

$\therefore$

$$\gamma = \sqrt{(RG)} + j\omega\sqrt{(LC)} = \alpha + j\beta$$

Equating real and imaginary parts, we find that when $LG = CR$

$$\alpha = \sqrt{(RG)}, \quad \beta = \omega\sqrt{(LC)}$$

So for $\alpha = \sqrt{(RG)}$, the attenuation coefficient is independent of frequency; as a result all frequencies are equally attenuated and the second cause of line distortion is also removed.

The third form of distortion, that due to varying time delays for different frequency components (or, what is the same thing, the variation in the velocity of propagation with frequency), can now be dealt with by a consideration of the familiar expression

$$\beta = \omega\sqrt{(LC)}$$

For since the velocity of propagation $v = \omega/\beta$, we have

$$v = \frac{\omega}{\omega\sqrt{(LC)}} = \frac{1}{\sqrt{(LC)}} \text{ km/s}$$

when $LG = CR$, and as ω has now vanished, the velocity (and hence the time delay) is independent of frequency.

8.1.2 The loaded line

The condition $LG = CR$ appears to be a most satisfactory one from which to set about the design of a transmission line, for under this condition the line will introduce no distortion. The signal as received will be an exact reproduction of the signal transmitted, reduced only in amplitude and delayed by a fixed time. Additionally, it can also be shown that this distortionless state gives us an added bonus: it is the condition in which the attenuation is a minimum. However, there are difficulties in the path of this seemingly hopeful future.

In the practical forms which a transmission line may conveniently and economically take, the ratio R/L is very much greater than G/C. The inductance is usually low and at very low frequencies is negligible, reactance-wise, in comparison with the resistance R. The line capacitance on the other hand is large and is not readily reduced. In order to approximate to the desired condition $LG = RC$, therefore, we must increase either L or G, neither C nor R being particularly amenable to alteration. Increasing G is not particularly wise because the attenuation and power losses will increase. Attempts have therefore to be made to increase L. Such artificial increases in line inductance is known as *loading*.

Loading may be carried out in either of two ways: by the insertion of coils at specified intervals along the line, a method known as 'lumped loading' and generally used on underground cable systems; or by a wrapping of iron or some other magnetic material in the form of a wire or tape around the conductors to be loaded so that the permeability of the surrounding medium is increased and hence the inductance of the conductors is increased. This method is known as 'continuous loading' and is employed on submarine cables where the problems of making water-tight seals at discrete points would be difficult and costly.

In the lumped-loaded line, provided that the spacing is uniform, the line performs at all frequencies up to the cut-off frequency more or less as if the added inductance were distributed uniformly along its length. Above the cut-off frequency there is a very rapid

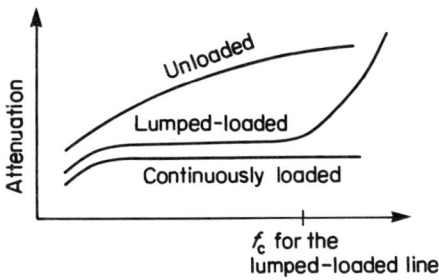

Figure 8.1

increase in the attenuation; but provided that f_c is greater than the upper frequency component of the transmission concerned, this is of no consequence and this method of loading is quite satisfactory. *Figure 8.1* shows typical curves of attenuation against frequency for unloaded, lumped-loaded and continuously loaded cables.

It is not practicable, however, to add *sufficient* inductance in this way to satisfy the distortionless criterion $LG = CR$. Quite apart from considerations of cost, any increase of inductance leads to an increase in resistance, and this situation is further aggravated by hysteresis and eddy current losses in the loading coils. The design of a loaded line is therefore, as in so many other things, a matter of compromise, governed by questions of the maximum permissible attenuation and the cut-off frequency.

8.1.3 The cut-off frequency

A lumped-loaded line behaves as a low-pass filter since the bulk of the inductance is lumped instead of being uniformly distributed throughout the length of the line. It can be shown that, below the cut-off frequency, the effect of the loading coils may be taken into account by considering it to be uniformly spread over the length of line between successive coils. If the inductance of a line *plus* its loading coil(s) is L_s henries per unit length and the capacity of the line is C_s per unit length, then

$$f_c = \frac{1}{\pi\sqrt{(L_s C_s)}}$$

Now let d be the spacing interval of the loading coils in kilometres and let L and C be the line inductance and capacitance per kilometre; then $L_s = dL$ and $C_s = C$, so that

$$f_c = \frac{1}{\pi\sqrt{(dLC)}}$$

whence

$$d = \frac{1}{\pi^2 f_c^2 LC} = \frac{4}{\omega_c^2 LC} \qquad (8.1)$$

Hence the coil spacing d must not be greater than $4/(\omega LC)$, where ω is the highest frequency to be transmitted. We observe the following factors:

(1) The cut-off frequency is unchanged if the loading coil inductance is multiplied by a factor N and the spacing is divided by the same factor N.

(2) If the loading coil inductance is unaltered, the cut-off frequency is inversely proportional to the loading coil spacing.

Example 1. An underground cable has the following primary constants per km length of run: $R = 30\ \Omega$, $L = 1$ mH, $G = 2 \times 10^{-6}$ S, $C = 0.05\ \mu\text{F}$. Find the value to which L would have to be increased to satisfy the condition of minimum distortion and comment on the answer.

The distortionless condition is that $LG = CR$

$\therefore$

$$L = \frac{CR}{G}$$

For the cable in question, this gives a value of inductance

$$L = \frac{0.05 \times 10^{-6} \times 30}{2 \times 10^{-6}} = 0.75 \text{ H/km}$$

This is 750 times the unloaded inductance of the line. Suppose such a loading coil could be obtained without additional resistance, then, if it was placed into the line as a single lumped component at 1 km intervals, the cut-off frequency would be

$$f_c = \frac{1}{\pi\sqrt{(LC)}} = \frac{1}{\pi\sqrt{(0.75 \times 0.05 \times 10^{-6})}} \text{ Hz}$$
$$= 1644 \text{ Hz}$$

This frequency is much too low even for speech circuits where a transmission range up to 3400 Hz is the minimum normally accepted. In order to get the required increase in inductance while doubling the cut-off frequency, it would be necessary to load the line with coils of inductance 0.37 H at intervals of 0.5 km (see note 2 immediately before this problem). Again, such heavy loading is unnecessary, and in practice the length of the loading section is standardised.

Loading is less important nowadays than it used to be, for present-day utilisation of cable circuits is by multi-channel systems employing v.h.f. carriers, and at these high frequencies even an unloaded cable has characteristics which approach the distortionless condition. In any event, even the lightest loading is not permissible on such lines owing to the extremely wide frequency band being transmitted and the consequently high cut-off frequency required.

Nevertheless, on account of the considerable growth of trunk networks as a whole, there is still a continuing and very substantial need for loading coils for use in main-cable trunk systems.

Example 2. A line has the following primary constants: $R = 40\ \Omega/\text{km}$, $L = 1\ \text{mH/km}$, $C = 0.1\ \mu\text{F/km}$, with G negligible. It is loaded with 80 mH loading coils, each of 4 Ω resistance, at 1 km spacings. What is the characteristic impedance of the line and what is the line loss before and after loading at an angular frequency of 5000 rad/s.

$$\text{Before loading } Z_0 = \sqrt{\left(\frac{R + j\omega L}{G + j\omega C}\right)} = \sqrt{\left(\frac{40 + j5}{j5.10^{-4}}\right)}$$

$$= \frac{40.3\underline{/7.1°}}{5.10^{-4}\underline{/90°}}$$

$$= 100\sqrt{(8.06)}\underline{/-41.4°}$$

$$\therefore \qquad Z_0 = 284\underline{/-41.4°}\ \Omega$$

Also
$$\gamma = \sqrt{[(R + j\omega L)(G + j\omega C)]}$$
$$= \sqrt{\{[40.3\underline{/7.1°}][5.10^{-4}\underline{/90°}]\}}$$
$$= 0.14\underline{/48.6°}$$
$$= 0.14\ [\cos 48.6° + j\sin 48.6°]$$

$$\therefore \qquad \alpha + j\beta = 0.093 + j0.105$$

The attenuation loss is therefore 0.093 népers/km.

Considering now the loaded cable, here $R = 44\ \Omega$/km, $L = 81$ H/km. Then

$$Z_0 = \sqrt{\left(\frac{44 + j405}{j5.10^{-4}}\right)}$$

$$= \frac{407.4\underline{/83.8°}}{5.10^{-4}\underline{/90°}}$$

$$= 903\underline{/-3.1°}\ \Omega$$

Also $\quad \gamma = \sqrt{(407.4\underline{/83.8°})(5.10^{-4}\underline{/90°})} = 0.451\underline{/86.9°}$

$$= 0.451\ [\cos 86.9° + j \sin 86.9°]$$

$$\alpha + j\beta = 0.024 + j0.45$$

The attenuation has now fallen to 0.024 népers/km.

We notice from this example that the characteristic impedance has become almost purely resistive at about 900 Ω and that the attenuation has been reduced by a factor of about four times. Also, the phase constant β has increased from 0.105 to 0.45 rad/km, so reducing the velocity of propagation from

$$v_1 = \frac{5000}{0.105} = 47\ 600\ \text{km/s}$$

to $\qquad v_2 = \frac{5000}{0.45} = 11\ 110\ \text{km/s}$

At this last value the velocity is practically constant over the entire frequency band.

8.2 APPROXIMATE EXPRESSIONS

We found in the previous chapter approximate formulae for the attenuation and phase constants α and β in the case of unloaded lines. Such approximate expressions can be deduced for loaded lines, for at low frequencies, i.e. speech frequencies, it is permissible to treat ωL as being very much greater than R and ωC as very much greater than G; compare the ωL values before and after loading in the previous example, and in general G is negligibly small.

Now $\gamma = \sqrt{[(R + j\omega L)(G + j\omega C)]}$

$$= \sqrt{\left\{ j^2\omega^2 LC \left[1 + \frac{1}{j\omega L}\right] \left[1 + \frac{1}{j\omega C}\right] \right\}}$$

$$= j\omega\sqrt{(LC)} \sqrt{\left\{\left[1 + \frac{1}{j\omega L}\right] \left[1 + \frac{1}{j\omega C}\right]\right\}^{\frac{1}{2}}}$$

Since $R/(\omega L)$ and $G/(\omega C)$ are small compared with unity, the last expression may be expanded by the binomial theorem, i.e.

$$\alpha + j\beta \simeq j\omega\sqrt{(LC)} \left[1 + \frac{R}{2j\omega L} + \dots\right] \left[1 + \frac{G}{2j\omega C} + \dots\right]$$

ignoring second and higher order terms;–

$$\alpha + j\beta \simeq j\omega\sqrt{(LC)} \left[1 + \frac{R}{2j\omega L} + \frac{G}{2j\omega C}\right]$$

$$\simeq \frac{R}{2}\sqrt{\frac{C}{L}} + \frac{G}{2}\sqrt{\frac{L}{C}} + j\omega\sqrt{(LC)}$$

Separating the real and imaginary terms we obtain the useful approximations for a loaded cable:

$$\alpha \simeq \frac{R}{2}\sqrt{\frac{C}{L}} + \frac{G}{2}\sqrt{\frac{L}{C}} \text{ népers/km} \qquad (8.2)$$

$$\beta \simeq \omega\sqrt{(LC)} \text{ rad/km} \qquad (8.3)$$

This last result is clearly what we should expect. In the case where G is negligible, the attenuation expression simplifies still further:

$$\alpha \simeq \frac{R}{2}\sqrt{\frac{C}{L}} = \frac{R}{2Z_0} \text{ népers/km} \qquad (8.4)$$

Example 3. Using the information given for the loaded line of Example 2, find approximate values for Z_0, α and β.

The total inductance per km is 81 mH and the total resistance per km is 44 Ω. Then

$$Z_0 \simeq \sqrt{\frac{L}{C}}\angle 0° \simeq \sqrt{\left[\frac{81 \times 10^{-3}}{0.1 \times 10^{-6}}\right]}\angle 0°$$

$$100\sqrt{(81)}\angle 0° \qquad 900\angle 0° \text{ Ω}$$

Since $G = 0$, we may use the formula

$$\frac{R}{2}\sqrt{\frac{C}{L}} = \frac{R}{2Z_0}$$

$$= \frac{44}{2 \times 900} = 0.024 \text{ népers/km}$$

Also
$$\beta = \omega\sqrt{(LC)}$$
$$= 5000\sqrt{(81 \times 10^{-3} \times 0.1 \times 10^{-6})}$$
$$= 5000 \times 9 \times 10^{-5}$$
$$= 0.45 \text{ rad/km}$$

These results should be compared with those obtained earlier.

Example 4. A certain line has a resistance of 25 Ω and a capacitance of 0.05 ΩF per loop kilometre, the inductance L and leakage resistance being negligible. Obtain a figure for the spacing of loading coils, each having an inductance of 50 mH and a resistance of 5 Ω to give a cut-off frequency of 5 kHz. For this value of spacing, calculate the line attenuation constant, in dB, at a frequency of 1 kHz.

From equation (8.1)

$$d = \frac{4}{\omega^2 LC}$$

Inserting values $d = \dfrac{4}{(2\pi \times 5000)^2 \times 0.05 \times 10^{-6} \times 50 \times 10^{-3}}$

$$= \frac{4 \times 10^3}{4\pi^2 \times 25 \times 0.05 \times 50} = 1.62 \text{ km}$$

Since the attenuation constant is required at a frequency well below cut-off, it is possible to use the approximate formula given in (8.2):

$$\alpha = \frac{R}{2}\sqrt{\frac{C}{L}} + \frac{G}{2}\sqrt{\frac{L}{C}}$$

As G is negligible, this reduces to

$$\alpha = \frac{R}{2}\sqrt{\frac{C}{L}}$$

Keep in mind when using this formula that L, C and R are the effective values of the primary constants per unit (km) length of the loaded line, i.e.

$$L = \frac{50}{1.62} = 30.9 \text{ mH/km}, \qquad R = \frac{5}{1.67} + 25 = 28.1 \ \Omega/\text{km}$$

$$C = 0.05 \ \mu\text{F/km}$$

Hence $\quad \alpha = \frac{28.1}{2} \sqrt{\left(\frac{0.05 \times 10^{-6}}{30.9 \times 10^{-3}}\right)} = 1.79 \times 10^{-2} \text{ népers/km}$

$$= 1.79 \times 10^{-2} \times 8.686 = 0.155 \text{ dB/km}$$

8.3 REFLECTION

We have so far considered current and voltage relationships in either infinite lines or in lines properly terminated in their characteristic impedance. In practice all lines are, of course, finite and their terminating impedances are never exactly the ideal values required by theory. In general, therefore, the termination will absorb only part of the energy of the forward travelling wave which finally reaches it, the remainder being forced to return along the line towards the generator. A *reflected travelling wave* is then said to exist on the line.

It is common in nature for reflections to occur whenever a change in the transmission medium is encountered. A reflection of light energy always occurs when a transition is made from, say, air to glass, part of the incident light being transmitted through the glass and the remainder reflected from the surface. Similarly, in the propagation of sound energy, reflections are produced in the form of echoes whenever a surface is presented to the forward movement of the sound wave.

A transmission line is the medium for the propagation of electrical energy and any such energy arriving at a termination differing in value from the characteristic value Z_0 encounters a more or less abrupt change in the impedance of the medium. Some reflection of the incident energy consequently takes places, and this reflected energy is lost to the receiving load.

The reflection is complete and the whole of the incident energy is returned along the line when the termination is the extreme case of either an open- or short-circuit. Between these extremes and the ideal case of the perfectly absorbing termination Z_0, all degrees of reflection are possible. In dealing with reflection it will be best to consider first these totally reflecting terminations.

8.3.1 The open-circuited termination

Let the length of line be open-circuited at the termination. No power can be absorbed by this termination as the current in it must be zero; this condition can be satisfied if we infer the existence of a reflected current emanating from the termination whose *amplitude is exactly equal to, and whose phase is exactly differing by 180° from,* the incident current. Further, since no power is absorbed but is returned in full back along the line, the reflected wave of voltage at the termination must be equal to that of the incident wave, and hence the voltage at the end of the line must be *doubled* by the open-circuit. Quite apart from this increase in voltage, however, which is due solely to the disappearance of the current, the forward voltage wave at all points on the line, and at the termination, must continue to follow the instantaneous variations of the sinusoidal input to the line. As reflection from the open-circuit is total, we may consider the termination itself to be acting as a generator, powered by the energy of the forward wave and transmitting in turn energy back along the line.

The phenomenon of reflection may be difficult to comprehend at first, but a convenient way of visualising the mechanics of the effect is to treat the travelling wave on the line as made up of electric and magnetic components; half of the energy is stored in the magnetic field due to the current, and half is stored in the electric field due to the voltage. It is the continual interchange of energy from the magnetic to the electric fields and vice versa which, as we have noted earlier, causes the propagation of the total electromagnetic energy along the direction of the wires making up the transmission line. When the wave reaches the open-circuited termination, the magnetic field must collapse, the current then being zero. But this energy cannot just disappear; the energy associated with the field must be converted to some other form. It does this by being converted into electrical energy and getting itself added to the electric field. The voltage at the termination will consequently double, and this increased voltage will initiate the movement of a reflected wave back along the line. As the electric field starts this movement, a magnetic field will be set up and the total energy of the reflected wave (which is in this case equal to that of the forward wave) will again be shared between the magnetic and electric field components. The resultant current and voltage at any point on the line and at any instant of time is therefore the sum of the currents and voltages due to the individual forward and reflected travelling waves.

8.3.2   The short-circuited line

Consider now the case of a line with the termination short-circuited. As the impedance is zero, the voltage developed across it must be zero and once again no power can be absorbed by the termination. For the condition of the zero voltage, the reflected voltage wave will have to neutralise the incident voltage wave at all times, and so must be *equal in amplitude but opposite in phase* to the incident wave at all times. This is comparable with the case of the current wave in the open-circuited termination.

At the termination the electric field must collapse and its energy be given to the magnetic field. This results in a doubled current. Thus the reflected current wave does not have a phase reversal. Also, as before, the resultant voltage and current at any point on the line is the sum of the voltages and currents due to the individual forward and reflected travelling waves.

Example 5. At a particular frequency a 100 Ω line is one-half wavelength long and when terminated in 100 Ω has an attenuation at this frequency of 3 dB. Using a phasor diagram to illustrate conditions, determine the sending end impedance of the line at the frequency concerned when the far end of the line is short-circuited.

At the shorted termination the current will be reflected without a reversal of phase but the voltage will suffer a phase reversal of 180°. To determine the sending end impedance we have to consider the termination as a generator and find the conditions at the sending end which result from the arrival there of the reflected wave. The question does not mention the point but we assume that the reflected wave will be totally absorbed by the generator effective load on its arrival there.

As the line length is λ/2 there will be a phase shift of 180° along its length. Considering first the current, let the forward current phasor I_f be represented as in *Figure 8.2(a)*; then at the termination it will be changed in phase by 180° because of its passage along the line and attenuated by 3 dB, phasor I_t. This current is reflected *without* phase reversal and so behaves as if it was generated at the short-circuited end of the line. As it travels towards the sending end it is attenuated by a further 3 dB and again changed in phase by 180°. At the sending end, therefore, the reflected wave is in phase with the incident (current) wave but reduced in magnitude by 6 dB (a ratio of 0.5). Hence the *total* effective current at the sending end is 1.5 I_f.

Considering now the voltage wave, let the incident voltage be represented by phasor V_f — see *Figure 8.2(b)*. At the termination this wave has been changed in phase by 180° and attenuated by 3 dB, phasor V_t. A further 180° shift is suffered on reflection (since the total effective voltage at the termination must be zero), and the wave starts its return journey *unchanged* in magnitude. A further 3 dB attenuation and 180° phase shift now occurs and at the sending end the reflected wave is 6 dB down on the incident wave and 180° out of phase with it. The *total* effective voltage at the sending end is therefore 0.5 V_i.

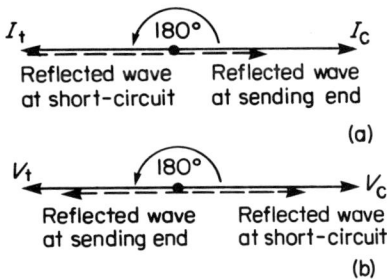

Figure 8.2

Now
$$Z_0 = \frac{V_f}{I_f} = 100 \ \Omega$$

∴
$$Z_{sc} = \frac{0.5 V_f}{1.5 I_f} = 100 \times \frac{1}{3}$$
$$= 33.3 \ \Omega$$

8.4 REFLECTION COEFFICIENT

Before going on to discuss specific current and voltage relationships along open- and short-circuited lengths of line, we will look at the magnitude of the reflected wave in a general case where a generator of impedance Z_0 is working into a line terminated in some finite impedance impedance Z_R, where Z_R is *not* equal to Z_0.

Energy flow along a line, either as a foward or as a reflected wave, has a particular direction and may accordingly be assigned a positive or negative sign to indicate that direction. *Figure 8.3* shows four possible interpretations of energy transfer. In cases (a) and (b) there is a positive flow of energy considered as acting from left to right; in (c) and (d) there is a negative flow of energy acting from right to left. These four cases clearly cover all the possibilities of circuit conditions for a forward and a reflected wave upon a line.

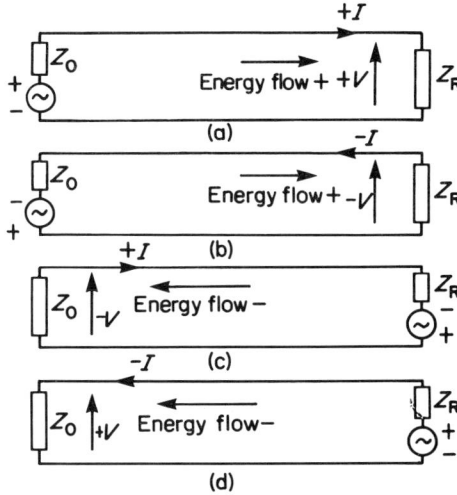

Figure 8.3

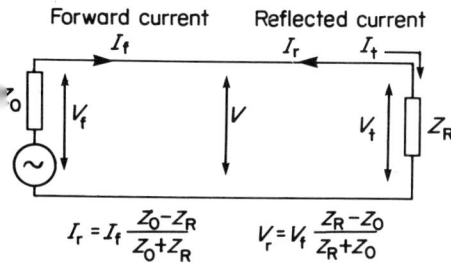

Figure 8.4

In *Figure 8.4* a generator of impedance Z_0 is working into a line terminated in Z_R. In our concept of reflection, an initial current I_f starts out from the generator and, *until* it arrives at the termination, behaves exactly as though the line was infinite or properly terminated in Z_0, that is, it is not aware of the position or nature of the termination until it arrives there. This incident wave consequently travels in the positive direction according to the convention just agreed, along a line whose characteristic impedance is Z_0. Thus

$$\frac{V_f}{I_f} = Z_0 \qquad \text{or} \qquad V_f = I_f Z_0$$

The transmitted wave reaches the termination and a current I_t flows in the load Z_R, developing a voltage V_t, hence

$$\frac{V_t}{I_t} = Z_R \qquad \text{or} \qquad V_t = I_t Z_R$$

This current produces a reflected wave which flows back from the load to the generator, i.e. in a negative direction. Hence

$$\frac{V_r}{I_r} = -Z_0 \qquad \text{or} \qquad V_r = -I_r Z_0$$

Now the voltage across the load is the sum of the incident and reflected voltages at the termination, and the current in the line is equal to that in the termination. Therefore, algebraically

$$V_f + V_r = V_t \qquad \text{and} \qquad I_f + I_r = I_t$$

Forward current I_f Reflected current I_r I_t

$$I_r = I_f \frac{Z_0 - Z_R}{Z_0 + Z_R} \qquad V_r = V_f \frac{Z_R - Z_0}{Z_R + Z_0}$$

Substituting for V_f, V_t and V_r from the previous equations we get

$$I_f + I_r = (I_f - I_r)\,\frac{Z_0}{Z_R}$$

$$\therefore \qquad I_f\left[\frac{Z_0}{Z_R} - 1\right] = I_r\left[1 + \frac{Z_0}{Z_R}\right]$$

so that

$$I_r = I_f\left[\frac{Z_0 - Z_R}{Z_0 + Z_R}\right]$$

and

$$\rho = \frac{I_r}{I_f} = \frac{Z_0 - Z_R}{Z_0 + Z_R} \qquad (8.5)$$

This is an important result, called the *reflection coefficient*. It gives the *ratio* of the reflected current to the incident current. In a similar way, we find that the reflection coefficient for the reflected and incident voltage waves is

$$\frac{V_r}{V_f} = \frac{Z_R - Z_0}{Z_R + Z_0} \qquad (8.6)$$

Notice that

$$\frac{I_r}{I_f} = -\frac{V_r}{V_f}$$

that is, both coefficients have the *same* magnitude but *opposite* signs. Notice also that when $Z_R = Z_0$, both coefficients are zero and there is no reflection.

We have taken the generator impedance to be Z_0. This properly terminates the line in the negative direction and so the reflected wave is completely absorbed on arriving back at the input end of the line.

8.4.1 Current and voltage at any point

Let the generator again be connected to a length of line terminated in Z_R at the far end. We require now to derive expressions for the current and voltage at any point Q at a distance x from the sending end. At such a point Q, see *Figure 8.5*, the total current at any instant is the phasor sum of

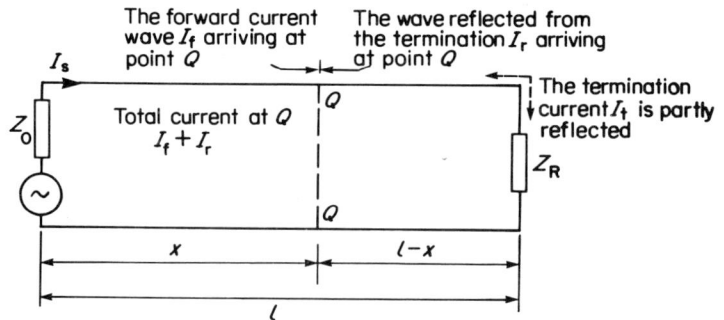

Figure 8.5

the forward and reflected waves at the point. Let the incident current at Q be I_f and the reflected current I_r. Then

$$I_f = I_s \cdot e^{-\gamma x}$$

where I_s is the sending end current and γ is the propagation constant. The current I_t at the termination is

$$I_t = I_s \cdot e^{-\gamma l}$$

where l is the total length of the line.

The amount of reflection occurring at the termination will depend upon the reflection coefficient. In the present case the current reflected from the termination will be I_t multiplied by the reflection coefficient, that is

$$\left[\frac{Z_0 - Z_r}{Z_0 + Z_r}\right] I_s \cdot e^{-\gamma l}$$

Treating this reflected wave as being generated at the termination, its magnitude on arriving back at point Q will be

$$I_r = \left[\frac{Z_0 - Z_R}{Z_0 + Z_R}\right] I_s \cdot e^{-\gamma l} \cdot e^{-\gamma(l-x)}$$

diminished by having travelled the distance $(l-x)$. Rewriting this we get

$$I_r = \left[\frac{Z_0 - Z_R}{Z_0 + Z_R}\right] I_s \cdot e^{-2\gamma l} \cdot e^{\gamma x}$$

The total current at Q is the sum of the forward and reflected currents:

$$I = I_f + I_r$$

$$= \left[\frac{Z_0 - Z_R}{Z_0 + Z_R}\right] I_s \cdot e^{-2\gamma l} \cdot e^{\gamma x} + I_s \cdot e^{-\gamma x}$$

This rather cumbersome expression can be written more compactly as

$$I = A \cdot e^{-\gamma x} + B \cdot e^{\gamma x}$$

where $A = I_s$ and $B = \left[\dfrac{Z_0 - Z_R}{Z_0 + Z_R}\right] I_s \cdot e^{-2\gamma l}$

The forward travelling wave is represented by the term $A \cdot e^{-\gamma x}$ and the reflected travelling wave by the term $B \cdot e^{\gamma x}$. When the reflection at the termination is total, the reflection coefficient is $\pm$ unity and B then reduces to $\pm I_s \cdot e^{-2\gamma l}$.

Many problems concerned with reflection can be dealt with by simple analysis, as the following example illustrates.

Example 6. A cable of negligible loss has a characteristic impedance of 80 Ω. It is terminated in a 200 Ω resistive load, and the voltage measured across this load is found to be 12 V. Calculate, for this line, (a) the values of the forward voltage and current waves, and (b) the values of the reflected voltage and current waves. The values, of course, will be in r.m.s. We find first the reflection coefficient:

$$\rho = \frac{Z_0 - Z_R}{Z_0 + Z_R} = \frac{80 - 200}{80 + 200} = -0.43$$

Thus the reflected voltage is 0.43 times the forward voltage and 180° out of phase.

(a) As the voltage across the load is 12 V, the load current $I_R = 12/200 = 0.06$ A. Now $I_R = I_f + I_r$ and $I_r = \rho I_f$:

$$I_R = I_f(1 + \rho)$$

Hence, the forward current

$$I_f = \frac{I_R}{1 + \rho} = 0.06 \times \frac{1}{1 - 0.43}$$
$$= 0.105 \text{ A}$$

The forward voltage $V_f = I_f Z_0 = 0.105 \times 80$
$$= 8.4 \text{ V}$$

(b) Now

$$I_R = I_f + I_r$$

The reflected current $I_r = I_R - I_f = 0.06 - 0.105 = -0.045$ A
The reflected voltage $V_r = -I_r Z_0 = -(-0.045) \times 80$
$$= 3.6 \text{ V}$$

8.5 STANDING WAVES

We now study in more detail, and qualitatively as far as possible, what happens as the result of the combination of the forward and reflected waves travelling simultaneously in opposite directions along a length of line. Consider a line, open-circuited at the termination and one-wavelength long (in terms of the frequency of the transmission it conveys), measured back from the termination. Then *Figure 8.6* shows the process of reflection for various instants of time t, from $t = 0$ to $t = 7T/8$, where T is the periodic time of the signal. In the diagram the thin line curves represent the forward travelling wave moving to the right, and the dotted line curves represent the reflected wave moving to the left. The resultant of these waves is indicated by the heavy lines. By assuming that the generator impedance is Z_0 we ensure that the reflected wave will be completely absorbed on arriving back at the sending end of the line, so only the single incident and reflected waves need concern us.

The two waves travelling in opposite direction along the line beat together, with the result that the current (and voltage) at any point varies with the position and is the sum of the forward and reflected waves at that point. *Figure 8.6* shows the forward and reflected current waves along the open-circuited line, and if these are added point by point it is seen that the resultant wave has the following characteristics:

(1) The amplitude of the resulting variations with time is different at different points along the line.
(2) At certain points the current is *always* zero.
(3) At certain other points the amplitude of the resultant wave is *double* the amplitude of either travelling wave.

The points at which the current is always zero are called *nodes*, and these occur at the termination and at distances $\lambda/2$, λ, $3\lambda/2$ and so on from it, that is, at all *even* multiples of a quarter-wave length *from* the termination. At all *odd* multiples of a quarter-wave length from the termination, $\lambda/4$, $3\lambda/4$ and so on, the amplitude of the resultant wave is twice the value of either

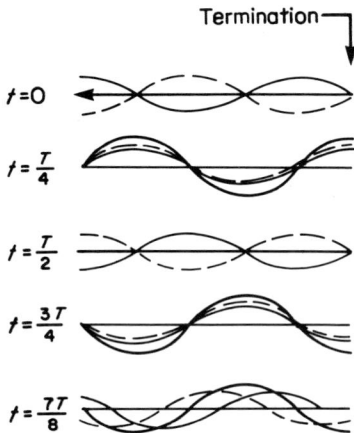

Figure 8.6. Current waves on an open-circuited line

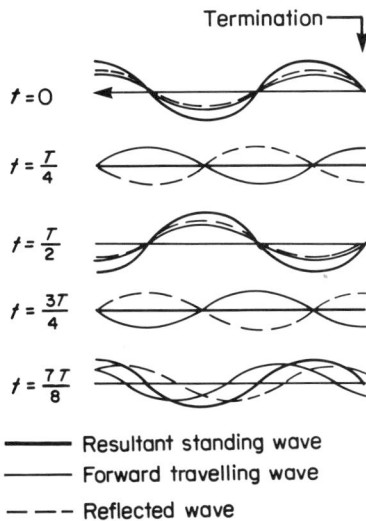

Termination

$t = 0$

$t = \frac{T}{4}$

$t = \frac{T}{2}$

$t = \frac{3T}{4}$

$t = \frac{7T}{8}$

——— Resultant standing wave

——— Forward travelling wave

— —— Reflected wave

Figure 8.7. Voltage waves on an open-circuited line

travelling wave, since at these points the two waves are equal and in phase. These points are known as *antinodes*, sometimes *loops*.

The voltage waveform for the open-circuited line may be deduced in a manner similar to that just outlined for the current wave. When the incident wave reaches the termination the magnetic field collapses and the amplitude of the voltage wave is doubled. Thus no phase reversal occurs and as before the voltage resultant is obtained by adding, point by point, the forward and reflected wave patterns. The result this time is shown in *Figure 8.7*. Voltage nodes occur at distances $\lambda/4$, $3\lambda/4$ and so on from the termination; voltage antinodes occur at distances $\lambda/2$, λ and so on from the termination. At any point the voltage and current waves are in quadrature.

The pattern of the resultant wave set up on a line as a result of the combination of forward and reflected travelling waves is known as a standing wave pattern, analogous in some ways to that of a vibrating string. The space patterns are sine waves in quadrature; the time variations are sinusoidal and in the same phase for both voltage and current. It must be carefully understood that for given termination, the nodes and antinodes of a standing wave do not move right or left *along* the line; a standing wave is a periodic variation taking place on the line without travel in either direction.

Figure 8.8 shows the r.m.s. values of current and voltage that would be recorded on a suitable r.m.s. instrument moving along the line with an open-circuited termination. These r.m.s. values are, of course, always indicated as positive, no note being paid to the reversals of phase in both current and voltage which occur at each successive half-wavelength. Notice that the termination is a current nodal point at all times.

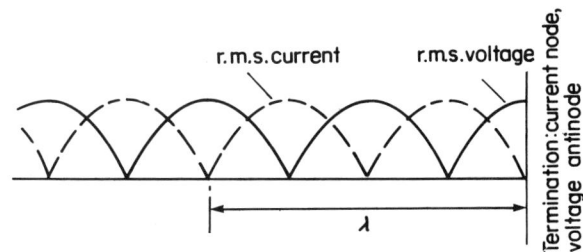

r.m.s. current   r.m.s. voltage

Termination: current node, voltage antinode

λ

Figure 8.8

Energy is transmitted along the line by the travelling wave only. Standing waves which result from reflection store energy. For the perfectly reflecting termination under discussion, the nodal points are true zero points and the standing wave pattern is known as complete. All the energy is stored energy and none is passed to the termination. In practice, for reasonably correct termination impedances, the standing wave is small, the nodal points are never zero and there is progressive attentuation along the length of the line. Little energy is now stored, and the bulk of the transmitted power is properly dissipated in the termination.

8.5.1   The short-circuited line

Consider now the termination of a line that is short-circuited. The impedance is now zero, $\rho = -1$, and total reflection again takes place, the termination being of necessity a voltage node. The type of reflection is, in fact, the reverse of that which we have already described for the open-circuited line: the current wave does not suffer a phase change but is doubled in amplitude, while the voltage wave changes sign by suffering a phase change of 180° to make the effective resultant at the termination zero.

We need not, therefore, draw a new diagram to illustrate these latest conditions, as voltage antinodes for the open line coincide with current antinodes for the shorted line, and voltage nodes for the open line coincide with current nodes for the shorted line. Consequently, *Figures 8.6* and *8.7* may be used to illustrate the voltage conditions and the current conditions, respectively, on a short-circuited line.

8.5.2 Standing wave ratio

Standing waves, as we have indicated, may be detected by a suitable r.m.s. instrument moving along the line. The measurement of such current and voltage maxima and minima not only provides a fairly accurate indication of the wavelength on the line but more importantly also reveals the amount of reflected energy relative to the incident energy that is absorbed at the termination.

Let the incident current at the termination be I_f and the reflected current I_r. Then if I_{max} is the sum of the forward and reflected currents and I_{min} is their difference, the standing wave ratio (SWR) on the line is defined as

$$\frac{I_{max}}{I_{min}} = \frac{I_f + I_r}{I_f - I_r}$$

whence

$$\frac{I_r}{I_f} = \frac{SWR - 1}{SWR + 1}$$

But the power absorbed in the termination $P_t = I_f^2 Z_0 (= V_f/Z_0)$, and the power is reflected $R_r = I_r^2 Z_0 (= V_r/Z_0)$. Hence

$$\frac{P_r}{P_t} = \left[\frac{SWR - 1}{SWR + 1}\right]^2 \tag{8.7}$$

The ratio of the reflected to the transmitted power is thus directly calculable from a knowledge of the standing wave ratio. For a perfectly terminated line the SWR is unity for there is then no reflected power. Consequently, in a practical line, the SWR should be kept as close to unity as possible. Ratios up to 2 are permissible before the loss of power exceeds some 10%, or 0.5 dB.

We note from a glance back to equation (8.5) that

$$\text{Reflection coefficient } |\rho| = \frac{SWR - 1}{SWR + 1}$$

from which

$$SWR = \frac{1 + |\rho|}{1 - |\rho|}$$

Example 7. A transmission line of negligible loss has a characteristic impedance of 600 Ω. What will be the standing ratio on the line when the terminating impedance has a value of $500 + j200)$ Ω?

$$\rho = \frac{Z_0 - Z_R}{Z_0 + Z_R} = \frac{600 - (500 + j200)}{600 + 500 + j200}$$

$$= \frac{100 - j200}{1100 + j200} = \frac{1 - j2}{11 + j2}$$

$$\therefore \quad |\rho| = \left| \frac{1 - j2}{11 + j2} \right| = \frac{\sqrt{5}}{\sqrt{125}} = 0.2$$

$$\text{SWR} = \frac{1 + |\rho|}{1 - |\rho|} = \frac{1.2}{0.8} = 1.5$$

Example 8. A low-loss transmission line has a mismatched load such that the reflection coefficient at the termination is $0.5\underline{/30°}$. If the line impedance is 80 Ω calculate the load impedance.

$$|\rho| = \frac{Z_R - Z_0}{Z_R + Z_0} = 0.5\underline{/30°}$$

$$\therefore \quad \frac{Z_R}{Z_0} = \frac{1 + 0.5\underline{/30°}}{1 - 0.5\underline{/30°}} = \frac{1 + 0.433 + j0.25}{1 - (0.433 + j0.25)}$$

$$\therefore \quad Z_R = 80 \times \frac{1.433 + j0.25}{0.567 - j0.25}$$

$$= 80 \times \frac{1.45\underline{/9.9°}}{0.62\underline{/-23.8°}}$$

$$= 187\underline{/33.7°}\,\Omega \text{ or } 154 + j104\,\Omega$$

8.6 PRACTICAL DESIGN OF LINES

The practical realisation of a line depends upon the functions that it may have to perform, and factors which have to be considered in the design and choice of a line for a particular job include the frequency bands concerned, the power to be conveyed, attenuation, problems of matching and coupling to other circuits, ease of maintenance, protection against weather and other natural hazards.

So far, a line has been imagined as being made up of two parallel wires and in many cases practical lines do take this form. Such lines have the advantage in economy, particularly for the transmission of high power at very low frequencies, their construction form is relatively simple and the maintenance straightforward, and standing wave distributions are easily measured. Parallel wire lines, however, have high radiation losses at high frequency operation and the problems of undesired coupling to adjacent circuits are considerable. Parallel wire lines are balanced systems.

The *coaxial cable,* familiar to most people in the form of the television feeder, consists of a central conductor surrounded by an earthed sheath throughout its length. This line has the advantage over the parallel wire line in that it is effectively self-screening and its radiation loss is very small even at high frequencies. Further, it is compact, can be run along walls or buried in the earth, and is little influenced by nearby circuits. It is, however, more costly to manufacture and is an unbalanced system.

In a parallel wire line where the radius of each wire is r and the wire separation is d, the characteristic impedance Z_0 is given by

$$Z_0 = 277 \lg \frac{d}{r} \ \Omega$$

A common value for such lines is 600 Ω. Other impedances are available, especially for television and f.m. use as aerial feeders in which the wires occupy the edges of a polythene-type tape. The wire spacing is thus automatically maintained and the line is flexible and easily positioned. Twisted pairs are also available, often enclosed in an outer screening, and many pairs can be found assembled together in large underground trunking systems where the covering is paper wrapped around the wires comparatively loosely so that the dielectric is effectively mixed paper and air. An accurate formula cannot be given for cables of this sort.

There are practical limits to the realisable values of Z_0 for parallel lines; with open parallel wires stretching over any reasonable distance it is not possible to have d/r much greater than 100 or less than 10, so that Z_0 is restricted to the range 250–600 Ω approximately. After the ratio exceeds 100, Z_0 remains fairly constant. If the wires are embedded in a dielectric a closer and more readily controlled spacing is possible and this, together with the increased line capacitance, enables Z_0 to be reduced to values below 250 Ω. For the minimum resistive losses it can be shown that the ratio d/r should be approximately 4, and for the maximum dielectric strength the ratio should be about 5.4.

The coaxial line has the inner conductor spaced away from the outer surrounding screen either by a solid dielectric, usually polythene, or by disc-like spacers set along the line at frequent and regular intervals. This latter construction gives a virtually airspaced cable and the characteristic impedance of such a line is given closely by

$$Z_0 = 138 \lg \frac{R}{r} \ \Omega$$

where r and R are the radii of the inner and outer conductors, respectively. Practical considerations again limit the values of Z_0 to a range of approximately 50–150 Ω as R/r ranges from 2 to 10, but most coaxial line is designed to an impedance of about 80 Ω. This impedance has certain advantages: minimum attenuation occurs when the ratio $R/r = 3.6$ and for this particular ratio $Z_0 = 77 \ \Omega$. For maximum dielectric strength, giving the least chance of a breakdown. R/r should be 2.72, which leads to Z_0 being 60 Ω.

To increase the effective radius of the inner conductor while retaining flexibility, a stranded inner conductor is often employed. This procedure is successful at the lower frequencies, but at higher frequencies such stranding increases the resistive loss in the innner. It is necessary to protect the outer conductor, which is often braided for flexibility, from the effects of weather, and such protection is usually afforded by an outer covering of polyvinyl. The fact that the outer conductor is braided so that the current paths along the strands of the braid are oblique to the axis, leads to a higher resistive loss than would occur with a solid outer.

PROBLEMS FOR CHAPTER 8

(1) An underground cable has the following primary constants per km length: $R = 20\ \Omega$, $L = 0.8$ mH, $G = 1.0\ \mu$S, $C = 0.03\ \mu$F. Find the value to which L would need to be increased to satisfy the condition for minimum distortion.

(2) How does coil loading decrease the attenuation coefficient of a low-frequency line. Why is such loading effective over only part of the frequency range? What do you understand by the cut-off frequency of a loaded line? Obtain an expression for the cut-off frequency of such a line.

(3) A cable has the following primary coefficients per km run: $R = 100\ \Omega$, $G = 1.5\ \mu$S, $L = 1$ mH, $C = 0.06\ \mu$F. Given that $\omega = 10^4$ rad/s, obtain *approximate* values for Z_0, α and β, if the cable is loaded with 88 mH coils each of resistance $4\ \Omega$ at 1.83 km spacing.

(4) A telephone line has a resistance of $20\ \Omega$ and a capacitance of $0.05\ \mu$F per loop kilometre, L and G being negligible. 80 mH loading coils, each having a resistance of $10\ \Omega$, are inserted at intervals of 1.125 km. Calculate the attenuation constant of the line at a frequency of 800 Hz, and the approximate value of the cut-off frequency.

(5) What do you understand by the term 'reflection coefficient'? If a long line having a characteristic impedance of $10(40-j2)\ \Omega$ is terminated in an impedance of $10(40+j2)\ \Omega$, what will be the value of the reflection coefficient?

(6) Explain what happens when a wave travelling in a line of characteristic impedance Z_0 arrives at the end of the line where it is terminated in an impedance Z. Derive expressions for the value of current and voltage in the reflected wave at the termination in terms of the corresponding values of the incident wave.

(7) Derive an expression for the coefficient of reflection for the case of a loss-free transmission line terminated in a pure resistance. A loss-free line has a characteristic impedance of $80\underline{/0°}\ \Omega$ and is terminated in a resistance of $220\ \Omega$. What is the ratio of maximum to minimum voltage along the line?

(8) Explain the meaning of the terms: (a) reflected wave, (b) standing-wave ratio, with reference to a transmission line.

A loss-free line of characteristic impedance $600\underline{/0°}\ \Omega$ is connected to a transmitting aerial of impedance $(400 + j200)\ \Omega$. Calculate the amplitude of the reflected wave relative to the incident wave. What is the standing-wave ratio?

(9) An infinitely long line has the following secondary coefficients at a frequency of 796 Hz: $\alpha = 0.1$ néper/km, $\beta = 0.1\pi$ rad/km.

If the sending end current is 1 mA, draw a polar spiral diagram to indicate the change in amplitude and phase of the current over the first 10 km of the line.

(10) Draw a graph of the relationship $Z_0 = 277\ \lg d/r\ \Omega$ for values of d/r ranging from 5 to 800. Use your graph to find the spacing necessary between two parallel open wires whose diameters are each 2 mm, to provide impedances of (a) $300\ \Omega$, (b) $550\ \Omega$.

If these lines are moulded into a solid dielectric, their spacing being undisturbed, and the relative permittivity of the dielectric is 2.5, what would their respective impedances become?

(11) The centre conductor of a concentric (coaxial) cable

feeding a television receiver has a diameter of 0.075 cm and the outer conductor has a diameter of 0.38 cm. If the relative permittivity of the solid dielectric used in the cable is 1.93, calculate the impedance of the cable.

(12) A low-loss line of characteristic impedance $500\underline{/0°}$ Ω is terminated in an impedance of $(500 + j577)$ Ω. The voltage across the termination is 1.32 V r.m.s. Calculate the amplitude (modulus) of the voltage across the line (a) at a point $\lambda/12$ distant from the termination, (b) at a point $\lambda/3$ distant from the termination.

(13) The standing wave ratio on a mismatched line is found to be 1.85. If the incident power *arriving* at the termination is 150 mW, what is the reflected power?

(14) A certain cable has $Z_0 = 60\underline{/0°}\Omega$. If the cable is terminated by an impedance $55\underline{/20°}$ Ω, what will be the SWR on the cable?

(15) A coaxial cable of characteristic impedance $80\underline{/0°}$ Ω is terminated by a component which may be represented at the operating frequency of 100 MHz as a resistor of 100 Ω in series with a 0.1 μH inductance. What will be the reflection coefficient on this line?

9 Resonant lines

For both open- and short-circuited lengths of line, the r.m.s. voltage and current distribution along the line have already been indicated in *Figure 8.8*. If therefore the length of a line is varied by moving the termination, the sending end impedance will depend, as far as magnitude is concerned, upon the ratio of the voltage and current amplitudes present at that end. When a line is correctly terminated, the input impedance is resistive and equal to Z_0 ($= R_0$ here); for every other condition of terminating impedance there is a standing wave set up along the line and the effective impedance is no longer Z_0. In a lossless line all nodes and antinodes for an open-circuited termination are positioned on the line (relative to the termination) at distances displaced by a quarter-wavelength from the positions they would occupy for a short-circuited termination. However, if the line length is changed the positions of the nodes and antinodes is changed also, but both move through equal distances and in the same direction so that they are always separated by a constant interval of a quarter-wavelength, i.e. the wavelength of the standing waves on a line is fixed by the frequency of the sending generator, *not* the length of the line. The voltage and current conditions at the sending end of a line *will*, however, depend upon the length of the line in so far as this is expressed in terms of the generated wavelength.

9.1 INPUT IMPEDANCE

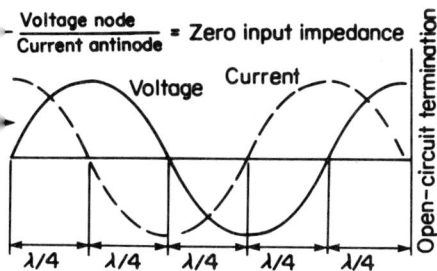

Figure 9.1

Let a line of length $5\lambda/4$ be open-circuited at the termination as shown in *Figure 9.1*. Then, standing waves of voltage and current will be established on the line as the figure depicts, the termination being always a current node and a voltage antinode. For the length concerned we see that the generator end of the line is a current antinodal and a voltage nodal point, i.e. a current maximum and a voltage minimum. Consequently, from the point of view of the generator, the open-circuited length of line behaves as a *zero impedance* load, and this will be true whenever a line contains an *odd* number of quarter-wavelengths and has an open-circuited termination. This may seem a very surprising result, but it is nevertheless true. Although in practice a line is not without loss and the nodal points are, as a consequence, never at zero level, the input impedance of such lengths of open-circuited line is still very small.

If the line is not an exact number of odd quarter-wavelengths long, the generator end will no longer be a point of current maximum and voltage minimum. From the diagram we see that the current and voltage along the line are everywhere in phase quadrature, the voltage leading the current by 90° when both curves are on the same side of the axis and lagging on the current by 90° when the curves are on opposite sides of the axis. But quadrature conditions of voltage and current represent pure reactance; so the impedance that a line presents to the generator when its length is not a multiple of quarter-wavelengths is therefore purely reactive. This reactance is inductive when the voltage leads by 90°, i.e. when the curves are on the same side of the axis; and capacitive when the voltage lags by 90°, i.e. when the curves are on opposite sides of the axis. A length of line consequently behaves as an *inductance or a capacitance* according as its length contains an even or an odd multiple of a quarter-wavelength *plus* a section of length less than a quarter-wavelength.

Similarly, a short-circuited line an odd number of quarter-wavelengths long will present a very high (theoretically infinite) impedance to the generator, for the generator end will see the conditions of a voltage maximum and a current minimum. Any impedance from zero to infinity may therefore be simulated at a given frequency by a suitable length of open- or short-circuited line.

This is an important property because it enables us to dimension a length of transmission line so that it will simulate the function of an inductance or capacitance of any desired value. Clearly, these effects are only of practical use at very high frequencies where the line lengths can be easily cut and handled. Such line lengths used for such purposes are known as *resonant lines*.

Example 1. Show by means of phasor diagrams that the input impedance of a short-circuited length of line less than $\lambda/4$ in length is inductive.

Consider the phasor diagrams for the forward travelling wave at the termination and the sending end, respectively; these are shown in *Figure 9.2(a)* and *9.2(b)*, the wave reaching the termination a short time after leaving the generator. This lag is represented by ϕ rad, where $\phi = 2\pi l/\lambda$, and has a value between 0 and $\lambda/4$, for a line length less than $\lambda/4$.

The reflected wave is obtained by reversing the voltage phasor V_t at the termination as shown at *Figure 9.2(c)*, and this wave now reaches the generator again after a further phase lag ϕ, as at *Figure 9.2(d)*.

The resultant voltage and current at the generator is obtained by combining *Figure 9.2(b)* and *9.2(d)*, and this is shown at *9.2(e)*. From this it is seen that the resultant current lags the voltage by 90°, that is, the line behaves, in the steady state, as a pure inductance.

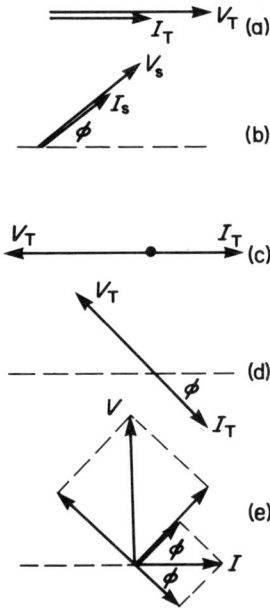

Figure 9.2

9.1.1 The general case

Referring to *Figure 9.1*, imagine that the open-circuited line there shown is of any length l measured from the termination. Then the current can be considered to be represented as a cosine expression and the voltage as a sine expression. On looking into the line at any distance x from the termination, the input impedance Z_i will be the ratio of a cosine to a sine expression, that is, a cotangent expression.

In the case of a short-circuited line of any length l, the current will be represented by a sine expression and the voltage by a cosine expression. The input impedance now at any point distance x from the termination will be a tangent expression.

It may be shown that for the general case of a lossless line of characteristic impedance Z_0 terminated in an impedance Z_R and of any length, the input impedance is given by

$$Z_i = Z_0 \cdot \frac{Z_R + jZ_0 \tan \dfrac{2\pi l}{\lambda}}{Z_0 + jZ_R \tan \dfrac{2\pi l}{\lambda}} \qquad (9.1)$$

From this, the following observations may be made, bearing in mind that Z_0 is resistive:

(1) If Z_R is a pure reactance, so is Z_i.

(2) If Z_R is a pure resistance equal to R_R, Z_i is resistive at all multiples of $\lambda/4$ from the termination; the corresponding values of Z_i are R_R at multiples of $\lambda/2$ from the termination and Z_0^2/R_R at odd multiples of $\lambda/4$. For all other lengths of line, the input impedance lies between these limits and is either purely inductive or purely capacitive.

(3) If $Z_R = Z_0$, $Z_i = Z_0$, as we have already seen.

(4) If $Z_R = 0$ which is the case for the short-circuited length of line, then

$$Z_i = jZ_0 \tan \frac{2\pi l}{\lambda}$$

and for $l = \lambda/4$, $3\lambda/4$, etc., Z_i is infinite.

(5) If $Z_R = \infty$ which is the case for the open-circuited length of line, by dividing numerator and denominator of (9.1) by Z_R we get

$$Z_i = Z_0 \cdot \frac{1 + j \dfrac{Z_0}{Z_R} \tan \dfrac{2\pi l}{\lambda}}{\dfrac{Z_0}{Z_R} + j \tan \dfrac{2\pi l}{\lambda}}$$

and for $Z_R = \infty$

$$Z_i = -jZ_0 \cot \frac{2\pi l}{\lambda}$$

For $L = \lambda/4$, $3\lambda/4$, etc., Z_i here is zero.

In *Figure 9.3* curves are drawn showing the effective ratio of voltage to current for both open- and short-circuited lines up to λ in length; the curves repeat indefinitely for greater lengths. These curves show the variation of Z_i as the distance x measured from the termination is varied and are therefore graphs of impedance against length. They are plotted from the equations noted under observations (4) and (5) above:

$$Z_i = jZ_0 \tan \frac{2\pi l}{\lambda} \quad \text{for the shorted line}$$

$$Z_i = -jZ_0 \tan \frac{2\pi l}{\lambda} \quad \text{for the open line}$$

and assume lossless lines and perfect open- and short-circuited terminations, conditions which are closely realised on very short lengths of line. Since the tangent term changes sign every $\pi/2$ rad (or every quarter-wavelength of line length), the sign of the *j* term is reversed accordingly; that is, the nature of the reactance takes either an inductive or capacitive form in the manner indicated in *Figure 9.3*.

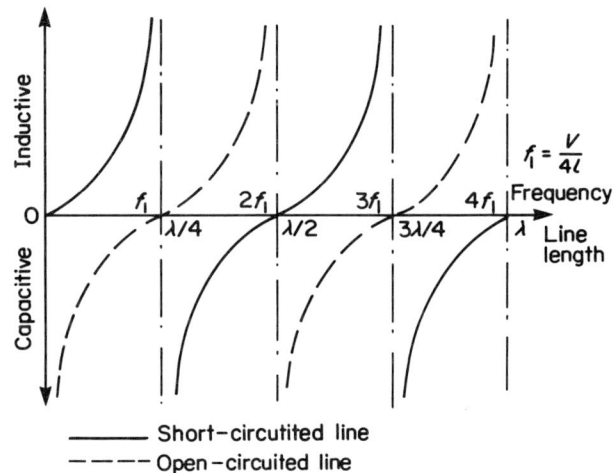

Figure 9.3

If instead of varying the *physical* length of the line, this is kept constant and the generator frequency is changed, the *electrical* length of the line will be altered just as effectively as when the frequency is kept constant and the physical length is adjusted. The curves of *Figure 9.3* may therefore be considered as having a *frequency* instead of a length plotted along the horizontal axis. Such an axis is indicated in the diagram. Hence, any reactance may be obtained from a length of line either by altering its length or by adjusting its input frequency.

9.2 GENERAL NATURE OF RESONANCE IN LINES

It follows from what has just been said that as the frequency of a signal applied to a short length of line is varied, the line exhibits the properties of series and parallel resonance at specific frequencies. For a uniform lossless line with either an open- or a short-circuited termination, the input impedance is either a pure reactance or else it is zero or infinite — the variations already illustrated in *Figure 9.3*. If *l* is the length of the line, series or parallel resonance occurs alternately, at all frequencies which are multiples of $v/4l$, where *v* is the velocity of propagation *on the line*.

There is little to be gained from speculation about the obvious physical differences between short lengths of transmission line and tuned circuits consisting of lumped coils and capacitors. A tadpole may not look much like a frog, but one will develop from the other and they are directly related, although not by much of a physical likeness. So with resonant lengths of line, the similarity of effect with tuned circuits lies in the identity of things *which happen at some particular frequency* when a piece of line is used in place of the more familiar coil and capacitor, not in appearance.

We have seen in the earlier chapters on resonant circuits that at frequencies above and below resonance, both series and parallel combinations of L and C exhibit either inductive or capacitive reactance. While at resonance the series circuit has the characteristic of low impedance and large current flow, and the parallel circuit has the characteristic of high impedance and small current flow. We bring forward from Chapter 3 the diagrams showing the frequency characteristics for both forms of resonance, these being shown in *Figures 9.4(a)* and *9.4(b)*.

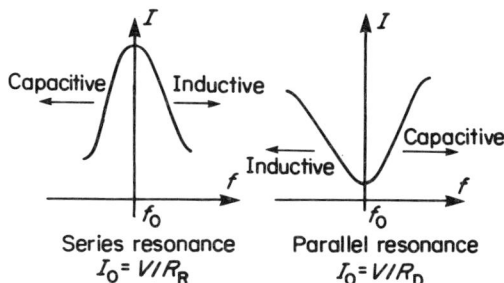

Figure 9.4

Below resonance the series circuit is capacitive, above resonance it is inductive. For the parallel circuit these conditions are reversed. The two conventional cases therefore correspond to those lengths of line which, respectively, present an impedance at the input terminals following the pattern

capacitive → low resistive → inductive

and   inductive → high resistive → capacitive

according as either the physical length or the input frequency is varied.

We have already seen that the effect of changing an open-circuit termination into a short-circuit termination on a length of line is merely to shift the nodes and antinodes from their original position through a distance $\lambda/4$ towards the end of the line. For a fixed length, therefore, a change of termination from open- to short-circuit produces at the input terminals a changeover in the relative positions of the nodes and antinodes. The two electrical lengths of line which interest us are (a) a quarter-wavelength or some exact multiple of it, and (b) a half-wavelength or some exact multiple of it. Considering these 'unit' lengths in turn we have the following four cases:

(1) *Line open-circuited and $\lambda/4$ long.* This condition is illustrated in *Figure 9.5(a)*. The standing wave pattern produces an input voltage node and a current antinode. The input impedance is therefore resistive and *low*.

(2) *Line open-circuited and $\lambda/2$ long.* This condition is illustrated in *Figure 9.5(b)*. The input is a point of current node and voltage antinode. The input impedance is therefore resistive and very *high*.

(3) *Line short-circuited and $\lambda/4$ long.* The input is a point of current node and voltage antinode — see *Figure 9.5(c)*. The input impedance is therefore resistive and very *high*.

(4) *Line short-circuited and $\lambda/2$ long.* Here the input is a point of voltage node and current antinode as shown in *Figure 9.5(d)*. The input impedance is therefore very *low* and resistive.

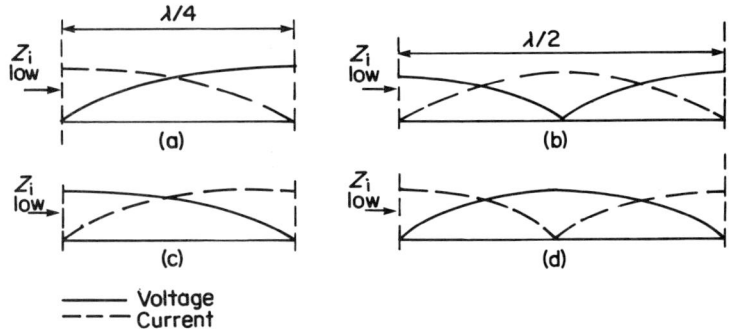

- —— Voltage
- – – – Current

Figure 9.5

An examination of these four cases and the figures illustrating them shows us that (1) and (4) can be compared with a series tuned circuit at resonance, while (2) and (3) can be compared with a parallel tuned circuit at resonance. So, for any given frequency, both series and parallel resonance can be obtained by using either $\lambda/4$ or $\lambda/2$ sections of line, or (as is more usual) by using $\lambda/4$ sections and terminating them appropriately. This use of a single length of line is possible because the addition of a $\lambda/2$ section, or any multiple of $\lambda/2$, simply causes the characteristics of the $\lambda/4$ section to repeat themselves. We note the following points.

(1) A quarter-wavelength of line has an input impedance that is the inverse of the terminating impedance; that is, an open-circuited termination gives rise to a low input impedance, and a short-circuited termination gives rise to a high input impedance. Such a length of line, therefore, may be used as a 'transformer' to match a low impedance to a high impedance circuit and vice versa.

(2) A half-wavelength of line has an input impedance which repeats that at the termination. Such a length of line may therefore be used as a 1 : 1 'transformer'.

It should perhaps again be pointed out that there is nothing to be gained from looking for a physical resemblance between such line section 'transformers' and those coupling transformers that were discussed earlier in the book.

We can expand on the question of matching transformers by considering again equation (9.1):

$$Z_i = Z_0 \cdot \frac{Z_R + jZ_0 \tan \dfrac{2\pi l}{\lambda}}{Z_0 + jZ_R \tan \dfrac{2\pi l}{\lambda}}$$

For $l = \lambda/2$ (or a multiple of $\lambda/2$), $\tan(2\pi l)/\lambda = 0$ and so

$$Z_i = \frac{Z_0 Z_R}{Z_0} = Z_R$$

irrespective of the value of Z_0. This principle is employed at very high frequencies to join two similar sections of transmission

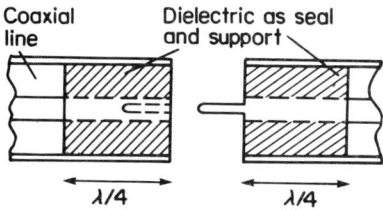

Figure 9.6

The effective wavelength of λ/2 forms a I:I transformer section between the two cables

line by a third section of different characteristic impedance. It is a principle that is often employed in the construction of plug and socket systems. These often introduce sections of line of different characteristic impedance and if their combined length is made equal to $\lambda/2$ a discontinuity at the connection point is avoided. A typical form of such a 1 : 1 transformer connector is shown in *Figure 9.6*.

Quarter-wave matching sections are impedance transformers used where the matching ratio is not 1 : 1. Dividing numerator and denominator of (9.1) by tan $(2\pi l)/\lambda$ we have

$$Z_i = Z_0 \cdot \frac{Z_R/\tan \dfrac{2\pi l}{\lambda} + jZ_0}{Z_0/\tan \dfrac{2\pi l}{\lambda} + jZ_R}$$

Making the substitution $l = \lambda/4$, we have tan $(2\pi l)/\lambda = \infty$ and

$$Z_i = Z_0 \frac{jZ_0}{jZ_R}$$

or
$$Z_0 = \sqrt{(Z_i \cdot Z_R)} \tag{9.2}$$

The quarter-wave section thus acts as an impedance transformer that may be connected between two other lines of characteristic impedances $Z_1 (= Z_i)$ and $Z_2 (= Z_R)$, provided that its own characteristic impedance satisfies the relationship

$$Z_0 = \sqrt{(Z_1 \cdot Z_2)}$$

A typical case arises in which a transmission line is to be connected to several similar lines which are in parallel at the junction point. An example is shown in *Figure 9.7*. The three 75 Ω lines present an effective impedance of 25 Ω and to match this to the outgoing 75 Ω line requires a $\lambda/4$ section of characteristic impedance $\sqrt{(25 \times 75)} = 43.3 \ \Omega$.

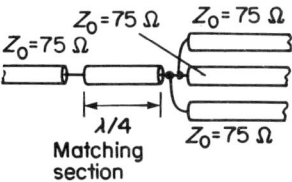

Figure 9.7

Example 2. An aerial of impedance 80 Ω is to be connected to a 600 Ω transmission line. How can this be done?

We cannot connect the 600 Ω line directly to the 80 Ω aerial as there would be a very serious mismatch and consequent loss of power. We can, however, use a quarter-wavelength of matching line between the aerial and the feeder to effect a match. This section should have an impedance

$$Z_0 = \sqrt{(80 \times 600)}$$
$$= \sqrt{48\,000} = 220 \ \Omega$$

PROBLEMS FOR CHAPTER 9

(1) After the manner of Example 1 in the text, show by means of phasor diagrams that the input impedance of a short-circuited length of line less than $\lambda/2$ but greater than $\lambda/4$ in length is purely capacitive.

(2) An aerial whose impedance is 75 Ω is fed from a 300 Ω transmission line. What impedance should a quarter-wave matching section have for insertion between the line and the aerial?

(3) Derive expressions for the sending end impedance of a

quarter-wave, low-loss length of transmission line when the receiving end is (a) short-circuited, (b) open-circuited, (c) terminated in its characteristic impedance.

(4) Obtain an expression for the sending end impedance of a line l km long having uniformly distributed primary constants when terminated in an impedance other than its characteristic impedance.

Thence prove that a quarter-wavelength of line can be used as a transformer for matching two different impedances. What will be the relationship between such impedances and the characteristic impedance of the section?

10 Complex waveforms

We have so far assumed that all alternating currents and voltages have been sinusoidal in waveform. Certain non-linear circuit elements, such as rectifiers and transistors, may modify a sinusoidal waveform or generate waveforms that are not sinusoidal, however, and such waveforms are said to be complex. Any repetitive complex wave of *fundamental*, or basic frequency f can be resolved into the sum of a number of sinusoidal waveforms having frequencies f, $2f$, $3f$, etc., the number of these component sine waves being finite or infinite. This is a statement of Fourier's theorem which you will encounter in mathematics. The component parts of the complex waves are *harmonics*, the second harmonic having twice the frequency of the fundamental, the third harmonic having three times the frequency of the fundamental, and so on.

Sounds produced by the voice or musical instruments always contain a large number of harmonics, and it is these which give musical notes their particular quality of 'timbre'. Electrical waves behave in the same way, so if a complex wave is acting in a circuit we can treat it as a series of sinusoidal alternating quantities, each acting independently of the others, for purposes of analysis.

10.1 COMPLEX WAVE EQUATION

Consider a complex wave, such as that shown in *Figure 10.1*. This wave is built up of a fundamental component having a period T seconds and frequency $1/T$ Hz plus harmonic components each of which will have its own phase angle with respect to time $t = 0$, and of varying amplitudes relative to the amplitude of the fundamental. The instantaneous value of the fundamental will be

$$\hat{V}_1 \sin (\omega t + \psi_1)$$

and of the successive harmonics

$$\hat{V}_2 \sin (2\omega t + \psi_2), \qquad \hat{V}_3 \sin (3\omega t + \psi_3)$$

and so on.

The resultant voltage will be, if it is acting in a linear circuit, the sum of the individual instantaneous effects of the fundamental and each harmonic by itself, that is

$$v = \hat{V}_1 \sin (\omega t + \psi_1) + \hat{V}_2 \sin (2\omega t + \psi_2) + \\ + \hat{V}_3 \sin ((3\omega t + \psi_3) + \dots \qquad (10.1)$$

and so on, up to the nth significant harmonic.

In the same way, the instantaneous value of a complex alternating current is expressed as

$$i = \hat{I}_1 \sin (\omega t + \phi_1) + \hat{I}_2 \sin (2\omega t + \phi_2) + \\ + \hat{I}_3 \sin ((3\omega t + \phi_3) + \dots \qquad (10.2)$$

If these voltage and current equations refer to the voltage across and the current through a particular circuit, then the phase angles

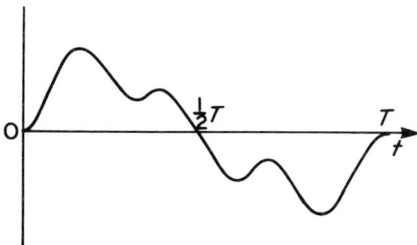
Figure 10.1

between the harmonic currents and voltages are $(\phi_1 - \psi_1)$ for the fundamental, $(\phi_2 - \psi_2)$ for the second harmonic component, and so on.

10.2 HARMONIC ANALYSIS

This subject, which can be dealt with in a precise manner by more advanced mathematics, is perhaps best covered here by way of worked examples, with the necessary discussion as we proceed.

Example 1. A complex current can be expressed as

$$i = 282 \sin 250 \pi t + 141 \sin 500 \pi t \text{ mA}$$

This current flows in a resistance of 200 Ω. Deduce an expression for the waveform of the voltage across the resistance.

Sketch on common axes these current and voltage waveforms, showing clearly the time-scales and maximum amplitudes, and by ordinate addition sketch in the complex wave of which they are the component parts. What power is dissipated in the resistance?

The current contains two separate a.c. components, the second being twice the frequency of the first. So there is a fundamental plus a second harmonic. Clearly the fundamental has a peak ampliude of 282 mA with an angular frequency $\omega_1 = 250 \pi$ (or $f_1 = 125$ Hz); and the second harmonic has a peak amplitude of 141 mA with an angular frequency of $\omega_2 = 500 \pi$ (or $f_2 = 250$ Hz).

Now in a resistance the current and voltage are in phase, so that the instantaneous voltage across the resistance is the product of the instantaneous current in amperes and the resistance in ohms:

$$V = 200[0.282 \sin 250\pi t + 0.141 \sin 500\pi t]$$

$$= 56.4 \sin 250\pi t + 28.2 \sin 500\pi t \text{ V}$$

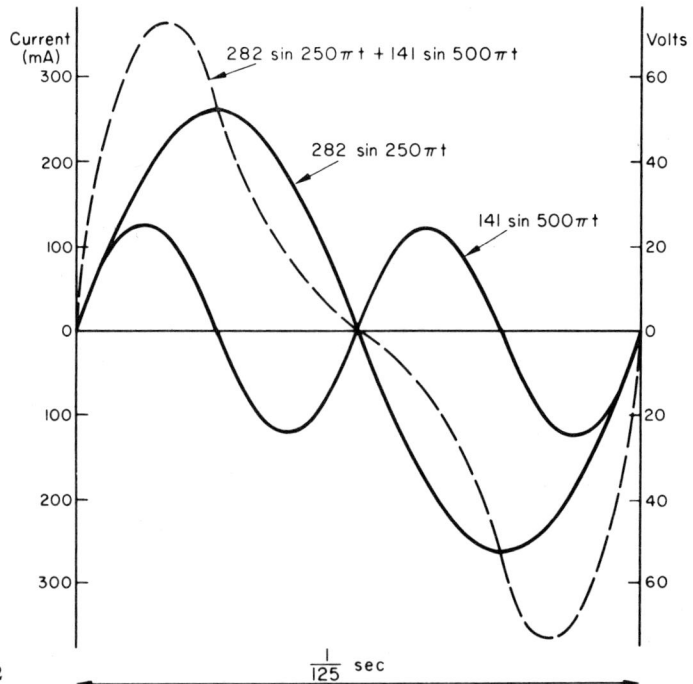

Figure 10.2

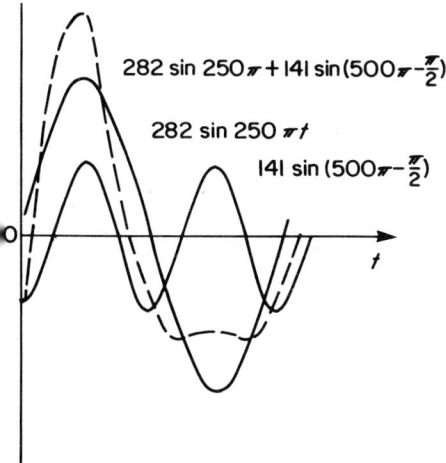

$$282 \sin 250\pi + 141 \sin(500\pi - \tfrac{\pi}{2})$$

$$282 \sin 250 \pi t$$

$$141 \sin (500\pi - \tfrac{\pi}{2})$$

Figure 10.3

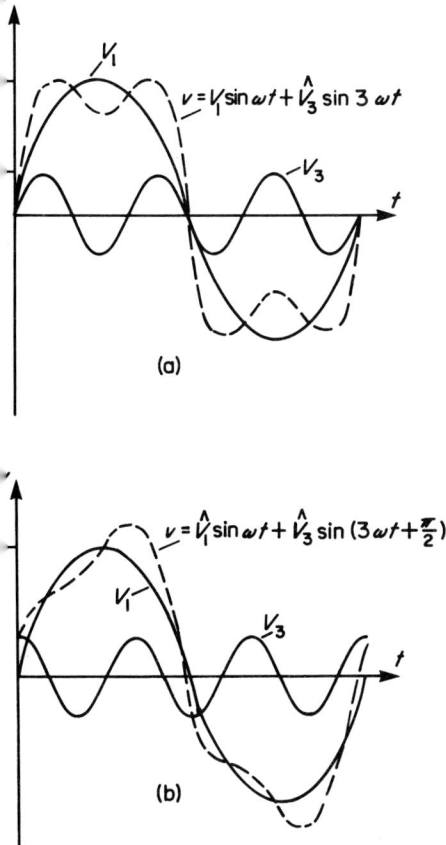

$$v = \hat{V}_1 \sin \omega t + \hat{V}_3 \sin 3 \omega t$$

(a)

$$v = \hat{V}_1 \sin \omega t + \hat{V}_3 \sin (3\omega t + \tfrac{\pi}{2})$$

(b)

Figure 10.4

The current and voltage waveforms for the two components can be sketched in the usual way to a common horizontal axis of time *t*, and vertical axes for both current and voltage values. *Figure 10.2* shows the waveforms for one complete cycle of the fundamental. The complex waveform actually present in the circuit is the sum of these two individual components; by ordinate addition the resultant wave is shown in broken line. This is the waveform of a sinewave with the addition of a second harmonic, in phase at zero time. Such 'distortion' from the true sinusoidal form is known here as *second-harmonic distortion*.

The power dissipated in the resistance is the arithmetic sum of the powers dissipated by each of the component waves, since power is not a vector quantity. So, power dissipated $= I^2 R$ for each component, where I is the r.m.s. current:

$$\therefore \qquad P_1 = (0.2)^2 \times 200 = 8 \text{ W for the fundamental}$$

$$P_2 = (0.1)^2 \times 200 = 2 \text{ W for the second harmonic}$$

Hence, the total power dissipated $= 10\text{W}$.

Since the amplitude of the harmonic in the previous example was 50% of the amplitude of the fundamental, the resultant complex wave is said to contain 50% second harmonic.

The effect of a phase change in the harmonic with respect to the fundamental can modify the resultant wave considerably, even if the percentage harmonic content is unchanged. *Figure 10.3*, for instance, shows the effect if the harmonic component of the above example is $141 \sin (500\pi t - \pi/2)$ instead of $141 \sin 500\pi t$. The resultant wave is completely changed from that shown in *Figure 10.2*.

Figure 10.4(a) shows a fundamental with a third harmonic one-third the amplitude of the fundamental. *Figure 10.4(b)* shows the effect when this same harmonic is shifted in phase with respect to the fundamental.

If we compare *Figures 10.3* and *10.4*, we notice that the complex wave produced by a fundamental and third harmonic components has identical positive and negative half-cycles, while the wave produced by the addition of a second harmonic component to the fundamental has dissimilar positive and negative half-cycles. In general, a complex wave with identical positive and negative half-cycles will contain no even-order harmonics.

Example 2. Write down the expression representing a current of 300 mA r.m.s. and frequency 10 kHz, together with a third harmonic having half the amplitude of the fundamental, the two being in phase at zero time. This current is flowing through a circuit made up of a resistance of 15 Ω in series with a 2 μF capacitor. Deduce an expression for the waveforms of (a) the voltage across the resistance, and (b) the voltage across the capacitor. What power is dissipated in each case?

For $I_1 = 300$ mA, $\hat{I}_1 = 300\sqrt{2}$ mA,

and for $f_1 = 10$ kHz, $\omega_1 = 2\pi \cdot 10^4$ rad/s.

Hence, for the fundamental

$$i_1 = 300\sqrt{2} \sin (2\pi \cdot 10^4 t) = 424 \sin (2\pi \cdot 10^4 t) \text{ mA}$$

The third harmonic has an amplitude half that of the

fundamental, that is, 212 mA peak and a frequency three times as great, that is, 30 kHz or $3 \times 2\pi \times 10^4$ rad/s. Hence, its expression is

$$i_3 = 212 \sin (6\pi \cdot 10^4 t) \text{ mA}$$

The complete expression for the complex wave made up of the fundamental and the third harmonic is therefore

$$i = i_1 + i_3 = 0.424 \sin (2\pi \cdot 10^4 t) + 0.212 \sin (6\pi \cdot 10^4 t) \text{ A}$$

(a) When this current flows in the 15 Ω resistance, the voltage across it is, as before, the product of the instantaneous current and the resistance, these being in phase. So

$$v_R = 15[0.424 \sin (2\pi \cdot 10^4 t) + 0.212 \sin (6\pi \cdot 10^4 t)]$$

$$= 6.36 \sin (2\pi \cdot 10^4 t) + 3.18 \sin (6\pi \cdot 10^4 t) \text{ V}$$

(b) When a current flows in a capacitor, the voltage across it will be the product of the current flowing and the reactance of the capacitor at the frequency of the current; so the voltage will be *smaller* at 30 kHz than at 10 kHz. Also, the voltage will lag the current by 90°. We consider the two components of the waveform independently of each other.

First, the reactance of the capacitor at the two frequencies concerned:

$$\text{At } \omega_1 = 2\pi \cdot 10^4, \ X_c = \frac{10^6}{2\pi \times 10^4 \times 2} = 7.96 \ \Omega$$

$$\text{At } \omega_3 = 6\pi \cdot 10^4, \ X_c = 0.33 \times 7.96 = 2.65 \ \Omega$$

At the fundamental frequency, the capacitor voltage is therefore

$$v = 7.96[0.424 \sin (2\pi \cdot 10^4 t - \pi/2)]$$

$$= 3.375 \sin (2\pi \cdot 10^4 t - \pi/2) \text{ V}$$

And at the third harmonic frequency

$$v = 2.65[0.212 \sin (6\pi \cdot 10^4 t - \pi/2)]$$

$$= 0.562 \sin (6\pi \cdot 10^4 t - \pi/2) \text{ V}$$

The power dissipated in the resistance is the sum of the I^2R terms for both component waves, hence

$$\text{Fundamental power } P_1 = (0.3)^2 \times 15 = 1.35 \text{ W}$$

$$\text{Third harmonic power } P_3 = (0.15)^2 \times 15 = 0.3375 \text{ W}$$

$$\text{Total power dissipated} = 1.65 \text{ W}$$

No power is dissipated in the capacitor.

We can now usefully generalise some of the points that have emerged from the worked examples above.

10.3 POWER IN A COMPLEX WAVE

In any circuit, the mean power dissipated is I^2R watts, where I is the r.m.s. value of the current, or V^2/R watts, where V is the r.m.s. voltage. The r.m.s. value of a complex wave is obtained from the square root of the sum of the squares of the r.m.s. values of the various components, that is,

$$I = \sqrt{(\text{Mean value of } i^2)}$$

Hence, for the current wave given by equation (10.2) earlier, since each component is sinusoidal

$$\text{Mean value of } i^2 = \left(\frac{\hat{I}_1}{\sqrt{2}}\right)^2 + \left(\frac{\hat{I}_2}{\sqrt{2}}\right)^2 + \left(\frac{\hat{I}_3}{\sqrt{2}}\right)^2 + \ldots$$

and so

$$I = \sqrt{\left(\frac{\hat{I}_1^2}{2} + \frac{\hat{I}_2^2}{2} + \frac{\hat{I}_3^2}{2}\right)} + \ldots$$

A similar expression is used for the r.m.s. value of a complex voltage wave.

The total power supplied by complex currents and voltages can therefore be calculated as the sum of the powers supplied by each harmonic component acting independently.

The true power supplied over each cycle of the fundamental is

$$P = V_1 I_1 \cos \phi_1 + V_2 I_2 \cos \phi_2 + V_3 I_3 \cos \phi_3 + \ldots$$

and the overall power factor will be

$$\frac{\text{True power supplied}}{\text{Apparent power}} = \frac{V_1 I_1 \cos \phi_1 + V_2 I_2 \cos \phi_2 + \ldots}{VI}$$

where V and I are, respectively, the total r.m.s. voltage and current. Since power is dissipated only in the resistive parts of a circuit, then if R is the effective series resistance, the total power is simply

$$P = I_1^2 R + I_2^2 R + I_3^2 R + \ldots$$
$$= (I_1^2 + I_2^2 + I_3^2 + \ldots)R = IR$$

Example 3. A complex voltage is expressed by the equation

$$v = 20 \sin \omega t + 10 \sin (3\omega t + \pi/6) + 5 \sin (5\omega t - \pi/2) \ V$$

When this voltage is applied to a certain circuit, a current flows which is expressed by

$$i = 0.5 \sin (\omega t - 0.15\pi) + 0.1 \sin (3\omega t - 0.08\pi) +$$
$$+ 0.05 \sin (5\omega t - 0.7\pi) \ A$$

Find the total power supplied and the overall power factor. Dealing with the fundamental and each harmonic separately:

(a) Fundamental power

$$P_1 = \tfrac{1}{2} \hat{V}_1 \hat{I}_1 \times \cos 0.15\pi$$
$$= \tfrac{1}{2} \times 20 \times 0.5 \times 0.9 = 4.46 \ W$$

(b) Third harmonic power

$$P_3 = \tfrac{1}{2} \times 10 \times 0.1 \times \cos (\pi/6 + 0.08\pi)$$
$$= \tfrac{1}{2} \times 10 \times 0.1 \times 0.71 = 0.36 \ W$$

(c) Fifth harmonic power

$$P_5 = \tfrac{1}{2} \times 5 \times 0.05 \times \cos (-\pi/2 + 0.7\pi)$$
$$= \tfrac{1}{2} \times 5 \times 0.05 \times 0.81 = 0.1 \ W$$

Total power $P = 4.36 + 0.36 + 0.1 = 4.83 \ W$

Also

$$I = \sqrt{\left(\frac{0.5^2}{2} + \frac{0.1^2}{2} + \frac{0.05^2}{2}\right)} = \sqrt{0.13125}$$
$$= 0.362 \ A$$

$$V = \sqrt{\left(\frac{20^2}{2} + \frac{10^2}{2} + \frac{5^2}{2}\right)} = \sqrt{262.5}$$
$$= 16.2 \text{ V}$$

$$\therefore \qquad \text{Overall power factor} = \frac{4.83}{16.2 \times 0.362}$$
$$= 0.824$$

10.4 HARMONICS IN PURE CIRCUIT ELEMENTS

An alternating complex voltage wave when applied to a single-phase circuit containing linear elements will cause a complex alternating current to flow. The question arises: is the percentage harmonic content of the current wave equal to that of the voltage wave? We consider the three cases of ideal circuit elements, pure resistance, pure capacitance and pure inductance.

(1) *Pure resistance.* Here the circuit impedance (resistance) is independent of frequency and current and voltage are in phase for each and every harmonic component. The expression for the circuit current will be

$$i = \frac{\hat{V}_1}{R} \sin \omega t + \frac{\hat{V}_2}{R} \sin 2\omega t + \frac{\hat{V}_3}{R} \sin 3\omega t + \ldots$$

and clearly this shows that the percentage harmonic content in the current wave will be the same as that in the voltage wave, this being given by

$$v = \hat{V}_1 \sin \omega t + \hat{V}_2 \sin 2\omega t + \hat{V}_3 \sin 3\omega t + \ldots \qquad (10.3)$$

Hence, the current and voltage waves will be identical in shape.

(2) *Pure capacitance.* The capacitive reactance $1/\omega C$ is dependent upon the harmonic frequency, but for every harmonic the voltage will lag the current by 90°. for the nth harmonic the ractance will be $1/n\omega C$ so that at this frequency

$$\hat{I}_n = \hat{V}_n \cdot n\omega C$$

The resulting current when the voltage wave of equation (10.3) is applied to the capacitor will be, therefore,

$$i = \hat{V}_1 \omega C \sin (\omega t + \pi/2) + \hat{V}_2 2\omega C \sin (2\omega t + \pi/2) +$$
$$+ \hat{V}_3 3\omega C \sin (3\omega t + \pi/2) + \ldots$$

Clearly in this case the percentage harmonic content of the current wave is *larger* than that of the voltage wave, the increase for each harmonic component being equal to that of the harmonic order.

(3) *Pure inductance.* The inductive reactance ωL is dependent upon the harmonic frequency, but for every harmonic the current will lag the voltage by 90°. For the nth harmonic reactance will be $n\omega L$ so that at this frequency

$$\hat{I}_n = \frac{\hat{V}_n}{n\omega L}$$

Hence, the general expression for the current will be

$$i = \frac{\hat{V}_1}{\omega L} \sin (\omega t - \pi/2) + \frac{\hat{V}_2}{2\omega L} \sin (2\omega t - \pi/2) +$$
$$+ \frac{V}{3\omega L} \sin (3\omega t - \pi/2) + \ldots$$

In this case the percentage harmonic content of the current is *smaller* than that of the voltage, the nth harmonic having only $1/n$th the corresponding harmonic content of the voltage wave.

Example 4. Write down an expression representing a voltage of 150 V peak value and frequency 1 kHz, together with its third and fifth harmonics having, respectively, one-tenth and one-twentieth the amplitude of the fundamental, all being in phase at zero time.

This voltage is applied to a 10 Ω resistor in series with a coil of inductance 1 mH. Find (a) an expression for the instantaneous value of the current, and (b) the power dissipated:

$$\text{For the fundamental } v_1 = \hat{V}_1 \sin \omega t = 150 \sin 2\pi \cdot 10^3 t \ V$$

$$\text{For the third harmonic } v_3 = \hat{V}_3 \sin 3\omega t = 15 \sin 6\pi \cdot 10^3 t \ V$$

$$\text{For the fifth harmonic } v_5 = \hat{V}_5 \sin 5\omega t = 7.5 \sin \pi \cdot 10^4 t \ V$$

The applied voltage is therefore expressed as

$$v = 150 \sin 2\pi \cdot 10^3 t + 15 \sin 6\pi \cdot 10^3 t + 7.5 \sin \pi \cdot 10^4 t \ V$$

At the fundamental frequency $\omega_1 = 6280$ rad/s

$$\therefore \qquad X_1 = \omega_1 L = 6280 \times 10^{-3} = 6.28 \ \Omega$$

The impedance at the fundamental frequency

$$Z_1 = \sqrt{(R^2 + X_1^2)} = \sqrt{(10^2 + 6.28^2)} = 11.81 \ \Omega$$

$$\text{Also} \qquad \phi_1 = \tan^{-1} \frac{6.28}{10} = 32° \text{ lagging}$$

At the third harmonic frequency, the inductive reactance will be $3 \times X_1 = 3 \times 6.28 = 18.84 \ \Omega$.

$$\therefore \qquad Z_3 = \sqrt{(10^2 + 18.84^2)} = 21.33 \ \Omega$$

$$\text{Also} \qquad \phi_3 = \tan^{-1} \frac{18.84}{10} = 62° \text{ lagging}$$

At the fifth harmonic frequency, likewise, $X = 31.4 \ \Omega$ and

$$Z_5 = \sqrt{(10^2 + 31.4^2)} = 33 \ \Omega$$

$$\text{and} \qquad \phi_5 = \tan^{-1} \frac{31.4}{10} = 72.3° \text{ lagging}$$

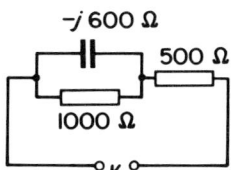

−j 600 Ω
500 Ω
1000 Ω
v

Figure 10.5

The expression for the total current is therefore

$$i = \frac{150}{11.81} \sin (\omega t - 32°) + \frac{15}{21.33} \sin (3\omega t - 62°) + \frac{7.5}{33} \sin (5\omega t - 72.3°)$$

$$= 12.7 \sin (\omega t - 32°) + 0.7 \sin (3\omega t - 62°) + 0.23 \sin (5\omega t - 72.3°) \ A$$

The power dissipated will be the product of the square of the r.m.s. current and the circuit resistance:

$$I^2 = \frac{12.7^2}{2} + \frac{0.7^2}{2} + \frac{0.23^2}{2} = 80.6 = 0.245 + 0.026$$

$$= 80.87$$

$$\therefore \qquad P = I^2 R = 80.87 \times 10 = 808.7 \ W$$

Example 5. The circuit of *Figure 10.5* is connected to a voltage given by

$$v = 20 \sin \omega t + 10 \sin (3\omega t + \pi/6) \ V$$

Find (a) the power dissipated, (b) an expression for the voltage across the 500 Ω resistor, and (c) the percentage harmonic distortion of the resulting current. The capacitive reactance shown is that at the fundamental frequency.

The circuit impedance at the fundamental frequency is

$$Z_1 = 500 + \frac{1000(-j600)}{1000 - j600} = 973 - j162 \ \Omega$$

$$= 986 \underline{/-9.45°} \ \Omega$$

At the third harmonic frequency the capacitive reactance will be $-j200 \ \Omega$, hence at this frequency the circuit impedance will be

$$Z_3 = 500 + \frac{1000(-j200)}{1000 - j200} = 538.5 - j192 \ \Omega$$

$$= 572 \ \underline{/-19.6°} \ \Omega$$

$$\therefore \qquad i = \frac{20}{986} \sin(\omega t + 9.45°) + \frac{10}{572} \sin(3\omega t + [30 + 19.6]°)$$

$$= 0.02 \sin(\omega t + 9.45°) + 0.017 \sin(3\omega t + 49.6°) \ A$$

(a) The total power is given by $V_1 I_1 \cos \phi_1 + V_3 I_3 \cos \phi_3$

$$= \frac{20}{\sqrt{2}} \times \frac{0.02}{\sqrt{2}} \cos 9.45° + \frac{10}{\sqrt{2}} \times \frac{0.017}{\sqrt{2}} \cos 19.6°$$

$$= 0.2 \times 0.986 + 0.085 \times 0.942$$

$$= 0.227 \ W$$

(b) The voltage drop across the 500 Ω resistor is $V_r = iR$

$$= 10 \sin(\omega t + 9.45°) + 8.5 \sin(3\omega t + 49.6°) \ V$$

(c) The percentage harmonic distortion of the current wave is

$$\frac{0.017}{0.02} \times 100 = 85\%$$

10.5 DIRECT CURRENT COMPONENT

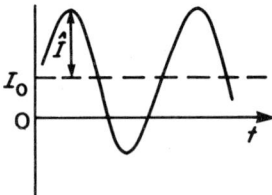

Figure 10.6

Up to this stage it has been assumed that all alternating current and voltage waveforms have operated about a zero base line, that is, the average current or voltage over a complete cycle has been zero. However, it is common practice to find that the complex alternating wave is superimposed on a direct current flow in the same circuit as illustrated in *Figure 10.6*. The effect of such a direct component is to shift the effective baseline of the wave by an amount depending upon the magnitude of the direct component. Calculations on such circuits are not affected to any great extent since the r.m.s. value of the d.c. component is the actual value of the component by definition. Hence, the r.m.s. value of a complex current wave containing a d.c. term, I_0, is found from the equation

$$I = \sqrt{\left(I_0^2 + \frac{\hat{I}_1^2}{2} + \frac{\hat{I}_2^2}{2} + \dots \right)}$$

$$= \sqrt{(I_0^2 + I_1^2 + I_2^2 + \dots)}$$

Where reactive elements are concerned, the direct voltage drop across a pure inductance is zero, and no direct current can flow through a capacitor.

Example 6. The current flowing in a certain circuit is made up of a d.c. of 3 A with a superimposed sinusoidal current of peak value 2 A. Wired into the circuit is a moving-coil ammeter and a moving-iron ammeter. What current reading will each of these meters indicate, and why?

The moving-iron meter will indicate r.m.s. values, while the moving-coil instrument will indicate average values. Hence the moving-coil instrument will read the d.c. component of current only, since the average of the sinusoidal wave itself is zero. The d.c. component will be simply 3 A.

The r.m.s. current as indicated by the moving-iron instrument will be

$$I = \sqrt{\left(I_0^2 + \frac{\hat{I}_1^2}{2} \right)} = \sqrt{(3^2 + 2)} = 3.317 \text{ A}$$

10.6 PRODUCTION OF HARMONICS

Harmonics are produced in the outputs of non-linear circuit elements. Such elements include rectifiers and any large-signal electronic amplifiers in which diodes, transistors, valves or iron-cored inductors are used. The characteristics relating current to voltage in such devices as diodes and transistors possess a marked non-linearity and any signal input transversing those parts of the characteristics that depart from linearity suffers varying degrees of distortion equivalent to the addition of harmonic components.

Two common forms of non-linearity are shown in *Figure 10.7*. The typical parabolic or *square-law* characteristic of diodes is shown in *Figure 10.7(a)*. For a sinusoidal input signal of sufficient amplitude, the output exhibits distortion in the form of either flattened or sharpened peaks, the wave being superimposed on a d.c. level indicated by current I_0. This resultant output can be shown to follow from the addition of even-order higher frequency components in which the second harmonic is dominant.

The typical cubic characteristic of transistors is shown in *Figure 10.7(b)*. For a sinusoidal input here, the output exhibits a flattening of both peaks, although not necessarily in a symmetrical form. This resultant output is obtained by the addition of odd-order higher frequency components in which the third harmonic is dominant.

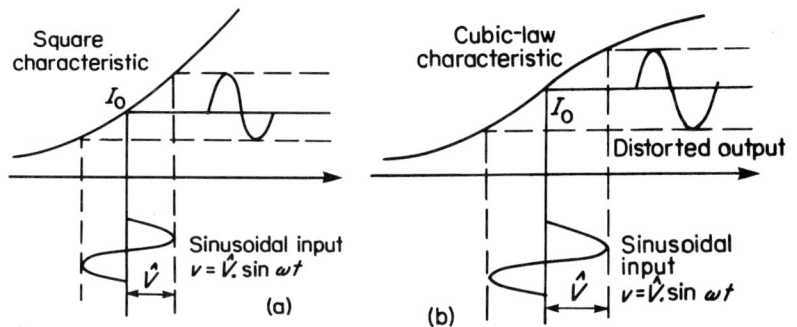

Figure 10.7

Iron-cored inductors are a source of distortion because of the non-linear characteristic of the *B–H* curve and hysteresis loop. The effect is seriously aggravated if saturation of the core is allowed to occur. *Figure 10.8* illustrates the form such distortion takes. For a sinusoidal voltage input, the flux density variation is also sinusoidal, but the resulting current (which is proportional to the magnetising force *H*) shows serious distortion as the hysteresis loop is traversed by the flux variation. It is seen that the current waveform has identical half-cycles so that odd-order harmonic components are dominant.

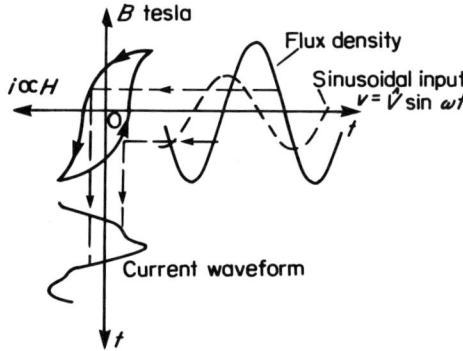

Figure 10.8

10.6.1 Second harmonic distortion

Consider a square-law characteristic having the current–voltage relationship

$$I = A + Bv + Cv^2$$

where v is the instantaneous input voltage amplitude. Clearly, the coefficient A represents the d.c. component of current, I_0, but the values of B and C have to be determined from an analysis of the complex output waveform.

Suppose the input signal to be $v = \hat{V} . \sin \omega t$, then

$$I = A + B\hat{V} . \sin \omega t + C\hat{V}^2 . \sin^2 \omega t$$

Writing $\sin^2 \omega t = \frac{1}{2}(1 - \cos 2\omega t)$, the amplitude of the term in $2\omega t$ (the second harmonic term) is $\frac{1}{2}C\hat{V}^2$.

The amount of distortion is expressed as

$$\% \text{ Second harmonic} = \frac{\text{Amplitude of second harmonic}}{\text{Amplitude of fundamental}} \times 100\%$$

which in this case is

$$\frac{\frac{1}{2}C\hat{V}^2}{B\hat{V}} \times 100\% = 50 \frac{C\hat{V}}{B} \% \qquad (10.4)$$

Clearly, for given values of B and C, the distortion will be small when the input peak amplitude $\hat{V}$ is small, as would be expected. To evaluate the coefficients it is sufficient to take suitable instantaneous values of the output waveform for given instantaneous inputs. Referring to *Figure 10.9*, for the instantaneous inputs at times $t = 0$ or π/ω, $t = \pi/2\omega$ and $t = 3\pi/2\omega$, the values are, respectively, 0, $+\hat{V}$ and $-\hat{V}$. Then $A = I_0$, and

$$\hat{I} = I_0 + B\hat{V} + C\hat{V}^2$$

$$\check{I} = I_0 - B\hat{V} + C\hat{V}^2$$

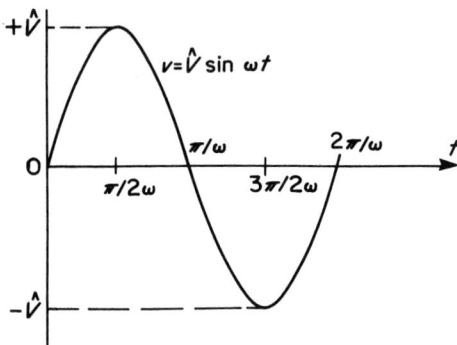

Figure 10.9

where $\hat{I}$ and $\check{I}$ indicate the maximum positive and negative peak current values, respectively. Hence

$$\hat{I}-\check{I}=2B\hat{V} \quad \text{and} \quad B\hat{V}=\tfrac{1}{2}(\hat{I}-\check{I})$$

$$\hat{I}+\check{I}=2I_0+2C\hat{V}^2 \quad \text{and} \quad C\hat{V}^2=\tfrac{1}{2}(\hat{I}+\check{I}-2I_0)$$

From equation (10.4), therefore, by substitution we have

$$\% \text{ Second harmonic}=\frac{\hat{I}+\check{I}-2I_0}{2(\hat{I}-\check{I})}\times 100\% \qquad (10.5)$$

This result can be illustrated as shown in *Figure 10.10*. Let the curve PQR represent the square-law characteristic of the circuit device and let the chord PR represent the desired linear characteristic. The sag CQ is then a measure of the departure of the characteristic from true linearity.

For a sinusoidal input signal the output waveform is distorted as shown. The output wave has an upward swing of i_+ where $i_+=\hat{I}-I_0$; and a downward swing of i_-, where $i_-=I_0-\check{I}$. Then, by addition and subtraction,

$$i_+ + i_- = \hat{I}-\check{I}$$

$$i_+ - i_- = \hat{I}+\check{I}-2I_0$$

Hence, equation (10.5) can be expressed in the form

$$\% \text{ Second harmonic}=\frac{i_+ - i_-}{2(i_+ + i_-)}\times 100\% \qquad (10.6)$$

In *Figure 10.10* the vertical projection of the chord PR is $i_+ + i_-$); at the centre C the curve droops by an amount CQ, that is

$$\text{(Height of centre)}-\text{(Height of } I_0)=\tfrac{1}{2}(i_+ + i_-)-i_-$$

$$=\tfrac{1}{2}(i_+ - i_-)$$

Substituting this into equation (10.6) expresses the second harmonic distortion in the form

$$\frac{\text{Vertical droop}}{\text{Vertical height}}\times 100\%$$

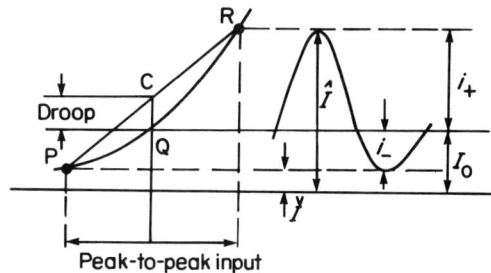

Figure 10.10

Example 7. Assuming that the instantaneous collector current i_c of a transistor amplifier is related to the source voltage $v=\hat{V}.\sin\omega t$ by the expression $i_c=I_0+Av+Bv^2$, where I_0 is the steady d.c. collector current and A and B are undetermined coefficients, show that the collector output current will consist of three frequency components. Distinguish between these components and illustrate their forms on a suitable diagram.

The effective collector load of the amplifier is 2000 Ω resistive. If the steady d.c. current is 6.3 mA and the collector current swings between the limits 11.8 to 1.2 mA over each input cycle, calculate (a) the percentage harmonic distortion, and (b) the fundamental power output in the load.

$$i_c = I_0 + Av + Bv^2$$

When $v = \hat{V} . \sin \omega t$

$$i_c = I_0 + A . \hat{V} . \sin \omega t + B . \hat{V}^2 \sin^2 \omega t$$

$$= I_0 + A . \hat{V} . \sin \omega t + B . \hat{V}^2 . \tfrac{1}{2}(1 - \cos 2\omega t)$$

$$= (I_0 + \tfrac{1}{2}B.\hat{V}^2) + A.\hat{V}.\sin \omega t - \tfrac{1}{2}B.\hat{V}^2 \cos 2\omega t$$

Hence, the output consists of three components: (i) a zero frequency (d.c.) component $I_0 + \tfrac{1}{2}B\hat{V}^2$; (ii) an a.c. component of fundamental frequency $\sin \omega t$ and of peak amplitude $A\hat{V}$; (iii) an a.c. component of twice the fundamental frequency (second harmonic) of peak amplitude $\tfrac{1}{2}B\hat{V}^2$.

The components are sketched in *Figure 10.11*. Notice that the d.c. component level shifts from its value of I_0 when the input signal is zero to $(I_0 + \tfrac{1}{2}B\hat{V}^2)$ when the signal is applied.

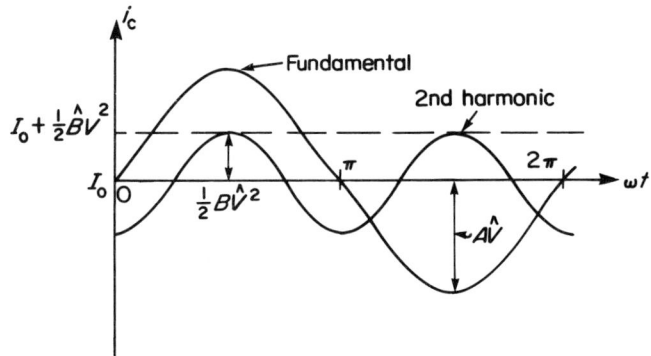

Figure 10.11

(a) From the given information $I_0 = 6.3$ mA, $\hat{I} = 11.8$ mA, $\check{I} = 1.2$ mA.

$$\% \text{ Second harmonic distortion} = \frac{\hat{I} + \check{I} - 2I_0}{2(\hat{I} - \check{I})} \times 100\%$$

$$= \frac{11.8 + 1.2 - 12.6}{2(11.8 - 1.2)} \times 100\% = 1.88\%$$

(b) Fundamental power = Mean squared fundamental current × load. Now the peak fundamental current = $AV = \tfrac{1}{2}(\hat{I} - \check{I})$

$\therefore$ Mean squared fundamental current = $\tfrac{1}{2} \left[\dfrac{\hat{I} - \check{I}}{2} \right]^2$

$\therefore$ Power = $\tfrac{1}{2} \left[\dfrac{\hat{I} - \check{I}}{2} \right]^2 . R = \dfrac{(\hat{I} - \check{I})^2}{8} . R$

Substituting values, we have

$$\text{Fundamental power} = \frac{2000}{8} [11.8 \times 10^{-3} - 1.2 \times 10^{-3}] \text{ W}$$

$$= 28.1 \times 10^{-3} \text{ W} = 28.1 \text{ mW}$$

PROBLEMS FOR
CHAPTER 10

(1) A current $i = 50 \sin 628t$ A flows in a circuit when a voltage $V = 250 \sin (628t + 2\pi/3)$ is applied across it. What is the frequency, peak and r.m.s. value of the voltage wave? What is the significance of the term $2\pi/3$ in the expression for the voltage?

Write down an expression for an alternating current that leads this voltage by $\pi/3$ rad.

(2) A complex current can be expressed as

$$i = 300 \sin 500t + 150 \sin 1000t \text{ mA}$$

This current flows in a resistance of 100 Ω. Find an expression for the waveform of the voltage across the resistance. What power is dissipated in the resistance?

(3) Write down an expression representing a current of 400 mA r.m.s. and frequency 10 kHz, together with its third harmonic having half the amplitude of the fundamental, the two being in phase at zero time.

This current is flowing in a circuit consisting of a resistance of 10 Ω in series with an inductor of 160 μH inductance having negligible resistance. Write down expressions for the waveforms of (a) the voltage across the resistor, (b) the voltage across the inductor.

(4) Write down the frequency, r.m.s. and peak values of a voltage wave expressed as $v = 14.1 \sin 1000 \pi t$. Write down also expressions for the current flowing when this voltage is applied across (a) a 5 Ω resistor, (b) a 1 mH inductor of negligible resistance, (c) a 150 μF capacitor.

(5) A voltage wave of 5000 rad/s fundamental angular frequency containing a third harmonic component, is applied to a circuit consisting of a 100 Ω resistor in series with a 2 μF capacitor. The amplitude of the fundamental is 100 V and that of the third harmonic 30 V. Calculate the fundamental and third harmonic voltage across the resistor.

(c) A current expressed as $282 \sin 200 \mu t + 141 \cos 400 \mu t$ mA is flowing in a 100 Ω resistor. Write down an expression for the voltage across the transistor.

Sketch the current and voltage waveforms on common axes over one cycle of the fundamental, indicating clearly the time-scale and peak amplitude values. What power will be dissipated in the resistor?

(7) A complex voltage wave is expressed as

$$v = 100 \sin \omega t + 20 \sin (3\omega t + 45°) + 5 \sin (5\omega t - 30°) \text{ V}$$

When this voltage is applied to a certain circuit, a current flows which is expressed by

$$i = 0.4 \sin (\omega t + 85°) + 0.8 \sin (3\omega t + 45°) + 0.05 \sin (5\omega t + 106°) \text{ A}$$

Find the total power supplied and the overall power factor.

(8) Explain why the current wave through a capacitor has a

larger harmonic percentage content than the voltage wave across it.

A complex voltage given by

$$v = 283 \sin 314t + 57 \sin 942t \text{ V}$$

is applied to a series C–R circuit. The resulting current has a peak value of 4·24 A with 30% harmonic content. Show that C is about 60 μF and R about 42 Ω.

(9) A power amplifier operating with a load resistor of 10 Ω has a steady collector current of 1.7 A. During one-half of the input cycle of signal the current rises to 3.1 A and during the second half-cycle it falls to 0.1 A. Calculate the fundamental power output of the amplifier and the percentage second-harmonic distortion.

11 Electrostatic fields and dielectrics

Electrostatic fields, magnetic fields and thermal fields are analogous in their characteristics and all of them may be examined by identical mathematical processes. We are mainly interested in electrostatic and magnetic fields in this course and this chapter will investigate the general concepts associated with the electrostatic field.

11.1 THE ELECTRIC FIELD

If an electron, representing a point negative charge, is placed in the region between two parallel metal plates charged to different potentials, there will be a force acting on the electron tending to drive it away from the negative plate towards the positive plate. A point positive charge would be acted on similarly by a force tending to move it in the opposite direction. Any region in which an electron or electric charge experiences a mechanical force is known as an electrostatic field, and the direction of the field is defined as that of the direction of the force acting on a positive charge placed within the field. Between the plates mentioned above, for example, the direction of the force and hence of the field's vector is from the positive towards the negative plate.

Such a field of force may be represented in magnitude and direction by vector lines of *electric force* drawn between the charged surfaces, where the closeness of the lines is an indication of the field strength — see *Figure 11.1*. An electric field will result at any time when a potential difference is established between two points or between two conductors and is a property of all electrical systems. *Figure 11.2* shows three typical field patterns.

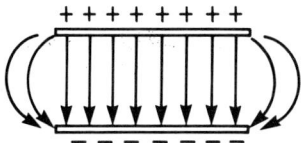

Figure 11.1

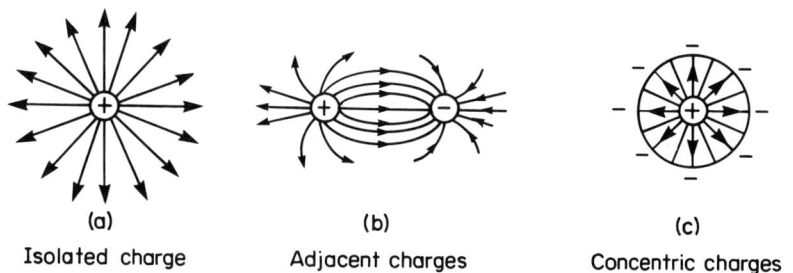

(a)	(b)	(c)
Isolated charge	Adjacent charges	Concentric charges

Figure 11.2

Electric lines of force, or *electric flux lines,* do not form closed loops but emanate from positive charges and terminate on equal negative charges. When a charged body is placed in proximity to an uncharged body, therefore, an *induced* charge of opposite sign appears on the surface of the uncharged body because lines of force from the charged body terminate on its surface.

Unit flux is defined as emanating from a positive charge of 1 coulomb, so that electric flux, ψ, is measured in coulombs and for a charge of Q coulombs the flux $\psi = Q$ coulombs. Electric flux density is measured in terms of the flux passing through unit area situated at right-angles to the direction of the field, hence flux density, or displacement, D, is measured in coulombs per square metre:

$$\text{Flux density } D = \frac{\psi}{A} \text{ C/m}^2$$

Clearly, the greater the charge Q, the greater will be the flux density at a particular location in the field.

Returning to the two parallel plates of 111, the mechanical force acting on a point charge of 1 coulomb placed between the plates is the electric force or the electric field strength E. It is measured in the direction of the field and its magnitude depends upon the potential difference between the plates and their separation. In moving from one plate to the other along a line of force we move from, say, $-V$ to $+V$ volts. The potential between the plates is therefore $2V$ volts and this potential changes linearly as we move from one plate to the other; we move, in other words, along a *potential gradient* which changes by equal amounts for each unit of distance covered. Lines may now be drawn connecting together all points within the field having equal potentials, and in *Figure 11.3* such *equipotential* lines have been traced for potentials of $-V$, $-\frac{1}{2}V$, 0, $+\frac{1}{2}V$ and $+V$. Notice that the zero equipotential line represents earth potential, and that the charges on the plates are, respectively, above and below earth potential.

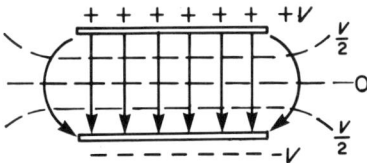

Figure 11.3

Equipotential lines are themselves part of an equipotential surface. Such surfaces, in the parallel plate examples, are parallel to the plates which are themselves equipotential surfaces. Since all points on an equipotential surface have, by definition, the same potential, there clearly can be no current flow between any given points on such a surface. Hence, a line of force must intersect an equipotential surface at right-angles, for if it did not there would be a component of current flow acting along the surface which would in turn imply that the surface was not equipotential.

By choosing two points in the field where the potential difference between equipotential surfaces if V volts and their separation is d metres, then the voltage gradient is V/d volts per metre. Let the electric force at any point P in the field be E newtons per coulomb, so that the force on a charge of 1 coulomb is E newtons and the work done by a positive charge of 1 coulomb at P moving in the direction of the field through a small distance δx metres is $E . \delta x$ joules. But the potential difference between two points is the work done, in joules, in transferring 1 coulomb from one point to the other. Hence, there is a drop in potential, δV, in the direction of the field and $E . \delta x = \delta V$. So

$$E = \frac{\delta V}{\delta x} \qquad \text{V/m or N/C}$$

or   Electric force = Voltage gradient

In general, then, for field strength E, voltage V and distance d

$$V = Ed$$

11.1.1 Dielectrics   At any point in an electric field, the electric force E maintains the electric flux and produces a particular value of flux density, D, at that point. E is the cause and D the effect. It seems reasonable to assume a linear relationship between E and D and it can indeed be verified experimentally that for a field established in vacuum (or for practical purposes, air) D, is everywhere directly proportional to E. Hence the ratio D/E is a constant, symbolised ε_0, and known as the *permittivity* of free space or the *free space constant*. It has the value $1/36\pi \times 10^{-9}$ or approximately 8.85×10^{-12} m.k.s. units (farads per metre).

When an insulating medium, such as polythene, mica or paper, is introduced into the region of the electric field, the ratio of D/E is modified from its free space value. Such an insulating medium is known as a *dielectric*.

When an electric field is established across a dielectric there is a distortion of the molecular structure of the dielectric material. The molecules of some dielectrics, like water, have permanent electric dipole moments and these dipoles tend to align themselves with the external field, as shown in *Figure 11.4(a)*. Because the molecules are in constant thermal agitation, the degree of alignment is restricted but will increase as the applied field strength is increased or as the temperature is decreased.

Materials which do not have molecules with permanent dipole moments do acquire them, however, by induction when placed in an electric field. This induced dipole moment is present only when the field is present, is proportional to the field strength and is created already lined up with the field — see *Figure 11.4(b)*. The overall effect of such alignment is to separate the centre of positive charge of the dielectric slightly from the centre of negative charge. Hence, the dielectric as a whole becomes *polarized*, although remaining electrically neutral. The result is a pile-up of positive charges on one face of the dielectric and of negative charges on the opposite face; within the dielectric itself no excess charge develops in any given volume element. Since the dielectric as a whole remains neutral, the positive induced charge at one surface face must be equal in magnitude to the negative induced charge at the other surface face. It must be kept in mind that the *displacement* of electrons in the dielectric is extremely small. There is no transfer of charge through the dielectric such as there is in metallic conductors.

Since the induced surface charges are always in such a direction that they *oppose* the applied field, the introduction of a dielectric into an electric field will tend to weaken the original field within the dielectric volume. This weakening of the field reveals itself as a reduction in potential gradient across the field boundaries (say the two parallel plates of *Figure 11.1*). The relationship $V = Ed$ established earlier holds whether a dielectric is present or not and demonstrates that the reduction in V is directly connected to the reduction in E. Hence, we may write

$$\frac{E_0}{E_d} = \frac{V_0}{V_d} = \text{a constant } \varepsilon$$

where E_0, V_0 are the field strength and potential difference established in a vacuum, and E_d, V_d are the field strength and potential difference established in the dielectric. We use the word

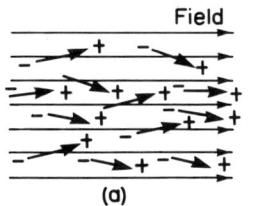

Field

(a)

Partial alignment of molecules with permanent electric dipole moments

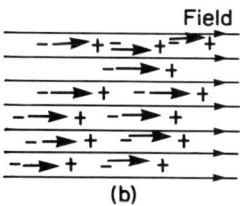

Field

(b)

Complete alignment of molecules by the effect of induction

Figure 11.4

'dielectric' here to mean a material substance between the plates, not the vacuum dielectric.

The constant $\varepsilon = \varepsilon_0 \varepsilon_r$, where ε_r is known as the *relative permittivity* of the dielectric material, and ε is the *absolute permittivity*.

The field system we are discussing is, of course, that on which the principle of the capacitor is derived. The introduction of a dielectric between two parallel metal plates causes a fall in potential and more charge may be added to the system to restore the potential to its former value. Hence, the effect of the dielectric is to increase the capacity of the plates by a factor equal to ε_r above that obtained with a vacuum dielectric, the relative permittivity of the vacuum being unity.

11.1.2 Dielectric strength

The molecular alignment and distortion of the electron orbits around the atoms of a dielectric when a field is established across the faces of the material produces a mechanical stress which in turn generates heat. Because the production of heat represents a dissipation of power, practical dielectrics always introduce a power loss, particularly in the case of capacitors employed in high frequency systems where there is a continual and rapid change in field polarity.

If the potential difference between the plates of a capacitor (or between the opposite surfaces of a sheet of dielectric material for that matter) is increased beyond a certain level, a point is reached where the insulation is unable to withstand the electric stress set up across it and there is a spark discharge which penetrates the material and destroys its insulating properties in the region of the breakdown. The potential gradient necessary to cause such a breakdown is a measure of the ability of a material to resist breakdown, and is usually expressed in kilovolts per millimetre. *Table 11.1* gives the permittivity and dielectric strengths of some commonly used materials.

Table 11.1

Material	Permittivity ε_r	Dielectric strength (kV/mm)
Vacuum	1.000 0	
Air	1.000 6	0.8
Paper	3.5–4.0	10–15
Mica	5.0–6.0	150
Porcelain	6.0–7.0	5
Quartz	3.5–4.0	8
Bakelite	4.0–5.0	12
Polystyrene	2.5	25
Titanium dioxide	100 +	6

Such figures are useful for relative comparisons of various dielectrics and are naturally a very important factor in the design of capacitors as well as other electrical equipment such as transformers, high voltage insulators, motors and generators. Dielectric strength depends upon many factors such as moisture content of the material, the operating temperature and the shape and size of the conductors associated with it. It is not directly proportional to thickness, so simply doubling the thickness of the dielectric does not double the working voltage. All practical capacitors have a safe working voltage stated on them, generally at a particular maximum temperature.

11.1.3 Energy stored in an electric field

Consider for a moment the parallel plate system of *Figure 11.1.* Treating this as a parallel plate capacitor, then from elementary theory, at any instant *t* the charging current

$$i = C \frac{dv}{dt}$$

and multiplying each side of this by $v.dt$ we obtain the energy equation

$$vi \cdot dt = Cv \cdot dt$$

For a final potential across the plates of *V* volts, the total energy input is

$$\int_0^V Cv \cdot dv = C[\tfrac{1}{2}v^2]_0^V$$
$$= \tfrac{1}{2}CV^2 \text{ J}$$

This energy is stored in the electric field, *not* on the plates. Now

$$\text{Electric flux density } D = \frac{Q}{A} \text{ C/m}$$

$$\text{Electric field strength } E = \frac{V}{d} \text{ N/C}$$

$$\text{Capacitance } C = \frac{Q}{V} \text{ F}$$

Then by substitution

$$\text{Energy stored} = \tfrac{1}{2}CV^2 = \tfrac{1}{2}\frac{DA}{V} \cdot V^2 = \tfrac{1}{2}DAV$$

But $V = Ed$, hence energy stored $= \tfrac{1}{2}DAEd$. But Ad = volume of the field.

$$\therefore \qquad \text{Energy stored} = \tfrac{1}{2}DE \text{ J/m}^3 \qquad (11.1)$$

This result applies to any electric field. Since $D = \varepsilon E$, this result may be expressed as $\tfrac{1}{2}\varepsilon E^2$.

Example 1. The parallel plates of *Figure 11.1* are 2 cm apart and each has an area of 0.025 m². If the voltage impressed across the plates is 20 kV and the dielectric is air, what energy is stored in the electric field?

Here $\qquad d = 2 \times 10^{-2}$ m, $A = 0.025$ m², $V = 20 \times 10^3$ V

$$\varepsilon = \varepsilon_0 \varepsilon_r = 8.85 \times 10^{-12} \text{ (taking } \varepsilon_r = 1.0)$$

Then $\qquad\qquad E = \dfrac{20 \times 10^3}{2 \times 10^{-2}} = 10^6 \text{ N/C}$

Energy stored in a field $= \tfrac{1}{2}\varepsilon E^2$ J/m

$$= \tfrac{1}{2} \times 8.85 \times 10^{-12} \times 10^{12}$$
$$= 4.43 \text{ J/m}^3$$

But the volume of field here is $0.025 \times 2 \times 10^{-2}$ m

$$\text{Hence, the energy stored} = 4.43 \times 0.025 \times 2 \times 10^{-2} \text{ J}$$
$$= 0.0022 \text{ J}$$

11.2 TRANSMISSION LINE PARAMETERS

Certain parameters relating to transmission lines can be derived from considerations of electric fields. The most commonly employed lines are the coaxial and the parallel-wire lines, already discussed in Chapters 7 and 8. The field inside a coaxial line takes the form shown in *Figure 11.2(c)*, while that between parallel wires is illustrated in *Figure 11.2(b)*. For the contents of this section we can clearly assume that the conductivity of the conductors in such cables is so high that any voltage drop within the conductors is negligible compared with the voltage difference across the intervening dielectric medium. As all points within the conductors are therefore at the same potential, the conductors form the bounding equipotential surfaces for the electric field.

11.2.1 The coaxial cable

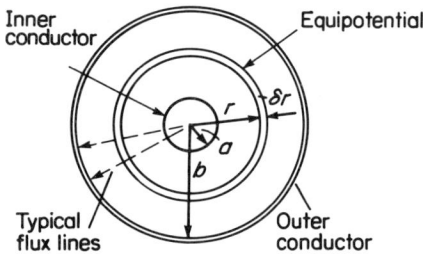

Figure 11.5

The cross-section of a coaxial line is shown in *Figure 11.5*. The inner and outer surfaces of the conductors form the bounding equipotentials, and all other equipotentials are concentric cylinders running the length of the cable. Let the radii of the inner and outer surfaces be a and b, respectively, and consider a unit length (1 m) of the cable run.

Let Q be the charge on unit length of the inner conductor, then the total flux χ, across the dielectric is Q C/m. Let a *thin* cylinder of radius r and thickness δ be situated within the dielectric. Then the total flux will pass through this cylinder and the flux density at radius r will be

$$D = \frac{Q}{A} = \frac{Q}{2\pi r} \text{ c/m}^2$$

since the cylinder length is 1 m. The electric field strength or stress at this radius is then

$$E = \frac{D}{\varepsilon} = \frac{Q}{2\pi\varepsilon r} \text{ N/C or V/m}$$

The potential difference across the cylinder walls is δV volts,

$$\therefore \qquad \delta V = \varepsilon \delta r = \frac{Q}{2\pi\varepsilon r} \delta r$$

The total potential across the dielectric is then

$$V = \frac{Q}{2\pi\varepsilon} \int_a^b \frac{\delta r}{r}$$
$$= \frac{Q}{2\pi\varepsilon} \ln \frac{b}{a} \text{ V}$$

From this, we can derive expressions for the capacitance per unit length and the dielectric stress in the cable.

$$\text{Capacitance per unit length} = \frac{Q \text{ per metre}}{V}$$

Hence

$$C = \frac{Q}{V} = \frac{2\pi\varepsilon}{\ln \frac{b}{a}} \text{ F/m} \qquad (11.2)$$

Rearranging this expression as

$$\frac{Q}{2\pi\varepsilon} = \frac{V}{\ln\dfrac{b}{a}}$$

and substituting this into the equation $E = Q/(2\pi\varepsilon r)$, obtained above, we have

$$\text{Dielectric stress} = E = \frac{V}{r\ln\dfrac{b}{a}} \text{ V/m} \tag{11.3}$$

There is an optimum value for the ratio b/a which depends upon the conditions imposed on the variation of a and b, that is, characteristic impedance, maximum permissible dielectric stress and maximum power transfer all depend upon this ratio.

From equation (11.3) the stress at any point in the dielectric is inversely proportional to r and will have a maximum value when r is a minimum. This clearly occurs when $r = a$. Hence

$$E_{\max} = \frac{V}{a\ln\dfrac{b}{a}}\text{V/m}$$

If $E_{\max}$ and V are fixed for a given dielectric material, then

$$\frac{V}{E_{\max}} = a\ln\frac{b}{a} = \text{a constant, say } K$$

and

$$\ln\frac{b}{a} = \frac{K}{a}$$

$\therefore$

$$b = ae^{K/a}$$

If b is to be a minimum to provide the most economical cable, then

$$\frac{\mathrm{d}b}{\mathrm{d}a} = e^{K/a} + a\,.\,e^{K/a}\left(-\frac{K}{a^2}\right) = 0$$

from which $a = K$. Hence, $b = k\,.\,e = a\,.\,e = 2.72a$.

This result was stated earlier in Chapter 8.

Example 2. Derive an expression for the capacitance per metre length of a parallel wire line, assumed isolated from earth, the radius of each wire being R and their separation d.

The field associated with the oppositely charged conductors A and B in a parallel wire line is shown in *Figure 11.6(a)*. Assume that conductor A carries a charge of $+Q$ coulomb per metre length, while conductor B is neutral. As in the case of the coaxial cable above, let a *thin* cylinder of radius r and wall thickness δr be situated in the field of conductor A, its length being 1 m and concentric with A — see *Figure 11.6(b)*. The flux density at radius r will be

$$D = \frac{Q}{\text{A}} = \frac{Q}{2\pi r} \text{ C/m}$$

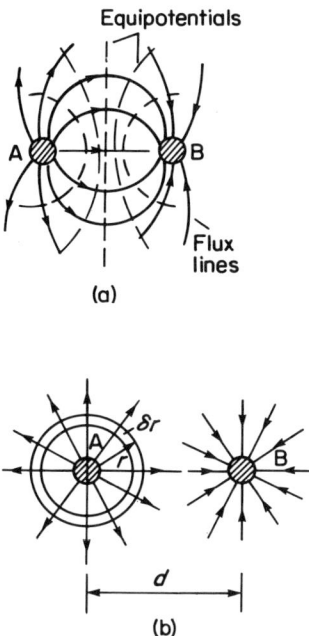

Equipotentials

A   B

Flux lines

(a)

δr

A   B

d

(b)

Figure 11.6

and, again as before, the electric field strength at this radius is

$$E = \frac{D}{\varepsilon} = \frac{Q}{2\pi\varepsilon r} \text{ N/C or V/m}$$

The potential difference across the cylinder walls is δV volts, therefore

$$\delta V = E \,.\, \delta r = \frac{Q}{2\pi\varepsilon r} \,.\, \delta r$$

We do not have an outer conductor now, as we had in the case of the coaxial cable, so we suppose that at a large distance from the conductor, R', the potential may be considered zero. Then the potential of A *above* zero will be

$$V_A = \frac{Q}{2\pi\varepsilon} \int_R^{R'} \frac{dr}{r}$$

$$= \frac{Q}{2\pi\varepsilon} \ln \frac{R'}{R} \text{ V}$$

Similarly, the potential at conductor B (since conductor B lies in the field of conductor A) is

$$V_B = \frac{Q}{2\pi\varepsilon} \ln \frac{R'}{d} \text{ V above zero}$$

Reversing the process, assume that the conductor A is neutral and conductor B carries a charge of $-Q$ coulomb per metre length. then, working as before, we obtain

$$V_B = \frac{Q}{2\pi\varepsilon} \ln \frac{R'}{R} \qquad V_A = \frac{Q}{2\pi\varepsilon} \ln \frac{R'}{d}$$

When both conductors carry equal and opposite charges, therefore, the total potential of A above zero is

$$\frac{Q}{2\pi\varepsilon} \left[\ln \frac{R'}{R} - \ln \frac{R'}{d} \right] = \frac{Q}{2\pi\varepsilon} \ln \frac{d}{R}$$

while the total potential of B below zero is

$$-\frac{Q}{2\pi\varepsilon} \ln \frac{d}{R}$$

The potential difference between A and B is therefore

$$V = 2 \frac{Q}{2\pi\varepsilon} \ln \frac{d}{R} \text{ per metre}$$

But the capacitance per metre length $C = \dfrac{\text{Charge per metre}}{\text{Potential}}$

$$\therefore \qquad C = \frac{1}{2} \,.\, \frac{2\pi\varepsilon}{\ln \dfrac{d}{R}} \text{ F/m} \qquad\qquad (11.4)$$

11.3 DIELECTRIC LOSS ANGLE

Because the power loss in dielectric materials is generally very small, it is usual to express such losses in terms of the phase angle by which the lead of the current on the voltage *falls short* of the ideal lossless case of 90°. In the case of a capacitor (although the same is true of any device having a dielectric in an alternating electric field) the dielectric loss arises from the work done in the continual orientation and reorientation of the molecular dipoles. Hence, there is a dissipation of energy within the dielectric which appears as heat and increases as the frequency increases. Such a dissipation can be represented as either a *small* resistance in series with an ideal capacitor or as a *large* resistance in parallel with an ideal capacitor. For a particular capacitor, values of a series or a parallel resistance can be found which lead to circuits identical in impedance and phase angle, and hence to equal power losses.

Figure 11.7 shows the impedance triangle for a series *C–R* circuit. We define δ as the loss angle of the capacitor; then $\delta = (90° - \phi)$ where cos ϕ is the power factor. For an ideal capacitor, cos ϕ is zero since no power will be dissipated, therefore $\phi = 90°$ and the loss angle is zero. As it is usually very small, capacitor loss angle is expressed in radians. A loss angle of 10^{-5} rad would be typical for a good quality paper capacitor at low frequencies.

A useful formula for the power factor of a capacitor may be derived as follows:

$$\cos \phi = \frac{\text{Resistance}}{\text{Impedance}} = \frac{\text{Power dissipated}}{\text{Apparent power supplied}}$$

which, referring to *Figure 11.7* is equal to

$$\frac{R}{Z} = \frac{R}{R + \dfrac{1}{j\omega C}}$$

Dividing top and bottom by *R*, gives us

$$\cos \phi = \frac{1}{1 + \dfrac{1}{j\omega CR}} = \frac{1}{\sqrt{\left(1 + \dfrac{1}{(\omega CR)^2}\right)}}$$

When the product *CR* of the equivalent series resistance and the capacitance is very small, ωCR is small also, and the expression reduces to the simple form

$$\cos \Phi = \omega CR = \sin \delta$$

If the equivalent resistance loss component of the capacitor is considered to be in parallel, the sides of the impedance triangle become $1/Z$, $1/R$ and $1/X$. The power factor cos ϕ is then Z/R and the approximate formula becomes

$$\cos \phi = \frac{1}{\omega CR} = \sin \delta$$

Example 3. An alternating voltage of 10 V r.m.s. at a frequency of 159 kHz is maintained across a capacitor of capacitance 0.01 μF. Calculate the capacitor current. If the power dissipated in the dielectric is 100 μW, calculate (a) the loss angle, (b) the equivalent series loss resistance, and (c) the equivalent parallel loss resistance.

At 159 kHz, $\omega = 10^6$ rad/s

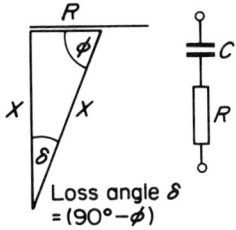

Loss angle δ = (90° − ϕ)

Figure 11.7

$$\therefore \qquad X_c = \frac{10^6}{10^6 \times 0.01} = 100 \ \Omega$$

$$\therefore \qquad \text{Capacitor current } I_c = \frac{V}{X_c} = \frac{10}{100} \ A = 100 \ mA$$

(a) Power factor $= \cos \phi = \dfrac{\text{True power}}{\text{Apparent power}}$

$$= \frac{100 \times 10^{-6}}{10 \times 100 \times 10^{-3}}$$

$$= 10^{-4}$$

But loss angle $\delta = 90° - \phi$, hence $\cos \phi = \sin \delta = 10^{-4}$. For small angles $\sin \delta = \delta$

$$\therefore \qquad \text{Loss angle} = 10^{-4} \text{ rad}$$

(b) Let the equivalent series resistance be r_s. Then since the power dissipated in it is 100 μW at a current of 100 mA

$$100 \times 10^{-6} = (0.1)^2 \times r_s$$

$$\therefore \qquad r_s = 10^{-2} = 0.01 \ \Omega$$

(c) Let the equivalent parallel resistance be r_p. Then since the power dissipated is 100 μW at a voltage of 10 V

$$100 \times 10^{-6} = \frac{10^2}{r_p}$$

$$\therefore \qquad r_p = 10^6 \ \Omega = 1 \ M\Omega$$

Example 4. A capacitor of 0.5 μF has a power factor of 2×10^{-4} and a capacitor of 0.75 μf has a power factor of 5×10^{-4}. Calculate the power factor of the combination of these capacitors joined (a) in series, and (b) in parallel.

Let the effective series resistance of C_1 be R_1 and of C_2 be R_2. Then, as shown in *Figure 11.8(a)*, the effective circuit of the capacitors in series is a capacitance of $C_1 C_2 / (C_1 + C_2)$ in series with a resistance of $(R_1 + R_2)$. Now

$$\cos \phi \simeq \omega CR$$

Hence, for the series combination

$$\cos \phi = \omega \ \frac{C_1 C_2}{C_1 + C_2} (R_1 + R_2)$$

$$= \omega \ \frac{C_1 C_2}{C_1 + C_2} \left[\frac{\cos \phi_1}{\omega C_1} + \frac{\cos \phi_2}{\omega C_2} \right]$$

$$= \frac{C_2 \cos \phi_1 + C_1 \cos \phi_2}{C_1 + C_2}$$

$$= \frac{1.5 \times 10^{-4} + 2.5 \times 10^{-4}}{1.25}$$

$$= 2.8 \times 10^{-4}$$

For the parallel combination, shown in *Figure 11.8(b)*, the total

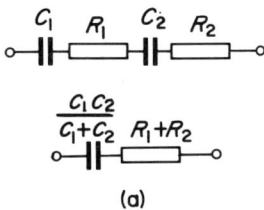

(a)

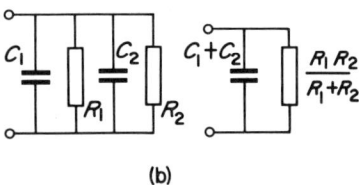

(b)

Figure 11.8

capacitance is $C_1 + C_2$ and the effective parallel resistance is $R_1R_2/(R_1 + R_2)$. Now

$$\cos \phi \simeq \frac{1}{\omega CR}$$

and so, for the parallel circuit

$$\cos \phi = \frac{1}{\omega(C_1 + C_2)} \left[\frac{1}{R_1} + \frac{1}{R_2} \right]$$

$$= \frac{C_1 \cos \phi_1 + C_2 \cos \phi_2}{C_1 + C_2}$$

$$= \frac{1 \times 10^{-4} + 3.75 \times 10^{-4}}{1.25}$$

$$= 3.8 \times 10^{-4}$$

PROBLEMS FOR CHAPTER 11

(1) The potential between two parallel metal plates spaced 1 cm apart, is 500 V. If the space between the plates is filled with polystyrene of relative permittivity 2.5, what is the flux density?

(2) A coaxial cable has an inner conductor of radius 0.5 mm and an outer conductor of internal radius 5 mm. If the dielectric is a material of relative permittivity 2.3, find the capacitance per metre run of the cable.

(3) The inductance per kilometre run of the cable of the previous problem is 1 mH. Ignoring line losses, what will be the characteristic impedance of this cable?

(4) Two parallel wires, each of diameter 0.5 cm are uniformly spaced in air at a distance of 4 cm apart. What is the capacitance of the line if its total length is 150 m?

(5) A concentric cable is to be used on a 10 kV, 50 Hz transmission system. The dielectric used has a maximum safe stress of 10^6 V/m r.m.s. and a relative permittivity of 3.5. What will be the dimensions of the most 'economical' cable?

Calculate also the capacitance of the cable per 100 m run and hence find the charging current for a line of this length. (Ignore the effect of inductance.)

(6) If the product of the equivalent series resistance and the capacitance of a capacitor at a certain frequency is 20×10^{-10} and the power factor is 10^{-3}, at what frequency was the measurement made?

(7) Explain the meaning of the term 'loss angle' as applied to a capacitor. State the main reasons for power loss in a capacitor with a solid dielectric.

A voltage of 1 V r.m.s. at a frequency of 1 kHz is applied to a capacitor whose reactance is 1000 Ω. Calculate (a) the capacitance, (b) the circuit current. If the loss angle of the capacitor is 10^{-4} radian, calculate (c) the power dissipated, (d) the equivalent series resistance.

(8) A a certain frequency a capacitor can be represented as a resistance of 0.1 Ω in series with a reactance of 1 kΩ. Calculate for this capacitor (a) the power factor, (b) the loss angle, (c) the Q-factor, (d) the equivalent parallel resistance.

(9) A capacitor of 0.4 μF with a power factor 2.5×10^{-4}, and a capacitor of 0.6 μF with a power factor of 5×10^{-4} are connected

(a) in series, (b) in parallel. Calculate the capacitance and power factor for each arrangement.

(10 Show that the Q-factor of a capacitor can be expressed as 1/(power factor)

A capacitor is equivalent to a reactance of 500 Ω in parallel with a resistance of 1000 Ω, the frequency being 2 kHz. What is the power factor of this capacitor? If the voltage across the capacitor at 2 kHz is 5 V r.m.s. what power will be dissipated.

(11) Derive from *first principles* an expression for the energy in joules in a charged capacitor, in terms of the capacitance and the voltage between the plates. Hence, show how the energy may be expressed in terms of the volume of the field.

(12) A parallel plate capacitor has plates with area A m^2 and separation d metres. The plates are charged to a potential difference V from a battery and the battery is then disconnected without loss of charge. A slab of dielectric material, thickness d, is then introduced between the plates. Show that the energy after the slab has been positioned is less than the original charge by a factor $1/\varepsilon_r$, where ε_r is the relative permittivity of the dielectric. Account for this effect.

(13)A metal sphere of radius R, in air, carries a charge of Q coulomb. Show that the energy stored in the field surrounding the sphere is given by

$$\frac{Q^2}{8\pi\varepsilon_0 R} \text{ joules}$$

(14) By calculating the work required to increase the plate separation of a charged parallel plate capacitor from x to $(x+\delta x)$ prove that the attractive force between the plates is $Q^2/(2\varepsilon_0 A)$ newtons.

12 Magnetic fields and materials

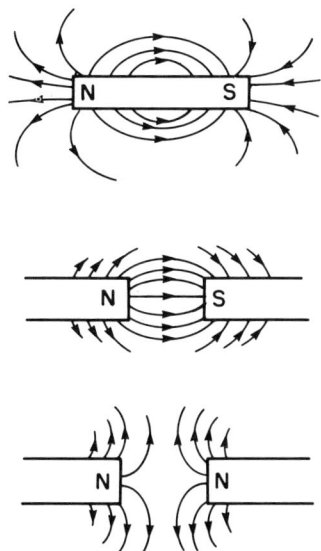

Figure 12.1

Any space in which a magnetic effect can be detected constitutes a magnetic field. If an isolated north pole is placed in a magnetic field, there will be a force acting on the pole tending to drive it away from the north pole of the field towards the south pole. The direction of the force on an isolated north pole defines the direction of action of the field. Such a field may be represented in magnitude and direction by vector lines of magnetic force, where the density of the lines is an indication of the field strength. *Figure 12.1* shows three typical field patterns. Such field configurations will be familiar to anyone who has plotted them at some time or other by means of compass needle or iron filings. Like poles will produce opposing fields and unlike poles attracting fields, and there will be a mechanical force of repulsion or of attraction, respectively, in either case. Magnetic fields, in theory, extend to inifinity, but their influence falls off inversely as the square of the distance, so for most practical purposes the fields are of importance only in the vicinity of the poles. Lines of force, or lines of magnetic flux, have no existence in reality and in a magnetic field one would not find forces acting along such discrete lines. The whole of the space occupied by the field, as is the case with the electric field, is *continuously* occupied; there are no regions of magnetic influence separated by regions where no such influence exists. Flux, then, is not measured in lines but in terms of the electromagnetic effect the field produces. The unit of flux is the weber: when unit flux of 1 weber links with a circuit of 1 turn in 1 second, or when a conductor cuts flux at the rate of 1 weber per second, an e.m.f. of 1 volt is induced in the conductor. Hence

$$1 \text{ weber} = 1 \text{ volt} \times 1 \text{ second}$$

The flux passing through 1 m² area normal to the field lines is the magnetic flux density B (or magnetic induction). For a magnetic field of cross-sectional area A m², with flux Φ threading that area

$$B = \frac{\Phi}{A} \text{ tesla}$$

where 1 tesla = 1 Wb/m².

Magnetic field strength is measured in amperes (or ampere-turns) per metre of flux path. When a current flows in a conductor, a magnetic field is set up in the space surrounding the conductor. The lines of flux take the form of concentric circles in planes at right-angles to the conductor, and the arrows in the diagrams of *Figure 12.2* show the line of action of the field as obtained from the direction of the force on an isolated north pole. The magnetic effect is directly proportional to the current in the conductor and for a given number of conductors, therefore, the current can be taken as a measure of the force which produces

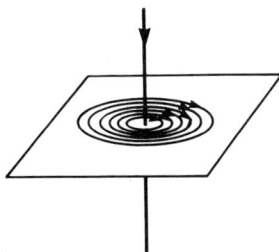

Figure 12.2

and maintains the magnetic field. This force is the *magnetic-motive force* (m.m.f.), and N conductors each carrying I amperes produce the same magnetic effect as a single conductor carrying NI amperes.

The available m.m.f. is the same for all the circular flux paths set up around the conductor. At greater distances from the conductor the length of the flux paths becomes greater and the less m.m.f. per unit length of *flux path* is available to maintain the flux in that path. The m.m.f. per metre length of *flux path*, the magnetising force or the magnetic field strength, symbol H; for a single conductor,

$$\therefore \qquad H = \frac{I}{l} \text{ A/m}$$

where l is the length of the flux path in metres. For circular paths the length of each path is $2\pi r$, where r is the radius of the ring, and so

$$H = \frac{I}{2\pi r} \text{ A/m} \qquad (12.1)$$

It is more usual to express the magnetising force in ampere-turns per metre (At/m) than in amperes per metre, although there is no fundamental difference between these two units, "turns" being a dimensionless number.

12.1.1 Permeability

At any point in a magnetic field the magnetising force H maintains the magnetic flux Φ and produces a particular value of flux density B. H is the cause and B the effect. Both H and B are vector quantities acting in the direction of the flux lines at a particular location. Clearly, if H changes, B will change, and if the field is established in air or in any non-magnetic medium, the ratio of B to H is found to be constant:

$$\frac{B}{H} = \text{a constant } \mu_0$$

The constant μ_0 is the *permeability of free space* or the magnetic space constant.

Consider a long straight conductor set in a non-magnetic medium and carrying a current of 1 A. The flux around the conductor will follow a concentric path. At a distance 1 m away from the conductor the length of the flux path is 2π m, and the m.m.f. producing this is due to the current of 1 A flowing in a circuit of 1 turn; hence m.m.f. = 1 ampere-turn and the field strength at the point is

$$H = \frac{NI}{l} = \frac{1}{2\pi} \text{ At/m}$$

Now if another long conductor, carrying a current of 1 A, is placed at the point 1 m distant from the first conductor, the force acting on this conductor will be BIl newtons. Hence for a 1 m length, the force will be B newtons. But from the elementary definition of unit current, the force per metre length between two parallel conductors each carrying a current of 1 A is 2×10^{-7} N. Hence

$$B = 2 \times 10^{-7} \text{ tesla}$$

and
$$\frac{B}{H} = \mu_0 = \frac{2 \times 10^{-7}}{\dfrac{1}{2\pi}} = 4\pi \times 10^{-7}$$

This unit is analogous to the permittivity of free space for the electric field, ε_0.

When iron or any other ferromagnetic material is used to provide the medium in which the field is established, there is a very large increase in the flux density B for a given magnetising force H. The factor by which the flux density increases, for the same value of magnetising force, is the *relative permeability* μ_r of the material. In the material, therefore

$$B = \mu_0 \mu_r H = \mu H$$

where μ is the *absolute permeability*.

All so-called non-magnetic materials, including air, do actually exhibit slight magnetic properties, but outside of the physics laboratory these can be neglected and non-magnetic materials can all be treated as having the same relative permeability as a vacuum, that is, unity. For ferromagnetic materials, notably iron, nickel, cobalt, dysprosium and gadolinium (and for a variety of alloys of these and other elements) μ_r is not constant and may even vary considerably for a given material, since the relationship between B and H, as revealed in the familiar magnetisation curves of such materials, is not linear. Further, relative permeability depends not only upon H but in many cases upon the past magnetic history of the specimen.

12.2 INDUCTANCE DUE TO INTERNAL LINKAGES

A flow current in a circuit is accompanied by magnetic flux linking with the circuit. A change in current causes a change in flux linkages and an e.m.f. is induced in the circuit, where the magnitude of the e.m.f. is proportional to the rate of change of flux linkages. In accordance with Lenz's law, the direction of the e.m.f. is such that the change in current is opposed. This opposition to change is the property of the self-inductance of the circuit.

The unit of inductance is the henry. A circuit has an inductance of 1 H when an e.m.f. of 1 V is induced in it by a current changing at the rate of 1 A/s. Alternatively, a circuit will have an inductance of 1 H if a current of 1 A flowing in the circuit causes 1 flux linkage with it. So

Inductance = flux linkages per ampere

We need to examine now in rather more detail the effect of flux linkages taking place *within* the conductor boundary. For extremely thin conductors or for conductors used at very high frequencies where the depth of penetration of the current is small (because of skin effect) compared with the cross-section of the conductor, the internal linkages are negligible and the circuit inductance is simply that due to the field in the space surrounding the conductor. When the conductor has an appreciable cross-sectional area and when the frequency is very low so that the current is uniformly distributed over the cross-section, the inductance due to the internal linkages has its greatest value.

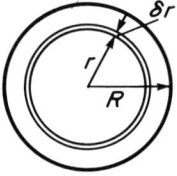

Figure 12.3

Figure 12.3 shows the cross-section of a conductor of radius R, carrying a current of I amperes uniformly distributed over the cross-section.

$$\text{Current density} = \frac{\text{Current}}{\text{Area}} = \frac{I}{\pi R^2} \text{ A/m}^2$$

For a thin ring element of radius r and width δr contained within the conductor, the current enclosed by this ring will be

$$\text{Current density} \times \text{Ring area} = \frac{I}{\pi R^2} \times \pi r^2$$

$$= \frac{Ir^2}{R^2} \text{ A}$$

So, if H_r is the field strength at radius r

$$2\pi r H_r = \frac{Ir^2}{R^2} \times 1$$

since, from equation (12.1), $I = 2\pi r H$ for a circular flux path and $N = 1$.

$$\therefore \qquad H_r = \frac{Ir}{2\pi R^2} \text{ At/m}$$

But the flux density at radius r

$$B_r = \mu H_r = \frac{\mu_0 \mu_r Ir}{2\pi R^2} \text{ tesla}$$

For a 1 m length of the conductor, the flux within radius r will be

$$\frac{\mu_0 \mu_r Ir}{2\pi R^2} \cdot \delta r$$

and this flux links the fraction r^2/R^2 of the whole conductor. therefore, the linkages due to the flux within radius r

$$= \frac{\mu_0 \mu_r Ir^3}{2\pi R^4} \delta r \text{ Wb-turns}$$

$\therefore$ total linkages per metre due to the flux in the conductor

$$= \frac{\mu_0 \mu_r}{2\pi R^4} \int_0^R r^3 \cdot dr = \frac{1}{4} \frac{\mu_0 \mu_r I}{2\pi} \text{ Wb-turns}$$

Then the inductance due to the internal flux = internal flux linkages per ampere

$$= \frac{\mu_0 \mu_r}{8\pi} \text{ H/m} \qquad (12.2)$$

Although the radius has disappeared and the inductance is therefore independent of this, the formula is modified slightly if the conductor is extremely thin.

12.3 TRANSMISSION LINE PARAMETERS

W are now in a position to derive values for the inductance of transmission lines per unit length as we did for capacitance in the previous chapter.

12.3.1 The coaxial cable

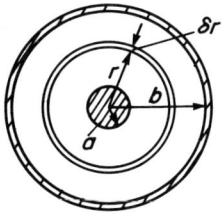

Figure 12.4

The cross-section of a coaxial line is shown in *Figure 12.4*. Let equal and opposite currents of *I* amperes flow in the core and outer sheath conductors. Consider a path of radius *r*, width δr, and let the radii of the inner and outer conductors be *a* and *b*, respectively.

Then the field strength at radius *r* is $H_r = I/2\pi r$ and the flux density at radius *r* is $B_r = \mu H_r = \mu_0 \mu_r H_r$ tesla

For a 1 m length of the cable, the flux within the cylinder of radius *r* and thickness $\delta r = B_r . \delta r \times 1$

$$= \frac{\mu_0 \mu_r I}{2\pi r} . \delta r \text{ Wb}$$

This flux links the loop of the cable formed by the core and the outer sheath, so that the flux linkages per metre length of the cable

$$= \frac{\mu_0 \mu_r I}{2\pi r} . \delta r \text{ Wb-turns}$$

The total flux is therefore

$$= \frac{\mu_0 \mu_r I}{2\pi} \int_a^b \frac{dr}{r} = \frac{\mu_0 \mu_r I}{2\pi} . \ln \frac{b}{a} \text{ WB-turns}$$

Hence, the inductance per metre $= \dfrac{\mu_0 \mu_r \ln b/a}{2\pi}$ H/m $\qquad$ (12.3)

At low frequencies the inductance due to the internal linkages must be added to this expression, giving a total inductance per metre of

$$\frac{\mu_0 \mu_r}{8\pi} + \frac{\mu_0 \mu_r \ln b/a}{2\pi} \text{ H/m} \qquad (12.4)$$

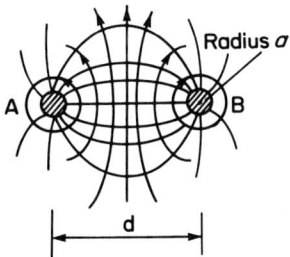

Figure 12.5

Example 1. Derive an expression for the inductance per metre run of a parallel wire line, assumed isolated from earth, the radius of each wire being *R* and their separation *d*.

Let the wires carry a current of *I* amperes in opposite directions. Then the magnetic field associated with the conductors will be as shown in *Figure 12.5* and there will be a force of repulsion between the conductors. As for the case of the electric field, we consider that only conductor A is carrying *I* amperes for purposes of analysis.

The field at radius *r* from A is

$$H_r = \frac{I}{2\pi r} \text{ At/m} \quad \text{and} \quad B_r = \frac{\mu_0 \mu_r I}{2\pi r} \text{ tesla}$$

The total flux in a 1 m length of the conductor is then

$$= B_r \delta r = \frac{\mu_0 \mu_r I}{2\pi r} \delta r \text{ Wb}$$

This is identical to the previous work. Now the linkages with conductor A due to this flux

$$= \frac{\mu_0 \mu_r I}{2\pi r} \delta r \text{ Wb-turns}$$

Let the field be assumed negligible at a distance R' from the conductor; then the total linkages with conductor A due to the current in A

$$= \frac{\mu_0 \mu_r I}{2\pi} \int_R^{R'} \frac{dr}{r} = \frac{\mu_0 \mu_r I}{2\pi} \cdot \ln \frac{R'}{R}$$

In the same way, the total linkages with conductor B due to the current in A

$$= \frac{\mu_0 \mu_r I}{2\pi} \cdot \ln \frac{R'}{d}$$

Suppose now that only conductor B is carrying a current of $-I$ amperes. Then by the same reasoning as used above

Total linkages with B due to current in $B = -\dfrac{\mu_0 \mu_r I}{2\pi} \ln \dfrac{R'}{R}$

Total linkages with A due to current in $B = -\dfrac{\mu_0 \mu_r I}{2\pi} \ln \dfrac{R'}{d}$

Hence, the total linkages with $A = \dfrac{\mu_0 \mu_r I}{2\pi} \left[\ln \dfrac{R'}{R} - \ln \dfrac{R'}{d} \right]$

$$= \frac{\mu_0 \mu_r I}{2\pi} \cdot \ln \frac{d}{R}$$

and the total linkages with B, similarly

$$= \frac{\mu_0 \mu_r I}{2\pi} \cdot \ln \frac{d}{R}$$

$\therefore$ for the two conductors, the total inductance

$$= 2\mu_0 \mu_r \frac{\ln d/R}{2\pi} \text{ H/m} \qquad (12.5)$$

Again, this result does not take account of the effect of the internal linkages. If these are included, the total inductance per *loop* metre (there and back)

$$= 2 \cdot \frac{\mu_0 \mu_r}{8\pi} + 2\mu_0 \mu_r \frac{\ln d/R}{2\pi} \qquad (12.6)$$

In most practical lines, u_r is unity.

Example 2. A coaxial television type cable has an inner core of radius 1 mm and an outer sheath of internal radius 5 mm, the dielectric being polythene of relative permittivity 2.5. Calculate the inductance and the capacitance of the cable per metre run, hence determine its characteristic impedance at low frequencies. Here

$$a = 1 \text{ mm}, \ b = 5 \text{ mm}, \ \varepsilon_r = 2.5, \ \mu_r = 1$$

From equation (12.4), by rearrangement and setting $u_r = 1$

$$\text{Inductance} = \frac{\mu_0}{2\pi} \left[\frac{1}{4} + \ln \frac{b}{a} \right]$$

$$= 2 \times 10^{-7} \left[\frac{1}{4} + \ln 5 \right]$$

$$= 3.27 \times 10^{-7} \ \text{H/m}$$

$$= 0.372 \ \mu\text{H/m}$$

From equation (11.2) earlier

$$C = \frac{2\pi\varepsilon}{\ln b/a} \ \text{F/m}$$

$$= \frac{2\pi \times 2.5 \times 8.85 \times 10^{-12}}{1.61}$$

$$= 86.3 \ \text{pF/m}$$

At low frequencies

$$Z_0 = \sqrt{\frac{L}{C}}$$

$$= \sqrt{\left(\frac{0.372 \times 10^{-6}}{86.3 \times 10^{-12}}\right)} = 10^3 \ \sqrt{\left(\frac{0.327}{86.3}\right)}$$

$$= 61.5 \ \Omega$$

12.4 LOSSES IN MAGNETIC CIRCUITS

Losses directly associated wtih the iron circuit of electromagnetic systems are divided into two parts: the hysteresis loss and the eddy current loss. In many devices such as relays, transformers and motors there are also resistance (copper) losses and frictional losses, but these do not concern us in this section.

12.4.1 Hysteresis

Hysteresis is the name given to the 'lagging' effect displayed by the flus density B whenever there are changes in the magnetising force H. Starting with a piece of ferromagnetic material in an unmagnetised state, when H is increased from zero the flux density follows the normal magnetisation or B-H curve, as shown in *Figure 12.6*. When the magnetisation reaches the point A and the magnetising force H is reduced to zero, the flux density does not follow the path of the magnetisation curve back to zero. The values of B and H follow the path AC and the material has a residual flux density or remanance given by OC. To remove this flux density requires the negative magnetising force OD, known as the coercive force. The lag of the values of B behind H is the hysteresis of the material.

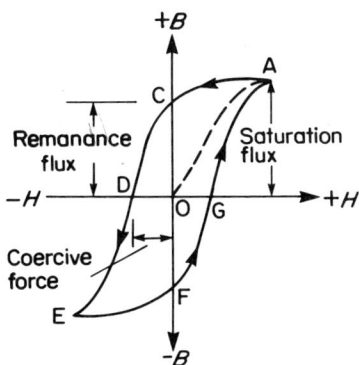

Figure 12.6

Further magnetisation in the reverse direction builds up the flux density to the value at E. Similar conditions occur as the magnetising is again reduced to zero, and then taken positive again. A full cycle of magnetisation in this way describes a curve known as the *hysteresis loop*.

The relative dimensions of the loop and hence its area depends upon the nature of the magnetic material and its magnetic properties. Hysteresis results in a dissipation of energy which appears in the form of heat, and the energy wasted in this way is proportional to the area of the loop. It represents the work done in orientating the magnetic domains to follow the variations in H.

In a specimen of iron subject to an alternating flux, the energy expended per cycle $= kB^n$ joules per cubic metre of the material, where B is the maximum flux density in tesla and k is a constant. The coefficient n lies in the range 1.6–2. Then at a frequency f Hz

$$\text{Energy dissipated per second} = kfB^n \ \text{J/m}^3$$

and

$$\text{Hysteresis loss } P_h = k_h fB^n \ \text{W/m}^3 \qquad (12.7)$$

The value of n is found by experiment for a particular material. The constant k_h, known as the hysteresis coefficient, can have values between about 200 and 5000 depending upon the material. For silicon steel, as used in transformer cores, its value lies between 200 and 400.

For devices which are subjected to rapid reversals of magnetism it is necessary to use a material having a small area hysteresis loop. *Figure 12.7* shows typical loops for hard steel, soft iron,

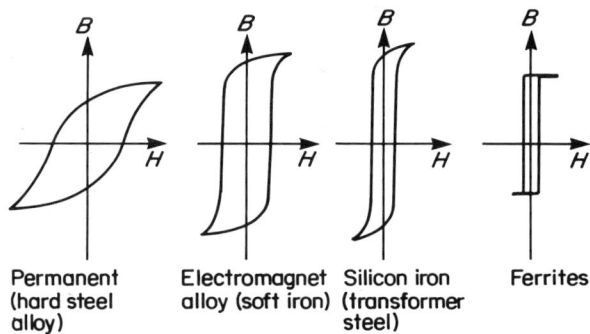

Permanent
(hard steel
alloy)

Electromagnet
alloy (soft iron)

Silicon iron
(transformer
steel)

Ferrites

Figure 12.7

silicon iron and ferrites. Hard steel has a high remanance and large coercivity; soft iron has large remanance and small coercivity. Ferrite materials exhibit small rectangular hysteresis loops which make them suitable as memory devices in computer systems. Ferrites are ceramic-like magnetic substances made from oxides of iron, nickel, cobalt, aluminium, magnesium and manganese. Although metallic, they are non-conductors and their hysteresis loss is extremely small.

Although the loops illustrated show that the specimen concerned has been carried into magnetic saturation, this has only been done for the sake of example. It is not necessary for the magnetising force to carry the specimen into saturation at the peak value before a hysteresis loop occurs; hysteresis result from reversals of flux of *any* value.

12.4.2 Eddy currents   In iron-cored components such as transformers, the alternating flux, as well as inducing voltages in the coil or coils where it is required, induces voltages in the iron core where it is not required. These induced e.m.f.s cause local currents to circulate in the core material, where their random distribution and magnitudes cause them to be known as eddy currents. Since they cannot be utilised they represent wasted energy which goes to heating the iron circuit.

To prevent these currents flowing, as much as possible, transformer cores, armatures and the like are built up of a large number of thin plates or *laminations,* separated from one another by very thin paper or varnish. This insulation offers a large resistance to any induced currents which would tend to flow across the laminations, and for any which flow in the plane of the laminations, the relatively high impedance of the iron path effectively reduces their magnitude. At the same time, the effectiveness of the magnetic circuit is not impaired to any great extent, as the flux can pass freely down each lamination without passing through the intervening insulation.

For low frequencies, the impedance of the iron path may reasonably be taken as constant. The eddy current loss is then proportional to f^2B^2 since the induced e.m.f. will be proportional to fB. Thus

$$\text{Eddy current loss } P_e = k_e f^2 B^2 \text{ W/m}^3$$

where k_e is a constant.

At high frequencies where it is impracticable to make laminations thin enough, core losses can be reduced by using so-called iron-dust cores or ferrite material. The iron dust, mixed with a binding material, is compressed in a mould which gives it a finished appearance comparable with a solid piece of iron. Such a core material has a very high electrical resistivity because of the non-conducting binder separating the iron grains. Eddy-current paths are, therefore, restricted to the size of the iron grains so that eddy currents are negligible even at high radio frequencies.

12.4.3 Separation of iron losses Since the hysteresis loss is proportional to frequency, whereas eddy current loss is proportional to the frequency squared, a method may be devised by which these losses may be separated out for individual evaluation. In a transformer or choke, for example, the total loss may be measured over a range of frequencies, while the other factors on which the losses depend, the core flux for example, are maintained constant. Then

> Hysteresis loss $= k_1 f$ watts, where k_1 is a constant
> Eddy current loss $= k_2 f^2$ watts, where k_2 is a constant
> Total loss $= P_t$ $= k_1 f + k_2 f^2$

and $$\frac{P_t}{f} = k_1 + k_2 f$$

If, now, P/f is plotted against f, a straight-line graph will be obtained having a gradient k_2 and a vertical intercept k_1 — see *Figure 12.8*. Hence, k_1 and k_2 can be found and hence the hysteresis and eddy current losses at a given frequency and maximum core flux density evaluated.

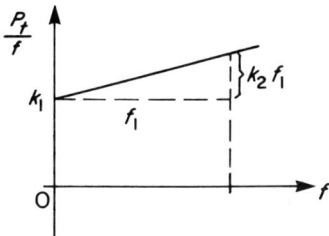

Figure 12.8

Example 3. A hysteresis loop is plotted to scales of 1 cm $= 0.005$ tesla, 1 cm $= 25$ At/m, and has an area of 120 cm². If the iron circuit associated with this loop has a volume of 10^4 cm³, find the hysteresis loss at a frequency of 50 Hz.

1 cm² of the loop represents $0.005 \times 25 = 0.125$ At–Wb/m³

Since the loop area $= 120$ cm², the energy loss per cycle
$$= 0.125 \times 120 = 15 \text{ J}$$

At 50 Hz, the power dissipated per *cubic metre* of the iron

$$= 15 \times 50 = 750 \text{ J/s}$$
$$= 750 \text{ W}$$

But the iron volume $= 10^4 \text{ cm}^3 = 0.01 \text{ m}^3$

∴   Hysteresis loss $= 750 \times 0.01 = 7.5 \text{ W}$

Example 4. The volume of a transformer core is 0.005 m^3 and the cross-sectional area of the winding limb is 100 cm^2. If the maximum value of the core flux is 9 mWb calculate the power loss due to hysteresis. Take the frequency to be 50 Hz, $k_h = 200$ and $n = 1.6$ in equation (12.7).

We require first the maximum flux density:

$$B = \frac{\Phi}{A} = \frac{9 \times 10^{-3}}{100 \times 10^{-4}} = 0.9 \text{ tesla}$$

From equation (12.7) hysteresis loss $= k_h B^n$

$$= 200 \times (0.9)^{1.6}$$
$$= 169 \text{ J/m}^3\text{/cycle}$$

Then the power loss $= k_h f B^n \text{ W/m}^3$

$$= 169 \times 50 \times 0.005 = 42.25 \text{ W}.$$

12.5 ENERGY IN THE MAGNETIC FIELD

In order to establish a magnetic field, energy must be expended, but once the field is established, the only expenditure of energy is that supplied to maintain the current flow against the circuit resistance, that is, the I^2R loss. This energy is dissipated as heat.

In a circuit made up of inductance and resistance, the applied voltage at any instant is

$$V = iR + L \frac{di}{dt}$$

Multiplying through by $i \cdot dt$ gives the energy equation

$$Vi \cdot dt = iR \cdot dt + Li \cdot di$$

where the component terms are, taken in order, the energy supplied by the source in time dt, the energy dissipated in the resistance, and the energy supplied in building up the magnetic field. The total energy stored in the field is, therefore, from this last term

$$= \int_0^I Li \cdot di = L \left[\frac{1}{2} i^2 \right]_0^I$$
$$= \tfrac{1}{2} L I^2 \text{ J}$$

This is an elementary result from an earlier part of the course. Substituting $L = \Phi N/I$ in this, we have

$$\text{Energy stored} = \tfrac{1}{2} \frac{N\Phi}{I} I^2 = \tfrac{1}{2} NI\Phi \text{ J}$$

But $NI = Hl$ and $\Phi = BA$, hence the energy stored

$$= \tfrac{1}{2} BHAl = \tfrac{1}{2} BH \text{ J/m}^3$$

since $Al =$ the volume of the field.

This gives us the energy in terms of the volume of the field, analogous to the case of the electric field. Since $B = \mu_0 H$, this result can also be expressed as

$$\text{Energy stored} = \tfrac{1}{2}\mu_0 H^2 \text{ or } \tfrac{1}{2}\frac{B^2}{\mu_0} \text{ J/m}^3$$

Example 5. Compare the energy required to set up in a cubic volume of 5 cm edge length (a) a uniform electric field of 10^4 V/m, and (b) a uniform magnetic field of flux density 0.5 tesla.

(a) For the electric field we have, from equation (11.1) earlier,

$$\text{Energy stored} = \tfrac{1}{2}DE = \tfrac{1}{2}\varepsilon_0 E^2 \text{ J/m}^3$$

Here the volume of the field $= (0.05)^3 \text{ m}^3$

$$\therefore \qquad \text{Energy stored} = \tfrac{1}{2} \times 8.85 \times 10^{-12} \times (0.05)^3 \times (10^4)^2 \text{ J}$$

$$= 5.53 \times 10^{-8} \text{ J}$$

(b) For the magnetic field, energy stored $= \dfrac{1}{2} \cdot \dfrac{B}{\mu_0} \times \text{volume}$

$$= \frac{0.5^2 \times (0.05)^3}{2 \times 4\pi \times 10^{-7}}$$

$$= 12.4 \text{ J}$$

Notice that considerably more energy is required to set up a magnetic field than is required to set up an electric field of similar reasonable magnitude.

12.6 IMPEDANCE OF SPACE

We have so far treated the wave travelling along a transmission line as being a current or a voltage wave. It is now possible to treat it as an *electromagnetic wave*, since both current and voltage set up their own electromagnetic and electrostatic fields, respectively, as energy is propagated along the line. In a uniform coaxial line the lines of electric flux are radial and the lines of magnetic flux are concentric, as illustrated in *Figure 12.9*. Notice that the two fields are mutually perpendicular to each other and to the direction of propagation, that is, the length of the line.

If an alternating signal is applied to the sending end of the line, then the distribution of E and H along the length of the line will be the same as the distribution of voltage and current. If the line is loss-free and correctly terminated in its characteristic impedance Z_0 (or if it is assumed to be infinitely long), it will have a purely resistive impedance and V and I will be in phase at all points. So E and H will be in phase at all points. Now

Electric flux
Magnetic flux

Figure 12.9

$$Z_0 = \frac{V}{I} = \sqrt{\frac{L}{C}}$$

and on substituting L and C for the coaxial cable from equations (12.3) and (11.2) respectively (assuming a vacuum dielectric)

$$Z_0 = \sqrt{\left(\frac{\dfrac{\mu_0}{2\pi} \cdot \ln \dfrac{b}{a}}{\dfrac{2\pi\varepsilon_0}{\ln \dfrac{b}{a}}} \right)}$$

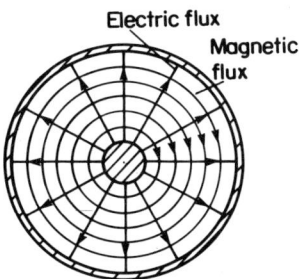

$$= \frac{1}{2\pi} \cdot \ln \frac{b}{a} \sqrt{\frac{\mu_0}{\varepsilon_0}} \qquad (12.8)$$

Now (from equation (11.3)

$$B = \frac{\mu \cdot I}{2\pi r} \text{ and } E = \frac{V}{r \cdot \ln \frac{b}{a}}$$

at any radius r within the cable. From these

$$\frac{E}{H} = 2\pi \cdot \frac{V}{I} \cdot \frac{1}{\ln \frac{b}{a}} \quad \frac{V/m}{A/m}$$

$$= 2\pi Z_0 \cdot \frac{1}{\ln \frac{b}{a}} \Omega \text{ (since } Z_0 = V/I)$$

Substituting for Z_0 from (12.8) above gives us

$$\frac{E}{H} = \sqrt{\frac{\mu_0}{\varepsilon_0}} \Omega$$

and putting $\mu_0 = 4\pi \times 10^{-7}$, $\varepsilon_0 = \frac{1}{36\pi} \times 10^{-9}$

$$\frac{E}{H} = 120\pi = 377 \ \Omega$$

This is the impedance of free space or the *electromagnetic wave impedance*.

When an electromagnetic wave is propagated through space, for example, a radio transmission, space is the transmission line, and such a 'line' must have a characteristic impedance. That impedance is 377 Ω.

From *Figure 12.9*, notice that while both E and H vary over the cross-section of the cable, the *ratio E/H* remains the same at all points. This is true of an electromagnetic wave propagated through space.

Since the transmission line simply acts as a guide for the electromagnetic wave, it is perfectly feasible to do away with the central conductor of a coaxial line and propagate the wave energy along the tube formed by the outer sheath. At very high frequencies such transmission systems are then known as waveguides.

12.7 FORCE BETWEEN MAGNETISED SURFACES

In *Figure 12.10*, let the cross-sectional area of the magnet by A m² and its distance from an armature x m. If a force of F newtons moves the armature through a small distance δx, then the work done on the armature is $F \cdot \delta x$ joules and this increase in energy is stored in the magnetic field. Before the armature moves the energy stored in the field is

$$\frac{B^2 A x}{2\mu_0} \text{ J}$$

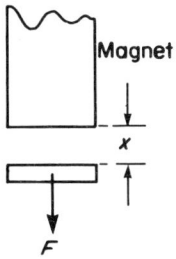

and after the armature has moved the energy stored is

$$\frac{B^2 A(x + \delta x)}{2\mu_0} \text{ J}$$

Hence, the *increase* in the energy is

$$\frac{B^2 A \cdot \delta x}{2\mu_0} \text{ J}$$

∴

$$F \cdot \delta x = \frac{B^2 A \cdot \delta x}{2\mu_0}$$

and so

$$F = \frac{B^2 A}{2\mu_0} \text{ N}$$

Figure 12.10

PROBLEMS FOR CHAPTER 12

(1) Show that the force between two long parallel conductors of length l m supported in air is given by

$$F = \frac{2I_1 I_2 l}{10^7 d} \text{ N}$$

where I_1 and I_2 are the currents in the wires and d their separation.

(2) A current of 5 A in a coil of 500 turns produces a magnetic flux of 5 mWb linking with the coil. What is the inductance of the coil?

(3) Wires carrying equal but opposite currents are often twisted together to reduce their external magnetic effect. Why is this effective?

(4) A light helical spring is hanging vertically with a small mass suspended from its lower end. If a current is sent through the spring what, if anything, will happen to the position of the weight?

(5) What must be the strength of a uniform electric field if it is to have the same energy as that established by a magnetic field of flux density 0.5 tesla?

(6) Show that the self-inductance of a length l of a long wire associated with the flux *inside* the wire only is

$$\frac{1}{2} \cdot \frac{\mu_0 l}{4\pi}$$

and independent of the wire diameter.

(7) A long coaxial cable has inner and outer conductors of radii a and b, respectively. The central conductor carries a steady current of I amperes, the outer conductor providing the return path. Show that the magnetic energy stored per metre length of such a cable is given by

$$\frac{\mu_0 I^2}{4\pi} \cdot \ln \frac{b}{a} \text{ J}$$

(8) The coaxial cable whose cross-section is shown in *Figure 12.11* carries a current of 10 A in the inner conductor and an equal but opposite current in the outer conductor. Calculate the stored magnetic energy per metre length of the cable (a) within the centre

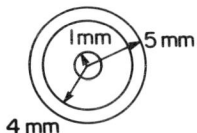

Figure 12.11

conductor, (b) within the space between the conductors, (c) within the outer conductor.

(9) The current in a circuit consisting of an inductor of 100 mH having a resistance of 50 Ω is increasing uniformly from zero at time $t = 0$ at a rate of 5 A/s. If at a time $t = 0.01$ s the coil is suddenly short-circuited, calculate the *total* energy which will have been dissipated in this period.

(10) A hysteresis loop is plotted to a scale of 1 cm = 0.008 tesla, 1 cm = 20 At/m, and has an area of 150 cm². If the iron circuit associated with the loop has a volume of 10^4 cm³ and operates at a frequency of 100 Hz, what is the hysteresis loss?

(11) The eddy current and hysteresis losses in a certain magnetic circuit are 3 W and 2 W, respectively. If the frequency of operation is reduced from 50 Hz to 40 Hz, the flux density remaining unchanged, calculate the new values of eddy current and hysteresis loss.

(12) What force is required to separate two flat metal surfaces having an effective area of contact of 100 cm² when the magnetic flux between them is 2 mWb?

(13) Show that the pull between two magnetic surfaces of effective area A can be expressed as

$$\frac{0.051B^2A}{\mu_0} \text{ kg-force}$$

Appendix I Transmission lines — hyperbolic forms

On page 111 we noted that for unit length of transmission line

$$I_S = I_R \cdot e^{\gamma}$$

where I_s and I_R are, respectively, the currents entering and leaving the section, and γ is the propagation constant.

Consider the T-sections of *Figure AI.1* to be part of a line

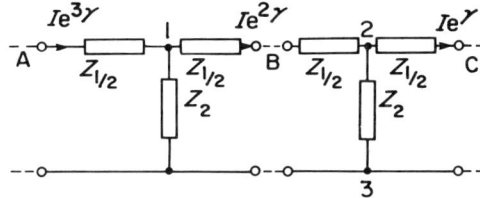

Figure AI.1

network. Let the currents at the points A, B and C be, respectively, $I \cdot e^{3\gamma}$, $I \cdot e^{2\gamma}$ and $I \cdot e^{\gamma}$. Applying Kirchhoff to the mesh 1–2–3–4 we have

$$(I \cdot e^{2\gamma})Z_1 + (I \cdot e^{2\gamma} - I \cdot e^{\gamma})Z_2 - (Ie^{3\gamma} - I \cdot e^{2\gamma})Z_2 = 0$$

Dividing by $I \cdot e^{3\gamma}$

$$Z_1 + Z_2 - e^{-\gamma}Z_2 - e^{\gamma}Z_2 + Z_2 = 0$$

$$Z_2[e^{\gamma} + e^{-\gamma}] = Z_1 + 2Z_2$$

$$\frac{e^{\gamma} + e^{\gamma}}{2} = 1 + \frac{Z_1}{2Z_2}$$

$$\therefore \qquad \cosh \gamma = 1 + \frac{Z_1}{2Z_2} \qquad (A.1)$$

Figure AI.2

Consider now a single terminated section— see *Figure AI.2*. Now $I_1/I_2 = e^{\gamma}$, and applying Kirchhoff to the output mesh

$$-(I_1 - I_2)Z_2 + \tfrac{1}{2}I_2 Z_1 + I_2 Z_0 = 0$$

$$\therefore \qquad I_1 Z_2 = I_2 \left[Z_2 + \frac{Z_1}{2} + Z_0 \right]$$

$$\therefore \qquad e^{\gamma} = \frac{I_1}{I_2} = \frac{Z_2 + \dfrac{Z_1}{2} + Z_0}{Z_2}$$

$$= 1 + \frac{Z_1}{2Z_2} + \frac{Z_0}{Z_2}$$

But $e^{\gamma} = \cosh\gamma + \sinh\gamma$, hence by substraction, since $\cosh\gamma = 1 + Z_1/2Z_2$ from (1) above:

$$\sinh \gamma = \frac{Z_0}{Z_2}$$

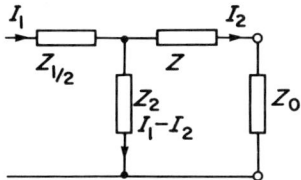

Hence

$$\tanh \gamma = \frac{\sinh \gamma}{\cosh \gamma} = \frac{Z_0}{Z_1 + \dfrac{Z_1}{2}}$$

Now

$$Z_0 = \sqrt{(Z_{oc} Z_{sc})} \quad \text{and} \quad Z_{oc} = Z_2 + \frac{Z_1}{2}$$

$$\therefore$$

$$\tanh \gamma = \sqrt{\left(\frac{Z_{sc}}{Z_{oc}}\right)} = \frac{e^\gamma - e^{-\gamma}}{e^\gamma + e^{-\gamma}}$$

$$= \frac{e^{2\gamma} - 1}{e^{2\gamma} + 1}$$

A worked examples will illustrate the use of the hyperbolic equations.

Example. A symmetrical network has open and short-circuit impedances of 600 $\underline{/-70°}$ Ω and 500 $\underline{/-10°}$ Ω. Find the attenuation and phase constants for the network

$$\tanh \gamma = \sqrt{\left(\frac{Z_{sc}}{Z_{oc}}\right)} = \sqrt{\left(\frac{500\underline{/-10°}}{600\underline{/-70°}}\right)}$$

$$= \sqrt{(0.833\underline{/60°})} = 0.913\underline{/30°}$$

Converting to rectangular form and rearranging, we have

$$e^{2\gamma} - 1 = [0.79 + j0.456](e^{2\gamma} + 1)$$

$$\therefore \qquad e^{2\gamma} = \frac{1.79 + j0.456}{0.21 - j0.456}$$

$$\approx 3.7\underline{/80°}$$

$$\therefore \qquad e^\gamma = 1.92\underline{/40°}$$

Hence

$$\alpha = \ln 1.92 = 0.65 \text{ néper}$$

$$\beta = 40° \ (0.697 \text{ rad})$$

Appendix II Curvilinear squares

This Appendix gives a method of solving certain field problems by a form of graphical estimation and is strictly applicable only to plane fields such as may exist, for example, between two long parallel conductors or parallel plates. The method is equally applicable to conduction fields established in a conductor carrying an electric current and to magnetic fields.

The basic of the method is perhaps best illustrated by examples, but the essential point is the division of the field (whether established in space or within a conductor) into a number of squares formed between the lines of force (or *streamlines* inside a conductor) and the equipotentials. In most cases, true squares will not exist since the field lines and the equipotentials will be curved; but because these respective lines will always *intersect* at right-angles, the approximate squares so formed by their intersections are known as 'curvilinear squares'. *Figure AII.1* shows a typical pattern, all the approximate squares having 90° vertices and approaching the form of true squares as the area we consider is further subdivided into smaller units.

Consider the electric field established between two parallel metal plates as shown in *Figure AII.2(a)*. The lines of force and the equipotential lines are sketched and form curvilinear squares. For a depth of field D metres behind each square, that is, looking into the page as *Figure AII.2(b)* shows, the capacitance of the flux tube formed will be, taking the equipotential surfaces $AA'BB'$ and $CC'DD'$ to be the effective parallel plates of an elemental capacitor,

$$\varepsilon_0 \varepsilon_r \frac{\text{Area of effective plates}}{\text{Separation}}$$

$$= \varepsilon_0 \varepsilon_r \frac{AB \times D}{BC} = \varepsilon_0 \varepsilon_r D$$

since $AB = BC$. Hence the capacitance of the flux tube whose end is a true square is *independent* of the size of the square. The total capacitance will then be

$$C = \varepsilon_0 \varepsilon_r D \times \frac{y}{x}$$

where x is the number of squares measured along each equipotential and y is the number of squares measured along each line of force.

It follows that for cases where the squares formed are not true squares (for example, beyond the end of the parallel plates), the capacitance per flux tube will still be $\varepsilon_0 \varepsilon_r D$. Thus the capacitance between the plates may be estimated by a field plot which divides the field into a number of curvilinear squares. It should be noted that neither x nor y need be integral.

It is difficult to illustrate the principles of the method in the

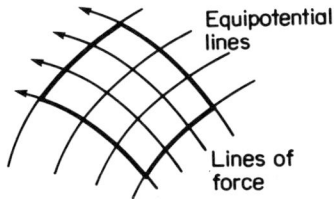

Figure AII.1

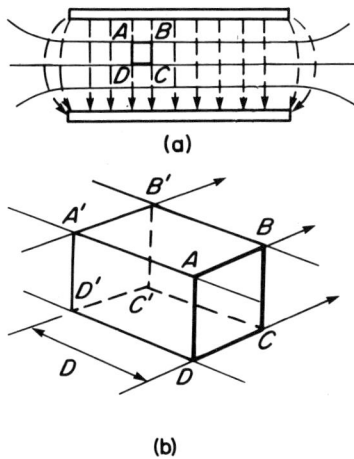

(a)

(b)

Figure AII.2

written word as the judgement necessary comes only from repeated practice and the proper procedure and techniques are best left in the care of the class lecturer. The following example should, however, provide the essentials of the method.

Example. Figure AII.3 shows the length of metal strip of uniform thickness *D* metres, which carries a current that is uniformly distributed throughout the strip. If a circular hole of diameter 5.0 cm is drilled centrally through the strip, estimate the percentage increase in the resistance of the strip resulting from the making of the hole.

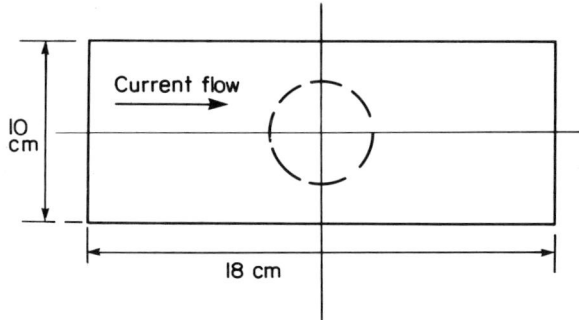

Figure AII.3

Since the hole is symmetrically drilled, the vertical central line will be an equipotential and the conduction field will be symmetrical about both the vertical and horizontal lines of symmetry — see *Figure AII.4*. Further, as no current can cross the edge of the strip, the edges must be conduction or streamlines. Hence, since equipotentials intersect streamlines at right-angles, the equipotentials must cross the strip edges at right-angles. As for the same reason, they must cross the edge of the central hole.

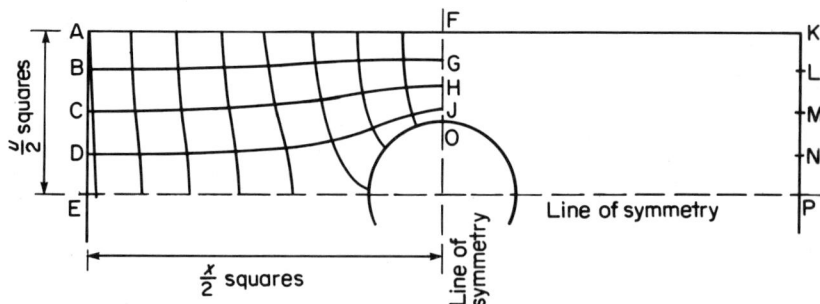

Figure AII.4

The area of the strip must now be divided into a number of curvilinear squares. A series of approximations (or inspired guesses!) have to be made to achieve this end, using a procedure rather as outlined below.

The current will divide equally on each side of the hole so we need consider only the upper half of the diagram, and for the purpose of drawing, only the top left-hand quadrant. Draw in the streamline CHM which divides the current on this upper side in

half. This line will start a little *higher* than the mid-point of the half-edge AE and cut the central equipotential F0 a little *below* its mid-point, say at H. The end end position M is symmetrical with C. The intermediate streamlines may now be estimated as BGL and DJN, again making the divisions such that
AB < BC < CD < DE and FG > GH > HJ > J0.

The equipotentials are now added, cutting the streamlines at right-angles and forming the best near squares over the area considered. This procedure requires some patience, and continual readjustment of both streamlines *and* equipotentials may be necessary before a reasonably accurate field diagram emerges.

From the diagram, let there be x squares in the streamline direction and y squares in the equipotential direction. Working similarly to the method outlined in the text above, the conductance per square element $= gD$ siemens, where g is the conductance of the material concerned and D is the thickness (or depth) of the element. Then the total conductance

$$= gD \times \frac{y}{x} = 0.49gD \text{ S}$$

since from the diagram plot we estimtate that $x = 16.2$, $y = 8$.

Before the hole was cut, the x number of squares would have been proportional to the strip length, 18 cm, and the y number of squares would have been proportional to the strip width, 10 cm. Then the total conductance without the hole

$$= gD \times \frac{10}{18} = 0.56gD \text{ S}$$

Hence, the ratio

$$\frac{\text{Conductance with the hole}}{\text{Conductance without the hole}} = \frac{0.49}{0.56} = 0.875$$

$$\therefore \quad \frac{\text{Resistance with the hole}}{\text{Resistance without the hole}} = \frac{1}{0.875} = 1.14$$

Hence, the percentage increase in resistance is about 14%.

Solutions to problems

CHAPTER 1

(1) (a) $2 \underline{/45°}$; (b) $5 \underline{/53.1°}$; (c) $13 \underline{/22.6°}$; (d) $5 \underline{/-36.9°}$;
(e) $6.7 \underline{/-116°}$; (f) $5 \underline{/53.1°}$; (g) $5.1 \underline{/11.3°}$

(2) (a) $3+j$; (b) $-2.5+j4.33$; (c) $-15.2-j1.74$ (i) $10 \underline{/150°}$;
(ii) $17.5 \underline{/45°}$; (iii) $0.4 \underline{/-90°}$; (iv) $0.875 \underline{/125°}$

(3) $Z = 9-j$ (series); $Z = 2.94-j0.45$ (parallel)

(4) (a) $100+j100$ Ω; (b) $10-j20$ Ω; (c) $10+j800$ Ω

(5) $444 \underline{/-45°}$ Ω

(6) $Z = 583-j750$ Ω; $|Z| = 950$ Ω, $\phi = 52.1°$ current leading

(7) $-0.366-j1.366$; $|Z| = \sqrt{2}$ Ω, $\phi = 105°$

(9) $Z = 342 \underline{/-7.7°}$ Ω

(10) $Y = 0.025-j0.043$ S

(11) 3.22 A; 5.29 A; 4.13 A

(13) $I = 0.307 \underline{/32.5°}$ A, leading

CHAPTER 2

(2) 26.9 mA; 38.6 Ω

(3) 67 mA

(4) 50 Ω

(5) 250 mA each row; 500 mA in the load. 375 mA, 938 mA, 563 mA

(6) 2.14 A; 2.93 A; 1.8 Ω

(7) 40 Ω

(8) 208 mA

(9) 2.95 V

(10) 12 mA; 45 mW. $200+j200$ Ω

(11) $E_{oc} = 100$ mV in series with $(5+j12)$ Ω. 5.4 mA

(12) $I = 1.28+j0.22$ A or $|I| = 1.3$ A; 50.6 W

(13) $R = 5$ Ω; 250 W

(14) $(88-j415)$ mA

(15) The source has zero impedance

(16) $E_{oc} = 5.6$ V, $Z_i = 191 \underline{/-11.8°}$ Ω

(17) $Z_{0T} = 30$ Ω; $Z_{0\pi} = 13.33$ Ω

(19) $6.1 \underline{/25.6°}$ mA; $6.1 \underline{/63.2°}$ mA

(20) $58.6 \underline{/-34°}$ V

(21) $2.5 \underline{/-96.7°}$ A; $1.18 \underline{/-29°}$ A

CHAPTER 3

(1) 1006 Hz; 10Ω; 63.3 V

(2) 8.84 kHz; 77.6

(3) 25 kΩ; 17.7 kΩ

(4) (a) 398 Hz; (b) 1 mA; (c) $\simeq 2$ V. (d) 0.45 mA

(5) (a) 159 Hz; (b) 100 Ω; (c) $141 \underline{/-45°}$ Ω; (d) 110 Ω

(7) (a) 14.6 A; (b) 43.4 µF; (c) 1390 Ω

(8) (a) 1.59 MHz; (b) 1 kΩ; (c) 100 kΩ; (d) 15.9 kHz. 10 Ω

(11) 36 kΩ; 16 kΩ

(13) (a) $0.5-j0.866$ mS; (b) 0.0866 µF; (c) 2000 Ω

(14) 5.1 mS; -0.99 mS; $192 \underline{/11°}$ Ω. $C \simeq 0.05$ µF

(16) $C_1 = \dfrac{C_B}{C_A}(C_A + C_B); \quad L_1 = \dfrac{1}{L_A\left(1 + \dfrac{C_B}{C_A}\right)}$

(17) $R = 298\ \Omega,\ L = 7.3\ \text{mH}$

(19) (a) $10.2\ \underline{/-78.6°}\ \text{mA};$ (b) 20 mW; (c) $R = 192\ \Omega;\ X = 962\ \Omega$

(20) (a) 27 mW; (b) $17.15\ \underline{/59°}\ \text{k}\Omega;$ (c) $-29.4\ \text{k}\Omega$

CHAPTER 4

(1) 318 µH

(2) 78.5 mV

(3) 0.122

(5) 14.2 µH

(6) (a) 0.025; (b) 12.6 µH; (c) 39.7; (d) 248 pF

(7) $130\ \underline{/-80°}\ \Omega$

(9) $2\sqrt{(2)}\ \underline{/-90°}\ \text{mA}$

(10) (a) $626 + j2.4\ \Omega;$ (b) 42 mW; (c) 40 mW

CHAPTER 5

(1) $R_1 = 2667\ \Omega,\ R_2 = 833.3\ \Omega$

(2) (a) 312 Ω, 422 Ω; (b) 419 Ω, 220 Ω; (c) 536 Ω, 68 Ω; (d) 589 Ω, 12 Ω

(3) $R_0 = 1333.3\ \Omega;$ 6 dB

(4) $R_1 = 39\ \Omega,\ R_2 = 56\ \Omega.$ Coaxial cable is unbalanced

(5) For each section the series value is given first: (i) 35 Ω, 5200 Ω; (ii) 69 Ω, 2580 Ω; (iii) 136 Ω, 1260 Ω; (iv) 258 Ω, 733 Ω; (v) 436 Ω, 196 Ω

(6) (a) 448 Ω, 1806 Ω; (b) 1120 Ω, 1000 Ω; (c) 2970 Ω, 733 Ω; (d) 5325 Ω, 672 Ω

(7) 110 Ω; 0.5 mW; 9.35 dB

(8) $R_0 = 106\ \Omega,$ (a) $E/160$ A; (b) $E/70$ A; (c) $E/106$ A, where E is the battery e.m.f.; -6.9 dB

(9) 100 mA

(11) $R_1 = 520\ \Omega,\ R_2 = 173\ \Omega;$ -11.4 dB; 9.5 dB

(12) $R_0 = 160\ \Omega;$ 0.33 V and 160 Ω; no

CHAPTER 6

(1) $L = 63.7\ \text{mH},\ C = 0.177\ \mu\text{F}$

(2) $L = 127\ \mu\text{H},\ C = 0.71\ \mu\text{F}$

(3) $f_c = 14.23\ \text{kHZ}$

(4) T-section: $L = 13.6\ \text{mH},\ C = 0.076\ \mu\text{F};$ π-section: $L = 27.3\ \text{mH},\ C = 0.038\ \mu\text{F}$

(5) 8.22 kHz, low-pass

(6) 7.1 µs

(7) $L = 212\ \mu\text{H},\ C = 53\ \text{pF each secton. 94 sections}$

(8) $L = 66.7\ \mu\text{H},\ C = 103\ \text{pF}$

(9) 3.19 kHz

(11) 1085 Ω

(13) Low-pass: $L = 31.8\ \text{mH},\ C = 0.088\ \mu\text{F};$ high-pass: $L = 13.6\ \text{mH},\ C = 0.038\ \mu\text{F}.$ Band-pass filter

(14) See *Figure A.1*

(15) To keep the reactance $Z_1\left[\dfrac{1-m^2}{4m}\right]$ of the same type as Z_1 at all frequencies $(1-m^2)$ must be positive, hence $0 < m < 1$

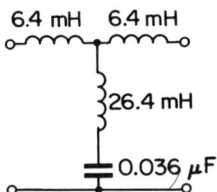

6.4 mH   6.4 mH

26.4 mH

0.036 µF

Figure A.1

CHAPTER 7
(1) 0.45 V
(2) 565 Ω
(3) 37.4 km, 59 500 km/s
(4) 126.5 Ω, $j0.047$, 134 km, 106 380 km/s
(5) 476 $\underline{/-30}$ Ω; 0.036 $\underline{/-3.4°}$ or 0.036 $\underline{/-195°}$
(6) $Z_0 = 290$ Ω, $\alpha = 0.18$ néper/km, $\beta = 0.21$ rad/km
(9) $R = 36.5$ Ω; $L = 52$ mH, $C = 0.069$ μF, $G = 12$ μS, all per km run
(10) 0.5 mA
(12) $\gamma = 1.06 + j233$
(13) (a) 3 m; (b) 10 mW; (c) $0.014 \sin (2\pi \cdot 10^8 t - 8/3 \pi)$ A; (d) 8/3 π; (e) 600 mJ
(14) (a) 80 Ω; (b) 4 m; (c) 12.5 mW; (d) $2 \cos (\pi \cdot 10^8 t - 3/2 \pi)$ V
(15) 0.42 $\underline{/-4.4°}$ mA

CHAPTER 8
(1) 600 mH
(3) $Z_0 = 1125$ Ω; $\alpha = 0.47$ néper/km, $\beta = 0.7$ rad/km
(4) 0.93 dB/km, 4.5 kHz
(5) $-j0.05$
(7) 2.75:1
(8) 0.277:1; 1.77:1
(10) (a) 12.1mm; (b) 97 mm. 190 Ω, 348 Ω
(11) 70 Ω
(12) (a) 1.5 V; (b) 0.5 V
(13) 14.6 mW
(14) 1.44:1
(15) 2.06

CHAPTER 9
(2) 220 Ω
(3) (a) $jZ_0 \tan \beta l$; (b) $-jZ_0 \cot \beta l$; (c) Z_0

CHAPTER 10
(1) 100 Hz; 250 V; 177 V. $i = \hat{I} \sin (628t + \pi)$
(2) $30 \sin 500 t + 15 \sin 1000 t$ V; 5.6 W
(3) $v_R = 5.66 \sin (2\pi \cdot 10^4 t) + 2.83 \sin (6\pi \cdot 10^4 t)$
$v_L = 57 \sin (2\pi \cdot 10^4 t + \pi/2) + 84.5 \sin (6\pi \cdot 10^4 t + \pi/2)$
(4) 500 Hz, 9.97 V, 14.1 V. (a) $v_R = 2.82 \sin 1000 \pi t$ V; (b) $V_L = 0.0045 \sin (1000\pi t - \pi/2)$ V; (c) $V_c = 6.64 \sin (1000\pi t + \pi/2)$ V.
(5) 71 V; 28.5 V
(6) $v = 28.2 \sin 200 \pi t + 14.1 \cos 400 \pi t$ V. Power = 5 W
(7) 6.83 W; 0.15
(9) 11.25 W; 3.3%

CHAPTER 11
(1) 1.1 $\mu C/m^2$
(2) 56 pF
(3) 134 Ω
(4) 1506 pF
(5) Inner radius 1 cm, outer radius 2.72 cm; 19.5 nF, 61 mA

(6) 80 kHz
(7) (a) 0.16 μF; (b) 1 mA; (c) 0.1 μW; (d) 0.1 Ω
(8) (a) 10^{-4}; (b) 10^{-4} rad; (c) 10^4; (d) 10 MΩ
(9) (a) 0.24 μF, 3.5×10^{-4}; (b) 1.0 μF, 4×10^{-4}
(10) 25×10^{-5}; 12.5 μW

CHAPTER 12   (2) 0.5 H
(3) The opposing fields overlap and become self-cancelling
(4) The spring will contract and the weight will rise. (You need a very light spring and a fairly heavy current to demonstrate this effect convincingly)
(5) 1.5×10^8 V/m
(8) (a) 2.5 μJ; (b) 14 μJ; (c) 0.8 μJ
(9) 542 μJ
(10) 24 W
(11) 1.92 W; 1.6 W
(12) 31.8 N

Checkbooks

Checkbook General Editors

for **Mathematics, Sciences, Electrical, Electronic, Telecommunications, Mechanical, Production, Marine and Motor Vehicle Engineering**

John O Bird and Anthony J C May of Highbury College of Technology, Portsmouth.

for **Building Construction, Civil Engineering, Surveying and Architecture**

Colin R Bassett, lately of Guildford County College of Technology.

Mathematics

•Mathematics 1 Checkbook

J O Bird and A J C May

1981   168 pages approx   186 × 123 mm
0 408 00609 9   Limp   Illustrated
0 408 00632 3   Cased

•Mathematics 2 Checkbook

J O Bird and A J C May

1981   272 pages approx   186 × 123 mm
0 408 00610 2   Limp   Illustrated
0 408 00633 1   Cased

•Mathematics 3 Checkbook

J O Bird and A J C May

1981   196 pages approx   186 × 123 mm
0 408 00611 0   Limp   Illustrated
0 408 00634 X   Cased

•Mathematics 4 Checkbook

J O Bird and A J C May

1981   240 pages approx   186 × 123 mm
0 408 00612 9   Limp   Illustrated
0 408 00660 9   Cased

•Engineering Mathematics and Science 3 Checkbook

J O Bird, A J C May and D Ayling

1982   128 pages approx   186 × 123 mm
0 408 00625 0   Limp   Illustrated

Sciences

•Chemistry 2 Checkbook

P J Chivers

1981   144 pages approx   186 × 123 mm
0 408 00622 6   Limp   Illustrated
0 408 00637 4   Cased

•Chemistry 3 Checkbook

P J Chivers

1982   144 pages approx   186 × 123 mm
0 408 00658 7   Limp   Illustrated
0 408 00662 5   Cased

•Building Science and Materials 2 Checkbook
and
Environmental Science 3/4 Checkbook
(see under Building Construction, etc. below)

- **Electrical Science 3 Checkbook**
  (see under Electrical, Electronics, etc. below)

- **Engineering Mathematics and Science 3 Checkbook**
  (see under Mathematics above)

- **Mechanical Science 3 Checkbook**
  (see under Mechanical, Production, etc. below)

Building Construction, Civil Engineering, Surveying and Architecture

---

- **Construction Drawing 1 Checkbook**

  J Greening and A Bowers

  1982   144 pages approx   186 × 123 mm
  0 408 00646 3   Limp   Illustrated

- **Construction Technology 1 Checkbook**

  R Chudley

  1981   160 pages approx   186 × 123 mm
  0 408 00602 1   Limp   Illustrated
  0 408 00642 0   Cased

- **Construction Technology 2 Checkbook**

  R Chudley

  1981   160 pages approx   186 × 123 mm
  0 408 00603 X   Limp   Illustrated
  0 408 00664 1   Cased

- **Building Science and Materials 2 Checkbook**

  M D W Pritchard

  1981 116 pages approx   186 × 123 mm
  0 408 00607 2   Limp   Illustrated
  0 408 00640 4   Cased

- **Environmental Science 3/4 Checkbook**

  M D W Pritchard

  1982   144 pages approx   186 × 123 mm
  0 408 00608 0   Limp   Illustrated
  0 408 00663 3   Cased

- **Geotechnics 4 Checkbook**

  R Whitlow

  1982   144 pages approx   186 × 123 mm
  0 408 00631 5   Limp   Illustrated

- **Building Law 4 Checkbook**

  A Galbraith

  1982   112 pages approx   186 × 123 mm
  0 408 00588 1   Limp   Illustrated

- **Economics for the Construction Industry 4 Checkbook**

  V J Seddon and G B Atkinson

  1982   144 pages approx   186 × 123 mm
  0 408 00655 2   Limp   Illustrated

- **Building Services and Equipment 4 Checkbook**

  F Hall

  1981   144 pages approx   186 × 123 mm
  0 408 00613 7   Limp   Illustrated
  0 408 00641 2   Cased

- **Building Services and Equipment 5 Checkbook**

  F Hall

  1981   144 pages approx   186 × 123 mm
  0 408 00614 5   Limp   Illustrated
  0 408 00651 X   Cased

Electrical, Electronic and Telecommunications Engineering

- **Microelectronic Systems 1 Checkbook**

  R Vears

  1981   84 pages approx   186 × 123 mm
  0 408 00552 1   Limp   Illustrated
  0 408 00638 2   Cased

- **Microelectronic Systems 2 Checkbook**

  R Vears

  1982   160 pages approx   186 × 123 mm
  0 408 00659 5   Limp   Illustrated

•**Electrical and Electronic Principles 2 Checkbook**

J O Bird and A J C May

1981   168 pages approx   186 × 123 mm
0 408 00600 5   Limp   Illustrated
0 408 00635 8   Cased

•**Electrical and Electronic Applications 2 Checkbook**

D W Tyler

1982   144 pages approx   186 × 123 mm
0 408 00616 1   Limp   Illustrated
0 408 00661 7   Cased

•**Electrical Principles 3 Checkbook**

J O Bird and A J C May

1981   160 pages approx   186 × 123 mm
0 408 00601 3   Limp   Illustrated
0 408 00636 6   Cased

•**Electrical Science 3 Checkbook**

J O Bird, A J C May and J R Penketh

1982   128 pages approx   186 × 123 mm
0 408 00626 9   Limp   Illustrated
0 408 00657 9   Cased

•**Electronics 2 Checkbook**

S A Knight

1981   144 pages approx   186 × 123 mm
0 408 00615 3   Limp   Illustrated
0 408 00639 0   Cased

•**Electronics 3 Checkbook**

S A Knight

1982   144 pages approx   186 × 123 mm
0 408 00623 4   Limp   Illustrated

Mechanical, Production, Marine and Motor Vehicle Engineering

•**Engineering Drawing 1 Checkbook**

L N Jeary

1982   160 pages approx   186 × 123 mm
0 408 00647 1   Limp   Illustrated

•**Workshop Processes and Materials 1 Checkbook**

R L Timings and T E Savage

1982   144 pages approx   186 × 123 mm
0 408 00621 8   Limp   Illustrated

•**Mechanical Science 3 Checkbook**

D Ayling

1982   144 pages approx   186 × 123 mm
0 408 00649 8   Limp   Illustrated
0 408 00665 X   Cased

Waterfield's School

A preparatory school in its Victorian heyday

SIMON WRIGHT

To Colin

with grateful thanks for your support
in the publication of this book and
with very best wishes
from
Simon Wright

THE HERONS GHYLL PRESS

The catalogue record for this book is available from the
British Library

ISBN 0 – 9523276 – 0 – 0

All profits from the sale of this book
are being donated to the Bursary Fund of
Temple Grove with St Nicholas School

Jacket design by Rob Grainger

Typeset in ITC Garamond by
Brenda Alexander, Wilmington, Sussex

Printed and bound by Biddles Ltd.
Walnut Tree House, Woodbridge Park, Guildford.

First published in 1994 by Herons Ghyll Press

Acknowledgments

The heading is conventional, the inference – an obligation to be inserted in any convenient gap – unavoidable. And it is not always easy to make the distinction between genuine gratitude and perfunctory acknowledgement apparent to those who are to be thanked, or indeed to the reader. But anyone who has written for publication will know that he or she owes an immense debt to people who have helped, advised or encouraged at the various stages.

David Crowley, Pam Allen and Gilbert Wheat all read the first draft of the book and made comments that I found extremely helpful. David Brewer and Raymond Gill put books at my disposal which I could not have had on extended loan from a library. Raymond Gill, furthermore, has guided me along the highways and byways of East Sheen history, of which he has unrivalled knowledge. Mike and Marie Smith welcomed me on my first visit to East Sheen and have answered numerous enquiries since then. The staff of the public library in Uckfield have been both patient and helpful in tracking down obscure volumes for me. Mary Ellingworth and her brother, John Waterfield, have given invaluable assistance in unravelling for me the tangled skein of the Waterfield clan. Margaret Wilson brought to my attention two very early school bills among the Knatchbull papers in the Kent Record Office. Sarah Lough (a Bursary Fund Trustee) has spared precious time to take care of the financial side of the whole enterprise. To her name I must add those of the anonymous sponsors whose interest-free loans have made the undertaking possible.

I have benefited more than I can say from the technical advice, constant encouragement and material help I have received from Peter Gillies of the Windmill Press, Hadlow Down. He did more than anyone to convince me that private printing was the best way to produce the book at an affordable price, a price moreover that still yielded a worthwhile dividend to the Bursary Fund. He has also tried, not always successfully, to call a halt to the constant 'tinkering' to which the manuscript has been subjected. I also want to thank, most warmly, Brenda Alexander who, as typesetter, has tackled an unwieldy manuscript with great care, good humour and admirable efficiency. Her professionalism has made the task of the printers considerably easier. To Julia Sentance and her colleagues at Biddles I owe a real debt. The care and trouble they have taken over a very small print run have been most impressive.

I approached the task of writing to publishers and literary agents with some trepidation. Knowing how strict were the copyright laws, and yet conscious of 'grey areas', I wondered how my letters would be received, and what sort of fees would be charged for the passages I was asking permission to quote. Almost without exception the replies I got were prompt, friendly and helpful and it is with gratitude that I acknowledge the kindness of those publishers, agents and private individuals who have allowed me to include extracts from their books without charge: Constable and Co: *Mrs Gatty and Mrs Ewing* by Christabel Maxwell; Faber and Faber: *When I was a boy* by Sir Ian Hamilton; David Higham

Associates Ltd: *The Scarlet Tree* by Osbert Sitwell; Hodder and Stoughton *A History of Education in England from 1760* by H.C. Barnard and *The Public School Phenomenon* by Jonathan Gathorne-Hardy; John Johnson Ltd: *The Happiest Days* by G.F. Lamb (originally published by Michael Joseph); Charles Knight: *Very Superior Men* by Alicia Percival; Lawrence and Wishart: *Studies in the History of Education* by Brian Simon; Macmillan London: *My Early Life* by Winston Churchill, and Curzon, *A Most Superior Person* by Kenneth Rose; Manchester University Press: *Schools for the Shires* by David Allsobrook and *Public Schools and Private Education* by Colin Shrosbree; John Murray: *Godliness and Good Learning* by David Newsome; Thomas Nelson and Sons: *The Rise of the Public Schools* by T.W. Bamford; Oxford University Press: *Thomas Arnold, Headmaster* by Michael McCrum; Penguin Books: *The Public Schools* by Brian Gardner (originally published by Hamish Hamilton); Peters, Fraser and Dunlop: *Boys Together* by J. Chandos (originally published by Hutchinson); Quiller Press: *Rowland Hill* by Colin Hey; Random House U.K. Ltd: *Changing Eton* by L.S.R. Byrne and E.L. Churchill and *Dreadnought* by Robert Massie (both originally published by Jonathan Cape), *Autobiography* (Volume 1) by Leonard Woolf (originally published by Hogarth Press), and *A Rough Record* by W.E. Goodenough, *The Life of E.F. Benson* by Brian Masters and *The Dukes* by Brian Masters (the last three originally published by Hutchinson); Routledge and Kegan Paul: *New Trends in Education in the Eighteenth Century* by N. Hans and *A History of Secondary Education in England, 1800-1870* by J. Roach; Taylor and Francis Ltd: *The Rise of the English Prep School* by Donald Leinster-Mackay (originally published by Falmer Press); A.P. Watt Ltd (on behalf of Althea Hannay and Susan Harper): *Pleasant Places* by George A. Birmingham (J.O. Hannay) and (on behalf of the Trustees of the Robert Graves Copyright Trust) *Goodbye to all that* by Robert Graves (originally published by Jonathan Cape).

The Royal Society for the encouragement of Arts, Manufactures and Commerce has kindly given permission for a summary of the Society's *Inquiry into the Existing State of Education in Richmond, Twickenham and Mortlake, 1870* to appear in Appendix 2. I am also extremely grateful to the following people for allowing me to use extracts from books by themselves or by members of their family: Michael Asquith for *Remember and be glad* by Lady Cynthia Asquith; Michael Freyer for *Bogey Baxter* by Dermot Freyer; Rupert Hart-Davis for his own *Hugh Walpole*; Lady Dorothy Heber Percy (Chairman of the Berners Trust) for *First Childhood* by Lord Berners; Charles Hollis (who most kindly sought the permission of all the co-beneficiaries of his grandfather's literary estate) for *Eton* by Christopher Hollis; James Sabben-Clare for his own *Winchester College*; Brian Simon for *The Victorian Public School* which he and Ian Bradley edited. Every effort has been made to trace the copyright owner of *Tom Brown's Universe* by J.R. de S. Honey (published by Millington Books) without success. Anyone claiming copyright should get in touch with the author.

Finally I wish to express my gratitude to Dr S.J. Gurman of the University of Leicester who has allowed me to read and make use of his expanded article on William Pearson, prior to publication.

Contents

List of Illustrations

The author wishes to express his gratitude to Colleges, Schools and other organisations, together with the individuals named, who have allowed him to make use of prints and photographs in their possession.

Dr. William Pearson, founder of Temple Grove

Introduction

The genesis of *Waterfield's School* lies in an exhibition organised, in 1988, by a newly-formed local history group in Uckfield. I was told that a number of schools still retained their Victorian log-books and I was asked to mount a section to display and explain them. Without actually revealing that I didn't know precisely what a log-book was I approached two or three schools and received permission to borrow and exhibit their log-books. When I came to look at them I began to realise that here in my hand I held fascinating documents of Victorian educational and social history. Concentrating on the most informative of them, the one kept by William Rollison of Uckfield Parochial School for 32 years, I started to trace the background, both nationally and locally, of such schools. The outcome, two years later, was a small volume entitled *Payment by Results*. By now my interest in Victorian education had been thoroughly aroused. I had retired from teaching and had become the unofficial curator of all the books and papers relating to the history of Temple Grove where I had taught for many years.

Ten years earlier, Meston Batchelor had published an excellent short history of the school, *Cradle of Empire*. But long before that, in fact as soon as he became headmaster, he had begun writing to the most senior surviving Old Boys of the school. Many had responded by letter but half-a-dozen of them had done better than that: they had compiled lengthy manuscript accounts of their experiences at Temple Grove in the days of Queen Victoria. Meston himself had assiduously tracked down most of the published accounts which contained references to the school, and the result was that when, in retirement, he came to write the history of Temple Grove he had a store of information from which to work. He received valuable help from Raymond Gill who probably knows more about East Sheen and its history than anyone else; he himself had been a boy at Temple Grove's second home in Eastbourne; on becoming headmaster he had moved the school from Eastbourne to Herons Ghyll; and finally he had remained on close terms with Harry Waterfield (Headmaster, 1902-35) and members of his family. He was therefore uniquely well equipped to write the history of the school.

Could there then be any justification for another person to attempt something similar only ten years later? There was, in the first place, the very considerable amount of material which Meston Batchelor had assembled, much of which he had been unable to use through lack of space. On top of that further information had become available, including the valuable discovery of O.C. Waterfield's *Name Book*; and finally *Cradle of Empire* had gone out of print and was unlikely to be republished.

Over and above these reasons, I felt there was room for a more detailed study of a Victorian prep school in its heyday. In writing *Payment by Results* I had attempted to answer such questions as: 'What did they actually teach in a National School and how did they teach it?'; 'How did they keep order and enforce attendance?'; 'How were teachers trained and what were they paid?'; 'How did the system known as Payment by Results operate?' After the book was published it occurred to me that a parallel study of a school catering for children (in this case only boys) of the same age, living in the same country and at the same period, but coming from homes of wealth and privilege, might be of value to people interested in educational and social changes. The questions I attempted to answer would, in many cases, be similar but they would range more widely since the boys lived at the school for three-quarters of the year. And the answers, in the main, would be supplied not by the headmaster's logbook but from the memories of the pupils themselves.

The histories of dozens of prep schools have been written in recent years. I have read, and enjoyed, a number of them. Nearly all of them give an account of the school's development from its foundation up till the day before yesterday, and they reveal a deep love of the institution they are describing. That is only natural, but it does make an objective view difficult to take. One cannot be expected to tell one's readers what a pig Mr So-and-so was when the man, or his widow, is still alive. One can merely hope that a reference to 'a warm heart within a somewhat crusty exterior' will evoke knowing smiles. There is, too, the danger that the writer's absorption in the affairs and achievements of, say, St Philibert's will tend to exclude any consideration of wider issues.

Before 1800 prep schools did not exist and up till the mid-century they were still comparative rarities. Thanks to the pioneering work of Donald Leinster-Mackay we now know much more about how and why they came into existence. What I have sought to do is to take the study of prep schools in the 19th century a stage further. To try to understand what the schools were attempting to achieve and how they set about actually doing it, one needs to have some knowledge of the development of public schools during the same period, and – even if more sketchily – an idea of the growth of state education. I have therefore started by giving an overall picture of education in England in the early part of the 19th century. From there I have gone on to discuss the 'great' schools (as the nine or ten oldest public schools were then known) and I have tried to give a more detailed account of Eton since it was there (in the 1840s) that Ottiwell Waterfield himself was educated and it was at Eton that his conception of the type of schooling suitable for sons of well-to-do families was formed. I have also attempted to show how the new breed of public schools came into existence from about 1840 onwards. At nearly all these schools the classics (Latin and Greek) formed the core of the curriculum. At some great store was set on academic excellence, others seem merely to have been content to keep to a wearisome routine of 'gerund-grinding' of little long-term benefit to their pupils. The great majority were boarding schools, run by headmasters who were usually clergymen of the Church of England; all enforced discipline by what today would be considered ferocious methods but paid scant attention to hygiene, adequate feeding or domestic comforts. Attitudes to the boys' 'free-time' varied considerably but in general there was much less individual freedom and much more emphasis on organised games in the later part of the century. Beneath the surface, too, much went on in schoolboy sub-culture – concerning prefectorial powers, bullying and immorality – of which the authorities were largely unaware.

It seemed to me important to discuss these matters in some detail since most prep schools emulated the practices and the ethos of the senior schools (indeed a prep school run on the lines of those today would have been a poor preparation for the harsh world which a boy entered at the age of 13). But it would be wrong to suggest that the schools themselves were impervious to change. Two Royal Commissions, known as the Clarendon and the Taunton, revealed – in the 1860s – widespread deficiencies, and indeed abuses, at schools of every type, and energetic measures were put in force to correct many of them. These reforms would scarcely have percolated through to prep school level until after Waterfield's day but their effects were beginning to be felt by his successors. What I have not attempted in this book is any considered assessment of the education provided by the state in the second half of the century. I have merely contented myself with adding (Appendix II) a summary of a report issued by the R.S.A. in 1870. It describes the education available to children from poor homes in Twickenham, Richmond and Mortlake (thus including East Sheen). The comparison with the schooling offered at Temple Grove is enlightening.

Every writer of the history of an individual school has to guard against the temptation to hint – by subtle or unsubtle means – that the one he is describing is, in some way, pre-eminent among schools of its type either in social standing or by means of academic excellence or sporting prowess or whatever. Temple Grove happens to be older, and probably better documented, than most prep schools, but its strengths and weaknesses, its virtues and failings, were, I feel, typical of scores of such schools.

Mention has had to be made of scholarships won, but I have avoided printing long lists of them, nor have I included tables of sporting and athletic triumphs. Such emphasis could all too easily become mere trumpet-blowing and would detract from the purpose of the book: to show a typical Victorian prep school in all aspects of its life.

To try further to avoid excessive parochialism I have attempted to put the school in its niche on the local, national and even international stages. I have even ventured to give an outline of world events for the year 1859 to try to remind readers that no institution exists in a vacuum, impervious to outside influences. The work of the Barnes and Mortlake Historical Society, and especially of Raymond Gill, has provided us with a clear picture of the development of East Sheen from the aristocratic locality of 1800 to the fast-growing outer suburb of a century later. Temple Grove, in spite of being a boarding school, is also an integral part of that process, and sections 2 and 3 of the same chapter describe the house, the estate and the neighbourhood.

The core of the book is contained in Chapter VI, in which, using as far as possible the published and unpublished evidence of former pupils, I have built up a detailed picture of the Headmaster and his staff, the curriculum and most aspects of the school at work and at play. I have also tried to consider the standards of behaviour – what was 'done' and what was 'not done' – in such a school. No account books have survived but there are enough figures, I believe, to hazard reasonable estimates as to running costs etc. Remember, please, that whereas prices largely remained stable for the second half of the 19th century, one now needs to multiply by about 40 to get a modern equivalent for prices and by anything up to 200 for salaries and wages, and property values.

Chapter VI continues the story up to the end of the century and includes some aspects of school life that lie just outside Waterfield's own period, though essentially the school continued to be run on the lines he had laid down. I fully intended to stop there but it occurred to me that readers who reached that point might well be asking – in so many words – 'What did all those boys you've talked about do in later life?' 'Can one trace any connection between their prep school grounding and the sort of men they turned out to be?' And so, using a variety of sources – and with valuable help from the archivists at a number of public schools – I have compiled in Appendix I brief biographies of almost every boy mentioned by name in Chapters V and VI who was a pupil at the school between 1859 and 1902 and, somewhat rashly, I have attempted to draw certain conclusions from that slender evidence.

Sources

In the first two chapters the main authorities consulted and quoted from have been:

A. Ainger *Eton Sixty Years Ago*
D.I. Allsobrook *Schools for the Shires*
T.W. Bamford *The Rise of the Public Schools*
H.C. Barnard *A History of Education in England from 1760*
L.S.R. Byrne and E.L. Churchill *Changing Eton*
J. Chandos *Boys Together*
A.D. Coleridge *Eton in the Forties*
Gavin de Beer *Charles Darwin*
F.W. Farrar *Eric, or Little by Little*
J. D'E. Firth *Winchester College*
B. Gardner *The Public Schools*
B.H. Garnons Williams *A History of Berkhamsted School*
J. Gathorne-Hardy *The Public School Phenomenon*
A. Gatty *A Life at one Living*
N. Hans *New Trends in Education in the Eighteenth Century*
C. Hey *Rowland Hill*
Christopher Hollis *Eton*
J.R. de S. Honey *Tom Brown's Universe*
Thomas Hughes *Tom Brown's Schooldays*
J. Hurt *Education in Evolution*
A.W. Kinglake *Eothen*
Charles Lamb *Essays of Elia*
G.F. Lamb *The Happiest Days*
John Leese *Personalities and Power in English Education*
Michael McCrum *Thomas Arnold, Headmaster*
E.C. Mack *Public Schools and British Opinion (2 vols.)*
G. Melly *School Experiences of a Fag at a Private and a Public School*
David Newsome *Godliness and Good Learning*
A.C. Percival *Very Superior Men*
J. Roach *A History of Secondary Education in England, 1800-1870*
James Sabben-Clare *Winchester College*
Colin Shrosbree *Public Schools and Private Education*
B. Simon *Studies in the History of Education*
B. Simon and I. Bradley (ed.) *The Victorian Public School*
F. St J. Thackeray *Memoir of Edward Craven Hawtrey*
D'Arcy Thompson *Daydreams of a Schoolmaster*

1 Education in England, 1800-1840

1 The General Situation

In the long and often inglorious history of English education the nadir was reached during the first forty years of the 19th century. Beacons of light there undoubtedly were, flashing from isolated outposts: the appointment of Thomas Arnold to Rugby, Samuel Butler's long headship at Shrewsbury, the lesser-known work of the Hill family at Hazelwood and the founding of two new Universities, London in 1828 and Durham in 1832. To this tally may be added the Mechanics' Institutes, initiated by George Birkbeck and also the Christmas lectures at the Royal Institution which Michael Faraday first gave in 1827. Admirable as was the work of these pioneers their light shed hardly more than a glimmer in the all-enveloping miasma of sloth and stagnation that surrounded English education. The situation appears all the more curious in view of the vitality that infused so many strands of life in the early 19th century. Beginning in the late 18th century and gathering pace in the early 19th were two movements which, in their different fashions, were to transform both the physical appearance of England and the ways of life and thought of the English people. The power of steam, and later of electricity, turned the country from an agglomeration of rural communities into the world's first industrial nation in the space of a single lifetime. At the same time the power of preaching and of expounding the Bible brought about a religious revival that was no less transforming. Why, then, did not the excitement aroused by new inventions and the zeal for spiritual growth combine to generate an educational revolution that would have put England's schools far ahead of those of any other country? Part of the answer lies, perhaps, in a third revolution to which England did not succumb but whose menace was apparent from 1789 onwards. Even after Waterloo the contagion of the French Revolution was perceived as a threat to English society and may account for the caution with which all attempts at reform, political, social and educational, were viewed in the first half of the 19th century. The rest of the answer may be deduced from the range and nature of English schools at which one must now take a closer look.

One may conveniently deal first with education provided by the State. Very simply the State made no provision for education at all, nor, until 1833, did it even allocate funds for the purpose. What the Whig government that came to power in 1830 did do was to institute an enquiry into the state of education in England and Wales. As a consequence of that enquiry, Lord Althorp, wisely awaiting the month when most Members of Parliament had withdrawn to their estates, introduced (August 1833) a proposal to vote the sum of £20,000 a year for the purposes of education. Even with the opposition of the redoubtable William Cobbett, who declared that all that such a measure would do would be to create a new race of idlers viz. schoolmasters and schoolmistresses, Althorp's motion was passed by 50 votes to 26. The money had been voted but carefully planned proposals for its use did not begin to appear until Dr James Kay (later Sir James Kay-Shuttleworth) was appointed secretary to the Select Committee of the Privy Council which was responsible for allocating the funds. He was convinced that the monitorial system, to which reference will shortly be made, was a disastrous failure and he wanted monitors replaced by a system of pupil teachers. For this to be possible there had to be an effective system of teacher training and Dr Kay concentrated his efforts in this direction, with the result that – in 1841 – two teacher training colleges were opened in London. Dr Kay further helped to raise standards by sending inspectors to schools which were receiving state grants and ensuring that such grants went only to schools where voluntary contributions were forthcoming. And, until the Newcastle Commission of 1858, that was as far as State support for education went. England, by the mid 19th century, was still a long way from providing free education for all children or from ensuring that all parents sent their children to school.

What sort of schooling, then, was available for children from poor homes around the year 1800? Effectively there were four types of school where the education was either free or very cheap. They were dame schools, Sunday schools, charity schools or those supported by ancient endowments. An illustration and brief description of a dame school appears in nearly every account of rural life. Some dames conscientiously attempted to instil the rudiments of reading and writing into their charges, even though their methods might not find favour, or customers, today. William Shenstone describes: 'A matron old, whom we Schoolmistress name,
              Who boasts unruly brats with birch to tame'.
More often, one supposes, the dame might be:
              'a deaf poor widow sits
              and awes some thirty infants as she knits' (George Crabbe).
Even so a great many working mothers must have been only too grateful to find someone who would look after their children for a penny or two a week. Older children sometimes received instruction in what were known as common day schools, usually from a man unfitted for more active employment.

The charity school movement flourished principally in the 18th century. The impetus behind it lay with the Society for Promoting Christian Knowledge, founded in 1699. Typically an Anglican clergyman left property in his will specifying that the rents were to be employed by the charity's trustees to provide a building, some basic equipment and the services of a schoolmaster or schoolmistress and that the school should offer free education to certain poor children of the parish. The master or mistress would

'A matron old . . . who boasts unruly brats with birch to tame'

'The lesson of the Sunday-school teacher is empty unless he or she is touched with the spirit of Christian love and earnestness'

be required to be a practising Anglican (though there were some dissenting charity schools as well) and the parson would be expected to provide religious instruction, especially the catechism. In this way a limited number of children did receive at least a grounding, although the formidable Sarah Trimmer had trenchant criticisms to make, complaining that the Bible and Prayer Book were taught by rote, so that the children seldom understood what they were learning.

The splendidly titled 'Society for the Bettering the Condition and Increasing the Comfort of the Poor', founded in 1796, sought to promote schools of industry where children from impoverished homes could be taught useful crafts. The idea was praiseworthy but, in practice, most such children were already in paid employment, working long hours in mills, in factories, on farms or in their own homes, with no leisure or energy for school work.

It was in order to try to combat the almost total ignorance of Christianity displayed by children of the very poor that the Sunday School movement came into existence. Robert Raikes, the owner of a Gloucester newspaper, may not have thought of the idea himself but he did more than anyone to popularise it, 'using his resources to open schools for the undisciplined and illiterate children who were employed in local factories all week and let loose on Sundays' (Barnard). And so yet another society came into existence, 'The Society for the Establishment and Support of Sunday Schools'. The chief aim of the schools was to enable children to read so that they could follow the text while the teacher was explaining a Bible story. By the year 1800 about half a million children were receiving this very basic form of schooling, and J.R. Green, in his *Short History of English People*, was probably right to claim that, 'the Sunday Schools established by Mr Raikes were the beginning of popular education'.

The next development came early in the 19th century when two very different men intoduced remarkably similar methods for instructing large numbers of children at minimal cost. Andrew Bell had been responsible for the Madras Male Orphan Asylum while working in Southern India as a chaplain. Employing a boy of eight to teach even younger children he gradually evolved the Madras System. Not long afterwards Joseph Lancaster, a Quaker, opened a school for poor children in Southwark, at first teaching all the pupils himself. When numbers rose from a hundred to five hundred, Lancaster, unable to afford to employ an assistant, began also to use older children to instruct the younger ones. From this genesis arose the two voluntary societies which achieved extraordinarily rapid growth. The British and Foreign School Society, based on Lancaster's system, was un-denominational and made its chief appeal to nonconformists, especially in urban areas. The Church of England responded to this challenge by launching the National Society for Promoting the Education of the Poor in the Principles of the Established Church. In both types of school the Master instructed the older pupils and they in turn, as best they could, passed on the instruction to the younger children. This was the essence of the monitorial system, elements of which persisted throughout the 19th century. The schools had the very great merit, as it then seemed, of being financed by private charity, by public subscription and by the tiny fees paid by the parents. No burden whatever fell on the tax-payer.

Both societies put up school buildings or adapted existing ones, the National Society enjoying a distinct advantage in small towns and villages where the local parson often had a controlling voice. As we have seen, it was not until 1833 that any government offered financial support. But, for all the shortcomings of the methods they employed, the British and National Societies were making determined efforts to reduce the level of illiteracy in England.

Just as these two societies were beginning to get into their stride, Henry Brougham, friend of Sydney Smith and associate of Jeremy Bentham and John Stuart Mill, and himself Member of Parliament for Winchelsea, succeeded in having appointed a select committee that instituted an inquiry into charity abuses, which Brougham was able to extend to the universities and to the major public schools. 'Some scandalous revelations were made and the governing bodies bitterly resented the inquisition' but when, in 1820, he introduced two bills, the one designed to curb abuses excluded Brougham's main targets (owing to political pressure) and the one intended to provide for the building and maintenance of parochial schools was defeated owing to nonconformist opposition. Over thirty years passed before the establishment of a Charity Commission which had the power to remedy abuses, but in the meantime the Parliamentary Committee had uncovered flagrant abuses in the administration of many schools: endowments misapplied, schoolmasters in receipt of salaries for which they performed no duties, and much of the teaching that was carried on of a very low standard.

It was in the grammar schools that the most numerous examples of corruption and sloth were to be found and it is hard to disagree with the Lord Chief Justice in his description of them as 'empty walls without scholars and everything neglected except the receipt of salaries and emoluments'. Here the judgment of Lord Eldon (Lord Chancellor) in 1805 needs to be borne in mind. He apparently laid down that it was illegal for a school, endowed by its founder to teach the classical languages, to do anything else. However, in the particular case at issue (concerning Leeds Grammar School) the case was never finally decided. Later commentators have claimed that Eldon was merely 'defining the primary purpose of the foundation, not denying that this purpose might be modified' (Roach). When the grammar school at Kirkleatham (N. Riding) was examined in 1822 it was revealed that for many years no boys had been educated there, the squire apparently objecting to the proximity of the school to the manor house. Yet the master and usher had been regularly paid their salaries from the school's endowment. Leicester Grammar School had once had 300 pupils. By 1820 the total had sunk to one boarder and four day boys. At Coventry, when the inquiry took place in 1833, the master was 82 and had held office for 54 years. He and the usher, appointed in

1794, had quarrelled and 'when the commissioners came there was only one boy left' (Roach). One of the best documented accounts is that of Berkhamsted, an ancient and well-endowed grammar school. It was, to all intents and purposes, 'purchased' by the Rev. John Dupré in 1788 and passed on by him to his son, Thomas, in 1805. 'Thomas Dupré's career was a scandal from start to finish; his record has no redeeming feature other than his devotion to the material interests of his enormous family'. As early as 1811 complaints were being made about his neglect of duty but it was not until 1841 that he finally resigned with a pension of £250 p.a. for life. During those thirty years 'Thomas Dupré had enjoyed the use of a comfortable residence . . . ; he had received upwards of £7,500; a succession of Ushers – his uncle, his closest old school friend and two of his sons in turn – had [further] received half that sum; and he himself was destined . . . to receive [in retirement] another £5,000. To earn all this . . . he had done literally nothing at all' (Garnons Williams).

These are some of the classic cases but at other schools, too, the children of local parents, for whom the founder had intended a free education, were neglected and the master, his position fortified by freehold tenure and guaranteed by the endowments, increased his income by taking in fee-paying boarders.

One's immediate questions, 'Didn't local people object?', 'Why didn't the trustees dismiss an idle or corrupt master?' etc. can best be answered by asking how easy it is in the late 20th century for parishioners to get rid of a parson they all dislike if he doesn't want to go. The overall problem facing scores of small grammar schools up and down the land was that their incomes had failed to keep pace with rising costs, they couldn't afford to pay the sort of salary that would attract a good teacher, nor could they bear the cost of broadening their curricula. 'The only way to make changes lay through legal action, either in the Court of Chancery which was expensive and slow [as readers of *Bleak House* will recall] or by private Act of Parliament, a measure contemplated by only a few schools' (Roach). When the wardens of the grammar school at Guisborough (N. Riding) in 1821 were anxious to get rid of an absentee master, 'they were advised that the master could not be charged with neglecting his scholars because he did not have any' (Roach).

The original purpose of a grammar school had been to provide free education for sons of poor parents to enable them to proceed to Oxford or Cambridge. But the term 'free' had become increasingly narrowly defined. At Manchester, an example of the best type of grammar school in the early 19th century, 'all the boys received a classical education free, but they were charged for writing, arithmetic and mathematics' (Roach). By then, of course, the demand for a classical education for the sons of tradesmen, manufacturers and engineers was greatly reduced. They wanted the kind of schooling that fitted their boys for the world in which they would earn their livings and at the same time gave them a little of the polish demanded by polite society. But the men appointed as masters of grammar schools had themselves received a purely classical education and were not equipped to teach anything else. To provide teachers of mathematics, science, drawing and dancing cost money and without a sufficiency of fee-paying pupils the necessary revenue did not exist. Dr Roach rightly sums up the problems when he explains that, 'if all these limitations – about subjects taught, numbers admitted, entrance fees and other payments – are taken together, the amount of free education provided by the early 19th century grammar school was severely limited'.

There *were* grammar schools, even then, that met local demand. Stafford is a good example, with fifteen classical and fifteen English scholars taught by the master, and a further seventy boys given more basic instruction by an usher, the school being free to inhabitants of the borough. But 'at the other end of the scale, particularly in rural districts, many so-called grammar schools were really elementary schools, and indeed it is difficult to believe that many of them had ever filled any other role' (Roach).

To conclude this survey of educational provision for the children of the poor, Alfred Gatty's account of the schooling available in the parish of Ecclesfield (near Sheffield) in the year 1839 may be taken as illustrating its variety – and its limitations.

'I will now relate the means for educating the poor which existed in the parish when I came to it. It was a subject of much importance and interest for there were five schools with endowments, and their condition was very bad. Before describing this, I would recall that public opinion had not yet decided that universal education was for the good of the country, so there were no trained or competent teachers. The post of schoolmaster was habitually given, by favour, to someone who was fit for nothing else; and he held it as a freehold, and voted for the county; and this easy, slipshod way of dispensing patronage had produced the following results.

'The Feoffees' School in the village had no scholars when I came to Ecclesfield, as the master was insane; but he drew an income from the charity, and locked himself in the house.

'The Infant School was kept by an old woman, who signed a cross when I paid her quarterly salary, because she could not write her name.

'Lound School was almost a ruin, in which a few scholars were taught by a man who was a cripple.

'Parson Cross School was under a man who had lost his right hand in a brawl and was placed there for a maintenance.

'Shire Green School was a complete ruin without a roof; but in the kitchen of the master's house sat a patriarch, eighty years old, who was making Aeolian harps, and apparently teaching two little children'.

A factor of increasing importance in the 19th century was the growing stratification of society. There had always, of course, been rich and poor but 'so much new wealth was being made and so many people were moving up socially because they were richer that society in general was more, not less, concerned about class distinctions. In the old stable rural society the natural leaders had a position so secure that they did not need to concern themselves about it. Their position was not threatened simply because (their children) sat on the benches of the local grammar school with boys from much humbler homes' (Roach).

The solution, for parents able and willing to pay fees, but anxious to avoid what they saw as the moral contagion of the public schools or the social contagion of the grammar schools was to send their sons to a private school. Some of these had come into existence as far back as the mid-17th century but it was in the 18th that they proliferated. 'Private schools were usually established in opposition to the old grammar schools and derived their tradition from other sources. Even the classical private schools tried to break away from the rigidity of old foundations and introduced new subjects and new methods' (Hans). Those that put mathematics and technical subjects in the forefront of their curricula tended to style themselves Academies and Hackney Academy, with an uninterrupted existence of over 120 years and a long string of distinguished alumni, provides an excellent example of the genus.

Probably over a thousand classical schools were founded in the 18th century, with at least two hundred in existence at any given time. Information about them is necessarily scanty and it is only from the records of the longer lasting ones that we can form an impression of their methods and their results. The owner of such a school was normally also the headmaster. He might be a professional teacher who had taken holy orders but who did not undertake parish work, or he might be a resident vicar who took boarding pupils into his own home to supplement a limited income. In either event they were small in scale and their lifetime ended when the proprietor retired or died or moved away. The atmosphere in the better ones was quite different from that of a larger establishment, the Master acting as 'more of a paterfamilias than the distant demi-god of a large public institution. Many aristocratic families, country squires and members of the clergy preferred private schools to public schools on these grounds' (Hans). The best known of them all, clearly, was Cheam. It was founded by George Aldrich in 1647 and, for a period of over fifty years (1752-1806), was run by the Rev. William Gilpin and his son of the same name, two outstanding schoolmasters whose aims and methods were far in advance of educational thinking of the day. So humane was their regime that when Henry Addington (later Prime Minister) went on to Winchester he was shocked at the bullying and wickedness he experienced there and he absconded. And when, on one occasion, Gilpin took three boys who had already spent time at a public school they soon began to corrupt the other pupils and he had to get rid of them.

An important point to make about classical schools is that they prepared boys for Oxford and Cambridge. Some parents chose to send their sons on to public schools, some boys went straight out into the world but their prime purpose was preparation for University, hence the importance of Latin and Greek. Since the syllabus the school followed depended entirely on the predilections of the owner, the classics might well be diluted with other –

more agreeable – subjects. Examples can be found of mapping and architectural planning, of land surveying and astronomy, of navigation and book-keeping, as well as the more obvious branches of science and mathematics. 'The number of ordinary private schools which taught science does show that these schools were a long way ahead of the public schools where science does not appear until the 1860s' (Roach).

One must not, clearly, paint too rosy a picutre of the private school world. It was riddled with snobbery 'since, when a free choice was available, no parent wished his children to consort with those who were socially beneath them, nor could he expect . . . to be admitted to the company of those above him' (Roach). The curricula followed, in spite of some praiseworthy innovations, was distinctly limited; and, however inspiring the headmaster might be, 'teachers in private schools were poorly paid, poorly qualified and lacking in social status' (Roach).

Nor is one ever likely to forget those schools which catered for parents who wished to see as little of their children as possible, or for guardians anxious to be rid of an unwanted encumbrance, the sort of school to which 'boys are sent out of the way to be boarded and birched at £20 p.a.' . The pages of *Nicholas Nickleby* and of *Jane Eyre*, once read, remain in one's imagination for ever, and Mr Squeers' prospectus, containing the ominous words, 'No vacations', made plain the purpose of such an institution. With no system of inspection and no regulations such schools were bound to exist. But at least the evidence from the better establishments 'goes a long way to destroy the claim that private schools were run only by dishonest charlatans, anxious only to make money from their helpless pupils' (Roach).

All too little has so far been said about the education of girls. Since, prior to the 19th century, the only schools they could attend were privately owned and – mostly – short-lived, records are necessarily scanty. The earliest may well have been the one established by Mrs Frankland in Manchester in 1680 for the daughters of dissenters. It survived for 35 years. During the 18th century such schools proliferated, some of them large seminaries with plenty of teachers, others small establishments where a handful of girls were taught in the proprietor's own home. An attractive picture of one such school is given in chapter 3 of *Emma*. Jane Austen herself had attended Abbey School at Reading in the 1780s and her description of Mrs Goddard's school at Highbury is closely modelled on it. A few years later Erasmus Darwin published *Lectures on Female Education* (1797), in which he advocated a curriculum to include English literature, French and Italian, History and Geography, Zoology and Botany, in addition to the 'polite' accomplishments, drawing, embroidery, dancing etc. He even considered it necessary for girls to be taught mathematics and applied sciences. Far-reaching these proposals may have been but they took little account of the prospects open to middle-class girls at that time. All too often, 'the teachers were untrained and . . . many of them had drifted into the work with little enthusiasm because there was nothing else they could do. Since middle-class girls looked forward to no career other than that of marriage there was none of the stimulus provided in boys' schools by the pressures of the job-market and the need to earn a living' (Roach). Much of the teaching was inevitably stereotyped and consisted of memorising the answers to questions on a wide range of unrelated topics. Richmal Mangnall (herself a headmistress) first published *Historical and Miscellaneous Questions for the use of young people* in 1798. It was reprinted again and again and became the great stand-by of the Victorian schoolroom.

However by 1864 the women's movement had made considerable progress and, when the Schools Inquiry Commission was appointed, Emily Davies (foundress of Girton College) and others urged that the girls should be included in the inquiry. 'The promise was made and fully kept and the Commission's papers give a valuable account of the position of girls' education at that time' (Roach). By then Queen's College (1848) in Harley Street and Bedford College (1849) in Bedford Square had been in existence for some years. Both had been founded as a result of the growing realisation that 'one reason for the poverty of many governesses was their ignorance and incapacity to teach' (Roach). Frances Buss (North London Collegiate School) and Dorothea Beale (Cheltenham Ladies' College) were early students of Queen's, together with Sophia Jex-Blake who later founded the London Medical School for Women.

Even then religious differences were a handicap to the development of girls' education. Queen's College was distinctly Anglican in tone and regarded Bedford College (founded by a Unitarian) as being tainted by heresy. Roman Catholics, meanwhile, were concerned to provide an education for their daughters that adhered closely to their own faith. Cardinal Wiseman, the first Archbishop of Westminster, in 1846 invited that remarkable woman, Cornelia Connelly, to found a teaching order in England. The Society of the Holy Child Jesus began in Derby, moved to St Leonards (Sussex) in 1848, and finally settled in Mayfield. 'The curriculum . . . was that of the superior girls' schools of the day but Mother Connelly . . . wanted her girls to have independence and self-government . . and ahead of most girls' schools her nuns introduced gymnastics and games' (Roach). As one of her early pupils wrote, 'there was a sense of freedom and broadmindedness about the school that was delightful'.

2   The Great Schools

The term 'Public School' is one that has caused considerable difficulty in addition to raising egalitarian hackles. Around the year 1800 the expression was hardly in use. One spoke of the Great Schools, meaning Winchester, Eton, Westminster and perhaps three or four more, and they were understood to mean those schools which educated boys up to 19 or 20 years of age in return for fees and which were largely boarding establishments, although it was recognised that a proportion of the pupils at each school would be accepted without charge or at reduced fees. Winchester and Eton had come into existence some four centuries earlier as the charitable foundations of William of Wykeham and King Henry VI and were intended principally for the education of poor scholars who would then proceed to New College, Oxford, and King's College, Cambridge, respectively, before becoming ordained. This was made possible by generous endowments. These consisted of estates, farms, villages or other property from which revenue in the form of rents could be guaranteed. At an early stage the admission of boys whose parents would pay fees was allowed for in the statutes and to these arrangements, as indeed to the syllabus, the intervening centuries had made little alteration. Even the Reformation, which sounded the death knell of most religious property-owning organisations, made little difference to Winchester and Eton. However Renaissance influences, at about the same time, did create a demand for the revival of classical learning, not as a training for priests but for secular purposes. Westminster owes its existence to Queen Elizabeth (probably the most learned of all our sovereigns) herself. The monastery, which housed a school, had been dissolved in 1539 and the great Abbey itself had survived only by fortunate circumstances. But the Queen, anxious to encourage a new generation of scholars, brought about the restoration of the school, in what had been monastic buildings, in 1560. For two centuries thereafter Westminster 'dominated in intellectual performance the great schools of England' (Chandos).

Elizabethan England offers numerous examples of men occupying high positions of state or making important contributions to scholarship and the arts who had come from relatively humble backgrounds. Such men would almost always have obtained a grounding at their local grammar school. The foundation of such a school, usually in his native town, was regarded as the crowning achievement of a successful merchant and his best hope of achieving immortality. Again there would be endowment, usually of land bequeathed in the founder's will, the revenue sufficient to pay for the upkeep of the building and, normally, of a single schoolmaster. Since the boys lived locally and probably brought some victuals with them, few other expenses were entailed. Again, however, provision was made for a certain number of fee-paying pupils and they would customarily board in the master's house. In this way scores of grammar schools appeared in England from around 1550 onwards, the term 'grammar' specifying their purpose: that of inculcating the grammar, syntax and vocabulary of the two classical languages, Latin and Greek.

The concept of boarding goes back further than the schools themselves. In noble households of the Middle Ages all the sons would be sent, from the age of seven onwards, to live in the households of other nobles and there to learn the deportment, military skills, musical attainments and other desirable attributes of a man who would never require paid employment but who would one day administer a great estate or serve his king in peace and war. The discipline and training necessary were more easily imposed if the boy were living among strangers. Long after the Middle Ages such an education was still customary for the sons of the wealthier families, though the provision of a tutor for the more academic accomplishments was also widely practised. But by the second half of the 18th century the custom, observed by the nobility and gentry, of sending their sons to one of the great schools had largely superseded the earlier practice. Eton, because of its royal foundation and proximity to Windsor, had already become the foremost school in England, socially if not academically. But it was certain of the old grammar schools, Harrow, Rugby and Shrewsbury, which were now becoming elevated to 'public school' status. The provision for free scholars shrank as the fee-paying aspect came to dominate a school's finances. This transformation came about because of the increasing need – and, indeed, desire – for younger sons, even those from wealthy families, to earn a living or pursue a career. The Services, the Law and the Church were the three most obvious openings and for each of them a modicum of education was necessary. Entry into Parliament was normally reserved for the eldest son unless a wealthy patron was prepared to help, since M.P.s were not paid and contested elections were often ruinously expensive. There was, as yet, no possibility of such a young man going into business. Indeed, if his father had made a fortune through mercantile enterprise, the son would be anxious to rise above such sordid commercial contagion.

Tom Brown's father may be taken as a not untypical early 19th century squire. Tom has spent eighteen months at a private school and is now to be sent away to an erstwhile grammar school that for some time has enjoyed public school status. The squire's aspirations for his son are well known but are worth quoting again: 'I don't care a straw for Greek particles or the digamma; no more does his mother . . . If he'll only turn out a brave, helpful, truth-telling Englishman, and a gentleman, and a Christian, that's all I want' (Hughes). As we shall see, the aims of Thomas Arnold were pitched more loftily than those of Squire Brown, but it seems likely that the average father was not too concerned about the academic content of his son's education.

By the time of Waterloo, then, the great schools had been joined by half-a-dozen others and when Henry Brougham was endeavouring to have all institutions of a charitable nature investigated the six leading public schools, Eton, Winchester, Westminster, Harrow, Rugby and Charterhouse, were specifically excluded. Dr Butler was aggrieved that Shrewsbury was not also included and, in the general view, the two leading London day schools, St Pauls and Merchant Taylor's, should be considered public schools. Christ's Hospital fitted in to no convenient category but the education it provided was similar to that of the other nine.

Before going on to examine one or two of these schools in some detail it would be as well to have a general idea of their aims and methods. The teaching was carried on in a large barn-like building over which the headmaster presided from his throne. He himself was normally responsible for all the senior boys (upwards of fifty) and ushers were seated along each side. The pupils occupied benches arranged all round the schoolroom, sometimes in tiers. There were no lessons as we understand the term in the sense of a class being instructed by a master and the pupils able to ask questions. Education consisted of endlessly learning by rote the grammar and syntax of the classical languages and long passages from Greek and Latin poets. In more or less regular rotation boys were summoned from their benches to stand before the master and recite perhaps twenty lines of Virgil, or the rules governing gerunds and gerundives. The principle enunciated by John Sleath may be taken as representative of most public schools: 'At St Paul's we teach nothing but the classics. If you want your son to learn anything else you must have him taught at home, and for this we give three half-holidays a week' (Gathorne-Hardy). At boarding schools it was sometimes possible to employ a private tutor if a pupil, or his parents, wished for instruction in non-classical subjects. Very little writing was required from the pupils. No desks were provided but boys learnt to balance a pad on their knees when necessary. When a boy was called on to 'construe', i.e. to translate into English, he stood up and worked his way through several lines of Greek or Latin until the next pupil was told to take over. Ushers, therefore, were seldom burdened with preparing lessons or marking written work. Some schools did expect their pupils to compose verses (according to strict classical rules) on a theme set by the master. The range, though wide, was not limitless and well-organised boys had access to a huge stock of ready-made verses to save them the trouble of fresh thought and composition each time.

Bearing this outline in mind we may now consider one or two schools, and the men who presided over them, in greater detail.

The figure of Thomas Arnold towers over the story of public school education in the first half of the 19th century. So much has been written about him, from the hero-worshipping of Arthur Stanley and Thomas Hughes to the debunking of Lytton Strachey, that it is difficult to adopt an impartial attitude and impossible, in a brief summary such as this, to put forward a challenging new view. The facts of his life can be briefly set out. Born in 1795 he entered Winchester in 1807, where, to judge from his surviving letters, he entered boisterously into the violent life of the college, describing 'dreadful sieges', sudden assaults and bloody encounters with considerable relish. From Winchester he went on to Oriel College, Oxford, where he became a member of a group known as noetics, deeply religious but not uncritically so. He took his degree in 1818 and was ordained the same year. He then settled at Laleham, married Mary Penrose (among their nine children was Matthew, the poet and critic) and set up a small school for private tuition, coaching pupils for university entrance. Nine years later, at the age of 33, he was appointed Headmaster of Rugby School, his selection in no small measure due to a testimonial from Dr Hawkins of Oriel who predicted that, if chosen, 'Mr Arnold would change the face of education all through the public schools of England'. Not long afterwards he was made Doctor of Divinity and seems to have been referred to as 'The Doctor' by staff and pupils alike. For two years he contrived to hold the Professorship of Modern History at Oxford (and was the author of a three-volume *History of Rome*) as well as the headmastership. But in 1842, still aged only 47, he died from a sudden and severe attack of angina. One cannot help wondering what he might have achieved, given another twenty years. He himself would not have thought of the headmastership of Rugby as the crowning achievement of a life's work. His aims went far beyond that. To Arnold England was a country in danger. Poverty and ignorance were the fuels of mob violence and eventually of revolution (the Chartist movement was seen as a real threat to the stability of the country). The fault lay not so much with the poor themselves, however. 'Arnold's fire was trained on the aristocracy . . . and on the whole of the upper classes . . . which had failed to do their duty to the poor' (Honey). Only the Church could stem the tides of unrest and only then if its leaders were truly religious. Church and State must work closely together, and 'the more closely they do so, the more nearly will human perfection be attained'. At the heart of his faith and his religious and political thought, as Michael McCrum emphasises, was his attitude to the Bible. While not unaware of the subversive claims of scholars and scientists, Arnold 'regarded the case for the Bible's plenary inspiration as indefensible'. From these beliefs it followed that 'Christianity should be the base of all public education in this country,' and indeed the chief function of the State *was* education, administered by the Church. Today such beliefs may be criticised, even ridiculed, but unless we understand them we fail to appreciate what Arnold was trying to achieve at Rugby.

Arnold began with clear-cut aims of what he intended to do. 'My object will be, if possible, to form Christian men, for Christian boys I can scarcely hope to make; I mean that, from the natural imperfect state of boyhood they are not susceptible of Christian principles in the full development'. He had the assurance of the Trustees that he would enjoy 'complete independence in all matters of school discipline and school routine' (Newsome), and he could afford to pay his assistant masters (nearly all clergy) salaries of £1500 p.a. (at least £50,000 today) which meant that they didn't need to hold benefices. His staff were amongst the first to do anything other than teach. They were encouraged to take a personal interest in their pupils, especially, of course, the housemasters, and to be on friendly terms with them in out-of-school hours and, as he made clear, 'his three main aims were to inculcate, first, religious and moral principles, secondly, gentlemanly conduct and, thirdly, intellectual ability' (McCrum). From the outset Arnold was determined to get the prefects on his side. He did not invent the prefect system but whereas most other heads used it merely to enforce discipline (and turned a blind eye on how that was achieved) Arnold looked to the Sixth Form to set the moral tone of the school. 'They were to be his lieutenants, endowed with very considerable powers and privileges, receiving his complete confidence, in return for which they were to serve him with loyalty and obedience' (Newsome). 'Rarely did he act in a matter of school discipline without consulting them, and about once a month he held a masters' meeting . . . to discuss school affairs in general' (McCrum). This close concern for the moral welfare of the school and its dependence on

Dr Thomas Arnold, Headmaster of Rugby, by T. Phillips

Rowland Hill (unknown artist)

James Kay, later Sir James Kay-Shuttleworth

the example shown by the Sixth Form extended into the holidays. Arnold had bought a house (Fox How) in the Lake District and quite often senior boys were invited to stay so that he could get to know them better without the constraints of school life. Even when a boy left Rugby he could still turn to Arnold for help. There is plenty of evidence that Old Rugbeians at University were helped to prepare for examinations (and sometimes given books to do so), received advice on the choice of career and, if in straitened circumstances, were given small sums of money.

In addition to being Headmaster, Arnold was also housemaster to the sixty-odd boys who boarded in School House and with them he took great trouble to get on close terms, preparing them for confirmation or just spending time in private conversation with individual boys. It may be true that he never really got to know the boys outside the Sixth Form or his own house, but he certainly knew how to influence them. In 1831 he was appointed to the chaplaincy (an unusual arrangement) and the power of his sermons in chapel is one of the best-remembered features of his regime. 'Stanley and Vaughan (the future headmaster of Harrow) used to nudge each other with anticipation of delight when they saw Arnold prepare to ascend the pulpit' (Chandos), and Thomas Hughes paints a vivid picture of 'a tall, gallant form, the kindling eye, the voice now soft as the low notes of a flute, now clear and stirring as the call of the light infantry bugle . . . What was it that moved and held us . . . three hundred reckless, childish boys, who feared the Doctor with all our hearts, and very little besides in heaven or earth? We couldn't enter into half that we heard . . . but we listened to a man who we felt to be with all his heart and soul and strength against whatever was mean and unmanly and unrighteous in our little world'. And it is worth noting in passing that – even in printed form – Arnold's sermons proved an inspiration in later years to such different people as Queen Victoria and Mr Gladstone. It must also be remembered, however, that there was a quite noticeable undercurrent of anti-religious feeling at Rugby, even in Arnold's time. In some houses boys who knelt at their bedsides to pray were liable to be pelted with boots and slippers.

The theme of nearly all his sermons was the battle against sin and the struggle for the soul of each boy. In today's world what Arnold was attempting would be labelled indoctrination, even brainwashing, and if he had understood those terms he would probably not have refuted them; it was the essence of his method, the way to achieve his goal. What effect did it have? Like most questions this one is a great deal easier to ask than to answer but it is worth quoting David Newsome's view that 'this manliness – not a masculine muscular ideal, the converse of effeminacy, but an adult sense of duty and a mature determination to develop their moral character and intelligence so as to become Christian *men* – was the direct result of Arnold's teaching and preaching'.

It is time to consider the academic side of Arnold's headmastership. He had inherited a small school of 136 boys, organised in six forms which were taught by six full-time assistant masters, all clergymen, while others of lower social standing taught drawing, writing, dancing and French. By 1841 the number of pupils had risen to 360, divided into ten classes, with a total staff of fourteen. Himself educated at Winchester and Oxford he could not envisage Latin and Greek occupying anything but a central position. Thucydides, Herodotus and Xenophon were among his favourite authors and as he once wrote, 'Greek and Roman History are a living picture of things present, fitted not so much for the curiosity of the scholar as for the instruction of the statesman and the citizen'. Military history was his chief hobby; there was nothing he enjoyed more than a visit to a classical battlefield and an investigation on the spot of the tactics employed. But Arnold's view of the value of a classical education was both wider and deeper than that of most contemporary schoolmasters. In one of his essays he declared that a man's education had two aspects. The first, the professional, fitted him for his daily work. The second, the liberal, fitted him for citizenship and, if that aspect were neglected, he would not be educated. A liberal education was one based on but not limited to the study of the classics. 'A classical teacher should be fully acquainted with modern history and modern literature no less than those of Greece and Rome. What is, or perhaps what used to be, called a mere scholar cannot possibly communicate to his pupils the main advantages of a classical education . . . the study of Greek and Latin, considered as mere languages, is of importance mainly as it enables us to understand and employ well that language in which we commonly think, and speak, and write'. Consider, for a moment, some of the subjects set by Arnold for composition *in Latin or Greek*: a conversation between Thomas Aquinas, James Watt and Sir Walter Scott; a description of Oxford through the eyes of a resurrected Herodotus; or a history of the province of Africa (Tunisia) from the Roman Conquest till modern times. The great value of the classics, he used to say, lies in training the judgment and the memory and in teaching the pupil to learn for himself.

It is claimed that Arnold set his face against the teaching of Science and to an extent this is true. But whereas other headmasters, as we shall see, did so because they regarded practical subjects as unworthy of the attention of a gentleman, Arnold thought the subject too vast to be accommodated in the timetable without severe detriment to other subjects. To natural history, however, he gave the greatest encouragement, regarding 'the search for an understanding of nature's laws as the most rewarding occupation of boys' leisure hours' (Newsome). Readers of *Tom Brown* may form the view that cricket and football were integral parts of education at Rugby. They certainly loomed large in the boys' estimation: to Arnold they were irrelevant. If Dr Newsome's thesis (*Godliness and Good Learning*) is correct, the cult of manliness and ultimately the veneration of athletic prowess stems more from Thomas Hughes than from Thomas Arnold. Arnold was in favour of as much freedom as possible for boys in their out-of-school hours. If they chose to organise cricket and football matches among themselves he would not interfere. But, as readers of *Tom Brown* will remember, a boy was free to go 'where he liked, with whom he liked, without hindrance'. 'In other words the unattended, untimetabled, unescorted hours of a boy's day were as much part of his education as the classroom itself. In those private hours he came across pleasures or, to use Arnold's word, evil. But education . . . was not sheltering the boy from evil but encouraging him to stand up to it and conquer it' (Simon and Bradley).

It is often held against Arnold that, for all his Christian principles, he was in the front rank of floggers, and this charge needs to be faced. Virtually all headmasters of the day flogged and some undoubtedly took pleasure in inflicting extreme agony on their pupils. As far as the evidence goes Arnold was not sadistic and did not thrash boys for mistakes in the classroom; instead he saw himself as waging war on sin, though we must recognise that what to others might have seemed mere naughtiness to him was wickedness. The sins of lying and deceit were special abominations and when a boy called March, who had been sent to him for bad work, not only claimed that the passage he had been set to translate was not one they had been told to prepare but persisted in the claim against the master's word, the Doctor repeatedly called March a liar and finally, 'in a passion', gave him *eighteen* strokes with the birch. When it transpired later that the master had made a mistake, 'Arnold's remorse was pitiful . . . and he made a public apology to the school' (Newsome). His treatment of March was inexcusable but, in general, it would be fair to say that he disliked the necessity for punishment, once saying to the assembled school, 'If I am to be here as a gaoler, I will resign my office at once'.

There remain two more aspects of Arnold's attitude to education that need to be considered. At bottom he did not really approve of public schools, or at least of boarding schools. One reason was that 'since school became the family for most of the year, he himself too much replaced the father' (Gathorne-Hardy). Arnold was a devoted family man and feared that the long absences of term-time would weaken family affections, and from the pulpit he deplored the tendency of boys 'to feel ashamed of indulging their natural affections and particularly of being attached to their mothers and sisters'. And he went so far as to advise his brother-in-law that, in the event of his own early death, 'the best thing would be for Mary to settle somewhere where they could go to school as day boys and be under her care at other times'.

Although schools like Rugby had been founded with the intention of providing cheap, or free, education for the sons of local people, the steady increase of fee-paying pupils meant that only a school in a large city could draw enough of them from a small area to be viable as a day school. But Arnold acted in direct defiance of the founder's wishes by excluding younger boys who lived in the town from attending his school. He had every right to be concerned for the moral welfare of small boys thrown into the company of teenagers but his efforts, which Bamford describes as 'a series of tricks incredible in their ruthlessness', culminating in a court action in 1839, do not make attractive reading. It was his clear intention (as described by McCrum in his account of the Wratislaw saga) to suppress the Lower School (where local boys from poorer homes could have reached the necessary standard to pass into the Upper School) by appointing incompetent teachers. It should be added that other public schools followed Arnold's example, though less ruthlessly, thus giving tremendous impetus to the growth of preparatory schools which could hardly have flourished in a world where public schools continued to accept boys of seven or eight.

It would be wrong to gloss over Arnold's faults. He could be hypocritical, he could be insensitive and rude, he could even be dishonest. To a certain extent his faults were symptoms of the age in which he lived; to a further extent they arose from defects of his personality and from the very single-mindedness with which he approached his task. But few would doubt that he was a great headmaster. One makes the judgment conscious that Rugby was far from being a totally reformed school even by 1842. Thomas Hughes and George Melly, both pupils at the school then, refer to poaching, drinking, smoking, fighting, bullying and cruelty all through Arnold's time, and W.C. Lake, although he admired the Doctor's idealism, had to confess that conditions at the school when Arnold died were much the same as when he took office.

Strangely enough Arnold's influence was more profound on other public schools after his death than at Rugby during his lifetime. George Moberly, headmaster of Winchester (1835-66), wrote: 'A most singular and striking change has come upon our public schools . . . This change is undoubtedly part of a general improvement of our generation in respect of piety and reverence, but I am sure that to Dr Arnold's personal simplicity of purpose, strength of character, power of influence and piety . . . the carrying of this improvement into our schools is mainly attributable'. And John Percival, a later headmaster of Rugby, called Arnold, 'a great prophet among schoolmasters rather than an instructor or educator . . . the secret of his power consisting not so much in the novelty of his ideas or methods, as in his commanding and magnetic personality'.

At the outset of this chapter certain 'beacons of light' were referred to and we must now turn our attention to two other remarkable schools of the early 19th century. Samuel Butler (Rugby and Cambridge) was appointed to the headship of Shrewsbury in 1798 at the age of 24. 'A famous school in the 17th century, [Shrewsbury] had been in decay for generations and [Butler's] predecessors had been at best listless mediocrities and at worst tipsy idlers who either stole or allowed others to steal valuable books and manuscripts from the school's library' (Chandos). Indeed an Act of Parliament was required as 'the only way of making changes sweeping enough to revive the school educationally and adminstratively' (Alicia Percival). By it St John's, Cambridge, retained the right to appoint the Headmaster and the Second Master, the Head alone to select the other masters. Butler, then, 'unlike Arnold who took over a school with a distinguished recent history, built up a great school from almost nothing' (Roach). And in total contrast to Arnold he had 'an almost morbid fear of religious enthusiasm' (Roach). A deeply religious man himself he aimed to make boys self-reliant in intellectual as well as in moral matters. He would, one feels, have stigmatised Arnold's methods, especially his preaching, as 'brainwashing' but he is himself open to the charge of failing to give his pupils adequate moral and religious leadership.

A brilliant teacher, Butler set about making new appointments and raising the academic standards of the school, and over the course of 38 years his pupils achieved an extraordinary record of success at Oxford and Cambridge. His most outstanding pupil (and successor as headmaster), Benjamin Hall Kennedy, actually won the coveted Porson Prize while still a Shrewsbury pupil. It was Kennedy who later wrote that Butler's 'crowning merit was the establishment of an emulative system, in which talent and industry always gained their just recognition and reward in good examinations'. Samuel Butler 'can be called one of the creators of the competitive spirit so characteristic of mid-Victorian England . . . The link between the public schools and university achievement . . . was one of the main reasons for their growing success as the century went on' (Roach). But if Butler's achievement in raising the academic standards, not merely of Shrewsbury but of public schools generally, is unquestionable, fair assessment of his standing as a headmaster is even more difficult to make than in the case of Dr Arnold. Three points need to be borne in mind. He was never well-off. He 'had sunk his borrowed capital in furnishing a boarding house and he could not move till he had got out of debt' (Alicia Percival). He did not enjoy good health. And most seriously of all he was saddled with a Second Master, John Jeudwine, with whom he soon ceased to be on speaking terms. For 37 years communication between the two men was by notes, written in the third person, and there is evidence that a Jeudwine party was formed in the town which must have greatly increased Butler's problems. Whether they also sought to blacken his reputation further afield is uncertain but when in 1806 the headship of Rugby fell vacant, the Trustees appointed Dr Wooll in preference to Butler.

It is hard to be sure how severe a disciplinarian Butler was. 'Cruel and unusual punishments' are referred to, but on the other hand several of the Shrewsbury Governors wrote to the Rugby Trustees asserting that he was 'as kind, humane and indulgent as any reasonable parent could wish'. Very possibly his absorption in classical learning and his total lack of interest in sport (he tolerated cricket but banned football as being 'fit only for butchers' boys') alienated him from the mass of his pupils and when trouble broke out Chandos is probably right in asserting that 'it was the more difficult to subdue because the headmaster lacked the allegiance and support of some of the most senior boys'.

Moreover it is undeniable that, however brilliantly the classics were taught, other subjects were largely excluded. Kennedy himself 'introduced Mathematics and French as subjects in the ordinary curriculum' (Barnard). The school's most famous alumnus, Charles Darwin (1818-25), summed up his education there as 'simply a blank' and, when he and his brother, Erasmus, improvised a laboratory in a tool shed at their father's home, 'he was publicly rebuked by the headmaster for wasting his time on such useless subjects' (de Beer).

As against these defects or limitations one must record 'letters stretching over three decades [which] bear witness to parents' gratitude not merely for the effectiveness of their boys' education, but for acts of kindness and generosity' (Chandos). In assessing Butler as a headmaster one must therefore take into account a mass of evidence, some adulatory, a good deal of it hostile, concerning his merits and his faults. The *Quarterly Review* said, shortly after his death, 'If the reformation which has been at work in our public schools . . attracts the notice it deserves . . . [men] will hardly fail to give honour due to that scholar who first set the example in remodelling our public education'.

Among the educational pioneers of the 19th century the name of Thomas Wright Hill is seldom mentioned. In 1787, at the age of 24, he became manager of a rolling-mill in Wolverhampton. His marriage to Sarah Lea yielded six sons (one of whom died young) and two daughters. They formed an exceptionally united family and although it was the third son, Rowland, who achieved world-wide fame, the support of his brothers was crucial to his own success. Influenced by Joseph Priestley, at whose Sunday School he taught for several years, Thomas Hill founded a small school in Birmingham which, in 1803, he transferred to larger premises at Hill Top, a mile from the city centre. His sons were educated at the school and each served in it as a pupil teacher. Four of them gradually moved into different professions so that on Thomas's retirement, Rowland and Matthew were left in charge of the school. The former had already gained practical experience in metal work through a part-time job at the Assay Office and was keenly interested in all branches of engineering. Even so one cannot but be astonished to find the young man acting as the architect of a much larger building to which the school – henceforth known as Hazelwood – moved in 1819. Instead of the usual huge schoolroom with masters and their classes spread along each side there was 'a large assembly hall with a stage and at least six ordinary classrooms, all heated by a central air-duct system . . . In addition there was a whole range of craft-rooms, a library, a laboratory, a gymnasium, studies for the older boys, an observatory, a swimming pool and a sports field' (Hey). The thinking behind these revolutionary school premies had evolved gradually during the years at Hill Top and it would be quite wrong to give Rowland the entire credit. Indeed it was Matthew who was largely responsible for the curriculum and for the actual teaching while Rowland concerned himself chiefly with administration and discipline. A much wider curriculum than any then practised was established at Hazelwood, including 'Penmanship, English, Elocution, Geography, History, Mathematics, French, Italian, Spanish, Latin and Greek (both optional), Gymnastics, Art, Music, Woodwork, Metalwork and Science'. Whenever possible both modern and ancient languages were taught by the direct method and short plays were staged in several of them. On Open Days 'the whole school was open for inspection'. In addition to the performances on stage there was a varied selection of the boys' work on view and actual experiments and demonstrations were conducted in the laboratory. As the result of an article in a French journal, *La Revue Encyclopédique*, visitors began to arrive at Hazelwood from abroad, while the publication of *Public Education* by Rowland and Matthew Hill in 1822 attracted wide attention. It was not merely the progressive curriculum which interested visitors. Corporal punishment had been abolished and a remarkable form of self-government by the senior boys had replaced it. If the various offices, including Magistrate of the Lower Court, Solicitor General, Custos Depositorum, Silentiary etc. sound faintly ludicrous to modern ears they must be set against the ferocious methods of discipline in force at most other schools. Rowland went so far as to introduce a complete range of token coins to reward academic achievement and conscientious fulfilment of duties. It would be wrong not to draw attention, at this point, to a dissenting voice, that of P.W. Bartrip. In *The Career of Matthew Davenport Hill* (unpub. thesis), he considers that the reforms arose chiefly from the Hills' desire to make Hazelwood 'a thoroughly good school'; that the first major improvement came when Matthew himself took on the teaching of Latin, and that subsequently the curriculum reverted to one in which Latin held a central position in school studies. W.L. Sargant, a pupil at Hazelwood for eight years and later a successful businessman and gunsmith, while remaining a friend of Matthew, was critical of the school's disciplinary system, considering it 'a moral hotbed which forced us into a precocious imitation of maturity'.

All the same one can hardly be surprised that distinguished visitors to the school included William Wilberforce, Thomas Malthus, Robert Owen (whose industrial and education experiments at New Lanark were equally revolutionary), Jeremy Bentham and Lord Clarendon (who later headed the Commission). Their admiration for Hazelwood encouraged Rowland to establish a sister school, Bruce Castle, in Tottenham in 1827. Arthur who had trained as a printer, came back to take over the headship of Hazelwood.

At Tottenham Rowland, assisted initially by Edwin, set out to repeat the earlier experiment but although successful at first he encountered increasing opposition from parents, specifically on the grounds that too much science (on which Rowland set great store) was being taught. Bitterly disappointed and 'with the sympathy and support of the whole family' Rowland abandoned what he had always intended to be his life's work and set off for a long holiday in France. Hazelwood School then closed and Arthur came south to take charge of Bruce Castle, a post he held for 34 years (1833-67) before passing it on to his son, George. When the latter gave evidence to the Taunton Commission he told its members that the teaching of science had long been discontinued at his school.

Rowland's later career, involving the master-plan for the colonisation of South Australia, the invention (with Edwin) of the rotary printing press, the development of the London and Brighton railway and, above all, the complete reorganisation of the Postal Service with the introduction of the famous Penny Black in 1840, entitles him to an eminent place in 19th century history. But one cannot help feeling that the short-lived Hazelwood experiment is one of the great might-have-beens of English educational history.

With these varied examples of schools and schoolmasters in mind we may begin to consider, in a little more detail, how the teaching was carried on in public schools of the early 19th century, together with various other aspects of school life. For a first-hand account of classroom procedure and practice we turn to chapter 8 of *Tom Brown's Schooldays* in which Thomas Hughes recalls the methods of teaching and learning as he himself experienced them at Rugby.

'The lower fourth, and all the forms below it, were heard in the great School, and were not trusted to prepare their lessons before coming in, but were whipped into School three-quarters of an hour before the lessons began by their respective masters, and there, scattered about on the benches, with dictionary and grammar, hammered out their twenty lines of Virgil or Euripides in the midst of Babel. The masters of the lower school walked up and down the great school together during this three-quarters of an hour, or sat in their desks reading or looking over copies, and keeping such order as was possible'.

In chapter 3 (part 2), by which time Tom has advanced to a higher form, we learn how lessons were prepared, out-of-school, by the older boys.

'The Vulgus . . . is a short exercise [four to six lines] in Greek or Latin verse, on a given subject, the minimum number of lines being fixed for each form. The master of the form gave out at fourth lesson on the previous day the subject for next morning's vulgus, and at first lesson each boy had to bring his vulgus ready to be looked over; and with the vulgus, a certain number of lines from one of the Latin or Greek poets then being construed in the form had to be got by heart. The master at first lesson called up each boy in the form in order, and put him on in the lines. If he couldn't say them . . . he was sent back and went below all the boys who did . . . say them; but in either case his vulgus was looked over by the master, who gave and entered in his book, to the credit or discredit of the boy, so many marks as the composition merited. At Rugby, vulgus and lines were the first lesson every other day in the week, on Tuesdays, Thursdays, and Saturdays; and as there were thirty-eight weeks in the schol year, it is obvious to the meanest capacity that the master of each form had to set one hundred and fourteen subjects every year, two hundred and twenty-eight every two years, and so on. Now to persons of moderate invention this was a considerable task, and human nature being prone to repeat itself, it will not be wondered that the masters gave the same subjects over again after a certain lapse of time. To meet and rebuke this bad habit of the masters, the school-boy mind, with its accustomed ingenuity, had invented an elaborate system of tradition. Almost every boy kept his own vulgus book and these books were duly handed down from boy to boy till [most] boys . . . are prepared with three or four vulguses on any subject in heaven or earth . . . which an unfortunate master can pitch on. The only objection to the traditionary method of doing your vulgus was the risk . . . that you and another follower of traditions should show up the same identical vulgus some fine morning, in which case, when it happened, considerable grief was the result'.

That his 'traditionary' method of coping with the problems of classical verse was not confined to Rugby is supported by examples from all the Great Schools. At Eton, around 1820, the captain of a Dame's house had inherited or acquired a collection of 3000 old copies, suitable for every occasion. The requirement to produce a verse marking the death of George III was met by supplying the one used sixty years earlier 'which, with only one small adjustment, met the demand handsomely' (Chandos).

At Winchester special emphasis was placed on repetition. 'The school had a ceremony, known as Standing Up, which took place towards the end of one term. Everybody below the two top forms had to learn and recite as many lines of Latin or Greek as he could remember . . . Marks were awarded and remarkable feats of memory were achieved. One boy, it is recorded, recited the whole of a Sophocles play without a single mistake [while another achieved] 16,000 lines of verse. It took eight lessons to accomplish the feat' (G.F. Lamb). And whereas today exceptionally gifted children of 10 or 11 may be found taking 'A' level mathematics, or confronting a Grand Master across the chess board, his or her equivalent 150 years ago could rate as a prodigy only in the classics. John Corington, at the age of 8, could already repeat a thousand lines of Virgil and 'amused himself by comparing different editions of the poet' (Chandos).

Since almost all the teaching staff at a public school had had a classical education, the narrowness of the curriculum made the headmaster's dispositions correspondingly easy and, in the event of sickness or unexpected shortage of staff, the headmaster was quite capable of taking the entire school himself. It is not easy to be precise about the allocation of teaching time because clearly there were variations, but a total of thirty hours a week would perhaps represent the average. 'At least three-quarters, and in some cases four-fifths, of the time was spent in class on Latin and Greek (with relevant ancient history and geography included)' (Bamford). Mathematics was given two or three hours at Rugby, left as an option at Eton, and entrusted at Harrow to 'that much-ragged Frenchman, M. Marillier, who taught his pupils to call the subject 'Teek' from his pronunciation of les Mathématiques' (G.F. Lamb). This valiant man, who endured an incredibly long period of service (43 years), 'combined the disadvantages of being a Frenchman with those of teaching Maths in what was to him a foreign tongue. He lived the life of a dog being received with hallooing and hooting whenever he appeared' (G.F. Lamb). French itself was normally taught by refugees and English, as such, was scarcely taught at all, 'but was expected to emerge as a by-product of classical studies' (Honey). And although a considerable array of options was available at most schools, such as German, Italian, fencing, natural philosophy, music and dancing, 'they were unmistakably frills and headmasters took no responsibility for them whatever' (Bamford).

The number of boys in a class varied from school to school, but almost everywhere setting was the rule. 'Under this scheme a boy progressed from one set to another by passing the appropriate test or examination, usually at the end of a term or a year. If he failed he was kept down while a bright boy might climb through the forms of a school with extraordinary rapidity' (Bamford). Since there were no external examinations until it came to university entrance, such a system was not utterly disastrous but it did have the effect of placing boys of widely differing ages in the same form.

Why did the classics remain paramount throughout the 19th and well into the 20th century? Why was the teaching of science largely ignored?

Headmasters are to be absolved, in the main, of blatant hypocrisy though not of failure to examine their own motives impartially. A headmaster who was also an inspired teacher, an Arnold or a Benson, could so widen and deepen the usual limits of classical education that a young man of at least average intelligence could emerge from the Sixth Form imbued with a love of learning, with an appreciation of classical and modern literature and with the cultural values of a Christian gentleman. If for the moment all scientific knowledge is excluded it could be claimed that he was as well educated as the product of any school or college anywhere, at least up to the mid 20th century. But the qualifications are only too apparent: the teacher himself must be a man of wide cultivation and sympathies; the pupil must be of at least average intelligence and have a receptive mind; and, of course, the exclusion of science is a huge escape clause. What of the thousands of pupils at public schools who were not specially bright or who never sat under an inspiring teacher? Their evidence needs to be considered. To A.F. Leach at Winchester 'Charles Wordsworth's Greek Grammar, written in Latin, remains . . . the ideal of all that is hateful and hideous in learning'. Indeed 'primers which had grammatical rules stated in English . . . were a concession introduced by Dr Arnold, of which he later repented' (Honey). D'Arcy Thompson (afterwards a perceptive and kindly schoolmaster himself) fared even worse at Christ's Hospital. 'The day after my entry into this colossal institution a Latin Grammar was placed in my hands. It was a bulky book of its kind . . . Over a space of years we went systematically through that book, page after page, chapter after chapter. It was all unintelligble, all obscure'. Edward Lockwood gave similar evidence when he wrote: 'I had not the faintest idea what the Latin Grammar was all about, and as no one made the faintest attempt to explain anything, I gave up all hope of understanding it'.

Nor was it merely uncomprehending pupils who assailed the aridity and uselessness of much of the classical teaching then practised. In the *Edinburgh Review* Sydney Smith wrote in 1809: 'There are few boys who remain to the age of nineteen at a public school without making above ten thousand Latin verses – a greater number than is contained in the *Aeneid*; and after he has made this quantity of verses in a dead language he never makes another as long as he lives . . . In classical learning it seems to be sufficient if the least possible good is gained by the greatest possible exertion'. H.H. Milman, Dean of St Paul's, termed the emphasis on classical verse 'a Cinderella slipper: it fitted one foot in a hundred, and the rash attempt to put it on might cripple the other ninety-nine for life'. Frederick Temple (Headmaster of Rugby, 1858-1869, and later Archbishop of Canterbury) gave a realist's view of the matter when he described the classics as the peculiar discipline for those who are to govern others and do not happen to possess such genius as to dispense with discipline'.

The *Westminster Review* was uncompromising in its denunciation: 'From eight to eighteen, nine or ten months in every precious year of youth are occupied for six or eight hours of every day in learning, or trying to learn, a little Latin and less Greek; in attempting, in fact, not to read and understand the matter of a classical author, to know the history, the poetry, the philosophy, the policy, the manners and the opinions of Greece and Rome, but the grammar, the syntax, the parsing, the quantities, and the accents; not in learning to write and speak the languages, but in getting by rote a few scraps of poetry, to be again forgotten, and in fabricating nonsense verses . . . In ten years of this labour, privation, punishment, slavery and expense, what is gained even of this useless trash? Nothing'.

And when, a good many years later, the Royal Commission on Public Schools came to issue its Report (1864) it concluded: 'If a youth after four or five years spent at [a public] school, quits it at nineteen, unable to construe an easy bit of Latin or Greek without the help of a dictionary, or to write Latin grammatically, almost ignorant of geography and of the history of his own country, unacquainted with any modern language but his own, and hardly competent to write English correctly . . . his intellectual education must certainly be counted a failure'.

Curiously enough the many critics of excessive concentration on the classics do not appear to have directed their fire at a conspicuously vulnerable part of the target. When one recalls the extremes to which prudery could go (Percival, at Clifton, ordering boys to keep their knees covered when playing football) it is at least surprising that no attack was mounted on the study of authors who were not merely pagan but often unashamedly licentious. At Westminster, for example, a favourite annual event, enjoyed by the boys and their parents, was the performance of Terence's *'Eunuchus'*, the most indecorous of all his comedies. William Sewell (Radley) at least aimed at the target when he wrote: 'A clean mind in a healthy body should be our ambition; but how can we hope to attain either when we place in the hands of boys, at the most critical period of their lives, a knowledge of language which unlocks the door to the lewdest literature in Western Europe'.

As far as opposition to the teaching of science is concerned, headmagisterial opinion is perhaps best summed up by Frederick Temple when he said: 'The real defect of mathematics and physical science as instruments of education is that they have not any tendency to humanise. Such studies do not make a man more human, but simply more intelligent'. 'The idea of a liberal education was fundamental to much of the 19th century discussion of school curricula . . . an education only qualified for the title 'liberal' if it developed all important aspects of the mind and the character. The significance of this concept for science teaching was that the sciences were commonly supposed to be of little value in the development of personality, whereas the traditional classical curriculum was regarded as nearly perfect' (Meadows and Brock in Simon and Bradley). The attitude of many parents to the teaching of science can be drastically but not perhaps unfairly represented by evidence submitted to the Devonshire Commission: 'Parents exhibit complete indifference to the whole subject, with the exception that they occasionally object to their sons devoting any time at all to it'. Lurking behind this attitude one may perhaps detect the fear of contamination. Science, in their minds, equated with smoke, steam and oily dungarees. Parents who had not themselves attended public school were understandably anxious to dissociate themselves from plebeian origins. The classical curriculum 'rested upon the belief that in the Classics were to be found, perfectly expressed, all the principles and mental discipline required to train the statesman, the divine and the gentleman; for travel on the continent, for coping with 'those damned dots' or for the solution of practical problems, the courier, the clerk or the carpenter would always be available' (Firth). Travel on the continent was becoming increasingly popular, indeed many families took a house in Italy or on the Riviera or at a spa for months at a time but it was no part of public school's task to teach conversational French or Italian. Such a study was considered to lack academic rigour and 'because French [in particular] was the native language of people who could not be counted on to be 'gentlemen' it was openly derided by [public school] masters as a 'tinpot' subject' (Honey).

There was, too, the feeling that a classical education bestowed of belonging to a secret society, one that spoke a language outsiders could not understand. When the Cabinet was discussing a financial deal with the Turks, Gladstone clinched the discussion with 'an extraordinarily apposite quotation . . . from Virgil' (Honey); while much later Lord Curzon warned his colleagues about the dangers of retreat from the Dardanelles (1915) by 'quoting at length Thucydides' account of the disastrous Athenian retreat from Syracuse' (Honey). It manifested itself, too, in 'exchanges in the House of Commons in which an epigram from a classical author quoted by one speaker might be capped by another from a speaker on the other side, or in which a false quantity was liable to be hissed' (Honey). And whether true or apocryphal the reported announcements of the conquest of Sind (Peccavi) and of Oudh (Vovi) provide good examples of both Victorian humour and cultural snobbery.

'What a classical education was designed to do was to forge a bond of shared thought, sensibility and manners between gentlemen of all ranks, uniting in one caste . . . nobility, gentry and the professional classes, requisite to the ordering of society. But what classical education [also] did in practice was to deny the designation 'gentleman' to increasing numbers of Englishmen who aspired impatiently to acquire it' (Chandos).

Two centuries ago the view of Dr Johnson, a notably humane man, was widely approved. Reproving a schoolmaster for his leniency with the rod, Johnson remarked that 'what boys gain at one end they lose at the other'. The question of punishing children is one that inevitably arouses strong feelings. The champions of 'Never, ever strike a child', 'A good hard slap when he knows he's done wrong', or 'Caning never did me any harm when I was a boy', are all equally firm in their conviction of the right method for dealing with troublesome children. What few now would question is that the infliction of pain often gives pleasure to the inflictor and that many men are sexually aroused in the process. What most reasonable people would accept is that attitudes to punishment – as to other aspects of social interaction – have changed very considerably in the past two hundred years and that one must not judge early 19th century attitudes with late 20th century minds. Consider this legal judgment of the 18th century: 'Where a schoolmaster, in correcting his scholar, *happens to occasion his death*, if in such correction he is so barbarous as to exceed all bounds of moderation, he is at least guilty of manslaughter; and if he makes use of an instrument improper for correction, as an iron bar or sword, or if he kick him to the ground and then stamp on his belly, and kill him, he is guilty of murder' (William Blackstone). The implications of this judgment, read slowly and pondered over, are truly frightful though specialists in black humour might relish the word 'happens'.

These matters need to be borne in mind when one comes to consider the question of punishment in public schools of the pre-Victorian era. The range of possible misdemeanour was considerable, especially as boys enjoyed so much freedom outside the classroom. But aside from poaching, stealing and vandalism, normally perpetrated on outsiders, crimes within the school were commonly lying, cheating, swearing, insolence and disobedience, while running away from school was the most serious offence of all. Within the classroom, of course, almost any mistake or failure could be regarded as punishable. The punishments available to a Headmaster were lines, solitary confinement (thankfully rare), expulsion and flogging. 'Even in the mildest schools the minimum punishment was . . . ten lines, to be copied out from a classical author in the best copperplate, but with a run of bad luck these penalties could mount up horribly, being doubled or even redoubled. Imagine the state of mind of a boy faced with 500 lines to be written out to a close deadline' (Bamford). Solitary confinement is best remembered from Charles Lamb's account in *Essays of Elia*:

'The sight of a boy in fetters, upon the date of my first putting on the blue clothes, was not exactly fitted to assuage the natural terrors of initiation . . . I was told he had run away . . . I was soon after taken to see the dungeons. These were little square Bedlam cells, where a boy could just lie at his length upon straw and a blanket . . . Here the poor boy was locked in by himself all day without sight of any but the porter . . . or of the beadle who came twice a week to [administer] his periodical chastisement'. Expulsion seems to have been reserved for sexual offences (though it is not easy to be dogmatic about this as schools are understandably unforthcoming). It normally took place after the boys had gone to bed (to avoid the possibility of a hero-worshipping send-off) and the boy might well arrive home before notification reached his parents.

But to most readers of school stories punishment meant corporal punishment and their assessment was correct. 'The use of corporal punishment was part of the fabric of nineteenth century education, not just in public schools but in all types of school' (Honey), to which one might add 'and in prisons and in the armed forces as well'. The birch was part of the standard equipment of the Victorian schoolmaster, as central to the work of teaching as is the textbook or the blackboard of today' (Honey). But the birch was by no means the only implement used. 'For simple offences they relied on sticks, rods or canes; for the severest cases split canes, thongs or a tightly bound mass of switches specially and freshly made up each day. Occasionally the weapon was weighted with lead, or else, for extra effect, the ends were left supple and pliant to lick around the edge to the tender regions where the sting was really felt. Misdemeanours were dealt with in private, but major offences became a public spectacle with the entire school looking on' (Bamford). Again it is hard to be dogmatic but generally it would seem that the cane was applied to the backside and the birch to the back as a whole. Caning on the hand seems to have been rare in public schools. In some schools the victim knelt over a block, being held down if necessary by other boys. At others a senior boy, or a master, 'takes hold of the victim, hoists him on his back by the wrists and keeps him suspended while the inflictor delivers the blows with a birch on the victim's bare back'.

Instances of sadistic floggings can be found in the annals of almost every public school. Christ's Hospital is perhaps unfortunate in having had unusually brutal cases recorded for posterity. In one instance a boy called Blount had been compelled to steal sugar for an older boy and was caught and flogged. 'That night', records Andrew Drew, 'poor little Blount could not sleep, and at last he begged me to help him. I accordingly took his shirt off, and found his back from the shoulders down to the waist, one mass of lacerated flesh, the blood sticking to his shirt so as to cause agony in getting it off. I then, with my finger and thumb, pulled out of his back at least a dozen pieces of birch-rod, which had penetrated deep into the flesh'. In another instance a boy 'was hoisted by one beadle, and another birched him, his back being raw for some days afterwards; and the agony he endured at night, when he took off his shirt, was fearful as the shirt stuck to his back, clotted with blood. The boy was then under twelve years of age'. By comparison Oliver Twist had a pampered childhood.

In earlier times nearly all beating was done by the headmaster or by the undermaster, but as Arnold's entrustment of authority to prefects, though without his firm guidance, spread to other schools, so beating by prefects (or monitors) for a host of petty infringements became the normal practice. Winchester, where it is known as 'tunding', had an especially black record. The Rev. W. Tuckwell, a boy there in the 1840s, described 'the worst tunding I ever saw inflicted by a college prefect on a commoner, and, though nominally official, was accentuated by personal resentment. The boy made no cry, but repeated audibly the 150 cuts as they fell; then was for some days in danger'. Another boy, Walter Hook, wrote home to his brother: 'I hate this place more and more every day. I was licked yesterday more severely than ever before . . . and am to be licked again today for cutting football to write this . . . If I am killed, as I think I shall be, tell the school butler to send you my books'. Charles Oman, later a famous historian, talks of the reign of terror he experienced as a new boy. Soon afterwards the 'great tunding row' broke out after a boy of seventeen received from a prefect thirty cuts across his shoulders, in the course of which beating five ground-ash sticks were broken. Angry letters to *The Times* revealed that this was not a solitary instance.

Prefect power was closely linked with fagging. Within reasonable limits this was a system that had some value to it, though it has now disappeared. An assertive thirteen-year old, whose father might be titled or a bishop, learnt the need for obedience and for being (for perhaps the only two years of his life) at the beck and call of others. With no supervision from above, however, it became a continuous and compulsory form of bondage. Outside class a junior was likely to be pounced upon for almost any task – cleaning shoes, making fires, sweeping up, shopping. He might be posted as a general lookout to warn of the approach of authority, or be sent out fishing or poaching. Cooking was a normal chore, and so was warming beds by lying in them'. At Winchester and at Rugby fagging is reckoned to have been severe, at Shrewsbury generally mild, while at St Pauls and Merchant Taylor's (day schools) it did not exist.

Bullying, too, was rife in schools (the majority) where the headmaster and his staff failed to exercise constant vigilance. Reminiscences of school days, and almost all school boy fiction, are studded with instances and, in real life, it was seldom that the perpetrator met his match as satisfactorily as when Tom Brown and East dealt with the appalling Flashman. A particularly unpleasant example was called 'Lamb Singing' and was prevalent at Rugby. 'New boys were required to stand on a table in turn and sing a song of their own choice . . . [and] if a majority of judges gave thumbs down, the singer was penalized by having to drink a brimmer of muddy water crammed with salt . . . Boys forced to swallow the noxious potion could be ill for days' (Chandos). On other occasions small boys were thrown, naked, into Anstey's Hole, 'notorious for its depth and mud'.

We have seen that Dr Arnold took almost no interest in the boys' games. Football at Rugby was boy-organised and boy-centred. Old Brooke's exhortation to the boys of school house epitomised the spirit and the approach. And as we have seen most public schools of the early 19th century allowed a remarkable amount of freedom to the boys in out-of-school hours. 'A boy had to attend at meals, prayers, classes and bedtime. Otherwise, at most schools, he could do what he liked and go where he pleased, if he could only evade fagging duties. Organised games . . . did not exist, . . . Walking, fishing, poaching, hunting, fighting . . . were favourite pastimes at all seasons of the year. It was common to find boys many miles from the school gates at one or other of several ale-houses forming traditional . . . focal points. Hobbies ranged from model-making, natural history and chemistry to ballooning' (Bamford). Older boys might be involved in the writing or acting of plays, in editing school magazines or in attending race meetings. Those interested in team games organised their own matches and, sometimes, employed a professional to coach them at cricket. The earliest photographs show casual assemblies of friends who look as if they had said something like, "Who wants to play a game today? What shall it be?" and then changed into a miscellany of old clothes, torn shirts, ancient jerseys and battered caps, tucked their trouser ends into the tops of their socks and got on with it' (Chandos). At Harrow Arthur Haygarth (in 1843) was prevailed upon by the captain of the Eleven to remain at school a further term in order to play cricket against Eton and Winchester and we are also told of 'one big boy, better at cricket than at Latin composition, who used to go round every Saturday evening with a bag containing slips of paper. Each scholar was required to draw a slip, on which was written 'Latin Prose', 'Algebra' etc. These were the assignments required for the coming week by the cricketer' (G.F. Lamb, quoting C.H.P. Mayo). One doubts whether those were isolated instances.

There is a danger that we may regard such freedom as an idyllic existence corrupted by late Victorian head- and house-masters whose zeal for the athletic prowess of their schools or houses ruined once and for all the concept that games are played for fun, and there is something to be said for this theory. Dr Newsome, in *Godliness and Good Learning*, supplies a useful corrective to this sanguine view:

'The root of the trouble lay in the deeply respected tradition that boys at school should be left to govern themselves . . . They might – and often did – choose jungle laws; but, provided that their follies did not lead to such wild excess that either the public in general or parents in particular felt obliged to demand magisterial intervention, the cruelties and injustices of boy life were studiously ignored. It was possible to defend such negligence on the grounds that the boys were learning the obligations and inconveniences of community life and that certain qualities of toughness and endurance were engendered by its rigours. Two important consequences came about. In the first place, masters and boys lived in a perpetual state of war . . . [and] in their particular domain, the boys formed a republic with their own code of laws and their own crude methods of punishment. Secondly, the recognition of mob rule led to the acquiescence in mob values. How could the principles of godliness and good learning have any hope of surviving in a community ruled by jungle law? Work within the republic too often consisted in organised and ingenious cribbing; recreation was frequently the baiting of the weak and the eccentric; conscience and the individual sense of decency and honour had to be subordinated to loyalty to the community, and thus one had to learn to lie boldly and never to betray any member of the group by sneaking into the enemy's camp'.

The freedom, and the brutality which often accompanied it, had not merely its apologists but also its ardent advocates. The Duke of Wellington's aphorism is too well-known and too often misquoted, but fifty years later one finds Thomas Hughes still enunciating much the same principles, though put more chivalrously:

'Fighting with fists is the natural and English way for English boys to settle their quarrels . . . Learn to box, then, as you learn to play cricket and football . . . As to fighting, keep out of it if you can, by all means. When the time comes . . . that you have to say "Yes" or "No" to a challenge to fight, say "No" if you can – only take care you make it clear to yourself why you say "No". It's a proof of the highest courage, if done from true Christian motives . . . But don't say "No" because you fear a licking . . . and if you do fight, fight it out; and don't give in while you can stand and see'.

Information about sexual attitudes in public schools at this period, what behaviour was commonly accepted, what was winked at, what was regarded with genuine aversion, and how prevalent any of it was is extremely difficult to ascertain and impossible to be dogmatic about. 'In the first half of the century, when a wide range of sexual maturity in a given school was guaranteed by a very wide age-range, the schools showed a marked lack of concern over the opportunities for homosexual temptation which this presented. The practice of two or more boys sharing a bed was common even in the most expensive schools including Eton and Rugby . . . and separate beds were boasted as a special advantage by Harper at Sherborne in the 1850s' (Honey). Sydney Smith, himself a Wykehamist, gave his customary forthright opinion: '. . . premature debauchery that only prevents men from being corrupted by the world by corrupting them before they enter the world'.

The difficulty of being sure is increased by the deliberate obscurity of the language employed. 'Open speech on sex was taboo, not only in school but throughout life; tracts on the social evils of sexual indulgence were couched in words that would not offend the most sensitive spirit, and so vague that it almost needs a suspicious mind nowadays to realise that they are about sex at all' (Bamford). In a letter written to a boy at Westminster by his father in 1824 he says, 'With regard to the Vice which you speak of as taking place in College, God grant that you may feel the same abhorrence of it as you express in your letter'. The boy's letter, presumably specifying the Vice, has not been preserved. 'Even Dr Arnold, usually extremely outspoken in reference to sin in his sermons, speaks only of sensual wickedness, such as drunkenness and other things . . . forbidden in the Scriptures' (Newsome). And Frederick Farrar, educated at King William's (Isle of Man) and describing the life there in *Eric or Little by Little* simply refers to wicked deeds which he is unable to describe. 'When Eric first heard indecent words in dormitory 7 he was shocked beyond bound or measure', and Farrar goes on, 'Now, Eric, now or never . . . Speak out boy! Tell these fellows that unseemly words wound your conscience'. And at the end of the chapter he exhorts his readers: 'Ah, young boys, if your eyes ever read these pages, pause and beware. The knowledge of evil is ruin, and the continuance in it is moral death'.

'Arnold himself, who could be desperately severe with any offence involving a breach of trust, showed himself sympathetic and understanding in his dealings with boys guilty of immorality' (Newsome). 'At times, on discovering cases of vice he would, instead of treating them with contempt or extreme severity, tenderly allow the force of temptation, and urge it upon them as a proof brought home to their own minds, how surely they must look for help out of themselves' (Stanley quoted by Newsome). Most headmasters, confronted with cases of immorality resorted to flogging or expulsion, reinforced by thunderous – though often obscure – denunciations from the pulpit.

3   Eton

It seemed better to attempt an overall picture of teaching, and life, at the great schools in general before turning our attention to the greatest of them all. In this way it may be possible to see how far the College exemplified the virtues – and vices – of a boarding, classical education in the early part of the 19th century.

More has been written about Eton than about any other public school. Some of that writing has been excessively laudatory, some hysterically critical; few authors have approached their task in a truly impartial state of mind. Conscious of that fact the present writer nevertheless feels he must attempt to give an accurate picture of Eton at that time since it is his contention that it was the overriding influence of the College that determined the kind of education offered at Waterfield's Temple Grove and, no doubt, at many of the other leading preparatory schools.

'The College that Henry VI founded (in 1442) was intended to provide free education for boys of moderate means, drawn preferably from the neighbourhoods of Eton and Cambridge, but, if boys from those neighbourhoods were not available, from elsewhere' (Hollis). The King's intention was that the boys, in term-time, should live in rooms round a large quadrangle but, since it was never built, all seventy (the number fixed by Provost Lupton in 1540) were huddled into Long Chamber. The conditions there have sometimes been described as 'medieval', but 'The Middle Ages never dreamed of anything so barbarous and Henry VI never intended his scholars to live in the humiliating conditions under which they were compelled to exist in the early years of the nineteenth century' (Hollis). For the first three hundred years the living standards, though spartan by modern reckoning, were no worse and no better than those of most other schools. Until the early 18th century the Lower Master lived at the east end of Long Chamber 'and was able to impose some sort of order upon its inhabitants' (Hollis), but in 1716 he moved into a separate house and his rooms were turned into supplementary dormitories, so that only 52 of the Collegers actually lived in Long Chamber, enabling each boy to have a separate bed. There were two fireplaces but 'most of the windows had no glass and a boy might easily awake in winter to find his bed covered with snow' (Hollis). There were no detailed regulations as to how Collegers should spend their time in Chamber. Each boy had a tallow dip, the 'candlestick' being an improvised school-book bent in two with a hole punched in the corner to take the candle. The senior boys, who alone had wash-basins, occupied one end of Chamber. The smaller boys fagged for them and the plea that he had been sent on an illegal errand by his fagmaster did not absolve the fag from punishment. 'I was not the only one sent . . . to fetch bottled beer from the *Christopher*. Apart from the doom impending if I met a master . . . the pocket of a Colleger's gown was not a convenient receptacle for quart bottles knocking against each other. But we had no choice' (Ainger). Although Long Chamber itself was locked for the night at 8 p.m., 'the gates of School Yard were never closed and at 9.30 every evening a messenger came to the window of Long Chamber and handed in the suppers which the boys had ordered from the *Christopher* . . . and the exercises which masters had corrected' (Hollis).

Life in Long Chamber clearly had its attractive side. 'On a winter night Long Chamber, with all its draughts and leaks, lit from top to bottom with fifty dips and two roaring fires, presented a vital and atmospheric picture. Some boys would be at their desks writing or lying down on their beds with candles standing up in their own wax, some reading, some walking up and down in laughing or earnest talk, others boxing, or fencing with single-stick' (Chandos). But the picture painted by Edward Thring, a colleger from 1835 to 1841, is less rosy. 'Who can ever forget the wild, rough, rollicking freedom, the frolic and the fun of that land of misrule, with its strange code of traditional boy-law, which really worked rather well *so long* as the Sixth Form were well disposed or sober? Rough and ready was the life we led; cruel at times the suffering and wrong; wild the profligacy. For after eight o'clock at night no prying eye came near till the following morning; no one lived in the same building; cries of joy or pain were equally unheard; and excepting a code of laws of their own there was no help or redress for anyone.' The various forms of torture practised can be read in their revolting detail in accounts such as those by Coleridge and Ainger. Evidence of sexual immorality is notoriously unreliable: the reader must form his own estimate on the basis of probability. On food, however, there is plenty of evidence. The authorities provided one meal a day: dinner in hall consisting (invariably) of mutton, bread and beer. The fags existed, for the most part, on whatever they could scrounge but the Sixth Form boys were served with cold supper in Chamber, waited on by the fags who afterwards 'cleaned the plates, knives and forks on their surplices or any other of their garments that they might find handy' (Hollis). In these conditions vermin could well have become a serious problem. 'What was left over from these suppers was used as bait for the rats by which Chamber was constantly infested . . . The boys caught them in their stockings and then banged them to death against their bed-posts. When in 1858 the floor of Old

Chamber was taken up, two large cart-loads of mutton bones had to be removed' (Hollis).

The only amelioration of this life-style was provided by the Dames who kept private houses in the town. In return for a fee a colleger could go round to his Dame's after early morning school, wash and have breakfast and even, for a further fee, have the use of a room in her house when Chamber became intolerable. And so a boy for whom, in theory, the College provided free board and lodging could, for about £80 p.a., enjoy some of the amenities of an Oppidan.

Oppidans, fee-paying pupils who lived in the town, had been admitted at an early stage of the College's history. Provost Roger Lupton in his will bequeathed enough money to provide for seventy scholars and a hundred Oppidans and, while the number of scholars was fixed by statute, the number of Oppidans steadily increased. Although Eton's early connections (pace Henry VI's foundations) had been closer with Oxford, 'in the 18th century the formula, Eton College to King's, Cambridge, became established as a rule and thereafter a place on the foundation carried with it a chance of substantial material benefits. The traffic was reciprocal and mutually advantageous. The Headmaster, Lower Master and all the assistant masters at Eton were former Eton Collegers and Kingsmen. Eton was a closed shop, providing jobs for its Old Boys' (Chandos). On the other hand the immediate advantages of being an Oppidan were so apparent that a parent, who had the means to do so and who wanted his son to go on to King's, often deferred the boy's taking of a scholarship – the necessary means of entry – until he had reached the Sixth Form. Oppidans lived in houses run by Dames or by masters (the last Dame's house closed in 1905) and – outside the classroom at least – led reasonably carefree lives.

Christopher Hollis is right to claim that 'there was really little at Dotheboys Hall which could not be paralleled in Long Chamber' and he asks, 'What was the reason for all this? Who was to blame? It was not that the College was poor. The College was far richer than it had been in the early 17th century . . . The reason was that the Provost and Fellows consistently misappropriated the College's money to line their own pockets . . . For generations Provost and Fellows had behaved as if all the estate of the College was their private possession to be disposed of in accordance with their pleasure. [Meanwhile] the Headmaster and Lower Master, regardless of their duty to supervise the Collegers, had moved from the rooms assigned to them at the end of Long Chamber into more convenient and distant quarters' (Hollis). They also enriched themselves by levying fees, contrary to the statutes, and 'a custom was allowed to grow up by which boys on leaving made presents of considerable value to both of them' (Hollis).

In earlier times the Headmaster himself had taken the whole of the Sixth and Fifth forms in a single division and, even when the Fifth Form had been split into two divisions, the Headmaster found himself responsible for 170 pupils, a total which rose to almost 200. It must have presented a remarkable scene with the Headmaster presiding, and teaching, from his desk at the north end of Upper School and the assistant masters conducting their divisions from the other desks in the body of the room. 'The boys had no desks, only benches. When they had any written work to do, they had to use their hats as desks' (Hollis). Such a system could only have worked with a curriculum that was 'not only almost exclusively classical but narrowly and monotonously so, a constant recapitulation of the *Iliad*, the *Aeneid* and Horace, together with books of extracts known as *Scriptores Graeci* and *Scriptores Latini*' (Hollis). All construing (translation) was done in oral fashion. 'Written translations from Greek or Latin into English, or vice-versa, were unknown . . . as were also written answers on history or divinity' (Maxwell Lyte). The entire staff consisted of classical masters, themselves the products of Eton and King's, Major Hexter, the 'Writing Master' (and the only teacher of mathematics), not being regarded as a proper 'beak'. It must be borne in mind, however, that much of the real teaching went on in the pupil rooms of the masters and not in the main schoolrooms. Although the custom of having one's work supervised by 'm'tutor' was evolving only slowly, a master with wider interests (like William Johnson, Arthur Benson or Oscar Browning later on) could stimulate a pupil's understanding and his desire for knowledge. Wealthier parents, moreover, provided and paid for external tutors who could instruct their sons in a variety of subjects in out-of-school hours. But not till the foundation of the three Newcastle Scholarships in 1829 was there any general examination, either in classical scholarship or religious knowledge, and although much time was spent in Chapel we find nothing of the reverence and awe which were present in Rugby Chapel.

'There, as the boys in the long row of benches knelt down, pencils were taken out and the odds, given and taken, were duly set down in the blank leaves of the prayer books: "Five to three against Vyne's stumping four in the match" (Eton v. M.C.C., 1813); "Done, Barker, if you will lay four to one against Barnard with catching".'

And can they be blamed when one finds a boy writing home, 'A clergyman read prayers for about twenty minutes but was perfectly inaudible on account of the noise made by the boys'; or when another recalls that 'the sermons were intolerably long . . . mumbled by old men with weak, smothered voices, not one word of which could be heard except by those immediately under them' (Hollis)? Nor must the contribution of the organist be overlooked. He was, it was said, a most charitable man 'for his right hand knoweth not what his left hand doeth'.

When one turns from the sterility of much of the academic and religious components of an Eton education to the freedom of life beyond the College walls, the contrast is a remarkable one. Here, too, there were rules and regulations but generations of Etonians had mastered every possible way to circumvent them, and beyond taking note of flagrant breaches of discipline and suppressing the occasional riot the authorities turned a blind eye to most of the boys' activities. The correct term for this procedure was 'shirking'. 'If a boy upon an unlawful journey saw a master approaching he simply dodged into a shop or down an alley until the coast was clear. The master might see the boy, but he was not supposed to see him, and it was contrary to the rules of the game for him to take cognisance of the boy's presence. The act of shirking was accepted as an adequate gesture of respect for the law, and an admission of its power' (Chandos). 'The only two things that are clear about boating at Eton . . . are that it was technically forbidden and that it was freely practised' (Hollis). And it was forbidden not because of the risk to life since swimming was seldom taught (and boys were not infrequently drowned) but because, in order to reach the Thames, it was necessary to violate the statutes of Henry VI. The procession of boats on the Fourth of June was frequently witnessed by distinguished visitors, and masters often went to the fireworks but took care not to be recognised.

Cricket was the only team sport indulged in, but again it was organised by the boys themselves who also employed professional coaches at their own expense. Here at least it is pleasing to note that the authorities marked Etonian success (especially against MCC) with approval. When Sir Christopher Willougby, after a successful innings, came into school, 'all construing ceased . . . and the Headmaster greeted him . . . with delight' (Hollis). The Wall Game, on the other hand, having enjoyed a turbulent history, was banned in 1827 as a result of a general affray. Poaching continued to be immensely popular, both for the risks involved and for the chance of supplementing a meagre diet, and there were several instances of boys being fired at by gamekeepers.

But if such sports seem the very epitome of schoolboy fun and daring, there were also many unattractive features of life outside the schoolroom. Relentless bullying of the eccentric and the vulnerable was all too common. Vandalism and drunkenness were far from rare, and bare-fist fighting was the accepted way of settling disputes. The most notorious of such fights took place in 1825, the opponents being Charles Wood and Anthony Ashley (youngest son of the sixth Earl of Shaftesbury). It arose from an accidental kick by Ashley which Wood chose to regard as deliberate. For a few rounds Ashley (who was smaller and younger) managed to hold his own. 'But after the fight had been going on for about two hours he became weakened and exhausted. His second asked Ashley major several times to stop the fight but he refused, each time urging his brother on to one more round for the honour of the Ashleys. For the last twenty minutes neither combatant was capable of striking a blow and finally Ashley collapsed and Wood, falling forward at the same moment, landed on top of him. Ashley in a state of coma was carried back to his house by two friends and placed on his bed. They did not inform the housemaster, thinking he would wake up later on but he never came out of the coma and around midnight he stopped breathing. Dr Ferguson was sent for but all he could do was to certify that Anthony Ashley was dead' (Chandos – shortened). Wood and his second were summoned to appear for trial but Lord Shaftesbury refused to prosecute (even though Wood's father was a personal and political enemy) and the charge of manslaughter was dismissed for lack of evidence. The Headmaster was not held to blame in any way and indeed his subsequent address to the school made a deep impression on the boys. 'He proceeded to express his sympathy with the bereaved parents in strains of genuine, because it was honest, eloquence and to urge upon us that, for the future, we should act in such cases with better judgment and under a deeper sense of responsibility' (Chandos). The Headmaster concerned was John Keate and some assessment of the man, and of his headship, must now be attempted.

Anyone reasonably familiar with the system of education and the methods of teaching employed in England today and invited to make a study of the country's educational history could scarcely do better than begin with an examination of John Keate's reign at Eton. At once one is transported into a different world having, it would seem, only a language and the physical laws in common with our own. Some idea of the circumstances of life at the College may have been gained from the preceding pages but the picture is incomplete without the central figure who was its ruler for 25 years. Born in 1773 John Keate entered Eton at the age of eleven, going on to King's seven years later. By 1797 he had been ordained and had returned to Eton as a master. In 1803 he was appointed Lower Master and at the age of 36 he became Headmaster. His predecessor, the mild-mannered ineffective but intensely conservative Joseph Goodall, held office as Provost throughout Keate's reign and 'offered obstinate opposition to every sort of change for 31 years', the statutes laid down by Henry VI remaining inviolate until the 1860s. These factors need to be borne in mind when it comes to assessing Keate's headship.

He was very short, but possessed strength, agility and courage out of all proportion to his size. According to Alexander Kinglake, 'He was little more if at all than five feet and was not very great in girth, but within this space was concentrated the pluck of ten battalions'. He had, we are told, 'a red face with a fiery eye' but his most unforgettable features were his red, shaggy eyebrows,'so prominent that he habitually used them as arms and hands for pointing out any objects whatever to which he wished to direct attention.' Clad always in a cloak, never without an umbrella out of doors and wearing an outdated cocked hat, Keate was a gift for caricaturists, a gift which Old Etonians the world over never spurned. 'When he approached a corner he gave the cough peculiar to him – Ba-a-affin – a sound that inspired the nickname by which he was known to all. After the cough came the point of the umbrella . . . to give further notice of his presence, and to prevent the outrage of possible collision 'by accident' with a boy hurtling round the corner; then followed the cocked hat and the sturdy little body in the widow woman's gown' (Chandos).

Keate was not merely caricatured, he was – inevitably – mimicked, most memorably by the Duke of Wellington's eldest son, the Marquess of Douro, who contrived to paint the Headmaster's green front door red one night, the watchmen swearing that they had observed Dr Keate himself armed with a paintbrush. And in addition he was the victim of innumerable practical jokes,

jokes that were 'hardly subtle . . . but [which] were served up with tireless gusto [so that] when one generation outgrew them another was on its heels ready to carry on the good work' (Chandos). More often than not Keate was quite capable of taking practical jokes, literally, in his stride. Locked out of his own classroom and compelled to make a long detour he arrived to find the doors to his own desk (which was raised on a dais and enclosed by panelling) screwed fast but the jubilant booing of his pupils changed to involuntary cheers when, 'his face glowing with rage like an angry meteor', he vaulted over the door and took his accustomed seat.

Keate was an outstanding scholar and could have been an inspiring teacher but the methods he felt himself forced to employ in personally conducting a class of up to 200 pupils (aged from 16-20) nullified scholarship and made good teaching almost impossible. Nor was his task made any easier by his assistant masters. They were, of course, Old Etonians themselves, men who had experienced only one system of education in their lives and who were 'extraordinarily unfitted for their positions'. Keate did contrive to make some well-judged appointments himself but to dismiss incompetent masters was beyond his powers. Only when he had the opportunity of teaching small groups of sixth formers in a separate room did Keate's real gifts show themselves. Gladstone long remembered the brilliance and stimulation with which he conducted such lessons giving his pupils the benefits of his deep understanding of the Roman and Greek poets.

For much of the time, tragically, these gifts remained hidden and Keate seemed to be doing no more than engage in a series of battles against an intransigent foe. 'When every allowance has been made . . . there is no denying that Keate's disciplinary methods were such that it is difficult to see how any sane man can have imagined that they would effect their purpose' (Hollis). For one thing he almost always assumed that a boy was lying. 'He abhorred above all things the tender of monotonous and overworked excuses . . . even more, if anything, when they were the truth' (Chandos). The expression, 'I see guilt in your eyes!' was familiar to every Etonian. Conducting, as he saw it, a lone battle against ignorance, idleness, untruthfulness and hooliganism, Keate's only weapon was the birch. In this he was no different from almost all 19th century headmasters: it was the extent to which he used the weapon that has preserved his reputation for flogging down to the present day. On one occasion defiance of his orders by the Lower Fifth led to his decision to birch all 80 of them – in public, as always. 'After he had birched about twenty without incident the spectators began to grow restive and to throw eggs at Keate. A more ridiculous and undignified spectacle than a Headmaster birching away [while] dodging eggs . . . it would be difficult to imagine' (Hollis). On another occasion 72 senior boys were birched as a result of widespread cribbing. The performance was described as 'a grand scene in the Library . . . the floor covered with victims and the benches and tables with spectators . . . jests and laughter accompanied the executions' (G.F. Lamb). An even more notorious episode occurred when a hundred boys deliberately

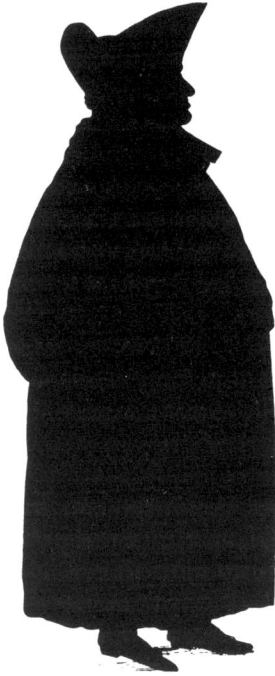

Dr Keate, the famous silhouette

'drawn from life in the cover of a Horace' by W. Harvey K.S.

Headmaster's Chambers, showing birches and block

Long Chamber, from Radclyffe's Memorials of Eton (1840)

Edward Craven Hawtrey, by Hélène Feillet

Dr Hawtrey

avoided a roll-call as a protest against one boy's expulsion. This time Keate employed the masters to drag them from their beds one by one and bring them to the block, 'and the floggings went on steadily through the night and up into the small hours of the Sunday morning' (Hollis).

'Yet despite appearances and his fearsome demeanour, Keate was not really a harsh or cruel man' (Chandos). Unlike many other heads, 'he was never known to punish any individual boy excessively' (by 19th century standards, that is) and the number of strokes was limited to five with a single birch, or ten with two birches for aggravated offences. It has been estimated that, on average, he flogged about ten boys a day. Alicia Percival draws attention to another aspect of Keate's disciplinary methods. During Goodall's headship, when Keate was Lower Master, he 'had the humiliation of sending up boys in the bill for punishment to the Headmaster and of finding their offences had been condoned or laughed off.' When he in turn became Head and assistant masters put down boys' names for punishment, he 'had the choice of either administering it . . . or of letting down his masters'. In the latter case several might well have decided to administer corporal punishment themselves.

This extraordinary man was also a person, 'when not on parade, of much geniality and kindness'. He and his wife, a very popular woman, used to entertain small groups of Etonians to breakfast and supper parties, occasions which were greatly enjoyed. And even his wrath could be turned aside quite unexpectedly. On one occasion when he had been thundering at one of his pupils, the boy stood up and said, 'You know, Sir, that I have always regarded you as a father'. And after a moment's silence Keate said quietly, 'I believe you do'.

Nor did Keate, himself obviously a Tory, attempt to stifle or censor opinions that ran counter to his own even, on occasion, when they might have been considered subversive. Young Gladstone, at 18, was clearly something of a firebrand and he was – one would have supposed – risking expulsion by writing verses in praise of the Cato Street Conspiracy (1820) which had aimed at slaughtering the entire cabinet. But Keate raised no objection to the publication.

When all these aspects of this remarkable man are considered it is perhaps not so surprising that Etonians – in the main – had a real affection for their Headmaster, and when his impending retirement was announced they subscribed £600 (say £25,000 today) as a leaving present and purchased with it some beautiful pieces of silver. 'Presented with them, Keate was so moved that he could do nothing except take off his cocked hat in recognition – the first time he had ever been known to raise it to anybody. Then, going into Chamber, he threw the (famous) hat upon the floor before the masters, saying, "This will not be used again" ' (Hollis).

On an earlier occasion, not long after Waterloo, he had been spotted by some Old Etonians on a visit to Paris. They decided to give him a dinner at Beauvilliers, the hosts including Lord Sunderland, who had been expelled for firing a small cannon on Keate's private lawn. Keate expressed his delight at finding that his old friends and pupils had not forgotten him, concluding a neat and appropriate speech with 'Floreat Etona'.

Even before Keate's retirement, criticism of Eton had been mounting, and numbers had been declining. The *Edinburgh Review* had printed articles 'criticising severely the education given at Eton, the text-books used and the manner of instruction' (Hollis). Numbers which had stood at over 600 in 1830 had dropped to about 450. Edward Hawtrey, who had been the senior assistant master, succeeded Keate and although he instituted some reforms there was not a great deal he could do until Goodall died in 1840. One immediate improvement was effected by Hawtrey removing himself from Upper School and teaching a mere 32 boys in the room next door. To the curriculum itself he made few changes but individual masters who now had charge of regular divisions were emboldened to introduce some improved methods of teaching which they had imbibed at Cambridge and which emanated from the great B.H. Kennedy of Shrewsbury. Hawtrey cannot have been unaware that this was going on but he always believed that 'the best work was generally got both out of masters and of boys by encouragement' (Hollis), and he did not interfere.

With the appointment of Archdeacon Francis Hodgson as Provost in 1840 real improvements were at last possible. He himself had survived several years in Long Chamber 'and the great task to which he felt himself called – by divine vocation – was to reform the life of College' (Hollis), both for humanitarian and academic reasons. ''Please God,'' he is reported as saying on his arrival, ''I will do something for those poor boys''. He began, working in close harmony with Hawtrey throughout, by reforming the diet with, inter alia, a proper meal served for supper at 8 p.m. Long Chamber was to become simply a dormitory, with separate cubicles, for the fifteen youngest boys. The New Buildings, of which the Prince Consort laid the foundation stone in 1844 and which were opened in 1846, provided bed-sitting rooms for the remaining 55 scholars. Arrangements were made so that Collegers, like Oppidans, should now be served breakfast and tea and accommodation was reserved for a resident Master in College.

The effect of these improvements was seen almost at once. Whereas the conditions for admission to College had previously been 'a grotesque scandal', with the opening of the New Buildings 'sixty candidates presented themselves for scholarships, prepared to do battle against one another in fiercely competitive examination' (Hollis). A Library was added at the far end of the New Buildings and a sanatorium built some distance away. Of equal importance was the coming of the railway to Windsor. It enabled the authorities to take the bold step of closing down the *Christopher*, the only inn where visiting parents could be put up for the night but also the focal point of many of the illegal activities that degraded Eton life. The advent of the railway also spelt the doom of Montem, a triennial event of much pageantry (well described in *Coningsby*), picturesque and largely harmless in itself, although the subsequent celebrations at the *Windmill* and the *Castle* tended to involve a good deal of vandalism. These in themselves would have been insufficient reason for suppressing an ancient custom, but cheap excursion tickets brought large crowds of casual spectators and every possibility of serious rioting. Even the reduction in scale of the 1844 Montem proved insufficient and Provost and Headmaster were agreed in deciding to make it the last.

Numbers started to rise again, reaching 777 in 1844. Numerous anomalies in the rules were abolished. The existence of the River Thames was officially recognised and boating was permitted though only to those who had passed a swimming test. Watermen were appointed to teach boys to swim. Cricket was positively encouraged and fives courts were built.

Although academic standards were undoubtedly raised the curriculum itself was not greatly widened. French, although tolerated, was scarcely encouraged in spite of the award of a prize for modern languages by the Prince Consort. The French master was not officially on the staff (he levied a fee on his pupils) and had no power to enforce discipline. Mathematics were taken on a purely voluntary basis until 1851 when Hawtrey's cousin, Stephen, was appointed Mathematics Master and a special building erected in South Meadow Lane. He hired several assistants to share the work but, again, their disciplinary powers were limited, they were not permitted to wear cap and gown and, unsurprisingly, were regarded as inferior beings. Geography continued to be treated as an extension of Ancient History and 'boys who did not know the locality of St Petersburg or Washington were expected to give the names of modern villages (such as Ramadan Oglu or Kissovo) occupying the sites of towns of the classical world' (Chandos).

The root of the trouble was the statute limiting the appointment of assistant masters to men who had been educated at Eton and King's. 'This was a very restricting choice as King's, at that time a College exclusively for Etonians, lived its own sheltered and indolent life and was hardly, except in name, a part of the University at all' (Hollis). Hodgson, so far-sighted in most other matters, in this respect would not agree to a change, and it was only when Hawtrey himself succeeded Hodgson as Provost in 1852 that the rule could be relaxed. By 1868 the number of boys had risen to over 900, the number of masters had increased from sixteen to thirty, and of those thirteen had been educated at other schools, enabling the College to employ six maths masters and one to teach science.

There is another aspect, too, and an important one, to the raising of academic standards. 'Once the gaining of scholarships was made dependent on a very stiff examination, it meant that only boys who had been at some exclusive and expensive private school which prepared for that examination had any chance of winning a scholarship and College, far from offering a career that was open to talents (which would appear to have been the aim of Henry VI) became . . . exclusively populated by . . . the sons of gentlemen' (Hollis).

No account of the changes wrought by Hawtrey in the 1840s would be complete without a brief description of the man himself. His attitude to his work is best expressed by a remark he once made: "Living here, I cannot feel the sadness of growing old, for this place supplies me with an unfailing succession of young friends." The best known historian of Eton concludes his description of Hawtrey's work with these words: 'Such was the man; not an accurate scholar, though versed in many tongues; not thoroughly well informed, though he had spent £30,000 on books; not able to estimate correctly the intellectual development of younger men, though he corresponded with the leaders of England and France; not qualified to train schoolboys in competition with a Vaughan or a Kennedy . . .; not one that could be said to organise well, for from first to last he dealt in makeshift and patchwork; yet for all that a hero among schoolmasters, for he was beyond his fellows candid, fearless and bountiful, passionate in his indignation against cruelty and ardent in admiring all virtue and all show of genius' (Maxwell Lyte). It is a noble tribute to a remarkable man. Like all the best schoolmasters Hawtrey was constantly mimicked. There was one boy (Money) who could simulate Dr Hawtrey to perfection. One night there was a grand masquerade in Long Chamber and the Headmaster found it necessary to go there in order to put an end to the uproar; but the Collegers had so often been taken in that they only shouted, "That won't do, Money. We know *you* well enough".

It was not long after Hawtrey's death in 1861 that the announcement was made of a Royal Commission to enquire into the condition of the nine 'great schools'. The findings of the Clarendon Commission will be discussed in due course. Here it is sufficient to remark that 'it was quickly established that the fines on the renewal of leases had never been audited and that over the years an enormous sum of money had found its way into the pockets of the Fellows in a furtive manner that was quite unethical' (Bamford). This was clearly scandalous but the Commissioners did not limit themselves to investigating the College's income and there is a nice irony about the questions directed to the Headmaster, Edward Balston, 'a great upholder of the classics', concerning the teaching of French at Eton. Considerably more damaging, however, was the evidence, much of it from university tutors, of the low standards attained by many Etonians in the one branch of learning in which they might be expected to excel – the classics.

When the Commissioners' report was issued it made a host of recommendations affecting the financial provisions, the governing body, the curriculum and other matters, but of course they were not binding. 'There was a long debate about what should be done. Balston quite clearly hoped that . . . nothing would be done' (Hollis), but seeing that changes were inevitable he resigned. The appointment of his successor broke tradition in a manner in keeping with the Clarendon Report. James Hornby was indeed an Old Etonian, but a mere Oppidan who had not only gone on to Balliol but had been teaching at Winchester where he was then Second Master.

The Public Schools Act of 1868 allowed Eton to conduct its own reformation. 'The old Fellows were swept away and in their place a new Governing Body was established' (Hollis). The statutes of Henry VI were repealed. The condition that scholars be 'poor and needy' was abolished. Changes in the curriculum included the introduction of 'extra studies' by which every boy had to devote a few hours a week to a subject outside the narrow range. And in 1869 Science, for the Fifth Form, made its appearance. Entrance to College must be between the ages of twelve and fourteen. Dames were gradually replaced by housemasters; and finally King's itself was to offer half its endowments to non-Etonians.

Space does not permit adequate treatment of Winchester. Many of the features of life at Eton could be paralleled at the older College, but the system of entry does merit a description of its own. 'Election to a Winchester scholarship could once guarantee a subsequent scholarship to New College, a benefice and security for life' (Sabben-Clare). For although William of Wykeham had made meticulous dispositions for most features of his College, 'the nature of the examination itself was left regrettably vague'. As a result 'the examiners . . . were able to bring forward their nominees and put them through the merest charade of an examination' (Sabben-Clare). One example will suffice: 'In July 1842 I went down with my father . . . to the Winchester election . . . The Candlesticks [candidates] were ushered into the Election Chamber where sat in awful state the two Wardens [of New College and Winchester], the Headmaster and three others. I had prepared with great care a hundred lines of Virgil but had not construed three before the examiner said, "That will do; can you sing?" I stared, and answered, "Yes". "Say," he continued, "All people that on earth do dwell." I recited the line. "Thank you, you may sit down". My examination was over and I was elected' (Sabben-Clare). The brief recitation's purpose was to satisfy the statutory requirement for proficiency in plainsong.

When Winchester was at length compelled to follow the lead of Eton in instituting a competitive examination the improvement was so rapid that in 1862 there were 137 boys competing for seven vacancies. Dr Moberly, who had earlier resolutely opposed change, could then claim, 'We do not know what it is to have a thoroughly stupid boy as a scholar' (Sabben-Clare).

4   Underlying Principles

The stage has been reached at which one must attempt some sort of judgment of English public schools in the early part of the 19th century, what might be termed the pre-Victorian age. Their defects, one hopes, are apparent from the foregoing pages and their virtues not entirely neglected. What principles underlay the system of education at the great schools and were those principles to be found in all, or only some, of them?

The 18th century, which bestowed great wealth through the ownership of land on hundreds of English families, also brought the civilization of Western Europe within reach of those families. For example Palladio had flourished in 16th century Italy but not until the mid 18th was his style of architecture adopted in England. Between them Lord Burlington, William Kent, the Adam brothers and others awakened wealthy Englishmen to an appreciation of the classical style, while Capability Brown put their houses into an Arcadian setting, and young Englishmen furnished them with paintings and sculptures from the Grand Tour. Nor was the interest in the classical world confined to art and architecture. Newton's 'Principia Mathematica' may have been the last important book to be written in Latin but the scholarship of men like the renowned Richard Porson ensured that the study of the classical languages was still in the forefront of learning. And from 1748 onwards the amazing discoveries at Pompeii awakened, or reawakened, the awareness of educated Englishmen to the culture of the classical world.

To the cultured Englishman of the late 18th and early 19th century, then, the attitudes to life and death of civilised Greeks and Romans two thousand years earlier made a strong appeal: the heroes of Homer and Virgil were admired; the philosophy of Plato and Cicero struck a chord of sympathy; and the indecorous satires of Aristophanes and Ovid at least cloaked less elevated feelings with scholarly wit. Englishmen nurtured on such a diet were not likely to be shocked by the affairs of the classical deities. Indeed it would probably be true to say that an English nobleman or gentleman of that period felt a closer affinity to cultured paganism than he did to Christianity.

That is not to say that Christian principles were not observed or that any open defiance or ridicule of them would have been tolerated (Chapter 15 of Gibbons' *Decline and Fall* had come in for severe criticism). In the main the Ten Commandments were observed and church attendance was part of the weekly routine. What was distrusted was enthusiasm, the religious enthusiasm of a Wesley, a Wilberforce or a Hannah More, which sought to put Christianity and its accompanying good works, together with its prohibitions, in the forefront of everyday life. And although Wesley and Wilberforce were the last people to approve of the atheistical principles of the French Revolution there existed a strong feeling that where enthusiasm, licence, novel ideas and a general questioning of the accepted standards were encouraged or permitted there pointed the way to the English Revolution. Hence the antagonism aroused by such varied reformers or innovators as Tom Paine, Percy Bysshe Shelley, Henry Brougham, Henry (Orator) Hunt or almost any of the would-be Trade Unionists, Chartists or Evangelicals who flourished in the early 19th century. Of course distrust of religious enthusiasm does not imply contempt for religion (the opposite is often true) but it may well tend to breed it. It is easy to make too much of the comparison between Rugby Chapel and the Chapel services at Eton but clearly Arnold felt himself fighting against the forces of anti-religion.

If then the education offered at public schools in the early 19th century was indeed based, partly consciously and partly as a reaction against revolutionary stirrings, on the classical world and its ideals, what principles should we be able to detect underlying that education? At bottom one finds conservatism, not in a narrow political sense but standing for the value of

tradition, of long-established institutions and customs as against innovation
and change. Here schoolmasters and schoolboys were united against a
common foe. 'Boys were tenaciously conformant to precedent and
established usage. At Eton, which was . . . the freest and least regimented of
. . . schools, compliance with certain forms of language and conduct were
effectively exacted, not by formal prohibition, but by the power of example
and the fear of ridicule' (Chandos). At Winchester, when Dr Williams sought
to improve the chaotic manner in which breakfast was served, 'he received a
disagreeable surprise. [When the] Wykehamist Commoners found breakfast
laid out for them in civilized style . . . with all needful crockery ready provided
. . . the boys rose en masse in protest and smashed the articles provided for
their comfort' (Chandos). Such an innovation was clearly copied from
despised private schools and as such was 'abhorrent to them'.

From the preservation of tradition and contempt for 'improvements' it is
only a short step to admiration for hardihood and endurance, for stoicism and
for what the 18th century called 'bottom'. Hardship is part of the life of all
classes and one is failing to prepare one's sons for life if one shields them from
hardship at school. A man (meaning a boy) must learn to stand up for himself,
to take – and give – knocks and to accept hardships without complaint. 'At
Westminster winter draughts of icy wind swept through cracks in the walls
and unrepaired windows [and], when the weather approached freezing point
juniors were roused in the middle of the night to pitch buckets of water down
the length of the dormitory for a slide in the morning' (Chandos). William
Barrow, in his *Essay on Education*, quotes an admiral as saying, 'Were it not
for the dormitory at Westminster and the quarterdeck of a man-of-war we
should be a nation of macaronis'. At Eton – and elsewhere – bare-fist fighting
was the accepted method of settling a personal quarrel. Lacy Yea, aged 13,
fought 'a very desperate battle' against a much bigger boy of sixteen and 'won
it by sheer pluck'. Thirty-three years later Colonel Yea distinguished himself by
his heroism and leadership at the battles of Alma and Inkerman, being finally
killed leading an assault on the Redan at Sebastopol.

And this leads us on to what was undoubtedly the most highly prized aspect
of public school life from the boys' point of view: their liberties. As we have
seen, apart from the obligation to attend lessons, meals, prayers and certain
roll-calls and, of course, to be in his house, dormitory or cubicle at night, a
public school boy enjoyed a freedom which would be unthinkable today. He
was free to go 'where he liked, with whom he liked, without hindrance' (*Tom
Brown's Schooldays*). Arnold himself realised full well the temptations that
such freedom offered and the 'evils' that a boy would encounter, but
education, he believed, was not sheltering the boy from evil but encouraging
him to stand up to it and conquer it. 'This, then, was what public school
traditionalists meant when they talked of 'freedom'; not the freedom of an
individual to flout or evade the customs and exactions of his peers, but the
freedom of boys, as a self-governing tribe, to live their lives and grow to
manhood without prying surveillance and interference by their titular

overlords' (Chandos). Put in another way, boys were expected to cope successfully, when the time came to enter the adult world, because they had been given liberty in their schooldays. Restrictions and excessive supervision merely prevented a boy from making the mistakes which were an essential part of his education.

Later in the century, when organised games had largely replaced the old freedoms, one finds a Headmaster of Eton, Dr Warre, to whom 'privacy was suspect and leisure (implying choice) an abomination', asking, "Do we know what they are doing?" of any boys 'ingenious enough to have evaded his autocratic clutch' (Chandos).

Very few boys, it is fair to say, and probably not many masters, worried their heads about the principles underlying the education they were getting or giving. A small minority of boys, coming under the influence of a Butler, a Hawtrey or an Arnold, might aspire to genuine scholarship and – at a later date – to the service of God and their fellow men. But the aim of most boys, one feels, could be expressed in simpler terms: to make the daily grind of Latin and Greek more bearable by a series of subterfuges (at the same time avoiding a flogging), to extract the maximum amount of enjoyment from out-of-school pursuits (both legal and illegal), to avoid all unnecessary contact with masters and certainly not to involve them in one's own private disputes; and finally (though largely unconsciously) to prepare oneself for a life of privilege, whether in the service of Church or State, or in the semi-feudal overlordship of inherited estates.

A schoolmaster, particularly a headmaster, would have phrased his aims for his pupils rather differently: that they should achieve a proficiency in the classics sufficient to enable them to proceed to University and subsequently to play a part in clerical or secular life befitting their status; and that the school should, at the very least, prepare him for his role as a Christian gentleman.

Arnold, and those who followed his lead, would naturally have aimed a good deal higher: to strengthen a boy's character, by academic teaching, by precept and by pastoral care, so that he entered adulthood fortified for the battle against temptation and sin and in doing so to take his share of the grave responsibility incumbent upon the ruling classes to give the lead to those beneath them by the uprightness of their personal lives and by their concern for those less fortunate.

In the end, perhaps, both learners and teachers would have been content to accept this aspiration as a fair one: 'If the effect of his schooling tended to make a subject brave, resourceful, confident and chivalrous, at ease with all ranks, truthful and honourable, yet shrewd in judging men and situations, and with such philosophic matter, great or small, in his head, culled from the noblest literature in the history of mankind – that might be accounted the education of an English gentleman, and who could ask for more?' (Chandos).

5   Universities and Academies

Even though a surprisingly small proportion of boys at the major public schools (roughly a third) went on to Oxford or Cambridge, the standards set – and the syllabus followed – at the two ancient universities inevitably influenced public school education – and at one remove that of preparatory schools – for most of the 19th century. The individual colleges had, for the most part, been founded between 1250 and 1550 for the benefit of indigent students and in order to preserve discipline. 'But by the 18th century there had grown up the practice of receiving large number of gentleman-commoners who paid fees and who greatly outnumbered the poor scholars' (Barnard). Since the fees they paid were considerable, these wealthy students expected privileges in return, principally that of receiving a degree without being required to work for it. Even for the serious students the standards required for obtaining a degree were absurdly low. 'An intelligent and well-prepared undergraduate, when he came into residence, often found that he had covered much of the work already; and that encouraged idleness' (Barnard). Not till 1780 was there any form of written examination at Cambridge; and twenty years later Oxford, while introducing written papers, retained the viva-voce. Cambridge, until 1824, offered only a mathematical tripos and when, in that year, a Classics tripos was established, 'it was open only to those who had already taken honours in the mathematical tripos' (Barnard). As it happened this innovation made it possible for Robert Peel to achieve one of the earliest 'double firsts' on record. The average undergraduate, however, was quite content to take a pass degree, enabling him to give 'the rest of his time to field sports in the country or to less desirable dissipation in London' (Barnard). These modest attainments should occasion no surprise. Most of the men who went up to university were not on the first rung of the career-ladder. They were, in the main, 'the sons of the upper class, far more interested in life and the gentlemanly pursuits' than in academic achievement. Many of the younger sons stayed on as Fellows until their college could offer them a clerical living which, with suitable patronage, would enable them to advance to an archdeaconry and possibly to a bishopric. Oxford and Cambridge did, however, have the unfortunate rule of celibacy (not relaxed until the 1880s) and it was the desire to marry that 'sent into the schools a supply of gifted academics' (Honey). Charles Wordsworth, while wandering round the Louvre, met an English girl, fell in love at first sight, abandoned a brilliant career at Oxford and was fortunate enough to obtain the Second Mastership of Winchester, worth £1000 p.a., with a house and other perquisites and minimal duties.

Reforms there were, in the first half of the nineteenth century, at both universities but they tended to stem from the efforts of a small group of men at a particular college, Oriel for example, rather than from the University as a whole. When, in 1833, 'a group of Oxford MAs put up a scheme for including some study of mathematics and science in the degree course, the measure was rejected' (Simon). Not unreasonably the *Quarterly Journal* commented: 'Bigotry and prejudice have doubtless had their share in leading to the formation of such a decision, but indolence and incapacity exercise an even wider and more pernicious influence; and an ascendancy of privileged inertness represses all attempts towards amelioration on the part of the more enlightened few.' Fifteen years later Cambridge did institute triposes in natural and moral sciences. 'Thus it was practically the middle of the century before the modern examination system was really under way at either university' (Barnard). Even then, at Oxford, the combined percentage of undergraduates reading modern history, botany and medicine was one percent. For many years to come the ancient universities continued to be mainly interested in Classics, Theology and theoretical mathematics. Moreover both lurked behind religious barricades. 'At Oxford no one could matriculate without subscribing to the 39 Articles; at Cambridge nonconformists might become members of the university, but they had not access to scholarships, fellowships, or university degrees'.

What academic openings, then, were there for young men without wealth or of the wrong religious persuasion? In the 18th century nonconformist academies had come into existence offering higher education to some of the most vigorous intellects of the day. The courses they offered normally lasted four years and were often superior, both in standard and range, to those at the two universities. Mathematics, medicine and natural science, philosophy and theology, modern and classical languages and also Hebrew all featured in their syllabuses. At Warrington Academy one of the tutors was Joseph Priestley, discoverer of oxygen. A man of wide-ranging interests he was far ahead of his time in his views on education, believing that 'the pursuit of truth and the practice of virtue' must underline all scholastic attainment. Such academies have been described as 'the forerunners of our modern universities' (Barnard), but it seems generally agreed that by 1800 they too were in decline.

The genesis of a new university may perhaps be found in a letter written by the poet, Thomas Campbell, in 1825. Published by *The Times* it was, in effect, a plea addressed to Henry Brougham for the establishment of a university in London. A meeting was held at the London Tavern, a prospectus issued and an appeal launched. A college was opened in Gower Street in 1828 with a curriculum embracing modern languages, mathematics, physics, law, history, political economy and medicine. Tuition fees of around £30 p.a. were to be charged. However there were no religious strings attached, a weakness which the Established Church was not slow to spot. Consequently a meeting took place, presided over by the Duke of Wellington (Prime Minister) with the two Archbishops on the platform, which led to the founding, in 1831, of King's College, London. The curriculum it offered was almost equally wide but was solidly based on a foundation of 'religion and morals'.

Not long afterwards the efforts of Bishop Van Mildert and others led to the foundation of the University of Durham. Unlike London it was organised on residential lines but at much lower cost than the ancient universities.

A brief summary of the state of education in England when Queen Victoria came to the throne might conclude that there had been little discernible improvement in the previous thirty or forty years. No more than the first stirrings of reform could be detected at the ancient universities. Those of London and Durham were still in their first decade: too soon for any judgment to be passed. The dissenting academies were nearly all in decline. The three 'great' schools, Eton, Winchester and Westminster, had been joined by half-a-dozen others which had undeniably attained public school status, but, except at Shrewsbury on a fairly narrow academic front and at Rugby on a wider spiritual and pastoral front, their aims and methods and their curricula had hardly progressed at all. Some long-established grammar schools were playing a valuable part locally on the educational scene; far too many were still in the grip of apathy or corruption. The State, for the past three years had voted £20,000 p.a. to assist the education of the poor, leaving it almost entirely to the British and National Societies to decide how the money was spent. Those societies were opening new schools (some in wholly unsuitable premises) but the education they were offering was of the most meagre and elementary kind, inevitable when so much of the actual instruction was being performed by older children. There were other charity schools, too, but without inspection their standards were just as low.

For the rest there were the private schools, ranging from child-minding dame schools to classical schools that, in syllabus at least, aped Eton and Westminster; from Dotheboys Hall to the high-minded (if somewhat priggish) Hazelwood; and, for girls, from Lowood to the Abbey School attended by Jane Austen. Finally one must not overlook the large numbers of children capably educated at home. For example the three Palmer brothers, Roundell, William and Tom, were taught entirely by their father, a clergyman. They started Latin at 5 and Greek at 6 and 'by the time Roundell was nine years old he was well grounded in Virgil and Horace . . . and had begun the Greek Testament. The works of Pope and Dryden were familiar and by the time [the boys] were approaching the age for dispatch to a public school they had made some progress in Homer . . . and were at least acquainted with Shakespeare and Milton' (Chandos).

2   The Growth of the Public Schools

1   The New Schools

The question as to what extent Arnold's influence spread to other public schools is one that is never likely to be resolved. Too many writers begin with a conclusion and then attempt to marshal all the evidence they can find in support of it. And a statistical weighing machine that can accurately measure the influence of a man's work, his example or his writing, has yet to be devised. The evidence of the Clarendon Commission (twenty years after Arnold's death) would suggest that he had had little direct influence on the older public schools. On the other hand by 1862 some twenty new public schools had come into existence. We need to look closely at them to see what effect they had on fee-paying education as a whole. First, why did so many come into existence in a period of twenty years? The 1840s were the decade of the railway boom. When Queen Victoria came to the throne there were some 500 miles of track; by the time of the Great Exhibition (1851) there were 6000. Not everyone, however, welcomed their relentless advance. In 1836 Dr Keate wrote to his former pupil, W.E. Gladstone M.P., asking for his help in opposing a project (the G.W.R.) which 'must be to the greatest degree injurious to Eton'. Two years later, however, when the London – Birmingham line, which passed close to Rugby School, was opened Dr Arnold remarked, 'I rejoice to see it and think that feudality is gone for ever'.

The advance of the locomotive was matched by the decline of the stage coach in, or on, which several generations of schoolboys had travelled at the beginning and end of term. But the railways themselves were simply the most spectacular manifestation of Britain's growing industrial might and prosperity. However much we deplore its more hideous aspects the Industrial Revolution ultimately brought greatly increased wealth (or the possibility of achieving it) to most classes of society. The middle classes experienced the benefit first and sought to bring up their children in a way that made them almost indistinguishable from the sons and daughters of the landed gentry. Improving health and consequent increase in the size of families was also to prove an important factor.

But 'mere demand for boarding education cannot produce new schools by itself. It may set people thinking and create a suitable climate for the production of a new venture, but fulfilment requires a great deal of planning and money' (Bamford). By the mid 19th century even the sovereign would scarcely have been wealthy enough to endow a public school i.e. give land and property whose annual revenue would provide, in addition to the fees, for the needs of the school, together with a large capital sum to enable the buildings to be constructed. In fact all the new public schools were founded by considerable numbers of well-to-do people combining their efforts to raise the necessary funds. The contributors might expect something in return for their money (though nothing comparable to investment in a business enterprise) or might simply make a donation 'with no strings attached'.

The first of the new public schools to open was Cheltenham. It was actually an amalgamation of several existing private schools with a complete range of new buildings. 650 shares at £50 each were put on offer. Each share entitled the purchaser to nominate one pupil (more often than not his son), and Cheltenham became the first of the Proprietary Schools. It catered mainly for families living locally, many of them having retired after service in India, and nearly all of the professional class. Tradesmen were specifically excluded for the following reason: 'Had we admitted tradesmen in the first place, we must have done so almost without limit, and in the confined circle of shops in Cheltenham, we should have had the sons of gentlemen shaking hands perhaps with schoolfellows behind the counter' (Bamford). But even without the dreaded tradesmen Cheltenham faced the fact that most of its pupils would have to earn their living, in the army, the Civil Service, the law and the Church. Separate Classical and Military departments were established from the outset, with entry to Sandhurst and Woolwich in mind.

'The development of railways ruined the old coaching inn on the London–Bath road through Marlborough and a fine building came on the market just at the time when a group of clergy, led by the Rev. G.H. Bowers . . . were looking for a site for a school which would supply first-class education at low cost for the sons of clergymen' (Honey). This, too, was to be a proprietary school but the need for economy presented grave problems from the very beginning. 'Having no experience of school management, the founders based their estimates of running costs on replies to their enquiries to a London Club and two Charity schools' (Honey). They nevertheless fixed the fees at 30 guineas a year for sons of clergy, 50 gns. for sons of laymen. The Rev. Matthew Wilkinson was a poor choice as the first Headmaster but the low fees encouraged an enormous response so that by 1848 there were 500 boarders. Sons of the clergy these boys may have been, docile and appreciative they certainly were not. Under Wilkinson's inconsistent leadership discipline deteriorated and in 1851 there took place what the school's official history, with considerable exaggeration, calls 'The Great Rebellion', but Wilkinson's replacement by G.E.L. Cotton, a disciple of Arnold, next year brought about a rapid recovery and a steady rise in Marlborough's status.

Rossall, founded in 1844, was in a number of ways the Marlborough of the north. Although it avoided Marlborough's disciplinary problems and was the first public school to have its Corps enrolled, it too suffered financial problems which were not really resolved for 30 years. Radley (1847) owed its existence to the Rev. William Sewell (already the founder of St Columba's College near Dublin), a member of the Oxford Movement, who failed to realise that his own views on fasting and on short commons generally would be unlikely to win his pupils' enthusiasm. 'Bulbs, flowers and acorns were taken from the grounds and eagerly devoured' (Gardner) but in spite of the economy the college ran into debt and was eventually rescued by Lord Addington.

The aims of the Rev. Nathaniel Woodard appear to have been clearly formulated at an early stage. 'Somehow or other', he wrote, 'we must get possession of the Middle Classes, especially the lower section of them, and how can we so well do this as through Public Schools . . . Education without religion is, in itself, a pure evil . . . Secular education makes Communists and Red Republicans [this was written in 1871] . . . Unless the Church, therefore, gets possession of this class at whatever cost, we shall reap the fruits . . . of an universal deluge' (Bamford). And, on another occasion, 'Till the Church do educate and train up the middle class, she can never effectually educate the poor'.

For a man with such high ideals Woodard proved also to be a remarkably successful fundraiser, a talent 'which he achieved by presenting the obligation as a duty rather than as charity' (Gardner). Money poured in and he began the task not merely of establishing and building a single public school but three tiers of schools which would eventually form chains extending across the country. 'The top tier was for the rich, the sons of gentlemen and of professional men . . . the next was for superior tradesmen, the semi-professionals and farmers. The third and lowest grade was to be designed for small tradesmen and artisans undergoing social mobility' (Bamford). The first three schools of the Woodard Foundation were St Nicholas College (1848), St John's Middle School (1849) and St Saviour's Lower Middle School (1858), now known as Lancing, Hurstpierpoint and Ardingly. By the time of Woodard's death in 1891, ' a network of schools had been created which included eight boarding public schools for boys and three for girls' (Honey).

The death of the Duke of Wellington (1852) sparked off almost a national debate as to how his memory should be enshrined, the family favouring an equestrian statue in every town. But, with strong support from the Prince Consort, the plan that was put into practice was one for a college devoted to 'the gratuitous . . . education of Orphan children of Indigent and Meritorious Officers of the Army'. In spite of the Duke's fame, and of the compulsory levy of a day's pay from all ranks in the army, only £105,000 was collected, more than enough, indeed, for the buildings and equipment but nowhere near enough to endow the school. Nevertheless work went ahead and Edward Benson was appointed Headmaster at the age of 29. He had been educated at King Edward's, Birmingham, under that inspiring teacher, James Prince Lee, and at Trinity, Cambridge, where he gained a First in classics and the Senior Chancellor's medal. 'In the space of fourteen years he succeeded in converting Wellington College into a great and important public school, despite the initial disadvantages of an inadequate endowment, unpromising material and some-what limited aims' (Newsome) and he achieved it by 'defying royalty without appearing to do so' (Bamford) by ensuring a steady flow of fee-paying pupils.

North West view of Marlborough College, Wilts., drawn by Isaac Shaw

The schoolroom at Lancing College (c.1870). Extract from College Rules: 'Any boy cutting desks or in any way destroying the property of the Society will be punished'.

Cheltenham College – Distribution of Prizes (c.1850)

Wellington College – North Front, 1873

Clifton, founded soon after Wellington, was also fortunate in the choice of its first Headmaster. John Percival, though briefly a master at Rugby was 'by no means a slavish adherent to Arnold's principles' (Gardner). Although mocked (nowadays) for his prudery, he was an astute selector of good teachers and Clifton soon built up a fine reputation for scholarship. He realised the importance of integrating day boys which he achieved by organising them into houses, with housemasters, and he founded a Jewish house so that those boys could practise their faith 'instead of just being exempted from Anglican rites' (Gardner). He was a pioneer in the spread of science teaching in public schools and taught the subject himself. Even more farsightedly Percival established the Clifton Ragged School in one of the poorer parishes of Bristol and encouraged Clifton boys to teach there as a form of social service.

Two more schools need to be looked at, since they illustrate another factor in the rapid expansion of public schools. In these cases small and insignificant local grammar schools were transformed by the vision and energies of two remarkable men. Edward Thring (Eton and King's) was 32 when appointed Headmaster of Uppingham in 1853, a school which then had 25 boys and two masters. Thring had a much wider view of education than most of his contemporaries. 'Every boy is good for something. If he can't write Iambics or excel in Latin prose, he has at least eyes and hands and ears. Turn him into a carpenter's shop, make him a botanist or a chemist, encourage him to express himself in music and, if he fails all round, here at least he shall learn to read in public his mother tongue and write thoughtfully an English essay' (quoted by Barnard). The traditional subjects must be retained but in afternoon school 'there was a wide range of optional subjects from which every boy could choose what interested him . . . and with the help of one of his masters Thring built up a strong musical tradition in the school' (Barnard). He had workshops and laboratories built and also a gymnasium, and he was a convinced believer in the value of organised games. Not surprisingly numbers quickly built up to 300 which he fixed as the limit so that he could know each boy individually.

Less famous than Thring was Daniel Harper but he achieved 'a scarcely less spectacular transformation' (Honey) at Sherborne, of which he was Headmaster from 1850 to 1877, as much by brilliant powers of organisation as by high educational ideals. Repton owed a similar debt to S.A. Pears who was Headmaster for twenty years from 1854.

John Henry Newman, the leader of the Oxford Movement, had been received into the Roman Catholic Church in 1845. He was the driving force behind the provision of Roman Catholic public schools at this period. Oratory, which opened in 1859, was unlike the others in being administered by laymen. Belmont Abbey was founded, near Hereford, in the same year, Blackrock College, Dublin, by the Holy Ghost Fathers in 1860 and Beaumont College by the Jesuits in the following year.

Thus, by 1861, almost twenty years after Arnold's death, a score of schools which would now be reckoned major public schools had either come into existence or had emerged from the ranks of moribund grammar schools. That year is a significant one in public school history: it marks the outset of the Clarendon Commission.

2   The Clarendon Commission

The Victorians were great people for enquiries, called Royal Commissions. They enquired into water supply, drainage and sewage, into hospitals and medical services, into army commissions and entry into the Civil Service, into mines and factories and the employment of children, and almost every enquiry revealed fresh horrors. Unlike those of today, however, the reports of Queen Victoria's Royal Commissions were frequently acted upon. And it is no surprise to find that education formed the topic of four Commissions in the space of twenty years. The enquiries into the two ancient universities had resulted in separate Acts of Parliament, for Oxford in 1854 and for Cambridge two years later. The Newcastle Commission of 1858 had recommended widespread reforms in elementary education but these were largely shelved for twelve years, apart from the introduction of the system known as 'Payment by Results'. (It is worth recalling that, by 1858, the 'decadent' Ottoman Empire had established 43 *secondary* schools).

As far as an investigation of the public schools was concerned, rumbles of unrest can be traced back to a journalist, M.J. Higgins, and a high court judge, Sir John Coleridge, two Etonians whose suspicion of financial malpractice at their old school had been aroused. But the spark that finally lit the fuse which was to explode beneath the nine 'great schools' was an article by Henry Reeve in the *Edinburgh Review* for April, 1861, 'which concentrated not so much on the outmoded education as on the illegalities exposed by Brougham's committee forty years before which, presumably, still went on unchecked. This was a challenge which accused the teachers of future Cabinet Ministers of embezzlement, and the issue could no longer be ignored' (Bamford). Even though the immediate issue was financial and administrative, the underlying concern about upper class educational standards had been growing for the past few years, ever since competitive examination had determined entrance to the Army and the Civil Service. The Royal Commission set up by Lord Palmerston was headed by Lord Clarendon (privately tutored and Oxford) who had already been Foreign Secretary for five years and was to be again for four more. Among its members were Sir Stafford Northcote (Eton and Oxford), Henry Halford Vaughan (Rugby and Oxford), Professor of Modern History at Oxford, Lord Lyttleton (Eton and Cambridge) and Lord Devon (Westminster and Oxford). Its brief was to enquire into the revenues, management and curriculum of those nine schools.

The opposition to setting up such a commission might have been stronger had it not been for the fact that, Eton and Harrow apart, the major public schools were suffering a disastrous decline in pupil numbers, while the newer schools were flourishing. Even so the headmasters of the leading schools were unanimous in their denial of the need for a Parliamentary investigation. Nevertheless 'the Commissioners conducted an extraordinarily thorough, even microscopic, examination of the position in each of the Schools' (Simon).

They began 'by sending round detailed questionnaires which went to the Head of each school . . . and, in May and June of 1862, Commissioners went to each school . . . They then called for verbal evidence and they interviewed 130 witnesses . . . [who] included all nine Headmasters but ranged also from the Dean of Christ Church and Professor Faraday to lads who had left their schools only two or three years before' (Alicia Percival).

As the enquiry proceeded, irregularities and abuses of privilege were laid bare day after day at school after school. 'No Headmaster came out of the revelations of the Commission with the honour of his school intact' (Bamford) but let a later judgment (by Firth) on Winchester serve to stand for the whole:

'But the law of the land, on whose letter the Warden and Fellows had relied for centuries to cheat the pupils and starve the ushers, now turned its bleak face against the exploiters themselves. They might whimper and snarl, like old dogs driven off a juicy and familiar bone; but their teeth were drawn'.

Although the checking of abuses was paramount some of the most interesting evidence submitted to the Commission concerns the curriculum and, in particular, the battle between the retention of the classics and the advocacy of science. An important witness was J.M. Wilson who was then teaching mathematics at Rugby (he was later Headmaster of Clifton). In addition to maths teaching he gave occasional lectures in science (attendance voluntary) which were conducted in the cloakroom of the town hall. While other scientists extolled the merits of their faculty in the abstract, 'it was left to Wilson to demonstrate the power and educative value of science with actual examples, showing science as a source of inspiration for the mind' (Bamford). Even more impressive was the evidence of Michael Faraday, who himself appears to have had no formal schooling, being apprenticed to a bookbinder and thereafter almost entirely self-taught. 'Asked whether science trained the mind, in the way the classics were supposed to do, Faraday launched a frontal attack on this whole approach to education: "Who are the men whose powers are really developed? Who are they who have made the electric telegraph, the steam engine, and the railroad? Are they the men who have been taught Latin and Greek? Were the Stephensons such? . . . It has only been those who having had a special inclination for this kind of knowledge have forced themselves out of that ignorance by an education and into a life of their own." ' (Simon).

And when they came to write their report the Commissioners showed themselves swayed by the scientists' arguments. 'Natural Science . . . is practically excluded from the education of the higher classes in England. Education is, in this respect, narrower than it was three centuries ago whilst Science has prodigiously extended her empire . . . This exclusion is, in our view, a plain defect and a great practical evil. It narrows unduly and injuriously the mental training of the young . . . With sincere respect for the opinions of the eminent Schoolmasters who differ from us, we are convinced that the introduction of the elements of Natural Science into the regular course of study is desirable' (Simon).

Although all the Commissioners had had a classical education themselves and were convinced of its value, they were far from blind to weaknesses in the way it was taught. They heard Charles Neate, fellow of Oriel College (Oxford) and Professor of Political Economy, claim that not only were the majority of public school boys 'almost incredibly ignorant' of English, modern languages, mathematics, natural history and modern history but also, after up to twelve years spent in exclusive study of Latin and Greek, were 'unable to construe off-hand the easiest passages'. Another Oxford don wrote: 'Very few . . . can construe with accuracy a piece from an author whom they profess to read . . . We never try them in an unseen passage. It would be useless to do so'. The Professor of Natural Philosophy at Cambridge declared: 'These latter [public schoolboys] can in many cases scarcely apply the *rules* of arithmetic, and generally fail in questions which require a little independent thought and common sense'. This ignorance of mathematics is confirmed by Matthew Arnold himself who confessed in later life that he had never thoroughly learned the rules of arithmetic until he embarked on a career as an H.M.I.

We can scarcely be surprised to find the Commissioners, in their summary of the evidence, stating that: 'Of the time spent at school by the generality of boys, much is absolutely thrown away as regards intellectual progress, either from ineffective teaching, from the continued teaching of subjects in which they cannot advance, or from idleness, or from a combination of these causes' (Simon).

A massive report in four volumes was issued in 1864 and when one has read only the briefest summary of the evidence one is surprised that the recommendations of the Commissioners did not go much further, surprised that is, until one reflects what sort of people they were and what was the cultural climate in which they lived and worked. They strongly advocated 'a ruthless and complete transformation of governing bodies' to be composed of 'trustees without pecuniary interest', in other words that the very considerable revenues of the older foundations should be directed to the purposes for which the founder had intended them. They put forward proposals to increase the efficiency of the schools: entrance and periodic examinations, prizes, the need for work done to be marked promptly, promotion from form to form etc. Their eleventh conclusion reads: 'The teaching of Natural Science should, wherever it is practicable, include two main branches, the one comprising chemistry and physics, the other physiology and natural history'. Some attempt, they felt, should be made 'to meet the case of that large class of boys, who are destined not for the universities, but for early professional life', proposing the inclusion in the curriculum 'not only of mathematics and a foreign language but also of music, drawing, history, geography and English composition' (Simon).

The Commissioners nevertheless averred that one of the great services performed by the schools had been the maintenance of classical literature as the staple of education and the commissioners were emphatic that 'the classics must be retained as the central core of education' though more emphasis should be placed on 'the content of the works studied and less on grammar and philology' (Simon).

'What really impressed the commissioners was that the public schools provided an excellent moral education and a sound character training for boys who had practical work to do in the world' (Roach), and they described the schools as 'the chief nurseries of our statesmen where men of all the various classes that make up English society, destined for every profession and career, have been brought up on a footing of social equality' (Simon). A classical education had proved its worth in shaping the minds and characters of the leaders of Britain for four hundred years and would be abandoned at peril to the future of the nation.

The claim regarding social equality was one put forward in direct contradiction to the facts. From Arnold onwards, headmasters had been at pains to exclude the children of tradesmen living in the neighbourhood, the very pupils for whose free education the founder had provided an endowment. To such an extent had fee-paying pupils taken over these schools that (for example) 'the children of Harrow parish cannot be sent safely or properly to the school on account of the . . . sons of the nobility and gentry of the kingdom, who constantly scoff at and ill-treat the other boys' (Simon). Anthony Trollope was one such victim. Apart from ostracism by the paying pupils, exclusion of local boys could be achieved by an increasing tendency to demand fees from their parents, Arnold setting the tone by immediately doubling the fees for supposedly 'free' scholars to twelve guineas per annum. Some compensation was offered by the provision of Sheriff's Grammar School at Rugby, by the Lower School of John Lyon at Harrow and by similar establishments; but a justifiable grievance remained in the minds of local people.

No exception to this process was taken by the Commissioners: rather they encouraged it by 'recommending the sweeping away of all such obsolete requirements (i.e. medieval statutes) and the opening up of the schools to competitive examination' (Simon), a solution 'entirely in accordance with the laisser-faire ideals of the Victorian upper classes and singularly effective for securing its end, since a good deal of expensive preparation was necessary to pass public school entrance examinations' (Mack). 'Expensive preparation' was, of course, the very thing preparatory schools were designed to provide.

The recommendations contained in the Clarendon Commissioners' Report encountered hostility and bitter opposition, both from reactionaries who 'thought interference dangerous and unnecessary' and from radicals who regarded it as 'one of the most delusive and reactionary measures ever put before the country'. There followed four years of controversy before the

Public Schools Act was passed in 1868. Brian Simon summed up what had been achieved when he wrote: 'The Clarendon Commission created an efficient and entirely segregated system of education for the governing class – one that had no parallel in any other country. The Commissioners had done what was required of them and had done it well'. Or, as Colin Shrosbree astutely commented (*Public Schools and Private Education*), 'The Public School Acts were in origin an expression of Whig reformist attitudes but, in effect and operation, were an expression of Conservative values which, however suitable for a gentleman's education, were likely to prove disastrous as a guide to secondary education in an industrial democracy'. Dr Shrosbree's book deserves to be studied by anyone interested in 19th century education in England.

3   The Taunton Commission

Thomas Arnold once observed: 'I have much to lose by revolution: I have nothing to dread by reform', and as David Allsobrook sagely observes, 'in [him] the principles of reform and guardianship of an aristocratic, gentlemanly tradition lived side by side'. In a letter to the *Sheffield Courant* in April 1832 he wrote, 'We are all aware of the growing power of the middling classes of society and we know that the Reform Bill will at once increase this power and consolidate it'. And he considered 'the education of the middling classes . . . a question of the greatest national importance'. One of his foremost concerns was the formation of a properly registered and certificated teaching profession. 'There is now no restriction', he wrote, 'upon the exercise of the business of a schoolmaster and no inquiry made as to his qualifications'. And, while the inculcation of Christian principles remained central to his concept of education for all classes, he also 'determined the path along which Broad Churchmen were to march towards a generous compromise with the demands of Dissent, in the interest of social and political harmony' (Allsobrook).

Nevertheless a further thirty-two years were to pass before, in 1864, the Schools Inquiry Commission was set up. It was entrusted with the colossal task of visiting and reporting on almost a thousand endowed grammar schools. The chairman was Lord Taunton (who, as Henry Labouchere, had been a successful President of the Board of Trade and Secretary of State for the Colonies) and it comprised a fine selection of the great and the good including: Lord Stanley, Lord Lyttleton (previously on the Clarendon Commission), Sir Stafford Northcote (also a Clarendon veteran and later Chancellor of the Exchequer), W.E. Forster (later responsible for the 1870 Education Act), Frederick Temple (Headmaster of Rugby, later Archbishop of Canterbury), Edward Baines (described as 'the statutory Dissenter') and Matthew Arnold (who was charged with investigating educational institutions in 'Foreign Countries' – effectively France, Prussia, Holland and Switzerland – a task that was highly congenial to him).

The enquiry was to prove 'a mammoth Victorian investigation of incredible scrupulousness, skill and industry' (Gathorne-Hardy), carried out, in the main, by a dozen assistant commissioners (including men of the calibre of T.H. Green and James Bryce). Their primary objective was to establish whether the endowments of the schools they investigated were being put to their proper purpose and it followed that, 'if the endowed schools were not doing good, they must be doing harm by standing in the way of better institutions' (Simon).

The work of the S.I.C. lies somewhat outside the scope of this book. Readers interested in more detailed analysis are referred to David Allsobrook's *Schools for the Shires* and Brian Simon's *Studies in the History of Education*. Here it will be sufficient to give as examples James Bryce's finding that only two schools out of nearly seventy in Lancashire were satisfactory while 'thirty-eight schools in Yorkshire and Durham, still claiming to provide a classical education in fact had no pupils at all' (Gathorne-Hardy). Elsewhere long drawn-out legal action drained completely the resources which the endowments provided.

Arnold's investigations are beyond the scope of this book but it is worth noticing that he considered that the reason the Germans were overtaking the British in industrial achievement was the emphasis placed on scientific education on the Continent. In his conclusion he observed that 'England possessed a few public schools which were excellent, but below that level there was nothing to compare with the state secondary schools of France and Germany' and, furthermore, England faced a period of rapid change 'with a working class not educated at all, a middle class educated on a second plane, and the idea of science quite absent'.

The Commission's enquiry resulted in a twenty-one volume report. It suggested that 'If all these endowments were put together and redistributed on a national scale, they could form the financial core of a great new national system of secondary education'. Detailed plans were proposed to encompass a modern curriculum, a national system of inspection and examination, and an overall financial strategy. 'The Commission's plans form one of the most fascinating might-have-beens of English educational history. They might have created a national system of secondary education before the public schools had established the massive prestige and authority which they gained in the next twenty years. They might . . . have produced an open system favourable to the poor but able boy or girl' (Roach).

The report was published in 1868 and the Endowed Schools' Act followed a year later but its provisions fell a long way short of the Commissioners' recommendations. H.C. Barnard simply says, 'It gutted the report'; but at least it gave wide powers to the three Endowed Schools Commissioners who, with the aid of seven assistants, at once set to work.

Almost inevitably the headmasters of the newer public schools reacted to the investigations of the S.I.C. with outraged horror, none more vociferously than Benson and Thring (the older public schools were, of course, excluded). And even to someone like Nathaniel Woodard who had championed the educational cause of the lower middle classes, the threat of state supervision caused considerable alarm since he feared it would undermine all he had sought to achieve in establishing education as the rightful province of the Church. John Mitchinson (King's, Canterbury and a pioneer in the teaching of science) and Daniel Harper (Sherborne) proposed a meeting of headmasters to confront the threat. Twenty-six attended, described by one of them, Edward Thring, as 'a very superior sort of men', and it was largely due to Thring's efforts that an ad hoc meeting was converted into a permanent organisation, the Headmasters' Conference; or, as Mitchinson neatly put it, 'I think that if I may fairly claim to have laid the egg . . . Thring did all the clucking necessary'.

At this stage the writer who has taken the whole of 19th century education as his field turns his attention to Forster's Education Act 1870. It may not, as many have supposed, have made schooling for the great mass of children either compulsory or free (though those aspects followed in the next twenty years) but it *was* the first great step forward in a national system of primary education. Secondary education, at least until 1902, had to depend, for a particular child, on obtaining a place at one of the revivified grammar schools. For the last thirty years of the century, then, attention is rightly focused on the development of State education. For the public schools these years witnessed no dramatic developments, rather a gradual infiltration of new ideas and techniques and a blurring of the distinction between the 'great' schools and the newer institutions. In some ways the most significant development was the increasing emphasis, documented by Dr Newsome, on athleticism and manliness, the effects of which are still felt today.

A deliberate gap has been left in this survey of education in Victorian England: there have been few references to the schooling available to the sons of well-to-do families in their earlier years. A good many, of course, were instructed by tutors in their own homes. But an increasing number of boys were being sent off to board at one or other of the scores of preparatory schools which had begun to flourish. It is time to assess the kind of education they provided, and the reaction of their pupils to it.

Sources

In chapter 3 the authorities consulted and quoted from have been:

Lord Berners *First Childhood*
Winston Churchill *My Early Life*
H. Freidrichs *Sir George Newnes*
Michael Gilbert *Prep School*
Robert Graves *Goodbye to all that*
T.A. Guthrie (F. Anstey) *A Long Retrospect*
Ian Hamilton *When I was a Boy*
Rupert Hart-Davis *Hugh Walpole*
Donald Leinster-Mackay *The rise of the English Prep School*
R. Meinertzhagen *Diary of a Black Sheep*
Kenneth Rose *Curzon, a most superior person*
Osbert Sitwell *The Scarlet Tree*
W.T. Sutthery *The Assistant Master, Past, Present and Future*
Leonard Woolf *Autobiography (vol.1)*
Hugh Walpole *The Crystal Box*
Eric Parker *Private Schools; Ancient and Modern*
Eric Parker *Preparatory School Assistant Masters*
Edward Peel *Cheam School from 1645*
R.G. Wickham *Shades of the Prison House*
Virginia Woolf *Life of Roger Fry*

3   Victorian Preparatory Schools

Donald Leinster-Mackay's research, embodied in his book, *The Rise of the English Prep School*, has clarified much that was vague and uncertain about the origins and development of this type of school and it seems unlikely that further substantial evidence will be added to the picture he has built up. This chapter, then, makes no secret of its indebtedness to Leinster-Mackay's work, though supplementing it by reference to other sources. He begins, quite rightly, by posing the question, 'What is a preparatory school?' The term 'private school' was the one more commonly used for the greater part of the 19th century and, while all preparatory schools were private (i.e. privately owned), by no means all private schools were preparatory (i.e. preparing boys for entry to public schools).

If Leinster-Mackay is correct hardly any true preparatory schools existed before c.1830. His criterion is simply that a school should exist *solely* for the purpose of preparing boys for their public schools or the Royal Navy and that the presence of older pupils whose education might continue to 16 or 17 disqualifies a school from 'preparatory' status. His assessment contains two further factors: separation and rustication. Separation is achieved by not allowing boys over 14 to be pupils at the schools in question; rustication is another word for boarding since early prep schools were usually established in semi-rural areas.

Having glanced briefly at schools run by dames (many of which 'were not preparatory schools in the classical sense: they prepared children of tradesmen for a life above that of the common herd who went to the National or the British schools') and that interesting group of Roman Catholic schools which operated illegally in England up till 1791, Leinster-Mackay focuses on the emergence, in the later 18th century, of the private classical school. 'By contrast with the private academies offering a wide utility-based curriculum, the private classical schools were concerned with a traditional curriculum and with entry to the universities. They differed from the typical local endowed grammar schools in having a more wealthy clientele and in being generally boarding in character. For those wanting a liberal education in the classics, these private classical schools formed an alternative to the endowed grammar, or public, schools of the day. They also exemplified the fourth characteristic which became an essential feature . . . they were run for private profit.'

Leinster-Mackay then goes on to formulate 'four main stages in the development of nineteenth century . . . preparatory schools'. The first stage takes one up to about 1830, 'by which date a few private classical schools had abandoned their previous practice of preparing some or all their boys for university and were concentrating largely, if not exclusively, on preparation for the great schools'. In the next 35 years a good many private schools followed their example, being recognised as 'preparatory schools' by the Clarendon and Taunton Commissions. By 1892 others (some of them run by assistant masters at public schools) had entered the field since the demand was steadily increasing and most public schools had hived off their younger (under 13) pupils. The fourth stage was 'the period after 1892 when the institutionalising of preparatory schools as a genre had been formalised by the setting up of the AHPS', but this stage goes beyond our period of study.

It is no part of the purpose of this book to pursue the claims of Twyford (1809) or any other candidate to be 'England's oldest preparatory school'. As Leinster-Mackay points out, the very title contains an inherent contradiction. Cheam, for example, was founded in 1645 but did not become a true prep school till 1855. How much does the change of (a) ownershp (b) site and (c) academic purpose affect the nature of a school? If a school, founded by Dr A at, say, Hitchin, is taken over by Mr B on the former's death and is moved to Royston under a new name, and – after Mr B's bankruptcy – is reopened in another building in Royston under Mr C, is it 'the same school'? Rather a far-fetched example, one may feel, but the early histories of Cordwalles (now St Piran's) or Fonthill are not altogether dissimilar and make a date for 'foundation' impossible to assess. York Minster Song School was established in 627, but over 1,250 years passed before it achieved prep school status. Let the discussion rest there.

What is important to our purpose is the existence in the early years of the 19th century of the private classical schools already mentioned. Out of 260 such schools recorded by Nicholas Hans, 240 were run by Anglican clergy. One of the most famous of them was Dr Nicholas's Academy, a large school with as many as 300 boys (1817) run on Eton lines. 'No private educational establishment enjoyed a higher reputation than did the Academy for Young Gentlemen carried on at Ealing by the Rev. George Nicholas, D.C.L., of Wadham College'. At Mitcham, too, the Rev. W. Roberts prepared boys specifically for Eton and had a 'great reputation'. Some impressions of Mr Roberts's establishment can be formed from the letters of James Milnes Gaskell. Mr Roberts himself seems to have been severe, but not – as so many were – capricious. He provided good food and he allowed his pupils more play time than at many schools. 'The boys rose at 6.30 and translated Ovid until 8, when there was breakfast of boiled milk and bread. Until 11 they learnt lines from Ovid; from 11 to 1 there was play. Dinner was at 1, and in the afternoon they read English and did sums. At 5.45 supper was served and after prayers there was play until bedtime at 8'. (Chandos). The establishments of Dr Curtis of Sunbury, Dr Hooker at Rottingdean (described as 'one of the most celebrated academies in England for the rank of its pupils, the comforts of the school and the superiority of its training'), Dr Horne at Chiswick (where the future Lord Shaftesbury spent five miserable years) and of the Rev. J.A. Barron at Stanmore enjoyed considerable (if not always merited) prestige. All have long since vanished 'despite the fact that several [of them] catered for the English nobility in their early years [aiming] to provide a complete classical education'. But, as Leinster-Mackay stresses, 'they were rivals to, rather than feeders of, the public schools . . . Only those classical schools which managed to adapt, such as Twyford, Temple Grove, Cheam and Eagle House', went on to become recognised prep schools. The first two of these are considered to have become 'preparatory in character' between 1815 and 1835, and it is Twyford which is believed to be the school to which young Tom Brown was sent for three 'half-years' before he went on to Rugby. The date would be around 1830 and so the account, brief as it is, is one of the few (in fact or fiction) of a private school in the first half of the 19th century. It is based on the experiences of Thomas Hughes himself who arrived at the school, with his brother, in 1830 and stayed three years. An imposition he was set (copying out part of the school rules) has been preserved and it is also recorded that, in 1833, he won a prize for reciting by heart 1200 lines of the *Aeneid*.

'It was a fair average specimen, kept by a gentleman, with another gentleman as second master; but it was little enough of the real work they did – merely coming into school when lessons were prepared and all ready to be heard. The whole discipline of the school out of lesson hours was in the hands of the two ushers, one of whom was always with the boys in their playground in the school, at meals – in fact at all times and everywhere, till they were fairly in bed at night'.

The picture of the two ushers is not a flattering one: '[they] were not gentlemen, and very poorly educated, and were only driving their poor trade of usher to get such living as they could out of it. They were not bad men, but had little heart for their work, and of course were bent on making it as easy as possible'. They achieved their objective 'by encouraging tale-bearing, which had become a frightfully common vice in the school in consequence, and had sapped all the foundations of school morality', and also by 'grossly favouring the biggest boys . . . [who] became most abominable tyrants'.

We learn little of the everyday life of the school, though we are told that 'Tom imbibed a fair amount of Latin and Greek', and that the half-holiday walks to Hazeldown (a huge area of woodland and waste ground) 'were the great events of the week'. We are hardly surprised to learn that when, midway through his third half, 'a fever broke out in the village' to Tom's delight he and the other fifty boys were sent home.

Since we have now reached the 1830s, it is time to reinforce Leinster-Mackay's disposal of another educational myth, namely that Dr Arnold of Rugby was the real instigator, if not actual founder, of the earliest preparatory school. We have seen already how he used every effort in his later years to run down and finally to close altogether the Lower School, believing that boys under 13 would become contaminated by close association with sinful adolescents. All that can be said is that in 1837 a school in Newport (I.O.W.), run by Dr Worsley since 1833, was taken over by a half-pay naval officer, Lieutenant C.R. Malden, and, after three moves, became Windlesham House and that, apparently, Dr Arnold gave Lt. Malden every encouragement. Certainly 'in the early years most of the boys went on to Rugby', though later Eton and Harrow became the most favoured schools.

Conditions at the early prep schools were seldom better than those at public schools in the first forty years of the century, and occasionally worse. Floggings of a severity to equal anything the public schools were capable of were not unknown. Indeed 'floggings at Eton were child's play' compared with those administered at Hampton by Mr Walton, while at Romanoff House (later Rose Hill) Mr Allfree 'thrashed a boy so cruelly that even now it makes me sick to think of it', wrote Sir Charles Rivers Wilson. It goes almost without saying that living conditions in many of those schools were spartan in the extreme. At the Manor House School, Chiswick (Dr Horne), there was 'no bathroom, no washing arrangements (that is, if one excludes the three tubs brought into the dining-room after breakfast), no proper lavatories and no matron to look after the boys' and, with few exceptions, the food was of the plainest quality and scantiest quantity. In most such schools there was little attempt outside the classroom to supervise the boys and, unsurprisingly, it was said of Dr Ruddock's school at Fulham, 'Whatever the alleged cruelties of public schools of that day, I [Lord Lytton] cannot believe that they equalled the atrocity of a genteel preparatory establishment in which the smallest boy was given up, without any check, . . . to the mercies of boys less small'. C.S. Lewis, writing of the cruelty and hardship prevalent at a preparatory school as late as 1908 commented: 'If the parents in each generation always or often knew what really went on at their sons' schools, the history of education would be very different'.

Not all children kept quiet throughout the holidays about life at school, not all children were disbelieved in principle by their parents, so that one must pose the question as to why English parents, almost alone in the world, were so anxious to send their children away to school at an early age. It is not an easy question to answer, but it is one that must be faced. There is, first of all, the historic answer, namely that since medieval times it had been customary for boys of the nobility to be sent away to receive discipline and training in the establishments of their peers. Then there is the fact that an increasing number of parents, in the reign of Queen Victoria, were serving their sovereign overseas and, in days when the voyage to India took many months, home leave was impracticable less than every five years so that boys perforce had to be left in England to be educated. Finally, and most importantly, there is the growth and increasing popularity of public schools, entry to which could be achieved only through the grounding in Latin and Greek which a prep school provided. This factor was reinforced by the increasing number of scholarships on offer from about 1870 onwards, the gaining of which not only enhanced a prep school's prestige but might well decide whether a bright child from an impecunious home (e.g. a vicarage) could go on to a public school at all. Even these considerations do not satisfactorily explain the fact that many parents must have known that their sons spent two-thirds of the year enduring harsh discipline – or worse – poor food and spartan conditions. Perhaps Charles Kingsley and other devotees of muscular Christianity have something to answer for here. It was Kingsley who criticised the education of the middle classes in that it had been lacking in that experience of pain and endurance necessary to bring out the masculine qualities. 'Now', as Professor Honey comments, 'as the cult of manliness spread, parents demanded even earlier access for their sons to the institutions which fostered it – which helps to explain the proliferation after 1870 of preparatory schools'. And he continues, 'This conception of the school as a forcing-house turning molly-coddled nurslings into *manly* young men is a vital clue to the problem of why parents defied the powerful deterrents of disease, immorality and cruelty in Victorian schools'. One would perhaps add the gloss that for 'parents' one might substitute 'fathers' since the mother's opinion would not often be consulted.

By about 1860, then, (in other words immediately prior to the Clarendon and Taunton Commissions) the preparatory school was a recognised feature of the educational world. There was still a wide variety in style and organisation between one prep school and another. An appreciable number of them were run by women, which might have been considered a handicap since few women in those days had received a classical education, but the difficulty was usually overcome by employing a man (sometimes a clergyman) to come in and teach Latin and Greek every day. Not all women, however, were so dependent. At a school in Cheltenham, kept by the two Miss Hills, Miss Bessie Hill, a first-rate Greek and Latin scholar, helped her sister keep about a hundred boys in 'terrific order'. And Winston Churchill (*My Early Life*) records with gratitude his transfer (in 1883) from St George's, Ascot, 'to a school at Brighton kept by two ladies. This was a smaller school than the one I had left. It was also cheaper and less pretentious. But there was an element of kindness and of sympathy which I had found conspicuously lacking in my first experiences. Here I remained for three years . . . At this school I was allowed to learn things which interested me: French, History, lots of Poetry by heart, and above all Riding and Swimming. The impression of these years makes a pleasant picture in my mind'. (The school, after two moves and several amalgamations, is now Stoke Brunswick, near East Grinstead.)

Almost inevitably, though, such schools were regarded with disfavour by all-male establishments. O.C. Waterfield's view may be taken as typical when (in 1866) he 'spoke out strongly against small preparatory schools opened in an amateurish fashion by . . . orphaned clergymen's daughters'. His views are understandable for these 'dame' schools were very real rivals to early 'professional' preparatory schools, since the gentle sex could become recognised as the more natural teacher of young boys. In the event the dame schools tended to concentrate on the lower age-range (5 to 8), becoming, in effect, pre-preparatory schools.

More typically the prep school of mid-Victorian days resembled an extended family. Between 30 and 40 pupils were boarded in a large private house (purpose-built schools were rare before 1900) which might have had additional classroom space built on. The headmaster, his wife and children (typically a son who might later succeed to the headship and two unmarried daughters) constituted the core of the staff. The headmaster, it goes without saying, taught classics to the senior pupils and he probably employed two poorly paid assistant masters to instruct the younger boys in Latin and Greek, and to supervise outdoor activities. Such schools and even larger ones, were so much a 'one man band' that even when they had official names they were more often spoken of as Evans' (Horris Hill), Tabor's (Cheam) or Waterfield's (Temple Grove), while Aldin House (then at Slough) quickly became, and has remained, Hawtrey's. Some of these establishments continued in the ownership of a single family for several generations. Salaried heads were unknown (except in the Junior departments of public schools) and, for that reason, it was unheard of for a prep school headmaster to transfer to another school.

The economic health of prep schools continued to be threatened by the habit of many public schools of taking boys at a very early age with Eton clearly the worst 'offender' in this respect. Henry Scott may have created a record by entering the College at the age of four, but examples of boys spending ten years there are by no means uncommon. In spite of Arnold's example at Rugby, when the Clarendon Commission began to take evidence, only one of the nine 'great' schools investigated (Harrow) contained no boys below the age of 13. 'It is not surprising, therefore, to find the Commissioners recommending the separation of the lower school from the upper in view of the tender years of some of the boys'. This recommendation led, in 1869, to a notice appearing in *The Times*, announcing that the Rev. John Hawtrey intended to continue his work of looking after young boys, but outside Eton, in fact at Aldin House, Slough. And within a year or two 'Hawtrey's' is on record as playing matches against some of the other famous prep schools of the day.

The majority of public schools followed the recommendations of the Clarendon Commission by raising the age at which new pupils could enter, and by instituting a form of entrance examination (chiefly, of course, in the classics), though agreement to produce a 'Common Examination for entrance to Public Schools' was not reached till 1904. The hiving off of younger boys from the senior schools gave added importance to these examinations and especially to the scholarship awards. These might be worth as much as £100 p.a. and were clearly a great inducement to prep schools to direct their efforts towards academic success. Even a small school could achieve surprisingly good results. Stoke House, (now a component of Stoke Brunswick) for example, with 50 boys, achieved fourteen scholarships, six of them at Eton, in 1878. The whole process had the effect, too, of binding a prep school to one particular public school which came to recognise the academic quality of its feeder e.g. Horris Hill and Winchester, Elstree and Harrow, Bilton Grange and Rugby and, of course, Hawtrey's and Eton.

The appearance of a growing number of preparatory schools in the last 30 years of Victoria's reign gave rise to increasing competition among them and this aspect of prep school history is well described in an article in *Longmans Magazine* in 1897 (Eric Parker: *Private Schools, Ancient and Modern*). Scholastic attainment, while it remained a priority, was by no means the only criterion for parental choice. Team games, which at most schools had been carried on in a haphazard fashion with little proper coaching and few inter-school matches, began to play a much more important role in school life. Indeed, by the end of the century some schools, following the example of their seniors, were making such a fetish of success on the games field that it dwarfed all other activities. Athletic prowess was, of course, closely linked to the cult of manliness whose spread David Newsome has so clearly documented. 'What they (i.e. the discerning schoolmasters) had discovered almost by accident', wrote Sir Cyril Norwood, 'was the team-spirit, which alone builds character . . . Team spirit is a commonplace to us [1929] but it is a recent arrival in the field of educational thought, and is indeed one of the present English contributions to methods of true education'. It would be difficult to take the cult of games further than it was taken by the Rev. Edwin Leece Browne, at St Andrews, Eastbourne, 'whose school magazine was given over almost completely to reports on matches and whose sermons were frequently illustrated by similes drawn from cricket'. Would it be considered irreverent to describe Browne's analogy of the Trinity ('three stumps, one wicket') as the apotheosis of Muscular Christianity? If Browne provides an extreme example of the cult, a more generally accepted assessment of the value of games was given by E.S. Dudding, Headmaster of Wolborough Hill in the 1890s, who observed: 'The boy who learns to play for his side at school will do good work for his country as a man'. Even if unstated, the corollary was clear.

Athletic activity, too, had the benefit of promoting health, another aspect of school life that was coming under increasing scrutiny. For those unfamiliar with the subject, England and Wales (from 1871 onwards) was divided into sanitary districts, each with a Medical Officer of Health and an Inspector of Nuisances. Four years later came the Public Health Act which provided, inter alia, heavy penalties for wilful exposure of persons to infectious disease. Newly founded schools had to take into consideration the supposed healthiness of the chosen site, gravel soil and bracing sea air being specially favoured. Mostyn House boasted a 'dry and bracing climate' in the Wirral 'where the rainfall is one of the lowest in the kingdom', resulting in 'remarkable freedom from the germs of epidemic or infectious diseases' according to the 1912 prospectus.

Fortunately the two most feared killers, cholera and small-pox, had largely disappeared, but diphtheria and scarlet fever were both dreaded. When three boys died at Eagle House in the 1850s during a double epidemic, no new pupils were entered for three years, the school then moving from Hammersmith to Camberley. At Twyford successive outbreaks of diphtheria caused the school to move to Westfields (Winchester) for a term and subsequently to Emsworth House (Copthorne) for a whole year, during which Twyford's own drainage system was completely reconstructed. Even epidemics of measles, mumps, whooping-cough and chicken-pox constantly disrupted school life and could have fatal results. On at least three occasions between 1866 and 1880 measles made devastating assaults on Cheam and several deaths resulted. The remedies advocated by the M.O. of Rugby School in 1887 were: 'instant isolation; perfect quarantine; perfect disinfection; plenty of cubic space; ventilation; efficient drainage; pure water'. And Dr Tatham, M.O. of Salford, suggested that 'it was a mistaken policy on the part of schoolmasters to urge, for economic reasons, the punctual return of pupils from vacation if they were recovering from infectious diseases'. The risk of illness was gradually reduced when the practice of bringing a doctor's clearance certificate by each pupil at the start of term began to be insisted on. Who, among today's senior citizens, does not remember desperately hoping to come into contact with a victim of, say, whooping-cough on the last day of the summer holidays?

Growing concern over health led to the provision of a sick-room and eventually to the addition of a sanatorium. Stubbington House even had two; Temple Grove had one at some distance from the school; St Ronan's, Aysgarth and Mostyn House were other pioneers in this important field. The latter, under its innovative head, A.G. Grenfell (1890 onwards), showed that prep schools need no longer be so restricted in their surroundings, their curriculum and their range of out-of-school activities. 'He launched a massive building programme and almost rebuilt the school, giving it at least three times the original floor area'. A large swimming bath (filled from an artesian well), a gymnasium and a chapel followed within a decade. Perhaps it was Mostyn House which inspired a contributor to Blackwood's to assert that 'a preparatory school should be a nursery for hardening young cuttings, not a hot-house to force exotic plants'.

We are approaching the era of the purpose-built school. Indeed Windlesham House had taken the lead when it moved into its new premises in 1846, a building which 'was but the nucleus of a school which expanded slowly throughout the century'. Later came Horris Hill, founded in 1888 and, in 1890, Aysgarth, 'the first great preparatory school of the north'. Aysgarth was the creation of the Rev. C.T. Hales and the whole campus was laid out, built and fitted up 'in one fell swoop'.

Various aspects of the prep school world in Victorian times must now be considered, beginning with the actual teaching. From the nature of the public school curriculum, the requirements of entrance examinations and, above all, the lure of scholarships, it seems clear that most prep schools concentrated on the classics, with mathematics in second place and other subjects relegated to the fringes. That being so, the time available for more aesthetic subjects was bound to be limited. There was also the question of finding people to teach them. The upbringing of the Victorian prep school master did not encourage him to exhibit artistic or musical gifts. And, inevitably, visiting teachers were accorded less respect than the resident staff. All too often, indeed, Frenchmen, with their comic accents, were ragged unmercifully and visiting women teachers would need to be exceptionally strong-minded to enforce discipline in a male-dominated society. But not all prep schools ignored 'the higher things of life'. Winton House, run by W.F. Rawnsley, was described by Richard Verney (later Lord Willoughby de Broke) in these terms: 'If education has indeed been properly described as "the equipment for a full life" we certainly got something very like education at Winton House'. A later headmaster of the same school, C.A. Johns, could even play the 'cello. But, in general, Leinster-Mackay is surely right to comment that 'music was regarded . . . as a study inappropriate for boys. The ability to play an instrument or to sing was regarded as an accomplishment suitable only for girls'.

From the very fact that they are scarcely mentioned one has the feeling that plays and concerts seldom featured in prep school life. The nearest equivalent would be the kind of entertainment at which two of the masters would sing rousing songs (the school joining in the chorus) or amusing ditties, interspersed with recitations of prose or poetry by selected pupils and with piano pieces played by the headmaster's daughter. Lectures there certainly would be, and occasional magic lantern shows, but of visits to concerts, to museums and art galleries or to places of interest it is hard to find any mention, transport being necessarily a restricting factor. The present writer's memory of art lessons (in the 1930s) may not be untypical, being restricted to pencil drawing of jugs and cups twice a week in the two lowest forms.

Food is probably the most emotive of all topics in the world of boys and since it features in later accounts of Temple Grove the evidence here will be confined to that of Thomas Guthrie (who, as F. Anstey, wrote *Vice Versa*). Here he describes the main meal of the day at 'Crichton House' (actually a private school in Surbiton): 'the midday meal was dreaded by myself and all those boys who were in the least fastidious. The meat was probably good in quality but, the cook being no artist, it often required an effort, even for healthy and hungry boys, to get it down and there were one or two soups and dishes which gave dismal notice of their imminence a good hour before we faced them . . . Any attempt on our part to leave unattractive morsels uneaten was a most serious offence in [the Headmaster's] eyes'.

Although prep schools were often small, with fewer than forty boys, even they had perforce to employ one or two assistant masters, while the larger ones might need ten or a dozen. Few, if any, of these men have recorded their impressions of teaching in preparatory schools. Indeed, if they had done so, one would necessarily be suspicious of axes being ground and grievances aired. *How I enjoyed my forty years of teaching Latin and Greek at St Philibert's* is not a title that would be likely to tempt even a Victorian publisher. It is, therefore, quite significant to find that the very first issue of *The Preparatory Schools Review* (1895) should carry an article, entitled *The Assistant Master, Past, Present and Future*, by W.T. Sutthery. It is most revealing concerning attitudes towards prep school teaching in Victorian times and bears quoting at some length.

'The world', he begins, 'is waiting for one of two things: either for a novelist who will give us a portrait of one of the many gentlemen . . . who become preparatory school masters . . . or for a race of schoolmasters who will so inspire novelists . . . that there will be a rush to make them the heroes of fiction'. He goes on, 'The fact is, we fear, that a false estimate of our profession is abroad which considers a gentleman in an absurdly false position if he teaches or looks after little boys'. The general opinion was that a university graduate, if he intended to teach, would naturally aim for a public school, where he 'rejoices that . . . he may at once take a high place among those who teach boys without the disagreeable necessity of having to know them out of school'. His college tutor would already have warned him, 'Never have anything to do with a preparatory school if you want to get on'. But all depended on the class of degree obtained. A first or a second opened the door to a public school post, but 'the man who takes a third or pass degree feels the nip of poverty at his heels . . . and after a wild look round for rescue which comes not – the public schools are busy reading the higher honours lists and the First Class averages – he plunges in at the door which bids him leave all hope behind, and here he is among us'.

Mr Sutthery's deep understanding of a prep school master's hopes, experiences and failures is evident when he goes on to describe the new master's early impressions: his failure to achieve an honours degree is no longer of importance – his Latin being quite as good as anyone else's; despite not achieving a cricket blue he is still a giant of the game in the eyes of the 1st XI; and the Headmaster is revealed as somewhat lower than the divinity he initially resembled. But it is the boys themselves who 'astonish him most of all [for] there are no wicked boys whom he may eloquently reprove or faithfully denounce . . . though for naughtiness they are beyond belief . . . He must be ever on the watch for new ways of ruling a mass of such boys and of securing order and respect from urchins who seem to enjoy disorder and respect nothing'. And perhaps after seven years or so our erstwhile novice will have learnt 'to teach out of school and in the playing fields, while in school he will sacrifice his literary taste in classics, or his scientific taste in mathematics, to the still higher pleasure of helping the lamest of dogs over the most hopeless of stiles'.

Now comes the rub. 'The salary which looked so attractive at three-and-twenty is no princely revenue at thirty; it has barely increased twenty-five per cent, and will increase no more'. Mr Sutthery then claims that younger men are 'pushing on to take his place' and that for all the experience he has gained, 'he will soon be no longer wanted'. Finally he 'either takes Holy Orders . . . or slips out of notice altogether'. One cannot help feeling that, in his anxiety to convince, Sutthery has somewhat overstated his case and that even in 1895 a man of 30 can scarcely have been considered middle-aged and not worth employing. But he concludes with a number of recommendations of which the main points are these: an insurance scheme enabling the AHPS (Association of Headmasters of Preparatory Schools) to pay pensions to assistant masters; a register of assistants who have worked in such schools; lower starting salaries and therefore larger increments; 'well paid posts without the ordinary supervision duty' for more senior staff; the appointment for short periods of 'supernumerary assistants to give a taste for the job to young men before university'; 'and lastly . . . can we not get the Headmasters' Conference to meet us and tell us what they are prepared to do for us in this matter?'

Three years later, Eric Parker, in another article in *Longmans Magazine* ('*Preparatory School Assistant Masters*') echoes some of the comments made by W.T. Sutthery, but is altogether more trenchant in his criticisms. Indeed he begins by giving 'utterance to a warning [which] amounts to a caution against entering the teaching profession' (he means, 'in preparatory schools'). He considers 'the professions open to a graduate' of some academic ability but of very limited means and he rapidly lists the attractive ones ('all these need capital') not available to them. 'But there are two professions which ask for no banker's reference: clergymen and schoolmasters'. And he then 'throws a searchlight' on those 'who are compelled to adopt a profession for which they have not knowingly an aptitude, or even perhaps an inclination; who become schoolmasters because they must have food and money'.

He goes on to paint an unflattering picture of the role of the scholastic agency. 'Is it generally known that this is the only possible method of entering the profession?' He describes the 'dingy little ante-room' and the interview with 'the man through whose hands pass I do not know how many appointments per week'. His comments on the kind of salary on offer are particularly valuable: 'I dare say the average amount offered is £80 or £90 a year, plus board and residence . . . but I have seen many of £40'. (Hawtrey at Aldin House was quite exceptional in paying his senior men £200 p.a.; at the other extreme Vincent van Gogh, teaching at a private school in Isleworth in 1876, was not paid a salary at all.)

He then considers the career of a classics graduate starting in his first post 'at a commencing salary of a round hundred'. 'It is not', he observes, 'an easy matter to teach small boys'. Latin and Greek may seem to present no problems, 'but could he teach sums to a boy of ten?' 'It takes years', he continues, 'to learn how to present the elements of Latin and Greek and French and mathematics to the mind of a child'. And further on he says that 'to succeed as a teacher . . . a man needs to spend his whole energies, not only in school . . . but also in the playground'. He advocates spending 'not more than six years under the same headmaster, and indeed . . . some headmasters make it a rule not to keep their assistants longer than three years . . . lest they fall into a groove'. By the age of thirty-six such a man is 'incalculably a better schoolmaster than he was twelve years ago. He has discovered methods of teaching, of influencing, of controlling boys . . . and he has discovered these himself in living experience'.

But what happens then? Without capital there is no prospect of our man becoming a Headmaster himself. Indeed his main concern will be to save enough money for his retirement. The whole tenor of Eric Parker's article implies that the man is necessarily a bachelor. At one point he says, 'Unless he has a home to go to . . .', and elsewhere he refers to 'a good holiday tutorship'.

And he declares that such a 'man is worse off at forty than at twenty-five', since the prospective employer is usually looking for an active young man able to coach cricket and football as well as to teach.

The various measures that Eric Parker discusses to make the teaching profession (in prep schools) a more inviting one bear a close resemblance to the proposals of Mr Sutthery: lower starting salaries with larger increments, less onerous duties for men of long service, the taking on of one or two senior masters as partners in the school by the Headmaster, and an insurance scheme for the payment of pensions. He goes into the proposals for such a scheme in some detail, pinpoints its weaknesses and concludes that 'the scheme is invertebrate'.

Having done, by his own admission, his utmost to deter unsuitable young men from entering the teaching profession he turns his attention finally to the plan put forward by Arthur Sidgwick for the training of teachers: 'Given that during the first year or so of his career the novice is worth very little considered as a teacher . . . let him anticipate this first year by a course of instruction in teaching'. And he advocates 'a course of lectures on the theory of education . . . lessons in voice-production and articulation; and, finally, genuine practice in teaching . . . under the superintendence of a competent mentor . . . Grant a diploma to the man who comes through it all successfully . . . and we are on the high road to a badly needed reformation'.

Eric Parker would surely have been encouraged to know that, within fifty years of the publication of his article, such a scheme was beginning to be put into operation.

If the facts about straightforward matters such as masters' salaries, music lessons and sanitation are not easy to establish, those concerning what might be described as the hidden side of prep school life are open to a variety of interpretations and unsafe generalisations. It must be remembered that evidence for life at these schools comes, in the main, not from headmasters, certainly not from assistant masters, not from parents, not even from letters home written by the boys (many were censored and, in any case, relatively few have survived) but from the reminiscences of former pupils penned many years after the event. These memoirs were often written by men who had achieved fame as soldiers and statesmen and whose theme tended to be: 'I had a good many floggings in my time – didn't do me any harm – never learnt anything, though'; or by authors, often men of great perception but who, by their nature, were untypical prep school boys and who may well have suffered more than most on that account. We need, at all events, to be cautious in assessing the evidence put forward, especially when it concerns fighting and bullying, beatings and sex and the varied miseries of boarding school life. Robert Graves, in *Goodbye to all that*, writes: 'Preparatory schoolboys live in a world completely dissociated from home life. They have a different vocabulary, a different moral system, even different voices. On their return from the holidays the change-over from home-self to school-self is almost instantaneous, whereas the reverse process takes a fortnight at least . . . school life becomes the reality, and home life the illusion'.

Osbert Sitwell, in *The Scarlet Tree*, makes two further points: 'In order for the reader to gain a correct picture of the scene . . . he must always bear in mind the continuity of the wretchedness, the underlying and enduring stratum of it. Being a boarding school, this establishment offered a horrible isolation from every warm current of life . . .' and later: 'Just as, when I was at the front (1915) I seldom knew . . . which part of the line we were occupying, so at school I never mastered the topographical essence of my place of internment . . . In both instances, I suppose, the surroundings were so laden with bitterness for me that their whereabouts seemed scarcely to matter'.

With these observations in mind we may consider some of the less attractive aspects of Victorian prep school life. Fighting between individual boys seems to have been endemic and one has the feeling that the authorities were at no great pains to put a stop to the practice which was closely bound up with the cult of manliness. Richard Meinertzhagen had his first fight at Aysgarth at the age of nine. 'There was a boy in the school called Brown. He was known as 'the Cannibal' and would boast that he loved eating little boys and that he did, in fact, do so at home, a story which we, of course, believed. He had a nasty habit of biting us in the leg and one day he bit Dan [Richard's brother]. I flew at him and a fight then ensued, but the Sergeant separated us before much damage was done. I had a tooth knocked out and a rib bruised and Brown was bleeding like a pig from some hits in the face'. A toned-down account of the fight in a letter home ends: 'We are both much swollen today. But Brown is worst'.

Less usual, but probably not too rare, were organised fights against outsiders. George Newnes recalls one such, in which his elder brother, Ted, was the ringleader. It occurred because the procession of Silcoates boys marching to chapel on Sunday was the inevitable target for the taunts and missiles of the local Halifax boys. 'One Sunday Ted found an opportunity to enquire of one of the tormentors, "Will six of you fellows fight six of ours?" The challenge was accepted and there and then it was arranged that six chosen warriors from Silcoates should meet six from Halifax at eleven o'clock next night. [The Silcoates boys] stole, unseen and unheard, to the appointed place chosen as far away from the beat of the police as possible. The Halifax lads fought pluckily, but those from Silcoates fought better and remained victors'. Hands were shaken and 'from that day forth no procession of Silcoates boys was molested in the streets of Halifax'.

Fighting was one thing, bullying was quite another matter. Again the school authorities cannot escape blame but the fault here lay in inadequate supervision rather than acquiesence. Of his two years at a school at Marlow, Hugh Walpole wrote, in *The Crystal Box*: 'I was frightened in the war several times rather badly, but I have never, after those days, thank God, known continuous increasing terror night and day. There was a period, from half-past eight to half-past nine in the evening, when the small boys (myself with them) were dismissed to bed but, instead, spread themselves in an empty classroom that is still, to me, when I think of it, damp green in retrospective colour. The bigger boys held during that hour what they called the Circus. Some of the small boys (I was always one) were made to stand on their heads, hang on to the gas and swing slowly round, fight one another with hair brushes, and jump from the top of the school lockers to the ground . . . I can feel again, as I write, the sick dizziness at my heart as I looked down at the shining floor . . . Worst of all was being forced to strip naked, to stand then on a bench before them all while some boy pointed out one's various physical deficiencies and the general company ended by sticking pins and pen-nibs into tender places ..' And he concludes: 'To those who would say: "We've all been through these private schools; a little roughing it does no one any harm. You ought to have stood up for yourself," I reply, "Quite so. But I did not stand up for myself then and I'm not trying to stand up for myself now". The point is exactly that I was a miserable child, and one month at [Marlow] was enough to make me sycophantic, dirty in body and mind, a prey to every conceivable terror'.

Against boy bullies there might, on occasion, be remedies. Against master and headmaster bullies there was none. Or rather, the only form of defence open to boys (as C.S. Lewis has pointed out) was that of lying. The gross abuse of power, of which a majority of Victorian headmasters were guilty, is surely the most sickening aspect of 19th century education. Examples of it occur in countless books. It is necessary here to give one or two samples if this is to be an overall picture of prep school life, but they will not be unnecessarily extended.

The first is taken from *First Childhood* by Lord Berners, a pupil at Elmley in the 1890s: 'Mr Gambril [Headmaster] had a stock of tortures. He would pull one up by the hair near one's ears. He would hit boys on the shins with a cricket stump. He had a way of pinching his victims that was positively excruciating. He excelled also in the administration of mental tortures'. These would frequently involve the examination of mark books, usually at lunch time. 'If one received a bad mark during the morning, the luncheon hour would be spent in an agony of fear. I can still remember that terrible devastating panic that seemed to paralyse the digestive organs and deprive one of appetite and if, as often happened the fatal summons was delayed till supper time, it was impossible to eat anything during either meal'.

High up the list of prep school tyrants comes Archibald Dunbar, assistant headmaster at Wixenford in the 1870s. One of his many victims was George Curzon, the future Viceroy of India: 'As a master he was for the most part detested by the boys, to whom he was savage and cruel. He practically shunted poor old Powles [the actual Headmaster] and ran the school by himself. He executed all or nearly all the punishments whether by spanking on the bare buttocks or by caning on the palm of the hand or by swishing on the posterior. He was a master of spanking . . . [and] I remember that it was at about the 15th blow that it really began to hurt . . . at about the 28th blow one began to howl. The largest number of smacks I ever received was, I think, 42'.

Most appalling of all these petty Hitlers was the Rev. H.W. Sneyd-Kynnersley

at St George's, Ascot. The evidence of Winston Churchill is supplemented by
that of Roger Fry. The former writes: 'Flogging with the birch in accordance
with the Eton fashion was a great feature in its curriculum. But I am sure that
no Eton boy, and certainly no Harrow boy of my day, ever received such a cruel
flogging as this Headmaster was accustomed to inflict upon the little boys who
were in his care and power . . . Two or three times a month the whole school
was marshalled in the Library, and one or more delinquents were haled off to
an adjoining apartment by the two head boys, and there flogged until they
bled freely, while the rest sat quaking, listening to their screams. This form of
correction was strongly reinforced by frequent religious services in the chapel
. . . I therefore did not derive much comfort from the spiritual side of my
education at this junction. On the other hand, I experienced the fullest
application of the secular arm'. Nearly thirty years later Mr Churchill, by then a
Member of Parliament, wrote to *The Times* in protest at an earlier letter from
the Headmaster of Sherborne who had apparently advocated corporate
punishment in certain cases. Churchill's letter ends: 'Does Mr Westcott flog his
boys in their corporate capacity?' The evidence of Roger Fry, himself one of
the two head boys at St George's, (quoted in Virginia Woolf's biography)
corroborates that of Churchill: 'In the middle of the room was a large box
draped in black cloth, and in austere tones the culprit was told to take down
his trousers and kneel before the block over which I and the other head boy
held him down. The swishing was given with the master's full strength and it
took only two or three strokes for drops of blood to form everywhere and it
continued for fifteen or twenty strokes when the wretched boy's bottom was a
mass of blood. Generally, of course, the boys endured it with fortitude but
sometimes there were scenes of screaming, howling and struggling which
made me sick with disgust'.

Clearly such sadistic tendencies had sexual undertones. As to sexual
practices among the boys themselves the evidence is scanty and
untrustworthy. One must remember that until 1958 sexual relations between
men were illegal and that, although certain actors and other men in public life
were recognised homosexuals, they contrived to avoid prosecution. And
certainly no one boasted of his homosexual experiences in print. Robert
Graves, in *Goodbye to all that*, claims that: 'In English preparatory and public
schools romance is necessarily homosexual. The opposite sex is despised and
treated as something obscene. Many boys never recover from this perversion.
For every one born homosexual, at least ten permanent pseudo-homosexuals
are made by the public school system: nine out of these ten as honourably
chaste and sentimental as I was'. It is an assertion easier to make than to prove,
or to deny.

H.M. Butler, Headmaster of Harrow (1859-1885) and writer of the hymn,
Lift up your hearts, was asked to lead a committee, appointed by the Purity
Society in 1888, to promote its objects in public and preparatory schools –
'for it was recognised [according to Professor Honey] that in the latter the
need for a campaign was probably even more urgent'. One wonders how much

evidence exists to sustain this claim. Butler's hymn, incidentally, was intended for school Confirmation services. The lines, 'The mire of sin, the slough of guilty fears . . . the deeds, the thoughts that honour may not name', are clearly intended to refer to homosexual practices. But it is all too easy for headmasters, and especially for clergymen, to suspect 'sin' where none exists or to magnify the lavatorial and lewd jokes of schoolboys into something much less desirable. It goes almost without saying that biology was not taught in prep schools and that a headmaster's talk on the subject of sex to a boy about to go on to his public school was couched in such obscure terms that those who already 'knew' would have had difficulty in keeping a straight face, while the innocent (one suspects the great majority) remained in that condition.

Leonard Woolf was a pupil at Arlington House in the 1890s. He writes: 'The only thing learned thoroughly [there] was the nature and problems of sex. They were explained to me, luridly and in minute detail, almost at once, by a small boy who had probably the dirtiest mind in an extraordinarily dirty-minded school. I was at the time completely innocent and I had considerable difficulty in concealing from him that it was only with the most heroic efforts that I was preventing myself from being sick . . . These facts are worth recording because they showed me, for the first time . . . the enormous influence a few boys at the top of a school exercise upon the minds and behaviour of the masses below them. By the time I left the atmosphere had changed from that of a sordid brothel to something more appropriate to fifty fairly happy small boys'.

Finally, for a truthful picture of some of the least attractive features of 19th century prep schools, one should read *When I was a boy* by General Sir Ian Hamilton. Chapter 3 (some fifty pages long) is devoted largely to his time at Cheam in the 1860s and it would be wrong to attempt to condense an experience so vividly recalled eighty years later into a mere paragraph. It simply needs to be said that the headmaster, R.S. Tabor, seemed, in young Hamilton's eyes, to have 'the character of a sadistic ogre' and that the boy endured untold misery in his early weeks there. Relief came in the form of a week-end visit to Uncle John and Aunt Gertrude at their estate near Carshalton.

'Coming from hateful, spy-ridden, pie-jaw Cheam, Carshalton with its jokes and bright, friendly, open welcome struck me all of a heap; and when they told me to run out and amuse myself till supper, I didn't quite know whether to laugh or cry, and was really as drunk as any lord with the boiling-up of queer thoughts within me as I danced and skipped down the smooth slope of the lawn towards the river, which had a punt moored to the bank, as if waiting there for me like a magic boat in a fairytale . . . I poled into the middle of the stream; floated down round a bend and out of sight of the house; anchored myself . . . and gazed into the crystalline current with my face about two inches above the surface. Life was wonderful, after all! And then, like the backwash of a wave, a terrible thought gripped me close as close, and seemed to try and pull me down under. Very soon I must go back to Cheam. Why go back? Why not stay by the river – in the river? An intolerable feeling of utter friendlessness and of my enemies in the dormitory shot like an arrow through my heart. The impulse to sleep enfolded by this haunted stream grew stronger – One brave leap; sleep, sleep, sleep. Now or never – In the river!

"Ian! Ian!! I-a-a-a-n!!! Where are you? What a fright you've given us. Come in, you naughty boy and get your supper at once." So now I know how and why boys do it'.

Since writing the above I have been able to borrow a copy of *Cheam School from 1645* by Edward Peel. Three hundred years of the school's history are readably chronicled, and the chapter on Robert Tabor (the contemporary of Waterfield and Edgar) deserves, in particular, to be read as a counterbalance to Sir Ian Hamilton's account. Cheam was undoubtedly a couple of rungs higher up the social ladder than Temple Grove: it was appreciably more expensive. But the points of similarity, and the differences, are enlightening and Peel offers a more sympathetic appraisal of Tabor himself.

Sources

In chapters 4, 6 and 7 the authorities consulted and quoted from (in addition to numerous letters) have been:

A. Anson *About Others and Myself*
A.A.M. Batchelor *Cradle of Empire*
A.C. Benson *Escape and other essays*
A.C. Benson *Memories and Friends*
E.F. Benson *David Blaize*
E.F. Benson *Our Family Affairs*
L.S.R. Byrne and E.L. Churchill *Changing Eton*
H.J. Coke *Tracks of a Rolling Stone*
Benjamin Disraeli *Coningsby*
P. Fletcher Jones *Richmond Park*
Dermot Freyer *Bogey Baxter*
R.C. Gill and F. Mattingley *Christ Church, East Sheen*
R.C. Gill *The Growth of East Sheen in the Victorian Era*
W.E. Goodenough *A Rough Record*
S.J. Gurman *The Rev. Dr. William Pearson*
T.A. Guthrie (F. Anstey) *A Long Retrospect*
J. Hannay (G.A. Birmingham) *Pleasant Places*
M.D. Hill *Eton and Elsewhere*
Pamela Horn *The Rise and Fall of the Victorian Servant*
Washington Irving *The Sketch Book*
M.R. James *Eton and Kings*
M.R. James *Collected Ghost Stories*
Bulmer La Terrière *Days that are gone*
A. Lawrence *Sir Arthur Sullivan*
A. Liddell *Notes from the life of an ordinary mortal*
Brian Masters *The Dukes*
Brian Masters *The Life of E.F. Benson*
C. Maxwell *Mrs Gatty and Mrs Ewing*
David Newsome *Godliness and Good Learning*
R.W. Pfaff *Montague Rhodes James*
R.E. Prothero *A Memoir of Henry Bradshaw*
R.E. Prothero *Whippingham to Westminster*
W.L. Rutton *Temple Grove in East Sheen, Surrey*
F. Shoberl *A Topographical and Historical Description of the County of Surrey*
Ronald Storrs *Orientations*
G.M. Trevelyan *Grey of Fallodon*
E.S. Turner *Boys will be Boys*

Three publications of the Barnes and Mortlake History Society:
Barnes and Mortlake as it was; Vintage Barnes and Mortlake; Alleyways of Mortlake and East Sheen

Unpublished recollections of:
C.C. Barclay, J.L. Brereton, A. Hailstone, H. Headlam, N. Macleod, J.C. Pitman, T.T. Pitman, T.V. Scudamore, C.E. Storrs, R.J. Yarde-Buller, W.R. Young

Also unpublished: R.C. Gill *Temple Grove and the Palmerstons*

Records held at Temple Grove:
Name Book (1859-1880); Name Book (1874-1893); First Class (1894-1912)
Library Withdrawals (1869-1903) (3 vols); Temple Grove Register (1905)
Biographical Notes of Old Boys (2 vols.); Elementa Latina
General Rules for the Conduct of Temple Grove School

4   *Temple Grove*

1   The Origins of Temple Grove

The career of the founder of Temple Grove provides a good illustration of the difficulties confronting a young man of ability whose family had very little money but who wished to better himself; but it also shows how, in the early years of the Industrial Revolution, someone who possessed obvious scientific ability could overcome handicaps and achieve considerable renown. William Pearson's father was a yeoman farmer who reared cattle and sheep and who was probably engaged in fishing since the farm, at Whitbeck in Cumberland, was close to the Irish Sea. It is possible that young William was instructed in the three Rs at a dame school in the village; otherwise he apparently received no formal education until he reached the age of eighteen. In 1785 he attended, presumably as a boarder, the free grammar school at Hawkshead, where William Wordsworth (three years younger than Pearson) had already been a pupil for seven years. Wordsworth made fun of the ill-educated 'clown' with his uncouth manners but recognised his natural ability and commented on the improvement in his manner and appearance 'by attrition with gentlemen's sons trained at Hawkshead'. Whereas Wordsworth went on to Cambridge University at the age of seventeen, Pearson remained at the grammar school for five years before being promoted (in 1790) to the post of Second Assistant Master. The date of his ordination remains uncertain but in February, 1793, he was appointed Under Master at the Free Grammar School in Lincoln (salary £50 p.a.) and curate of St Martin's in the city (£21 p.a.). By 1795, the year of his father's death (the farm passed to the elder brother, John), William had married Frances Low and had been admitted to Clare College, Cambridge, (although almost certainly never taking up residence). Through his own efforts, and largely self-taught, he had acquired a lifelong interest in, and deep knowledge of, astronomy, and he was becoming proficient in the construction of scientific instruments. He used to give public lectures on scientific topics and he wrote a series of eight scientific papers for Nicholson's Journal. He invented a small, portable electrostatic generator and he also designed an astronomical clock. This clock, in addition to the usual dials, contained a unique feature: 'a satellitian giving an accurate representation of Jupiter and the motion of the four satellites of it then known'. These satellites were important 'since observations of their eclipses were used to determine longitude' (Gurman). This clock now belongs to the Royal Astronomical Society. Later came an even more ambitious device, a large planetarium which was afterwards used by John Dalton (the propounder of the atomic theory) in his lectures. But his masterpiece is the instrument shown in the portrait of Pearson with his wife and daughter (also called Frances), painted by Thomas Phillips, which hangs in the Council Room of the Royal Astronomical Society.

He also designed simplified versions for use in schools and, in Rees' *Cyclopaedia*, he wrote: 'The purchase comes within the reach of every respectable master of an academy, which was the object the author had in view'. Pearson is rightly regarded as 'the outstanding builder of planetary machines of his time'.

In 1800, having accumulated some capital, Pearson moved his family to London where he became the junior partner in a small school, Elm House, at Parsons Green, which took in boys aged from four to ten. When the senior partner, Mr Sketchley, died in 1803, Pearson became the sole proprietor. 'He contrived', wrote Wordsworth, 'to manage it with such address, and so much to the taste of what is called High Society and the fashionable world', that he was able to spend £1000 on buying land at Grasmere and Rydal (is it possible to detect a note of envy in the impoverished poet's letter?). At all events 'the success of Elm House School enabled Pearson to consider expansion' (Gurman) or, alternatively, a move to larger premises.

Only a few miles away *Sheen Grove*, an estate of almost fifteen acres with a fair-sized mansion on it, was being offered for sale by Sir Thomas Bernard. The detailed history of this property, which has been clearly documented by Mr Raymond Gill, lies outside the scope of this book; only a brief account can be attempted here. Early dates are necessarily tentative but in 1617 a certain Ralph Treswell was commissioned by the Earl of Exeter to make a detailed survey of the manor of Wimbledon. In that survey one reads: 'One capital messuage and manor house called East Sheen otherwise West Hall with houses outhouses barns stables orchards and garden lying west of the King's Highway – 5 acres'. The owner was Thomas Whitfield who had bought the property some 23 years earlier and who owned a total of 217 acres scattered all over Mortlake. It seems fair to assume that he had pulled down an earlier building since, in *The Environs of London* by the Rev. Daniel Lysons, the date 1611 is given for the manor house. Somewhere about 1680, after several changes of ownership, the estate was bought by Sir John Temple (younger brother of Sir William Temple who was the husband of Dorothy Osborne, friend of Jonathan Swift and the owner of a large house at West Sheen). Sir John, as was not uncommon at the time, had a chequered political career, enjoying special favour under William III (for whom he served as Attorney-General of Ireland) and finally retiring to East Sheen in 1695. On Sir John's death in 1704 his son Henry inherited the property. Henry made two wise decisions: he married Anne Houblon, a wealthy heiress (the marriage, we are told, was 'one of great happiness') and he gave loyal support to Sir Robert Walpole. In due course he was ennobled as Viscount Palmerston but as the title was an Irish one its holders were not debarred from election to the English Parliament. In 1736 he purchased *Broadlands* at Romsey and handed over the East Sheen estate to his eldest son,

Henry John Temple, third Viscount Palmerston, by T. Heaphy (1802)

Sheen Grove, seat of Lord Palmerston (1798)

Henry. This young man, however, died suddenly in 1740, leaving an infant son, also called Henry, who became the second Viscount Palmerston. On coming of age (soon after his grandfather's death) he chose to live at *Broadlands*, his mother remaining at East Sheen until her death in 1789. The second Viscount was a member of Parliament for 40 years and in 1801 he unsuccessfully petitioned for a seat in the House of Lords but died in the following year. He had remarried in 1783 and his son, Henry John, was born in 1784. This young man therefore became the third Viscount Palmerston at the age of eighteen, but East Sheen was let (to Lord Castlereagh, later Foreign Secretary) until he attained his majority. By then he had decided to make *Broadlands* his principal residence; all the family portraits were moved there from East Sheen and the estate was broken up and sold off to neighbouring landowners. Last to be sold was the big house itself together with its immediate twenty acres. The buyer (in 1808) who paid £12,065 for the property was Thomas Bernard. (One branch of the family always pronounced the name 'Barnard', which has led to numerous errors.) He already owned a substantial house called *The Priory* at nearby Roehampton and it seems unlikely that he ever actually occupied his latest acquisition. He nevertheless rebuilt the Jacobean-style east front 'at considerable expense with corresponding taste'; he had a 'viranda' constructed on the west front; he sold off four or five acres and he cut down nearly all the fine horse chestnuts which had formed an avenue leading to the large pond. At this point (1810) Sir Thomas (who had inherited a baronetcy) decided to sell both *Sheen Grove* and *The Priory*, and the fact that he and William Pearson were both involved in the Royal Institution (Sir Thomas had been a joint founder in 1799 and Pearson became a Hereditary Proprietor in 1800) and were probably acquainted is at least a pointer to the reason for Dr Pearson's decision to move from Elm House. It is not known how much Pearson paid for *Sheen Grove*. If the figure was indeed around £12,000 he must have made a handsome profit out of the sale of Elm House, and even so he would have had to raise a private mortgage. At all events we are told that 'no expense has been spared in building school rooms, making gravel walks, forming a playground, draining wet parts by giving to springs a proper direction in their descent into the pond, planting ornamental fences and erecting outbuildings to correspond with the magnificence of the house itself' (Shoberl). A map of the estate 'surveyed and delineated' by James Wadmore jnr. (1811) shows the extent of Pearson's transformation. He later 'commenced the creation of an observatory over the roof, which, when finished, will have a semi-globular dome moveable on ebony rollers, so as to present its opening to any point in the heavens'. The design was entirely Pearson's work. He also designed and built an azimuth-mounted telescope and purchased four refracting telescopes and a Gregorian reflector. Bearing all this in mind it might be fair to assume that the running of the school was not necessarily uppermost in his mind. In 1817 when he sold the school to the Rev. John Pinckney, he was anxious to continue his astronomical observations and so (in 1818) he purchased a two-acre plot some 300 yards south-west and on it

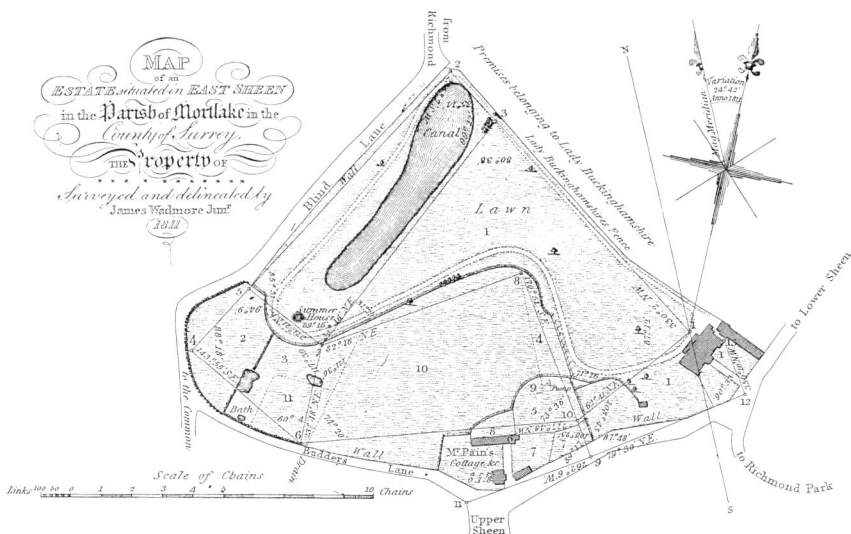

A Map of an Estate situated at East Sheen in the Parish of Mortlake in the County of Surrey, the Property of the Reverend William Pearson. Surveyed by Jas. Wadmore Junior A.D.1811 (SRO90/40/2)

Key to the Numbers

Freehold

1. The House, School Rooms, Fore Court, Lawns, Shrubberies, Canal etc.
2. Garden
3. Part of Pightle, formerly Woolley's
4. Adjoining Shrubbery and Lawn
5. Freehold part of Play Ground
6. Stable

In the occupation of Thos. Pearse

7. Cottage and 2 Gardens

Copyhold

8. Sheds and Yard
9. Copyhold part of Play Ground
10. Woolley's Three Acres
11. Late Stalls

**Total of Estate:
14a. 2r. 5p.**

erected *Observatory House*. In all probability he then transferred to it the 'semi-globular dome' and the astronomical equipment which he had installed at Temple Grove.

Even if Dr Pearson can hardly be described as a dedicated schoolmaster, the school appears to have prospered from the outset. Pearson was shrewd enough to associate it, at least by name, with the family which had owned the property for over a century by calling the school Temple Grove. Since Henry John Temple, the third Viscount Palmerston, held public office (with a few intermissions) for over fifty years, dying as Prime Minister in 1865, the tenuous connection must surely have been to the school's advantage.

It is possible that Pearson had already made the acquaintance of members of the aristocracy and of fashionable society before the move from Parsons Green. A son of General Sir Alexander Hope may have been among the boys whom Pearson brought with him to East Sheen. At all events two sons of John Hope (later 4th Earl of Hopetoun) were among the earliest positively identified pupils of Temple Grove. A great friend of the Hopes was General Sir George Murray who served under the Duke of Wellington in the Peninsular War, and it is therefore of considerable interest to find the Duke's wife writing to Murray (probably in February, 1813):

'My Dear Sir

You will oblige me very much if you will give Lord Wellington your opinion of Mr Pierson's [sic] school. I like it so much as to be very anxious our boys should go there, and I am sure your opinion of it would be most satisfying to him. With my best wishes for your happiness and success, I remain yours truly obliged.

D.C. Wellington'

General Murray was then just about to sail from England to rejoin the Duke for the culminating campaign of the Peninsular War. It is intriguing to think that among all the discussions of strategy and logistics, time could have been found to pass on a recommendation for a preparatory school. And on May 18 (1813) we find the Duchess writing:

'My dear General Murray

I am extremely obliged by your compliance with my request. The consequence has been exactly what I had hoped. Lord Wellington's confidence in your judgment has induced him to decide for Mr Pearson's school [endorsed: 'At East Sheen'] and I am quite happy. My poor little Douro has had a violent attack of bile, but his recovery has been so much quicker and he has been so much less weakened than by any former illness that we see he has gained considerably in strength. Charles is in perfect health.

Yours truly obliged
D.C. Wellington'

The Observatory, built by Pearson in 1818, and sold to Pinckney in 1823

ROTATIVE ROOF.

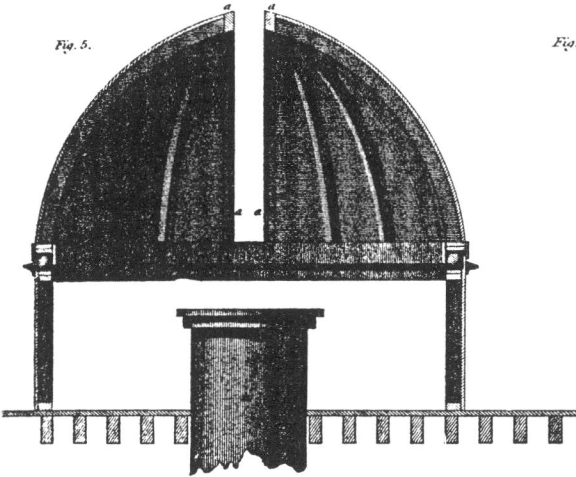

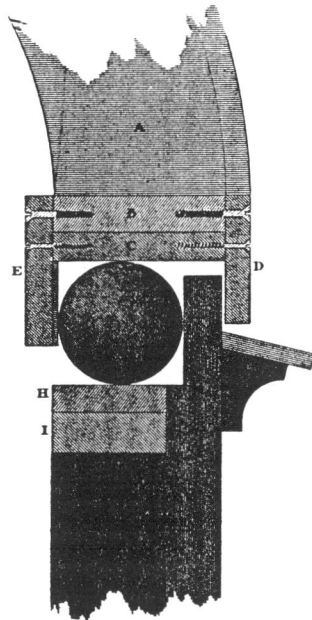

Design for a rotative roof by Pearson, originally on the roof of Temple Grove, transferred to The Observatory in 1818

Hardly any greater advertisement for a new school can be imagined than that the most famous man in Britain should send his two sons there. They entered Temple Grove in 1813 (aged six and five) and went on together to Eton in 1817. One hopes that their heads did not become too swollen when the news of Waterloo reached England: the Duke would not have approved.

General Murray had no sons himself, but his brother, Sir Patrick Murray, sent his son, John, to the school and in July 1816 the General wrote from Paris to his sister:

'As to Johnny, I wish chiefly that he would hold up his head, for when we saw him at East Sheen his poke was really frightful'.

Another pupil at that time who can be identified with certainty is Richard Twining (the fifth generation of the famous tea family). Three entries from his father's diary for 1816 read:

'Feb.12.   Went to Dr Pearson's Temple Grove, East Sheen, to settle for Richard's going thither.
'April 29.   Dear Richard went to school at Dr Pearson's at East Sheen. We had walked over the unfinished Waterloo Bridge.
'June 13.   Fetched Richard home 1st Holidays.'

Richard, then 8½, went on to Rugby in 1818 at the age of 10. He became a partner in the firm in 1829 and shared joint control with his cousin, Samuel Harvey Twining, from 1857 until their retirement forty years later. Richard's lifetime spanned all but a few months of the period that Temple Grove was at East Sheen as he died in March, 1906, aged 98.

By the time that Richard left Dr Pearson had handed over to Dr Pinckney, though the engraving, dated 1818, still accords Pearson precedence in its wording: 'To the Rev. Dr Pearson and the Rev Dr Pinckney and the Several Noblemen and Gentlemen educated in their establishment, This View of TEMPLE GROVE, East Sheen, Surrey, is most respectfully dedicated by their Obedient Humble Servant, Edmund Dorrell'. About twenty small boys (some very small) are shown walking across the lawn immediately south of the main building with a young woman bending solicitously over two of the youngest ones. A junior master walks behind this little procession; a married couple stand just outside the verandah surveying the scene; and another young woman holds the hands of two small boys in the far corner. In the right foreground cows sit chewing contentedly. This, in effect, is a 19th century prospectus. It is saying what almost every prep school prospectus ever compiled has sought to say: 'There are many good schools to choose from but *ours* is set in almost idyllic surroundings, where every possible comfort is provided for the well-being of the children. An atmosphere of kindness and tranquillity prevails, the grown-ups devoting themselves to the welfare of their charges etc. etc.'

In a book entitled *Speeches selected for the use of the Young Gentlemen of the Seminary at Temple Grove* the introduction strikes a similarly reassuring note. The first sentence is worth quoting in full:

William Pearson, his wife, Frances, and their only child, Frances, by T. Phillips

To the Revd Dr Pearson, the Revd Dr Pinckney, and the several Noblemen and Gentlemen educated in their Establishment, this View of Temple Grove, East Sheen, Surrey, is most respectfully dedicated by their Obedient Humble Servant Edmund Dorrell (1818)

'Dear Pupils

The short speeches with which you are here presented have been selected with an anxious desire that you should pronounce them in an audible, distinct and emphatical manner, and at the same time with a graceful attitude; and, we assure you, if your desire to learn be in proportion to our anxiety for your improvement, you will not only gain the love of your parents and the admiration of your acquaintance, now while you are young; but, as you grow in years, yourselves will daily experience those advantages which distinguish the polished gentleman from the rustic clown.

(Several more sentences follow and the introduction ends:)

Yours very affectionately,

W. Pearson, J.H. Pinckney.'

The booklet had originally been compiled for use at Elm House in 1809. Later editions added the name of Jonathan Thompson to those of Pearson and Pickney, so it clearly continued in use for a good many schoolboy generations. Among the speeches it contained were *Cato's Soliloquy on the Immortality of the Soul* and *Brutus on the death of Caesar*, titles which must have impressed parents with the school's concern for the moral and intellectual improvement of their children.

In 1817 William Pearson was appointed Rector of South Kilworth (Leics.) but he continued to live at *The Observatory* (while a curate attended to matters in Leicestershire), devoting himself to astronomical observations and the construction of new instruments. He was, by then, recognised as one of England's leading astronomers and when (in 1819) his name was put forward for the Royal Society, the Astronomer Royal was among the proposers. The following year, on January 12th, Pearson organised a dinner at the Freemasons' Tavern at which it was resolved to establish 'a Society for the cultivation of Astronomy'. An explanatory address was drawn up by John Herschel, with the assistance of Pearson and Babbage, and thus the Astronomical Society of London was founded. Ten years later the Society, receiving its charter from William IV, became the Royal Astronomical Society.

In 1819 Pearson's daughter, Frances, married W.P. Moffatt and he, together with John Pinckney, was recruited by Pearson as a founder member of the new Society. Two years later Pearson made the decision to leave East Sheen and move to Leicestershire though he did not actually sell *The Observatory* to Dr Pinckney (for £1,568) until 1823. Even after the move he contrived to attend nearly all the monthly meetings of the Astronomical Society (of which he was Treasurer for ten years) in London, travelling by stage coach from Rugby and staying with his sister, Hannah, in Islington. His visits to London continued to within a year of his death.

Pearson was now a wealthy man. He settled down to complete his great work, *Practical Astronomy*, in two volumes, for which he received the Gold Medal of the R.A.S. He continued his observations and his writing up to the time of his death in 1847 at the age of 80. His first wife died in 1831 and not long afterwards he married again, his second wife, Eliza Sarah, outliving him by many years. His son-in-law (who died in 1838) and daughter had no children and on Pearson's death the bulk of his estate passed to his nephew, William (John's son). *His* son, Colonel William Pearson, lived at North Kilworth Hall until 1908.

Let it also be remembered of this remarkable man that he paid for the rebuilding of the north aisle of his chruch, for a new organ and for a set of communion plate. In addition he built and endowed a school for the village and for many years acted as a magistrate.

Little seems to be known of the origins of John Pinckney, beyond the fact that he was born in 1777, married Harriet Lockton, who bore him four sons and a daughter, and was clearly a man of means. He bought up a number of properties in Mortlake at different times and redeveloped them and it may well have been he who purchased a substantial house in the Upper Richmond Road (just east of Milestone Green) which he named Temple Grove Cottage.

Clearly John Pinckney was an astute businessman and it is specially fortunate that evidence is available from several different quarters about life at Temple Grove in the 1820s and '30s.

Two documents, discovered by genealogist Margaret Wilson among the Brabourne family papers in the Kent Record Office, give valuable information about the running of the school in those early years. They are the midsummer bills for the two Knatchbull brothers, Norton and Charles, rendered in 1820, and in all likelihood are the earliest prep school bills to have survived. The basic fee was for the actual boarding of the pupil: £18.7.6. That is, of course, for a period of six months less a few weeks. Tuition, interestingly, is divided into three sections: Latin, Greek and Geography at £2.12.6; Writing, Arithmetic and English at £2.2.0; and French (optional) £2.2.0. Other, more or less unavoidable charges, were for copy books, paper, pens, exercise books, slate, pencils and ruler which amounted to 11/-. And, since the charges are the same on the two boys' bills, it is reasonable to infer that they were uniform throughout the school: Assistants and Servants 21/-; Chaplain 5/-3d (the weekly collection at Mortlake Church, presumably); Domestic Medicines (brimstone and treacle?) 3/-6d; Windsor Soap 6d. There is also a charge of £2 for Washing which one assumes to mean laundering.

Almost equally interesting are the charges for optional extras which neither boy incurred. They include Military Exercise, Dancing, Drawing, 'Wine daily, by Desire', 'Tea, Morning and Evening by Desire', 'Pure Milk, ditto', and 'Single Bed'. It is, sadly, impossible to guess what proportion of the pupils availed themselves of these assorted privileges.

The opposite page of the account is headed 'Disbursements' and this, too, contains some revealing items. Many of them relate to clothing: 'Hatter's Bill' (around 7/-); 'Taylor's Bill for repairing and buttoning clothes' (12/-6d for Charles); Mending, 5/-; Gloves, Shoe-strings, Shoes cleaning (around 2/-6d each); New shoes (14/-); Shoes repaired (in one instance, 17/-6d); together with 'Hosiers Bill', Combs and Brush and Tooth brush (not charged on either bill). Each boy, too, seems to have avoided the expense of dental extraction (see p.97) but each was charged 4/- for hair cutting. Norton's bill includes 4/- for 'Letters and Parcels', and both boys were required to pay for individual school books (eg. Dictionary 10/-6d, Hort's *Pantheon* 5/-6d, French Reader 4/-6d). One wonders whether 'Coach Hire' (5/-6d) covered the cost of journey by coach to and from school. There was also provision for 'Vacation', presumably for boys whose parents were working or serving abroad. Charles's broken window cost his father 6/-, and two of the most intriguing items – in addition to an Apothecary's Bill for £1.2.6 – also appear on his bill: 'Flannel etc.' (2/-6d) and 'Wine and Egg' (5/-6d) 'given medicinally by order of the Apothecary after the Mumps'.

Together with disbursements the total bill comes to £34.8.9 (Norton) and £33.11.6 (Charles) so that the total cost for the two boys in a full year would have been under £140. So much for the figures, it is time now to see what some of the pupils felt about the school and its Headmaster. The first account concerns the son of a London attorney.

'At the age of seven Alfred Gatty was taken in a post-chaise to his preparatory school, Temple Grove, at East Sheen, where he was consigned to Dr Pinckney's guardian care. The school was a distinguished one and numbered among its pupils sons of the nobility and rich merchants. One of the occupants of Alfred's dormitory of eight boys was James Disraeli, a younger brother of Benjamin.

The school had a high reputation for scholarship, but to young Alfred it was like plunging his hand into a hornet's nest. He had the unfortunate habit of snoring, with the consequence that he was continually disturbed by shoes being thrown at his head. The food was also unpalatable and ill-cooked and Alfred, who was very unhappy, confided his misery to Richard Burn, a youth with a prolific imagination.

A combination of the hard life, the nauseating food and no doubt excessive home-sickness induced the lads to run away and seek adventure. The boys had no money, but saved some crusts of bread from breakfast and with these packed in a small box together with a prayer-book they ran through the private grounds and dropped from a wall six feet high on to the road. The two then set forth for London, a distance of eight miles.

The old wooden bridge at Putney was their first obstacle, as they had nothing with which to pay the half-penny toll; so Alfred offered to pawn his small black-silk handkerchief, which offer the good natured man allowed without any security, and after a weary tramp the two tired boys reached London. Alfred went to his father's house and Burn to Russell Square, where his guardian lived, only to find at both places a charwoman in possession and the family out of town.

The adventure ended ingloriously for the boys were taken back to school in the charge of an emissary dispatched to find them. All romance and courage had faded out in weariness; but, as Alfred was very young, a relative had insisted before he was taken back that he should not be punished for his escapade. As Alfred was not to be punished someone had to stand proxy, and the Doctor exercised his birch on the person of his cousin William, who had no connection with the whole affair and who never quite forgave Alfred for this vicarioius whipping'. (Slightly adapted from the account in *Mrs Gatty and Mrs Ewing* by Christabel Maxwell). It is some consolation to know that by 1824, when he left to go on to Charterhouse, Alfred stood sufficiently high in Dr Pinckney's esteem to have been made one of the three head boys.

It is from Alfred Gatty, too, that we learn about the twice-yearly visit of the dentist, Mr Kidman, 'who received five shillings for every tooth he drew. It was a most sanguinary business. About half-a-dozen boys were admitted into the Doctor's study, where he sat, pen in hand, to note the number of extractions to which each of us was in turn subjected. I have sat through four double teeth being drawn at one time, and can now perfectly recall that chamber of horrors, the shrinking boys at the door awaiting the order to advance to the chair of execution; the little old butcherly dentist with his instrument of torture; the pail in the corner of the room to which we retired after the operation; and the calm Doctor with his powdered head, sitting through a morning as the dentist's numerator, unmoved by pity or remorse'.

Although the evidence is unsupported, the reference to James Disraeli as a fellow member of Alfred Gatty's dormitory makes it at least likely that Benjamin's younger brother *was* a pupil at Temple Grove in the early 1820s. The 1905 Register confidently claims the Prime Minister as an Old Boy of the school, adding in the third brother, Ralph, for good measure. Lord Blake's researches conclusively prove that Benjamin was never at Temple Grove, but leave his brothers still in some doubt. Even if they had not been pupils, Temple Grove was sufficiently well known for Disraeli, as a successful young author, to send the hero of his best-known novel, *Coningsby*, to school there. The book, written in 1844, is set around the time of the Great Reform Bill (1832) and in chapter 2 we read:

'The boy was recalled from his homely, rural school, where he had been well-grounded by a hard-working curate, and affectionately tended by the curate's unsophisticated wife. He was sent to a fashionable school preparatory to Eton, where he found about two hundred youths of noble families and connections, lodged in a magnificent villa that had once been the retreat of a minister, superintended by a sycophantic Doctor of Divinity, already well beneficed, and not despairing of a bishopric by favouring the children of the great nobles. The doctor's lady, clothed in cashmeres, sometimes inquired after their health, and occasionally received a report as to their linen.

'This change in the life of Coningsby contributed to his happiness. The various characters which a large school exhibited interested a young mind whose active energies were beginning to stir. His previous acquirements made his studies light; and he was fond of sports, in which he was qualified to excel'.

That is almost all we are told about his early shooldays since, 'when Coningsby had attained his twelfth year, an order was received (from his grandfather) that he should go at once to Eton'. Four chapters follow on his life at Eton, colourful but not entirely accurate.

Twice a year Mr Kidman ('the little old butcherly dentist') visited Temple Grove

Cornet William Bankes, posthumously awarded the V.C.

The Rev. Dr George Rowden

How much, then, can we rely on his brief pen-picture of Temple Grove? There were certainly never as many as two hundred pupils, nor is there evidence that Dr Pinckney was well-beneficed or that he coveted a bishopric, but the description of him as 'sycophantic' and 'favouring the children of the nobles' has a ring of truth about it. It is a characteristic that one has encountered elsewhere. Further evidence of Pinckney's quest for sons of the aristocracy may be gathered from a letter which Lady Charleville wrote to her daughter in 1831 in which she remarks:

'Lady Mountjoy did not come to London [from Dublin] with any views to advise or assist Countess d'Orsay. She came to place Richard at East Sheen, at a school his father selected, where Greek is first taught, and many other things in the new fangl'd mode'.

The most detailed picture of life at Temple Grove in the 1830s comes from the pen of Major-General Sir Archibald Anson in his memoirs, *About Others and Myself*. Anson was born in 1826 so he was eight when he arrived at the school where he spent three years. His memoirs were not published until 1921 (he died at the age of 99) but the details he gives of daily life (other than actual lessons) are sufficiently convincing to be taken on trust.

'On the 18th April, 1834, I was sent to the Rev. Dr Pinckney's school, at Temple Grove, East Sheen. It was then the great preparatory school for Eton. Three of my brothers had been at the school before me.

My grandfather drove me down from London to East Sheen, and I have a keen recollection of my feeling of desolation as he drove away, and the iron gates were closed, and I was left standing alone on the long, paved footpath leading up to the front door of the house. I was barely eight years old, and so was placed in what was called the little school, which was presided over by the lady teachers, Miss Field and Miss Evatt. My bedroom contained twelve beds, and commanded a view of the pagoda in Kew Gardens. My companions in the bedroom were two brothers, George and Gilbert Elliot, younger sons of the Earl of Minto; and two brothers Tuke. The names of the others I forget.

Dr Pinckney was a very skilful administrator of the cane and the birch. From the "little school" we used to be sent down to his table, where he held his class in the middle window of the "big school", to hold out a hand, which received one or more cuts with the cane. One punishment for a greater offence was named a "tight breech"; to administer this, the culprit was laid in a position bending over the Doctor's leg, while he pulled his trousers tight with one hand, and applied the cane with the other. For a still greater offence, the boy was laid face downwards across the Doctor's table, when two of the class held each a leg, and two others held each an arm, and the head boy of the class placed a Latin Grammar in his mouth to bite, to relieve his feelings during the operations. On more than one occasion, I assisted at this function. It was no easy matter to hold on to the arm or leg of a wriggling boy undergoing this description of torture.

At this school we had, for breakfast, oblong chunks of bread, about four inches long, one and a half wide, and one inch thick, the first pieces having a smudge of butter on them, the rest being dry. This was washed down by an allowance of milk and water served in a small white basin without a handle. A few boys, by special request of their parents, were given pure milk, and a few others, tea, in similar basins.

For dinner, there was a pudding, either rice with little milk, very dry, and served in round tin dishes similar to those used for soldiers in barracks, or a doughy sort of pudding, with a few raisins in it, and baked. My share of this, one day, was a lump of mortar, of which I got a mouthful. Then there was a joint of either roast beef or mutton, or sometimes a stewed aitchbone of veal. The pudding was served before the meat. The supper was the same as breakfast. Sometimes the boys at one table would have a match with those at the next, to see which could eat the greater number of pieces of bread.

The wife of a schoolmaster, whom I was lately telling how we were fed at this school, said, "If we could feed boys like that now, we could make some profit."

When a boy had game, or, at Michaelmas, a goose sent him, he made out a list of those of his friends he wished to share it with him, and this was handed up, at the dinner hour, to the Doctor, or whoever might be carving for him, at the sideboard. Cake or other goodies sent to the boys had to be delivered up, and a certain quantity for each owner was arranged on a long table, in the "little Schoolroom", at four o'clock each day, so long as it lasted. Jam was issued in like manner, spread on bread, at supper time to those who were lucky enough to possess it.

Dr Pinckney's custom was to come into the hall at supper time, and when we were supposed to have finished he would give the word "Rise", when all stood up. He would then call out the name of some boy, who would repeat the Lord's Prayer. He would then name another boy, who would repeat an evening hymn. After breakfast the Lord's Prayer and a morning hymn were repeated in the same way.

After supper, Dr Pinckney placed himself leaning against a dresser, at the end of the dining hall, nearest the exit, with one hand extended, which each boy shook as he passed. Should there be a boy under condemnation to be flogged, the Doctor would stop him, and five other boys, as they came up to him, and when all the rest had departed, these six boys were taken to the other end of the hall, to the sideboard where the meat was carved. The culprit was then prepared for punishment by the removal of that portion of his garment that would interfere with it. He was then laid upon the sideboard (a long stout table with four legs), and held, as described for the caning, by the five boys, whilst the Doctor vigorously applied the birch.

When the boys came down in the morning, there was, from time to time, a head-washing, carried out by the lady teachers. This took place in the writing-room, a room intermediate between the big and little schoolrooms. Each boy had his head washed over a basin of warm soap and water, and then a spongeful of rosemary and water squeezed over it. I can realize the smell of this, as it trickled down my nose, to this day. There was, in those days, no such thing as washing any other part of the body except the hands, face and feet, and perhaps the neck and chest. The feet washing was a great ceremony once a fortnight. A large oval tub was placed in the dining-hall, with a long form on one side of it, on which about half a dozen boys would sit at a time, with their feet in the water in the tub. One or two maids would kneel on the opposite side of the tub, and wash the boys' feet with soap.

In the meantime, one of the lady teachers would sit at the end of one of the dining tables, and one boy at a time, after his feet were washed, would sit on the table, with his feet almost touching her chest, when she would proceed to cut his toe-nails.

Every spring, it was the custom to give each boy a dose of sulphur and treacle. For this purpose there was brought into the writing-room at washing time in the morning, a large basin full of this mixture and each boy had a dessert-spoonful put into his mouth.

Our boots and shoes were cleaned only twice a week. The clean ones were given out on Wednesday and Saturday afternoons, when the dirty ones were taken away to be cleaned.

The walls of the writing-room were papered on canvas, between which and the brick wall there was a space. Here rats used to disport themselves, and Mr Dodd, the writing master, was very expert at spearing them with the leg of a pair of compasses, through the paper and canvas.

A dentist attended every half-year, and each boy was sent into him, to have his teeth examined, and it was the exception if he did not pull one out, and in those days a dentist appeared to consider it the correct thing to cause as much pain as possible. By instructions from my father, I was exempted from attending the dentist.

Now and then, a more adventurous boy would manage to evade the eye of old John, the man-servant at the lodge, and, under cover of darkness, break out of bounds and go to the tuck shop in the village.

Among the boys were Tom Coke (the late Earl of Leicester), and his brothers Edward and Henry. I remember their father, "Old Tom Coke of Norfolk", as he was then called, coming to see them, and my being introduced to him. His second daughter, having married my uncle, Viscount Anson, was my aunt. The last time I saw her was on a Sunday, shortly before her death in 1843, when I assisted her into her carriage when she was leaving our house in Devonshire Place. She married at the age of fifteen.

To please little boys who did not receive many letters from their friends, the Doctor sometimes addressed sham letters to them. No doubt this was kindly meant, but the result was disappointing to the recipient. I know I felt it so when I received one. As I received plenty of letters from home, it was unnecessary to have sent me one.

On the 16th October, 1834, I saw, from the school, the light from the burning of the Houses of Parliament.

At the end of my second half-year, at Christmas, I obtained a prize which I still have. It bears, on the outside of the cover, the following inscription: "The

gift of the Revd. Dr Pinckney to A. Anson for excelling his class in accidence".
The prize was a small book entitled *Footsteps to the Natural History of Beasts
and Birds, Designed for Children.* On arriving at home for the holidays,
almost the first thing I did, at tea-time, was to cut my little finger rather badly,
when I was immediately chaffed by my eldest brother, for having had a prize,
as he said, for accidents. I have the mark of this cut at the present day.'

It is worth noticing that Anson refers to 'my second half-year'. Until the
coming of the railways there were just two terms a year (as we have seen) and it
would have been no uncommon thing for a new boy to arrive while a term was
in progress. With reference to the birching described by Anson an article in a
magazine published in 1891 (in which the dedication of the new chapel is
briefly mentioned) describes Dr Pinckney as 'a man whose fame was as the
fame of Dr Keate himself. And indeed many a lad had the privilege of being
flogged by each of them, for the great majority of the boys at Temple Grove
were sent on from there to Eton College'.

The Henry Coke (a younger son of the famous Coke of Norfolk) who is
mentioned was himself the author of an entertaining autobiography, *Tracks of
a Rolling Stone*, published in 1905. It is more colourfully written than Anson's
book but it bears out the latter in certain details besides adding a brief
summary of the author's academic progress so that it seems worth quoting in
full:

'Soon after I was seven years old, I went to what was then, and is still, one of
the most favoured of preparatory schools – Temple Grove – at East Sheen,
then kept by Dr Pinckney. I was taken thither from Holkham by a great friend
of my father's, General Sir Ronald Ferguson, whose statue now adorns one of
the niches in the facade of Wellington College. The school contained about
120 boys; but I cannot name any one of the lot who afterwards achieved
distinction. There were three Macaulays there, nephews of the historian –
Aulay, Kenneth and Hector. But I have lost sight of all.

'Temple Grove was a typical private school of that period. The type is
familiar to everyone in its photograph as Dotheboys Hall. The progress of the
last century in many directions is great indeed; but in few is it greater than in
the comfort and the cleanliness of our modern schools. The luxury enjoyed by
the present boy is a constant source of astonishment to us grandfathers. We
were half starved, we were exceedingly dirty, we were systematically bullied,
and we were flogged and caned as though the master's pleasure was in inverse
ratio to ours. The inscription over the gates should have been "Cave Canem".

'We began our day as at Dotheboys Hall with two large spoonfuls of sulphur
and treacle. After an hour's lessons we breakfasted on one bowl of milk –
'Skyblue' we called it – and one hunch of buttered bread, unbuttered at
discretion. Our dinner began with pudding – generally rice – to save the
butcher's bill. Then mutton – which was quite capable of taking care of itself.
Our only other meal was a basin of 'Skyblue' and bread as before.

As to cleanliness, I never had a bath, never bathed (at the school) during the
two years I was there. On Saturday nights, before bed, our feet were washed by

the housemaids, in tubs round which half a dozen of us sat at a time. Woe to the last comers! for the water was never changed. How we survived the food, or rather the want of it, is a marvel. Fortunately for me, I used to discover, when I got into bed, a thickly buttered crust under my pillow. I believed, I never quite made sure (for the act was not admissable), that my good fairy was a fiery-haired lassie (we called her 'Carrots', though I had my doubts as to this being her Christian name) who hailed from Norfolk. I see her now: her jolly, round, shining face, her extensive mouth, her ample person. I recall, with more pleasure than I then edured, the cordial hugs she surreptitiously bestowed upon me when we met by accident in the passages. Kind, affectionate 'Carrots'! Thy heart was as bounteous as thy bosom. May the tenderness of both have met with their earthly deserts; and mayest thou have shared to the full the pleasures thou wast ever ready to impart.

'Did we learn much at Temple Grove? Let others answer for themselves. Acquaintance with the classics was the staple of a liberal education in those times. Temple Grove was the atrium to Eton, and gerund-grinding was its raison d'être. Before I was nine years old I daresay I could repeat – parrot, that is – several hundreds of lines of the Aeneid. This, and some elementary arithmetic, geography, and drawing, which last I took to kindly, were dearly paid for by many tears and by temporarily impaired health. It was due to my pallid cheeks that I was removed'.

During the time that Anson and Coke were pupils, Dr Pinckney, at the age of 58, handed Temple Grove over to his successor, Jonathan Thompson. Thompson, then 39, inherited what, in spite of its faults, was undoubtedly a flourishing school and yet within six years the numbers had dropped by 40%. We know this because the census of 1841 is the first to provide information about names, ages and types of employment. Spelling is not always to be trusted and ages (from 15 onwards) were usually given only in multiples of five. Nevertheless it is a valuable document, most especially for our purposes since it was taken during term-time.

In the main school we note Jonathan and Ann Thompson and their three young daughters. Then come no less than eighteen female school assistants, ranging from Amelia Smith, aged 60, down to Mary Roper, 15, though Mary and five of the other youngest ones are simply entered as female servants. There is also one male servant, John Creed, aged 40. Even with the minimal wages paid it does seem an excessive work force. The seventy-five pupils are headed by William Marriott, who, at 15, may simply have been a private pupil. No other boy was older than 12. Apart from a Horace Walpole, aged 10, and Henry Bradshaw, also 10, none of the pupils' names arouses immediate interest. Five families appear to have three boys each at the shool, including the three Thompson boys. From *A Memoir of Henry Bradshaw* by G.W. Prothero comes the only other evidence concerning Temple Grove pupils in the Thompson era: 'When Henry was eight years old he went to school at Temple Grove, East Sheen . . . Here Henry got his schooling for about four years. Not long after his arrival, Mr Thompson wrote to his mother as follows:

"Harry gains the first prize of the lower school, and with truth I never examined a little fellow with a sounder head or better memory. He is a boy of the greatest promise, so zealous and steady and unswerving from his point, that I will back him against any that have yet gone through my hands. He has made deep inroads upon his Latin Grammar, and has done in one quarter what costs the toil of four or five to most children".'. Bradshaw justified Thompson's judgment by becoming one of England's leading scholars and antiquarians.

Temple Grove Cottage housed the teaching staff: Charles Wilson (aged 40), Edward Pritchard (35), Jean Lefevre (35), John Hughes (25), Robert Bewglass (25) and Alexander Ross (20). It would be safe to assume they were all unmarried. Their qualifications, if any, are not divulged. That is all the evidence we possess for the Thompson era. It seems probable that the school had sunk in public estimation, though whether that was because the Headmaster was not in orders or whether he lacked the 'sycophantic' skills of his predecessor and was unable to attract the sons of the nobility, it is impossible to say.

His successor was a 23 year old unmarried clergyman, George Croke Rowden, born in Wiltshire (1820) and educated at Winchester and New College, Oxford, where he became a Fellow in 1840. In 1846 he married Emily Twining (sister of Richard). She was already 30 but nevertheless bore her husband three daughters in the next five years.

Rowden is perhaps better remembered as a musician than as a schoolmaster. He was a prominent member of the short-lived Musical Institute of London in the early 1850s, giving a lecture to the Institute, in February 1852, 'On the different character of keys in music'. His *Magnificat* and *Nunc Dimittis* were published three years later and he also composed glees to words by Mrs Hemans. He had some reputation, too, as a preacher, delivering three sermons, 'On the recent visitation of the cholera', at Mortlake Parish Church in 1849 in the presence of H.R.H. the Duchess of Gloucester, and, three years later, two sermons entitled 'The Christian's Retrospect of the year 1851'.

Rowden must have sought to build up the school's reputation and numbers. However, after seven years the total of pupils was only five greater than in 1841. The two witnesses whom we shall be calling, who were both pupils in the last years of Rowden's headship, describe it as a school of 'about a hundred boys', whereas the 'Name Book', started by O.C. Waterfield in 1859, lists only 64 pupils at the time when he took over. No clear interpretation can be put on that discrepancy.

The 1851 census is a good deal more informative than its predecessor, giving not only reasonably accurate ages and occupations, but each person's relationship to the head of the household, place of birth and marital status. At Temple Grove, in addition to Rowden and his family, there are listed Maria Surrey (47) and Eliza Adnams (46) matrons, both unmarried, nine female servants, all young (19-25) and unmarried except for Eliza Page, a widow of 45, and Edwin Baker (32), a solitary male servant, married. The eighty pupils deserve comment. The average age is considerably higher than it had been ten years earlier. There was a boy of 6, another of 7, but only twelve under 10 altogether, as against eight 13 year olds, five 14 year olds and one 15 year old. Does this suggest that fewer boys were going on to Eton than previously? The 15 year old was in fact Thomas Brassey, son of the famous railway engineer and himself to become a distinguished man in public life. Of the famous Bankes family (mainly associated with Dorset), two brothers, whose parents lived at *Sheen Elms* (quarter of a mile from Temple Grove), are listed in the school records. The elder, William, who left in 1849, was to become the first Old Boy to win the newly instituted Victoria Cross, while Wynne Albert appears, aged 10, in the census yet contrived to be serving in the Royal Navy in the Black Sea three years later and to have been present at the final surrender of Sebastopol in 1855. Montague Ommanney, aged 8, member of a noted East Sheen family is also on the list. Most significantly no less than fifteen boys (almost one-fifth) are recorded as being born in India. It is reasonable to assume that most of their parents were still living, and serving, there and that these boys spent their entire school-days separated from them in the care of guardians. In the 1861 and 1871 censuses (taken in school holidays) there are, respectively, seven and ten pupils resident in the school.

Temple Grove Cottage continued to provide accommodation for the unmarried masters. Jean Lefevre (French) was still on the staff; William Bewglass (Mathematics) *may* have been the same as the Robert listed earlier (the ages tally near enough) but the new names are all of young men: Henry Levander (24), B.A.(Oxon.) is given as First Classical Assistant Master, James Cowan (23) as Second ditto and James Jones (23) as Third. Did the latter two not have degrees? The last was Frank Rollins (English) who, at 18, may have been about to go up to University. It is not impossible, of course, that a married master lived elsewhere, but it seems unlikely.

Now for our witnesses. Adolphus Liddell writes in *Notes from the life of an ordinary mortal*:

'In the end of January, 1855, I went [at the age of 8 ½] to a school at Temple Grove, East Sheen, a large private school of about a hundred boys, kept by the Rev. Dr. Rowden. My cousin, Edward Liddell, accompanied me, being about a year my senior. It was the Crimean winter, and the day we went to school was bitterly cold and the ground covered with snow. When our parents left us we were taken to the schoolroom, feeling very bewildered with the noise and novelty.

'The next day school life began in earnest, and I came to know for the first time what it was to "eat my bread in tears". The cold of the great bare house, with its long windows and few fireplaces, was severe, and I soon was troubled with chilblains from getting my feet wet in the snow. But that was only a small part of my misery. The sudden change from a home where I was surrounded with care and affection to the indifference and occasional rough usage of school overwhelmed me, the more because I had never imagined anything of the kind.

'My new abode stood in extensive grounds, of which a large field and a gravelled square surrounded by walls were dedicated to the use of the boys. The field was only used in the summer months; in the winter we were confined to the dismal playground.

'The teaching of Temple Grove was not bad, but the food was coarse and the general arrangements rough. Every night and morning we filed before the matron – Miss Surrey – a kind, deep-bosomed old dame, who dealt two or three ferocious strokes with a hair-brush on to our heads as we passed under her. The good lady, oddly enough, had been a maid at my father's private school, so was kind to me in her way, tempering the blows of her hair-brush by saying that my forehead reminded her of my father's'.

One other remark that Liddell makes is worth noting: 'As for the fine arts, I attended the drawing class, which was instructed by an old gentleman called Wichelo, who had filled the same post at my father's private school, and an assistant known to the boys as "Bones".' And he concludes: 'At the end of 1858 Dr Rowden gave up Temple Grove, and I left the school'.

Four sons of the Rev. J Hailstone (of Bottisham, Cambs.) came to Temple Grove, Arthur (in 1855), Walter (1856), Herbert (1859) and Samuel (1861), so that only the two eldest were pupils of Rowden's and it is Arthur who has supplied further details.

'The Headmaster, when I first went to Temple Grove, was the Rev. Dr Rowden, an austere, bad-tempered man, and an awful bully. He used to wear a heavy ring on the little finger of his right hand and when he was going to box a boy on the ear, he would turn the ring round so that the stone was inwards, and didn't the blow raise a bump on the side of your head! I experienced this once when I had been stood on a form for punishment. The schoolroom was empty, and I was executing a pas seul, pour passer le temps, when the Doctor crept up behind and knocked me off the form with his favourite blow'.

He adds, charitably, 'Dr Rowden played the double bass and used to take part in the Handel festivals at the Crystal Palace', though he goes on, 'I remember our being all marched off to Mortlake Church every Sunday, where we sat in the gallery. It was a dreary service and never did me a bit of good'.

'Some of the assistant masters were very gentlemanly nice men. I remember Mr Smith and Mr Vernon, the latter a clever caricaturist. He drew a clever little sketch once for me of Achilles, who was only vulnerable in the heel, being shot at by a revolver as he was going upstairs and formed an easy prey. The French master, M. Alphonse, was a dear old chap. The butt of the school was the junior master. I forget his name but it began with Bu..., and a favourite amusement with us boys, when formed in line in the gloaming to march into tea, was to call out surreptitiously "Bu..., Bu..., Bu...".

'There was a gravelled playground with a fives court in one corner. Here we used to play "Battles", a game peculiar to this school. We formed sides and, provided with a large number of hard, leather-covered tennis balls, used to shy these at our opponents; when a boy had been hit three times, he had to retire from the Battle. This went on till all on one side or the other had been "killed" and there was one survivor left. It was a jolly good game, and took the place of football in the winter. We played cricket in the Summer.

'On the whole I spent a happy time at Temple Grove and got on well with my studies'.

Dr Rowden, in 1859, moved to Chichester where he was Precentor of the Cathedral and founder of the Chichester Choir Association. He died in 1863, his widow surviving him by 36 years.

2   East Sheen

Before the coming of the railway East Sheen, although only eight miles from central London, possessed an air of rural tranquillity which is hard to imagine today. Prior to 1846 there was a huddle of cottages around Milestone Green, including Colston's Almshouses (built in 1707), a second row of almshouses (Juxon's) facing on to Church Path, and another little cluster of cottages around *The Plough* in Upper Sheen. There were seven or eight substantial houses of 17th or 18th century origin, including *Sheen House*, which Earl Grey had leased from the Marquess of Ailesbury for two years during the struggle over the Great Reform Bill, *Sheen Elms*, the home of the Bankes family, *The Cedars* and, of course, *Temple Grove*. Their grounds were not extensive by 19th century standards, though those of *East Sheen Lodge* extended from the junction of Sheen Lane and Well Lane to the wall of Richmond Park, but with the River Thames nearby, commons and farmland adjacent and the 2500 acres of Richmond Park almost on their doorstep they enjoyed the advantages of seclusion and delightful surroundings coupled with that of a mere hour's drive to Westminster.

In many parts of England the Tithe Apportionment Map is often the first detailed survey of the area on a large scale. The one covering East Sheen was drawn up in 1838 and, as Raymond Gill comments, 'It would not have been much different had it been drawn at the beginning of the century'. The farm at the entrance to Sheen Common was sold off by Henry John Temple when he became the third Viscount Palmerston, along with *Sheen Grove* and the other Temple lands, but it continued as a farm until 1849.

In 1846 the London and Richmond Railway with its terminus of Nine Elms (Battersea) was opened and, although it was soon absorbed by the London and South-Western Railway (Waterloo becoming the terminus), the building of a station at Mortlake was to prove a crucial event in the development of the whole district. This was the era when not merely the aristocracy but gentry and professional men, businessmen and merchants could aspire to live a little distance away from London to allow their families to enjoy the untainted air of the countryside and yet be within easy reach of the business premises that sustained them. It was a relatively short period in English history, no more than 70 years, and when one looks at say, the O.S. map of East Sheen of 1913, it is hard to believe it ever happened.

O. S. map (scaled down from 25 inches : 1 mile) of East Sheen (1866). Crown Copyright.

East Sheen   Ordnance Survey Map 1866

Owners or Occupiers of named properties in that year

ROSE COTTAGE – Peter A.L. Muzy

PETRA HOUSE *(later Furness Lodge)* – **J. Barker**
(previously Sir Henry Durand)

STONEHILL LODGE – **William T. Atwood**

SHARON VILLA – **Mrs Earle** *(previously Joseph Wigley)*

TEMPLE SHEEN VILLAS –

   1. The Rev. Henry Venn

   2. Edmund C. Batten

   3. John Ball

   4. Mrs Elizabeth Beachcroft

SHEEN MOUNT – **H. Porter Smith**
(previously Thomas Hare)

UPLANDS – **(Sir) Henry Taylor**

THE PLANES – **Octavius Ommanney**

THE ORCHARD *(later Wedderlie)* – **Charles Ellis**

THE GABLES – **(Sir) Charles Bagot**

THE COTTAGE – **Arthur W. Blomfield**
(purchased by Sir Henry Waterfield in 1870)

THE LIMES *(formerly Spencer House)* – **Henry Kendall**

TEMPLE GROVE – **Ottiwell C. Waterfield**

THE OBSERVATORY – Unoccupied
(The Rev. J.H. Pinckney had died in 1864)

PERCY LODGE – **Mrs Graham**
(previously the Hon. Josceline Percy)

WORTLEY LODGE *(formerly East Sheen Lodge)* –
   James Stuart Wortley

THE VICARAGE – In course of construction

THE ANGLES – **Mrs Euphemia Scott**
   (widow of Dr John Scott)

THE FIRS – **Edward B. Meyer**

SHEEN HOUSE – **Edward J. Darley**

THE CEDARS – **Edward H. Leycester Penrhyn**

HOLLY LODGE – **(Sir) William White**

CLARENCE HOUSE – Leased by O.C. Waterfield
   for his assistant masters

Guided by Raymond Gill let us briefly examine what did take place in East Sheen between 1850 and 1890. The decision of Richard Smith to sell the farmland lying between Sheen Common and the Upper Richmond Road in twenty lots necessitated extending the roads later known as Christchurch Road and Temple Sheen Road westwards, and constructing new roads leading off them. One of the first of the new properties was *Uplands*, built for Sir Henry Taylor who was not only a senior civil servant but an esteemed poet. *Derby Lodge*, a villa in the Italianate style, was built by William Morley but his ambition outran his pocket and his miscalculation accounts for the appearance of the curiously shaped Stanley Road which gave access to 33 new building plots in 1866. Thomas Hare, well-known as an Inspector of Charities, designed for himself *Sheen Mount* in 1852. He was Archbishop Benson's brother-in-law and the house was described by Arthur Benson as 'a fantastic little chateau, with two pepper-box spired turrets, and overloaded with perky ornaments'. In fact Hare did not stay there long and the property was bought by H. Porter Smith. *Petra House* (later *Furness Lodge*) was built by Henry Goodale for Sir Henry Durand who was to become Lieutenant-Governor of the Punjab. Next door to it appeared *Sharon Villa* (later *Coval Lodge*) built by Joseph Wigley, a fire-hose manufacturer. On a more modest scale, immediately south, were built (in 1861) two pairs of semi-detached houses, known as *Temple Sheen Villas*, of which No.1 was occupied by the Rev. Henry Venn, Secretary of the Church Missionary Society.

James Stuart Wortley had served as Judge Advocate-General, Recorder of London and Solicitor-General (in the Palmerston ministry) but a riding accident crippled him so severely that he was forced to abandon public life. He bought *East Sheen Lodge*, a Georgian mansion with extensive grounds, paying £11,500 for it, and came to live there in 1858. He also purchased *Percy Lodge* and land in Stonehill Field, but sold off three plots, fronting on to Fife Road, one of which was later bought by the architect, Arthur Blomfield. Blomfield designed and built *The Cottage* for his own use and was also the architect of *Eastdale* and *The Halsteads*. The latter was built by Joseph Tall, the first contractor in England to develop the use of concrete. On the Stonehill land two new roads were laid out, extending eastwards from Sheen Lane, Vicarage Road and Stonehill Road, but the new properties there were not built until the 1870s. Finally, in the extreme south-east corner of East Sheen, come three very different but extremely interesting properties, each of which was occupied by a person of more than merely local importance. In point of time *Park Cottage* takes precedence. It was bought in 1849 by Adolphus Liddell and, as his son (who became a pupil at Temple Grove in 1855) records: 'One of the reasons which had induced my fther to take *Park Cottage* was the proximity of Richmond Park, of which his brother Augustus was Deputy Ranger'. The Liddells remained there for twenty years and young Adolphus paints an agreeable picture of the neighbourhood and some of its residents. He describes Mrs Julia Cameron (later famous as a photographer) as a 'clever and eccentric, but very kind, person. She belonged to the well-known Calcutta family of the Pattles, so remarkable for their individuality that Lord Dalhousie divided mankind into "men, women and Pattles". If she had once set her mind on any end, she never rested until she had attained it. The story of her forcing Tennyson to be vaccinated is well-known'. She was also known to be devoted to Henry Taylor.

Next door to *Park Cottage*, at *Sheen Lodge*, lived Professor Richard Owen, the famous anatomist. 'He was one of the best talkers I have ever met, and had a special gift of making things clear and interesting to children.' Not surprisingly he had many distinguished visitors, including Tennyson and Dickens who, apparently, had a special liking for Mrs Owen's home-made cream. Such visitors, however, impressed young Liddell far less than 'a stranger who had just returned from the North Pole' and who had shot through the heart a polar bear 'which afterwards ran for two hundred yards and swam for two hundred more'.

When the Liddells left *Park Cottage*, the property was taken by Edwin Chadwick. He had been largely responsible, in the earlier part of his life, for two of the most far-reaching pieces of 19th century legislation. For the first, the Poor Law Amendment Act of 1834, he was execrated for the remainder of the century. He had become a disciple of Jeremy Bentham and was convinced that there were neat, tidy, administrative solutions to all of society's problems if one took the trouble to gather all the necessary evidence, but he underestimated the greed and selfishness which can distort the best-laid plans and which converted the new workhouses into bleak prisons for the aged and infirm. For the second, the Public Health Act of 1848 (the first attempt by a British government to bring disease under control), he received belated recognition with the bestowal of a knighthood in his ninetieth year.

In 1862 Frederick Wigan bought eleven acres of land extending from Sheen Gate to Stonehill Road. He employed a relatively unknown architect, Robert Pope, to design a large house which was completed in 1866 and was occupied for the next forty years by the Wigans and their children. Wigan became an extremely wealthy man (and one of 'princely generosity') and in 1893 he engaged two of England's leading architects, Sir Aston Webb and Ingress Bell, to extend the house by building a picture gallery and a conservatory and to make other alterations. Frederick Wigan became a baronet at the time of the Diamond Jubilee.

Early in 1889 Princess Louise, eldest daughter of the Prince of Wales (the future Edward VII), became engaged to the Earl of Fife. Although the Earl had no royal blood, Queen Victoria was pleased. 'It is a very brilliant marriage in a worldly point of view as he is immensely rich'. The deficiency in blood was largely compensated by his creation as Duke of Fife on the day of their wedding (July 27th 1889). Nine years earlier he had purchased *East Sheen Lodge* and 'as the happy pair drove past [Temple Grove] after the wedding we schoolboys lined the wall and sang a charming lyric which had been composed for the occasion:

"Oh, What will the Germans do?
Oh, What will the Germans do?
The Duke of Fife has married a wife
The beautiful Princess Lou." (Hugh Headlam)'.

The lyric reflected the general satisfaction that 'the Princess had not gone abroad to choose some German princeling whom nobody had ever heard of, but had chosen instead a Scot, whom she loved, and the Queen had given her blessing' (Brian Masters: *The Dukes*).

This leads us on to a brief mention of *White Lodge*. The house (which lies in Richmond Park about three-quarters of a mile from East Sheen Gate) had been built around 1728 by George II as a quiet retreat for himself and his family. In 1850 Queen Victoria appointed her aunt, the Duchess of Gloucester, as Ranger of the Park and she lived at *White Lodge* for seven years, spending the summer months there. The house was presented by the Queen to the Duke and Duchess of Teck in 1869. Two of their sons entered Temple Grove ten years later, but their most famous child was May (born in 1867) who married the Duke of York (the future George V) in 1893 and who returned to *White Lodge* for the birth of her eldest child (Edward VIII, later Duke of Windsor) in 1894 (see p. 208). May of Teck had originally been engaged to the Prince of Wales's elder son, always known as Eddy, who died suddenly in 1892. May became engaged to Prince George by a pond in the grounds of *East Sheen Lodge* where George's sister, Louise (Duchess of Fife) had suggested they might like to go 'to look for frogs'. Should George have remained unmarried the throne would have passed eventually to Louise 'who, highly-strung and mouse-like, would scarcely have made an ideal Queen' (Brian Masters) and who dreaded the very possibility.

Up till the 1860s the people of East Sheen had no church of their own but attended the parish church of Mortlake, some three-quarters of a mile N.E. of Temple Grove. Not only did the growth of population after 1846 make a new church desirable, but the appointment of a new vicar, the Rev. J T Manley, whose 'eloquence was such that Mortlake Church was filled to its utmost capacity' made the need imperative. It was Edward Penrhyn, who had come to live at *The Cedars* in 1823 on his marriage to Lady Charlotte Stanley, who took the initiative, approaching the Queen to point out that a new church would be a great convenience for the occupants of White Lodge. Penrhyn's death in 1861, when only £1000 had been raised, led to the formation of a Building Committee. John Manley acted as Chairman but the main work was undertaken by James Stuart Wortley (of *East Sheen Lodge*) and Octavius Ommanney (of *The Planes*). Ommanney was a banker and he and Charles Bagot (of *The Gables*) were joint Honorary Treasurers. Most of the notables of East Sheen, who have already been mentioned, were on the Building Committee, including O.C. Waterfield, and by the end of the year sufficient funds had been raised to enable the Committee to instruct an architect. Arthur Blomfield who had recently designed his own home, *The Cottage*, was the obvious choice. He was the son of the late Bishop of London and had himself obtained 'a wide ecclesiastical connection through the quality of his workmanship'. His plans showed a church containing a nave, chancel, south aisle and tower and allowed for the addition of a north aisle at a later date. Land opposite Sheen Mount was purchased from the new owner, H Porter-Smith, for £266.13.4, and on June 17th, 1862, the corner stone of Christ Church was laid by Edward Penrhyn's son, E H Leycester-Penrhyn. Consecration was planned for April 16th, 1863, but early on the morning of Sunday, March 15th, 'the tower fell with a tremendous crash, carrying with it a portion of the roof'. Thomas Hardy was then a pupil of Blomfield's and when he arrived at the office the following morning the architect 'said slowly without any preface, "Hardy, that tower has fallen"'. Mercifully no-one had been hurt and the only witness had been the milkman, but no blame attached to Blomfield since it was discovered that, in the absence of a clerk of the works, the lower part of the walls of the tower had been 'packed with small stones and rubbish instead of being formed of solid masonry'. However the task of rebuilding was at once put in hand and on January 13th, 1864, the consecration was conducted by Dr Tait, the Bishop of London. Dr Pinckney (who died later that year) was one of a large gathering of clergy who were present. A month later came the first baptism at Christ Church, that of Wilfred Edgar, son of the Rev. and Mrs J H Edgar, of Temple Grove.

A brief paragraph must suffice to take the story of East Sheen from 1890 up to the outbreak of the First World War. The only development of note in the '90s was the opening up of Fife Road, leading from Sheen Lane to Christ Church Road, to replace Turtons Alley which had cut across the grounds of *East Sheen Lodge* (the story of the Alley and its eventual closure is clearly told in an intriguing booklet: *Alleyways of Mortlake and East Sheen* by Charles Hailstone). But in 1896 the sale of some 50 acres of the Palewell estate by the Gilpin family was to prove the precursor of a decade of hectic development and the transformation of an elegant locality of substantial houses into a typical outer London suburb. Herbert Shepherd Cross M.P. bought the Gilpin estate and the construction of Palewell Park led to the building of 150 houses. The new Headmaster of Temple Grove, Harry Waterfield, took fright and placed the estate on the market. It was sold at auction, in one lot, on April 11th, 1907, and the O.S. map of 1913 shows graphically what happened in the space of six years.

3   The House and Grounds of Temple Grove

Some description must now be given of the house as it stood in 1859 and of the estate. *Sheen Grove* appears to have been early 17th century in origin, though it underwent considerable alteration in the 120 years it belonged to the Temple family. Between 1730 and 1740 the first Lord Palmerston rebuilt the garden front but as late as 1798 the appearance of the entrance front was still markedly Jacobean. 'There are forward projections at either end, gabled at the roof and each showing an attic window; a third gable marks the centre of the elevation, where projects a two storied porch, having a flat roof with a railing round it. The circular-headed entrance-door in this porch, flanked by pilaster supporting an entablature is in character purely Jacobean' (W L Rutton). According to Daniel Lysons, 'The rooms are spacious and lofty. The drawing-room is hung with tapestries (which could well have come from the famous Mortlake factory) representing the four seasons. In the dining-room are the portraits of Sir John Temple, the younger, his brother . . . and others of the family'.

The estate was described by the Earl of Minto in 1802 as having 'pleasure grounds of seventy acres, pieces of water, artificial mounts and so forth'. Most of this was drastically altered by Sir Thomas Bernard during his brief tenure. The entrance front was rebuilt in a plain late-Georgian style although the actual entrance was contained in an unusual projection which resembled three sides of an octagon and extended to the full (three-storey) height of the house. He added the verandah on the garden front but he also sold a strip of land (four or five acres), immediately adjoining the house and extending right across to Blind Lane, to the Dowager Countess of Buckinghamshire, leaving *Sheen Grove* in the distinctly odd situation of standing in the extreme eastern corner of its estate and, as we have seen, he cut down a great many fine trees. The estate shown in James Wadmore's map may be described, then, as an inverted heart for want of a more accurate term. A curving path winds from the house across an extensive lawn in the direction of the summer house and then curls round the far side of a tongue-shaped lake, some 175 yards long, and there are two small ponds south of the summer house. Mr Pain's cottage is the only other residence on the estate, though three small buildings (not specified) are shown close to the wall on the S.E. side. At right angles to the big house is a long, rectangular building, almost certainly stables.

If Dr Pearson's school was soon to contain over a hundred pupils, even in the cramped conditions acceptable in the early 19th century, he must have speedily converted the stable block into classrooms and sleeping accommodation while linking it up to the big house, but unfortunately no detailed plans of Temple Grove survive from any stage of the school's hundred-year occupation. As already mentioned Pearson made other necessary alterations to the grounds to fit them for school use and in 1818 he built *The Observatory* as his own private residence. It seems likely, therefore, that the main building continued to house both pupils and staff while the school was still expanding.

Few further changes appear to have taken place in the next 45 years. All we know for certain is that, by 1841, the school had purchased or leased the building in the Upper Richmond Road known as *Temple Grove Cottage* (and later as *Ivy Cottage*) and that six unmarried masters lived there. Also, by that date, John Pinckney and his family were living at *The Observatory*.

When O.C Waterfield took charge he clearly intended to build up what had undoubtedly become a very run-down school and he must have had the full support, and financial backing, of the owner, Dr Pinckney. Work probably started almost at once to extend the former stable block in a north-westerly direction, giving an extra wing some 75′ in length and 25′ in width. When completed, as the photographs show, it blended in harmoniously with the existing building. The result, as the 1866 and subsequent O.S. maps show, was a T shaped building whose two component parts were of very similar cubic capacity, the one housing Mr Waterfield and his family, the other a superintendent, a housekeeper, four matrons, a dozen 'domestic servants' and about a hundred pupils.

Temple Grove Cottage proving too small to accommodate a larger staff, in 1865 Waterfield took a lease on a substantial house, also in the Upper Richmond Road but on the north side, known as *Clarence House*. Six assistant masters are noted as living there in the 1871 census, though as it was taken during school holidays that may not have been the full complement. A butler and his wife, two male cooks and five other domestic servants also lived there. When the house was first put on the market the sale particulars noted, on the ground floor, in addition to the usual reception rooms, a 'large school room'. One wonders whether this was where William Bewglass (on the Temple Grove staff in 1851) and his wife were running a day school (noted in the 1861 census). At all events no sale took place and Waterfield was able to lease the property.

As already mentioned no architectural plans for Temple Grove have survived. Our evidence for the ground plan of the school in c.1870 is based largely on the detailed sale particulars of 1907 which give the measurements of all the principal rooms supplemented by a plan drawn by T. T. Pitman and by various descriptions such as that of A.C. Benson in *Memories and Friends* and aided by a plan of the private quarters done (in 1948) from memory by Edward Graham who left the school in 1872. The latter is valuable as it shows that the use of some of the principal rooms changed after O.C. Waterfield left. The most important of these was the conversion of the Library (available to the boys) into the Headmaster's private dining-room. Benson describes it thus: 'The one delightful refuge was a big but cosy, carpeted and curtained room, many-windowed, with a verandah outside, called the Library, communicating with Waterfield's study, and therefore always orderly and quiet. Here there was a huge bookcase of readable books, and many comfortable chairs and sofas'. The adjacent room on the other side is described (in 1907) as the 'Smoking Room (now used as a Boudoir)' but it may have served a different purpose in 1870 or it may simply have been closed off at a later date.

The first floor of the private side is best described in the stately if not always grammatical language of the sale particulars: 'Ascending the fine old Oak Staircase with panelled oak wainscotting and Parquet Flooring to Landing, the First Floor is reached, and off the Principal Corridor (9′6″ wide), taking the entire Front of the Building, is the Private Bed-room Accommodation, comprising Six Principal Bed and Dressing Rooms, all communicating, well-lighted, pleasant rooms . . . all fitted with fire-places'. Above them would have

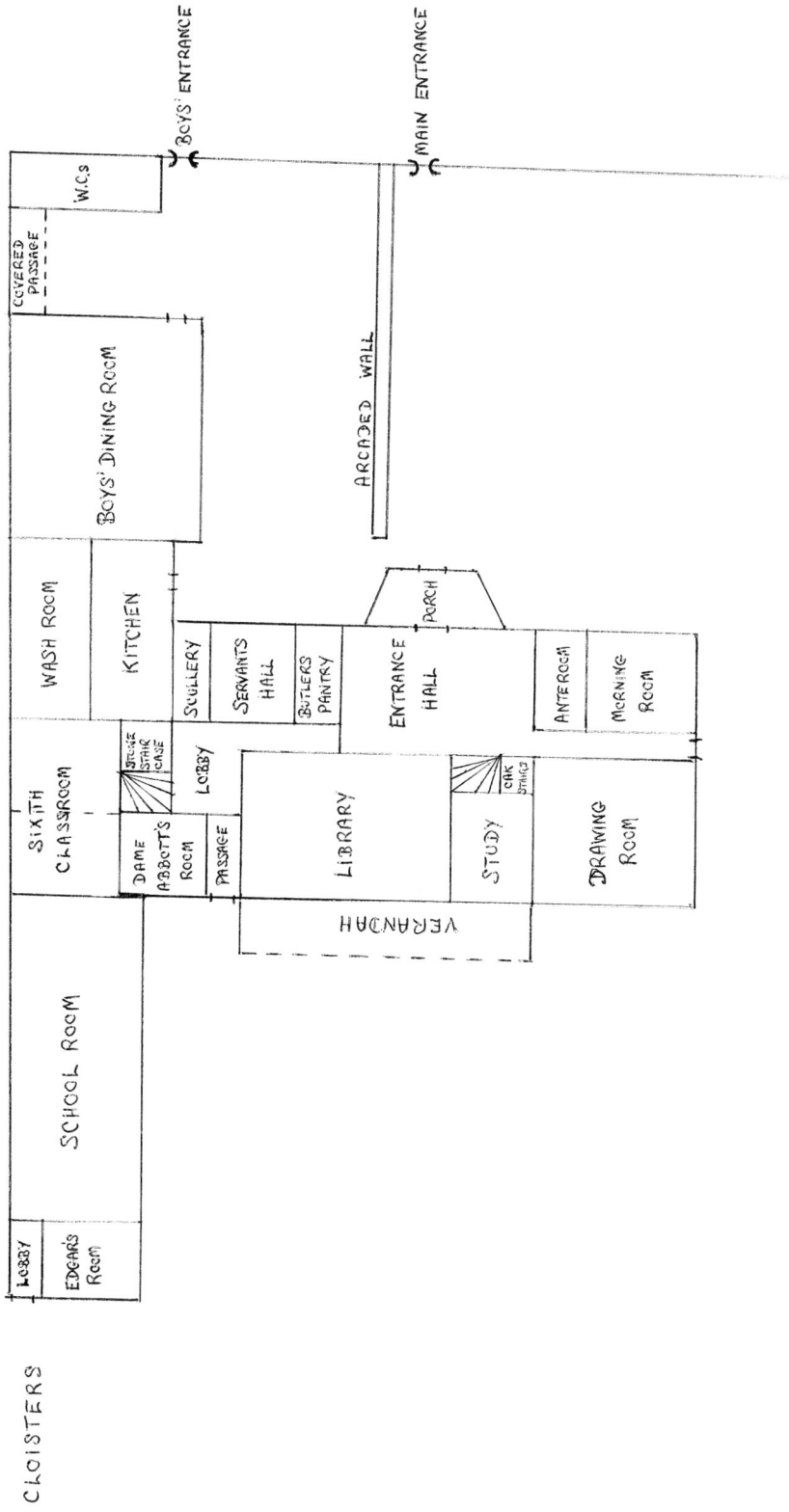

Ground Floor Plan of Temple Grove, 1861-1906. (Scaled down from 1" : 25')

The Entrance front (arcaded wall on the right) (Photo by A. Cheese)

The N.W. front (showing the verandah)

been the bedrooms of the domestic staff.

The boys' wing contained, on the ground floor, the main classrooms in the west wing and the dining hall and kitchen in the east wing. The sale particulars add (rather ominously), 'By covered Walk the School w.c's and latrines are reached'. Much of the teaching, in Waterfield's day, was done in what Benson describes as 'a big bare schoolroom, with much guarded fires, with five desks [raised up on small platforms] for masters surrounded by rows of deeply hacked and inscribed lockers'.

On the first floor were seven boys' dormitories, the largest (in the west wing), measuring 60' × 30', and being fitted with twenty-five cubicles, with a matron's room next door. The east wing also had a large dormitory, 38' × 36', with nineteen cubicles and a matron's room adjacent, and also a small dormitory, while the centre block contained four dormitories and a matron's room. 'Down stone staircase [vide Sale Particulars] on Half-Landing is the Large Bath Room, with five baths standing in lead sink fitted with hot and cold water supply'. Unfortunately it is not possible to discover when this amenity was first provided. The top floor contained another boys' dormitory, a changing room (fitted with hand basins) six bedrooms and three servants' bedrooms. Of all this Benson merely remarks, 'Upstairs there were airy cubicled dormitories, which we were not allowed to visit in the daytime'.

Benson, however, was much more forthcoming about the estate, and his description is worth quoting in full: '[The house] was surrounded by large and beautiful grounds, and there were signs everywhere of stately and leisurely occupation. I remember, for instance, a curious stone summer house, of rough rustic work, with a pediment like a Greek temple, of which the front had been boarded up, which ruefully surveyed the arid gravelled playground. In the grounds was a big artificial mound covered with trees, with a gazebo at the top. There had been a lake, which had been drained and turfed, but the sluice which had fed it still projected from a slope covered with elder bushes. There were big elms all about, and all round stood other fine suburban mansions in walled grounds. Indeed my idea of East Sheen is a place of high and secret walls, with great trees and solid facades discerned within.

'In the grounds at one place was a partly ruinous paling through which it was possible to peep. There was a dense shrubbery here, and some little mounds inside, with headstones, much overgrown with periwinkles, the graves no doubt of dogs, but which in my own mind I believed to be the graves of children – perhaps of boys who had died at the school, and which I regarded with a mournful pleasure . . .

'Altogether the demesne was both stately and romantic. If the place had been one's home, one would have regarded it with great delight and affection. We were allowed under certain conditions the run of these grounds and the hours spent there were some of the few really happy ones that I recollect at Temple Grove'.

Whereas Waterfield always insisted that the boys who were boarded out at Wedderlie should walk entirely along the road each time, his successor relaxed the rule and the walk – three times a day in each direction apparently – supervised by the drill sergeant, took the boys diagonally across the school grounds. A clear description of the walk is given by Dermot Freyer who arrived at the school in November, 1893.

'A door in a high brick wall [this would be the extreme S.W. corner of the grounds] led . . . into a combined kitchen garden and orchard which was traversed by a narrow cinder path between lines of old fruit trees early trained to grow with stems as if outstretched along a wall. At the first extremity of the orchard the path began to rise along a gentle incline and then, taking a bend to the right, wound gradually round, skirting the lower slopes of a large substantial hillock – a pleasant sort of mountain in miniature. This was known as 'The Mound' and was probably, to some extent at least, an artificial formation – the fruit of some landscape-gardening effort of a century or two earlier – for to the left of the track the ground continued to fall for some distance and then extended away in a long apparently scooped-out swampy hollow, christened familiarly 'The Dell'.

'These two features of the school domain had a romantic fascination in the eyes of the boys. The Mound to its summit was thickly wooded, and thus stood out as a landmark from every angle of view, the mystery of its hidden fastnesses being greatly enhanced and heightened by the fact that its slopes were at all times strictly out of bounds. The Dell and its surroundings, in like attractive manner left almost entirely in the wild state, presented a picturesque tangle of reeds and willow herb, of tall rank grasses interspersed here and there with alders and more substantial trees standing out high above the denser undergrowth. A path traversed the furthur end and in and out along the side, to join up eventually with the track along the lower slopes of the Mound.

'Quitting the Mound the path now turned sharp to the left and in a gradual curve again to the right, enclosed now on either hand by low wire railings, and passed between one of the larger playing fields – that devoted to the second and third 'clubs' – and a wide meadow that sloped towards the Dell and was used as a pasture for cows. Here again one or two fine trees were in evidence; in the centre of the playing field a luxuriant lime with a seat constructed round the foot of its bole; and at intervals beside the path and overhanging it, two old elms and a couple of fine Spanish chestnuts. In the meadow, providing in summer welcome shelter for the grazing cattle, a few more massive elms, and a single magnificent horse-chestnut just where this field adjoined the boundary of the first club pitches'. (*'Bogey Baxter'*).

The lake appears to have been drained within a few years of Waterfield's arrival. It is shown on the 1866 O.S. map but that, of course, had been surveyed earlier. By the time Arthur Benson arrived (1872) the Dell was already a feature and one former pupil conjectured that it might have been the inspiration for a rhyme 'which we used to help us scan pentameter and hexameter Latin verses':   'Down in a deep dark Dell
Sat an old cow munching a beanstalk;
Out of her mouth came forth
Yesterday's dinner and tea'.

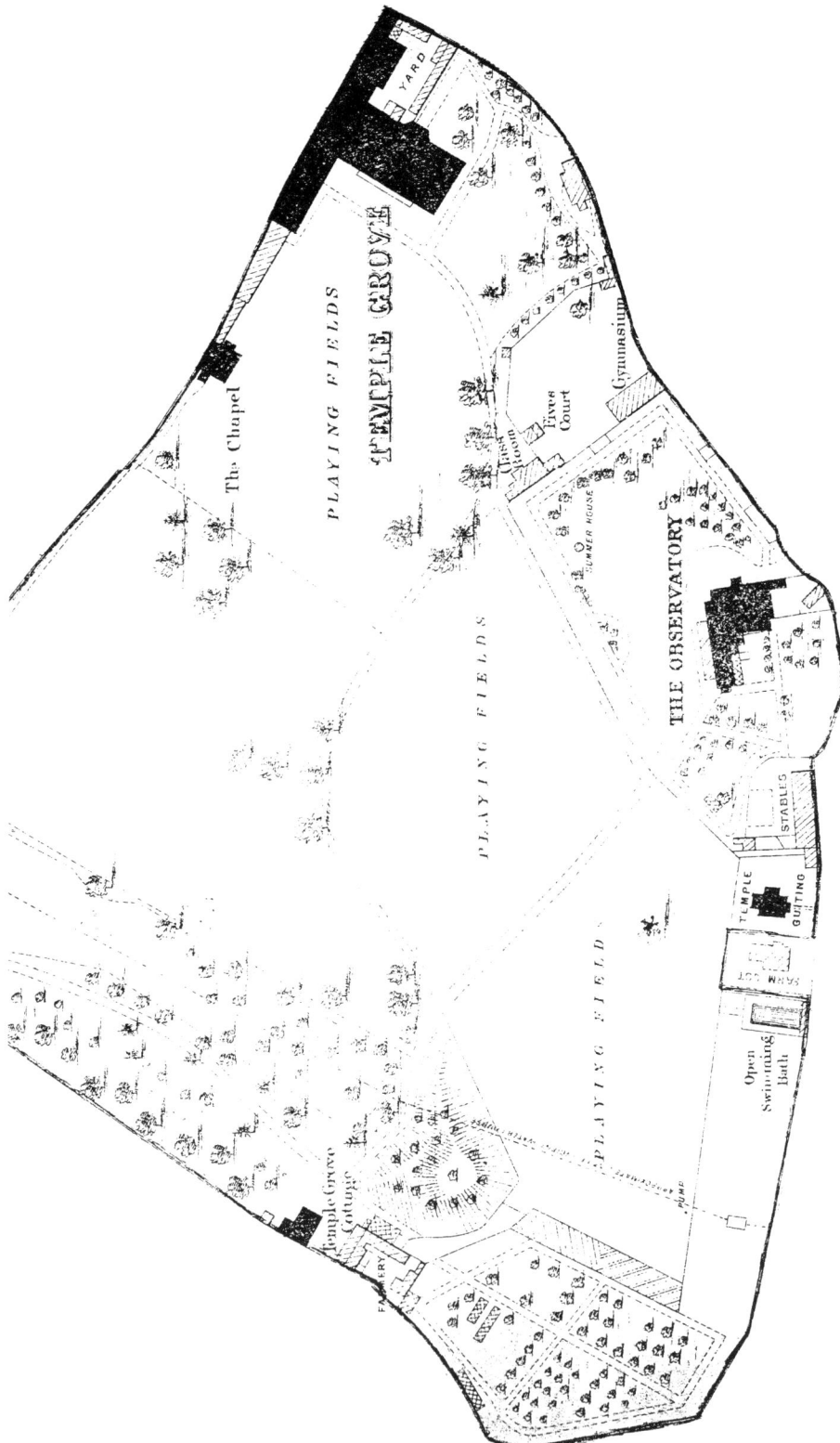

Plan of the estate prepared to accompany the sale particulars, 1907

The Chapel

TEMPLE GROVE

PLAYING FIELDS

YARD

PLAYING FIELDS

Gymnasium

Fives Court

Fifth Room

SUMMER HOUSE

THE OBSERVATORY

PLAYING FIELDS

PLAYING FIELDS

STABLES

TEMPLE

GUITING

FARM COTT

Open Swimming Bath

PUMP

Temple Grove Cottage

FARMERY

The Dell was created after the big lake had been drained

A view south across the grounds from a dormitory window

Throughout Waterfield's and Edgar's time only boys 'on the snore' were allowed, on half-holidays, to enjoy rambling along the paths around the Dell, 'a privilege which was highly esteemed'. But with the building of the school chapel in 1890 the walk to church – twice every Sunday – was replaced by a walk round the Dell which must have robbed it of much of its esoteric charm.

One further addition which probably dates from Waterfield's time was the arcaded wall (on the right of the photograph), running up from the road to the middle of the house, which served to separate the private garden from the school yard. The latrines already mentioned may well have backed onto this wall.

From 1823 onwards *The Observatory* was the home of Dr Pinckney and his family. The house was a substantial one with a large entrance hall and five reception rooms, including a conservatory, on the ground floor and on the first floor a 'handsome drawing room, in two parts, divided by archway supported by fluted wood columns and forming a Music Room', together with seven principal bed and dressing-rooms and a further four bedrooms on the top floor. 'Tasteful pleasure grounds' were screened from the view of the pupils by high brick walls. Here Dr Pinckney, who owned the entire estate, lived for forty years. By 1851 he and his wife, Harriett (both aged 73) were sharing the house with his sister-in-law and his unmarried son, Horace, aged 46 and described as an annuitant. They were looked after by a butler and three elderly female servants. Harriet died in 1855 and the widower was then joined by a younger son, George (a solicitor) and his 17 year old daughter; and by his daughter, Henrietta, and her husband another George (a physician) with their two children. (It was the physician's widow, Mrs George Wood, who visited the school on Sports Day in 1904 and, seated in her bath chair, presented the prizes.) Dr Pinckney died in 1864, aged 87, but the whole estate remained in the possession of his descendants until it was purchased by the Rev H.B. Allen in 1894.

5   The World in 1859

1859 – the year in which O.C. Waterfield took charge of Temple Grove – may fairly be judged the crucial year of the 19th century, the year indeed in which many of the determinants of our own century were laid down. Consider the events. France joined with the small Kingdom of Sardinia in launching an attack on Austria who controlled much of Northern Italy. Although only partially successful, it paved the way for Garibaldi's audacious enterprise in the following year which brought about the first unification of Italy since Roman times. That event, in turn, emboldened Prussia to challenge succesfully for the leadership of German-speaking Europe by defeating first Austria and then France. France's humiliation and loss of territory ensured that she would insist on the emasculation of Germany by the Treaty of Versailles in 1919, the treaty that – in retrospect – made a second World War inevitable.

In the U.S.A. two significant events took place. John Brown, in pursuit of his campaign to free Negro slaves, with eighteen followers seized the federal arsenal at Harpers Ferry. The tiny force was soon compelled to surrender by Robert E. Lee (and Brown was later hanged) but that episode – often wildly distorted in popular legend – probably did more than anything to create a climate favourable to Civil War which broke out two years later. When one considers all the implications of slavery – long after it had ended – in the confrontation of black and white in America, in Europe and in Africa it is hard to repress feelings of horror at 'Man's inhumanity to Man'.

In that same year a less publicised event had consequences that were very different but equally far-reaching. Petroleum was produced commercially for the first time at Oil Creek, Pennsylvania, and before long John D. Rockefeller had begun to build the first Oil Empire. In doing so he created a divinity whose worshippers today outnumber the devotees of Christianity, of Communism, of Fascism and of Islam.

The map of Africa drawn in 1859 would have shown a carefully charted coastline, well-established centres of population in Egypt and along the southern shores of the Mediterranean and an interior that was almost entirely blank. Three events in that year pointed the way towards the opening up of the Dark Continent. Richard Burton and John Hanning Speke, after incredible hardships, returned to Zanzibar, the latter claiming that a huge lake (Victoria Nyanza) which he had discovered was indeed the source of the Nile. During the course of an even more prolonged expedition through what is now Mozambique, Zambia and Malawi, David Livingstone discovered Lake Nyasa. And Ferdinand de Lesseps had begun the ten-year task of digging the Suez Canal. Within forty years every square mile of this vast continent, except for

Abyssinia and Liberia, had come under the rule of a European power.

In Asia changes were taking place more slowly. Alexander II did not abolish serfdom in Russia until 1861. China, torn apart by the greatest peasant uprising in history, the Taiping rebellion, was further subjected to humiliation by encroaching westerners. The legacy of this deep-seated resentment against outside interference is still in evidence today. In Japan, where western influence was of much more recent origin, the Tokugawa shogun, Ii Naosuke, in defiance of the Emperor, had authorised Japanese signature of trade treaties with U.S.A., Russia, Britain and France in 1858. The widespread opposition to these concessions resulted in his assassination in 1860.

The world's most successful trading company, the East India Company, ended its existence in November, 1858, so that 1859 was the first year in which the rule of that huge sub-continent passed officially to the British Crown. Lord Canning, formerly Governor-General, became Viceroy and worked hard to heal the bitterness resulting from the Mutiny. India was henceforth to present for 90 years a field of service, military, political, educational, missionary, medical and administrative, offering scope to some of the finest talents that Britain had to offer.

It would be unwise to pinpoint the events of a single year and attempt to see in them enough seeds of conflict, economic and ideological, religious and racial, to go on germinating well into the next millenium. Rather do isolated episodes suggest the underlying tendencies. Did nothing, then, happen, the reader may be asking, to point the way towards a better world, a greater hope for mankind? A Swiss citizen, Jean Henri Dunant, happened to come across the battlefield of Solferino in N. Italy and was so moved by the agonies of the wounded and the dying that he published a little book which led, in 1864, to the founding of the Red Cross, the world's first non-political humanitarian organisation.

For Britain, too, 1859 was to prove a crucial year. For the first half of it a minority Tory government, headed by Lord Derby and Benjamin Disraeli, was in power but on June 6th a highly significant meeting took place in Willis's Rooms, attended by Liberals, Peelites, Whigs and Radicals. It was addressed by Lord Palmerston, Lord John Russell, John Bright and others and it is not too far-fetched to claim that that gathering marked the beginning, for better or worse, of the present political system in this country with two, or even three, tightly organised parties resting on a solid platform of local organisations throughout the country. Appropriately enough Palmerston, the former owner of *Sheen Grove*, once more became Prime Minister, an office he held till his death in 1865.

In the industrial and economic sphere the 1850s had been a boom time for Britain. The railway system continued to expand until it covered the entire island; a new landmark in steamship construction had been reached with the launching of the *Great Eastern*; in the coal-mining, steel-producing and textile industries Britain still led the world. But by 1859 (the year of Brunel's death) the boom was over, foreign competition was eroding British supremacy and trade unions were becoming restive. A strike in the building industry led to the formation of the London Trades Council opening a new chapter in the history of British trade unionism. At this point a 40 year old German refugee became involved. Karl Marx had published his *Critique of Political Economy* earlier in the year and in 1864 he helped to found in London the International Working Men's Association for which he drew up the Inaugural Address and the Statutes. As one writer put it, 'one cannot help considering how much better off the world's capitalists might have been if the master builders had given their men a nine-hour day in 1859'.

For Britain, too, the clouds of 1859 had a silver lining. The Crimean War had not only drawn attention to British military ineptitude: it had focused on medical shortcomings. One outcome was the launching, by the third most formidable woman in our history (Boadicea not included) of the very first training scheme for nurses; another the founding of Britain's first Cottage Hospital, at Cranleigh. In the field of literature the lining may fairly be described as golden. Though Dickens's *Tale of Two Cities* was slated by some of the critics it quickly won popular approval, while *Adam Bede* by the almost unknown George Eliot ('evidently a country clergyman', opined one learned reviewer) met with both critical and popular acclaim. Meredith's *Ordeal of Richard Feverel* was rather too unconventional and daring but *Idylls of the King* by the Poet Laureate, Lord Tennyson, was regarded as one of his finest works. In contrast Edward Fitzgerald's *Rubaiyat of Omar Khayyam* was ignored until a copy reached D.G. Rossetti two years later. Two minor works (in the literary sense) were to achieve an abiding fame: Mrs Beeton's *Book of Household Management* and Samuel Smiles's *Self-Help*.

But the two most influential books of 1859 have still to be mentioned. One was *On Liberty* by John Stuart Mill. Mill's education (Greek at three but no religion till thirty) and his defiance of convention (he had lived with Mrs Harriet Taylor for twenty years before their marriage) made him an object of deep suspicion to most middle-aged Englishmen but the book 'had an electric effect on the ardent men and women of the younger generation' (R.B. McCallum). Mill was no egalitarian, he feared 'the brute mass of opinion' but he defended the individual's right to think for himself and indeed his brief membership of Parliament ended because of his support on principle for Bradlaugh (an atheist) whom he disliked in practice.

If none of the foregoing events had taken place, or books been published, 1859 would still be seen as a year in which thinking men and women were forced to re-examine their most strongly held beliefs. Charles Darwin's *The Origin of Species* was published on November 24. Evolution is today such an accepted part of scientific teaching and such an integral, if unconscious, factor in our attitude to the unfathomable mysteries of life, that it is difficult to recapture the shock it inflicted on civilised, conventional thought and belief in mid-Victorian England. Even though mankind was only fleetingly referred to in *The Origin*, the implication was there: mankind had evolved in the course of millions of years from sub-human species; the Garden of Eden no longer existed.

Do these mighty happenings have any bearing on the story of an English school in late Victorian times? In the 1860s probably very few of those parents who sent their sons to Temple Grove concerned themselves greatly with the Unification of Italy, the exploration of Africa, the Taiping Rebellion or the theories of Karl Marx, J.S. Mill and Charles Darwin. But the boys who were educated there – however well or badly that education fitted them for the world they were entering – those boys, every one of them, were affected directly or indirectly, by events in far-off lands and by the scientific, economic and political thinking of the day.

6   Waterfield's School

1   O. C. Waterfield

The Waterfield family can be traced back to William W. (1698-1763) of Peterborough who married Susannah Matley. It was his third son, Thomas (1738-1781), who continued the line and Thomas's elder son, another William (1779-1827), who became the progenitor of most Waterfields alive today. This William's third son, Charles (1806-1871), a barrister, married Katharine, daughter of the Rev. James Suttall Wood, in 1830. Their eldest son, Ottiwell Charles, was born in London on May 23rd, 1831, and was followed by William in 1833 and Donald in 1847. The name Ottiwell comes from Ottiwell Rider of Bolton Castle in Yorkshire and may have previously been the surname (corrupted from Attewell) of an ancestor on his mother's side. Of his early education nothing seems known. He may have had a tutor or he may have attended a private school. He entered Eton in 1844 by scholarship and was thus enabled to experience the joys and griefs of Long Chamber before the New Buildings came into use in 1846. An almost exact contemporary at Eton was Robert Cecil (later third Marquess of Salisbury and three times Prime Minister) who wrote to his father: 'I am bullied from morning to night . . . I am obliged to hide myself all evening in some corner . . . I am obnoxious to all of them because I can do verses but will not do them for the others'. He was later withdrawn from Eton and educated at home.

Admission to King's, Cambridge, nominally by scholarship, was still at that date a formality and Ottiwell was admitted to the College on July 3rd, 1850. The only letter of his known to have survived was written the following year from near Biarritz. It is addressed to Charles Booth and the opening sentence, congratulating his friend 'on your admission to the club', makes it likely that he was about to follow Waterfield to King's. Ottiwell and another friend, James, had been on a walking tour, exploring first the Auvergne and then the historic cities of southern France, Avignon, Nimes, Arles, before traversing the Pyrenees (with a guide) from east to west. 'We are now at a little sea-bathing place, close to Bayonne, where the customs with regard to bathing are decidedly primitive. Ladies and gentlemen all bathe together, in very scanty costume, while those who are not positively bathing sit on chairs at the water's edge and examine the ladies' costumes and the gentlemen's figures'. A bull fight he watched at St Esprit drew the comment: 'One of the most painful humbugs I have ever witnessed. The bulls would not fight and the men were afraid'.

Back at King's, Waterfield became a Fellow in 1853, retaining his fellowship

A simplified WATERFIELD family tree

Headmasters, and Assistant Headmasters of
Temple Grove are shown in CAPITAL letters.
Pupils at Temple Grove are shown in *italics*.

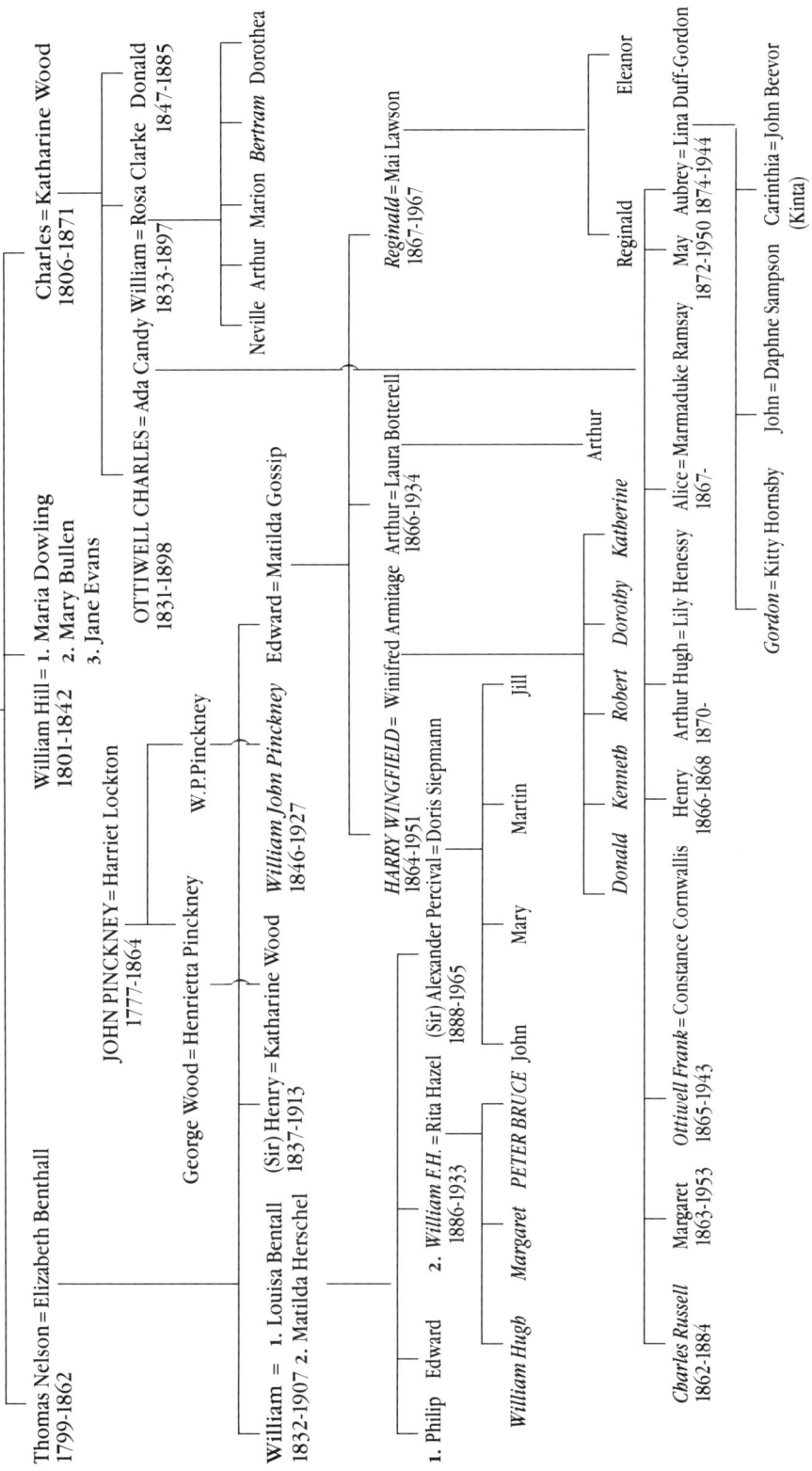

William = Susannah Matley
1698-1763 |
Thomas = Sarah Bothamley
1738-1781 |
William = Elizabeth Patey
1779-1827

Charles = Katharine Wood
1806-1871

OTTIWELL CHARLES = Ada Candy   William = Rosa Clarke   Donald
1831-1898   1833-1897   1847-1885

Neville   Arthur   Marion   *Bertram*   Dorothea

Reginald = Mai Lawson
1867-1967

Reginald   Eleanor

Aubrey = Lina Duff-Gordon
1874-1944

May
1872-1950

Carinthia = John Beevor
(Kinta)

John = Daphne Sampson

Gordon = Kitty Hornsby

Alice = Marmaduke Ramsay
1867-

Arthur Hugh = Lily Henessy
1870-

Henry
1866-1868

Katherine

Dorothy

Robert

Kenneth

Donald

Arthur

Arthur = Laura Botterell
1866-1934

HARRY WINGFIELD = Winifred Armitage
1864-1951

Jill

Martin

Mary

John

(Sir) Alexander Percival = Doris Siepmann
1888-1965

2. *William F.H.* = Rita Hazel
1886-1933

PETER BRUCE

Margaret

William Hugh

1. Philip   Edward

William = 1. Louisa Bentall   (Sir) Henry = Katharine Wood
1832-1907 2. Matilda Herschel   1837-1913

George Wood = Henrietta Pinckney

William John Pinckney
1846-1927

Edward = Matilda Gossip

W.P.Pinckney

JOHN PINCKNEY = Harriet Lockton
1777-1864

William Hill = 1. Maria Dowling
1801-1842   2. Mary Bullen
               3. Jane Evans

Thomas Nelson = Elizabeth Benthall
1799-1862

Charles Russell
1862-1884

Margaret
1863-1953

Ottiwell Frank = Constance Cornwallis
1865-1943

until his marriage eight years later. His B.A. degree followed in 1854 and two years later he left Cambridge to return to Eton as an assistant master, taking his M.A. in 1857. It is not impossible that he might one day have aspired to the headship of Eton, but an opportunity presented itself long before that was any more than a distant dream. His first cousin, Sir Henry Waterfield, had married Katharine Wood, grand-daughter of Dr John Pinckney, who, in the autumn of 1858, was looking round for a successor to Dr Rowden. The fine reputation Ottiwell had won at King's would have been a strong recommendation and financial support – one assumes – would have been forthcoming from his father (aged 52), so that, at the age of 27 and still a bachelor, O.C.W. became Headmaster of Temple Grove when Rowden departed to Chichester at the end of 1858. It is commonly, and not unreasonably, supposed that Victorian headmasters were born aged 50, heavily whiskered and exhibiting a stern expression which they had been practising in the womb. The likelihood is that the photograph (p. 130) shows him as an Eton master, a tall fine-looking man, not too stern and with only incipient whiskers.

Many tasks awaited him on taking up his new responsibility, not the least of which was the need to find a wife. In 1861 he married Ada Candy, the fourth daughter of William Candy of Chipstead Place, described as 'very typical of the rich businessmen of that day, very prosperous, very generous and very autocratic'. Ada was 18 or 19, handsome, good-natured and kindly. Charles Russell, the eldest of Ottiwell and Ada's eight children, was born the following year and there were to be four more sons, of whom Henry died in infancy, and three daughters. Margaret, the eldest girl, became a talented watercolour painter of gardens and landscapes and also a knowledgeable botanist (an article on her by Diana Baskervyle-Glegg appeared in *Country Life*, October 25th, 1990). Aubrey, the youngest child (b.1874), also became an artist, specialising in large flower pieces and distinctive murals.

One of Waterfield's first actions was to institute the *Name Book*. Earlier headmasters must have kept some record of their pupils but none has survived. Waterfield began (February 1st, 1859) by causing each of the 64 boys then in the school to inscribe his own name, the Headmaster adding the name and address of the father. Every term thereafter each new boy wrote his name in the book so that, in 266 pages, just over a thousand names are recorded though not, unfortunately, till 1864 did Waterfield think to add the date of leaving and the public school on the right-hand page. Scholarships, however, are listed at the back from 1860 onwards, the first eleven awards all being gained at Eton. On the very first page we find a Twining (Daniel) and a Pinckney (William John), grandson of Dr J.H. Pinckney. Arthur and Walter Hailstone, Frederick Ommanney and Claude Champion de Crespigny are others among the pupils Waterfield inherited from Rowden. The school is thus in the fortunate position of having an almost complete record of its pupils for the past 130 odd years.

What sort of man was O.C. Waterfield? For better or worse he ruled the lives of a thousand human beings during the most impressionable years of their

Ottiwell Waterfield from a photograph in the possession of King's College, Cambridge

boyhood and it would indeed be remarkable if he had not left a permanent imprint on the minds and characters of a good many of them. We must be grateful that five of his former pupils recorded their impressions of Waterfield in print and that a later Headmaster, Meston Batchelor, by writing – in the mid 1930s – to some of the then most senior Old Boys, elicited memories of Temple Grove and its headmaster from seven more. In all, then, we have impressions of the man from the standpoint of a dozen of his pupils. So far as one knows no parent of those days has left an assessment of Waterfield and it is reasonably certain that not a single member of his staff dared to do so. Those on whose evidence we rely comprise two authors, a Provost of Eton, a Privy Councillor, three soldiers, a sailor, a schoolmaster, two judges, and a clergyman, a fairly reliable jury (if one may vary the legal metaphor), one might feel. Indeed their verdict is nearly, if not quite, unanimous; and if the account that follows seems to rely too heavily on the Benson brothers that is because they wrote most fully about him. The assessments of all twelve jurymen have been carefully considered and quotations from most of them appear in this account.

Arthur Benson (*Memories and Friends*), speaking of Waterfield's time as an Eton master, comments perceptively: 'I should imagine that he was a masterful young man who liked a position of supreme authority'. Indeed he might well have aspired to the headship of Eton itself but that would have involved a lengthy wait; instead, at the age of 27, he was in charge of Temple Grove. Benson arrived at the school in 1872, in other words when Waterfield's methods and personality had been stamped on it for a good many years. His description is not only the longest but seems probably the fairest and deserves to be quoted at length:

'Waterfield himself was a tall, impressive looking man, with a vague resemblance to the portraits of Charles Dickens, wearing his curly hair rather long, and with a short grizzled beard. He was always dressed like a great gentleman, sometimes in a frock-coat, sometimes in loose, well-fitting grey. When he was arrayed in a full silk gown he was almost too majestic for words. A faint scent of Havana cigars hung about him. He walked with a slight limp, which gave him a swaying motion, and he had eyes of great brilliance which opened wide, if he was surprised or vexed, and struck terror into our souls. I have never in my life been so afraid of a human being as I was of him. I thought of him as wholly indifferent to us boys – that we were just more or less inconvenient adjuncts to his surpassing greatness. He seemed to live in a far-off atmosphere of fashion and high culture. As a matter of fact, he observed us very closely, and his reports were models of insight and penetration. He was a really great teacher, extremely clear and forcible, and it was from him that I first learnt that it was possible to be intellectually interested. He did not teach us regularly, but he used to send for a class to his study, or take us in Latin composition; the first-class schoolroom was in a separate building, some way from the house, and had a small study adjoining it, a sort of garden smoking-room. Waterfield used suddenly to open the door from this into the classroom, and ask Rawlings to send him in a set of boys. He had a curious habit, as he taught us, of breaking off in the middle of a sentence, or even in the middle of a word, and we believed that we were meant to supply the missing word or syllable if we could. He had a rich, distinct voice, with a pleasant laugh, and, if he chose, could make a lesson extraordinarily stimulating. I shall never forget his reading us some of Barnes's Dorsetshire poems to illustrate the difference between Doric and Ionic Greek. He had a singular, almost hysterical, trick whenever he read poetry aloud: his voice used to falter, a large tear brimmed his eyes, and fell on his waistcoat. I did not think of this as in any way emotional – it was just a habit he unquestionably chose to adopt.

'But he was too severe, and punished a good deal, believed in caning and flogging – I still shudder at the sound of his bunch of keys, when he unlocked a drawer in his writing-table and pulled out a cane. It was a thoroughly topsy-turvy affair, to beat a terrified boy for mistakes, not carelessly so much as stupidly made, and then to expect him to imbibe information which after all was utterly trivial. Waterfield really had not the slightest need, in most cases, to punish boys at all, for his words spoken in anger bit deep, and he was regarded with almost fanatical respect.

'[His] theory of discipline must have been, I make little doubt, a tradition instinctively imbibed and perpetuated. As an Etonian of the forties, soon after the departure of Dr Keate, though Hawtrey . . . had been a teacher of real inspiration . . . such rough-and-ready obedience as there might be was maintained by ceaseless flagellation, while the younger boys were, to use an ugly old phrase, remorselessly licked into shape by the upper boys. I do not imagine that a man like Waterfield – who had certainly absorbed much of the real culture of Hawtrey – ever got as far as thinking that discipline could be maintained in a big school, or a brisk and business-like attention to work secured, except by personal chastisement. Indeed if Waterfield had reserved his chastisements for serious moral offences instead of mere intellectual peccadilloes, I should feel that the place had been run on truly enlightened lines.

'When Waterfield was in a thoroughly good humour, in teaching-hours, he showed a degree of patience and sympathy which were remarkable. I remember some mathematical lesson in geometry, given to a class of which I was a member, when a jolly boy of the sturdy, puzzle-headed Saxon type, with a great shock of curly hair, proved entirely impenetrable to the simplest geometrical considerations. Waterfield made a rapid demonstration with three half-crowns to prove the doctrine of equalities; and I can see now the boy, under the impression that it was a conjuring trick of some kind, scrutinising the three half-crowns one by one, with anxious nicety, and at last handing one of them to Waterfield, saying, "I think this is rather smaller than the other two, Sir." Waterfield burst out laughing, rumpled the curly head, and said, "Ah, you don't understand the use of symbols! What's a symbol, Benson?" And there followed an ingenious little Platonic dialogue which I cannot reproduce now. One could hardly believe him to be the same man who would say to a dull boy, palpitating with terror and quite unable to recall the correct answer, "Don't bully me, sir – I'm not going to stand here half the morning to be bullied by you like this!" '

Bulmer La Terrière (*Days that are gone*) speaks of Waterfield always keeping himself very much aloof and never making himself cheap, so that an interview with him, whether for good or bad, meant something. And Monty James (*Eton and King's*) describes Waterfield as Olympian (he might have said Jovian) both in appearance and manner, someone who could 'discharge his bolts with stunning force and unexpectedness'. 'One day Grey and I, as first and second in the school, were persuaded by our friends to go to Waterfield and ask for the evening [i.e. remission of evening school]. Now we had both been reported to him for talking in the dormitory the night before; but we didn't know it, or hadn't taken the matron's threat seriously. And when, confronting Waterfield in his study, we said: "Please Sir, may we have the evening tonight?" his eye flashed (as nobody else's eye ever did) and he said: "No, you mayn't have the evening, but I'll tell you what you may do; you may both of you go on the Double Daily list till further notice". Blasted by the stroke we retired to a diet of two copies [pages of copybook] per diem and two hours' extra drill'.

All too often the punishment meted out was a good deal sterner than that. William Goodenough (*A Rough Record*) after speaking of 'the majestic figure of Mr Waterfield [which] pervaded all' goes on to say, 'The cane, or rather a very thick ruler that he used in place and with which he once broke a boy's finger, an act that caused even that tremendous being to shed tears, and the birch – more familiarly known as the swish – came easily to his hand and it was a real experience to see and to feel the manner in which, with his left hand, he drew the lapel of his frock coat across his chest and with his right hand, applied the weapon with a swing that would have done credit to a plus two golfer'.

Tom Pitman offers the testimony of another victim when he refers to 'Waterfield's study, a summons to which meant trouble and possibly pain. I only suffered the latter once but I fondly believe that the marks still remain on the place where I felt it. Waterfield had a pimple on the side of his nose which used to swell up enormous [sic] when he was really angry, but although we were all frightened to death of our Headmaster we had the greatest respect for him'. James Hannay (*Pleasant Places*) remembered all too vividly the summons to a birching: 'There was a bedroom known as 'number twenty-two' set apart for these executions, which were always carried out just before our mid-day dinner. Waterfield had a cold-blooded and rather gruesome plan of saving up his victims until he had half-a-dozen under sentence. Then the school porter went round to the various classes and informed those who were to suffer that they were to go up to room twenty-two immediately after morning school. By the time the birching was finished the rest of the school had sat down to dinner and we used to march into the dining-hall behind Mr Waterfield under the eyes of one hundred and twenty or so of our school-

'Don't bully me, Sir'

A pupil in Sunday attire (F.W. Scudamore, 1868)

Charles Geoghegan (Geege), the popular one-armed Irish master

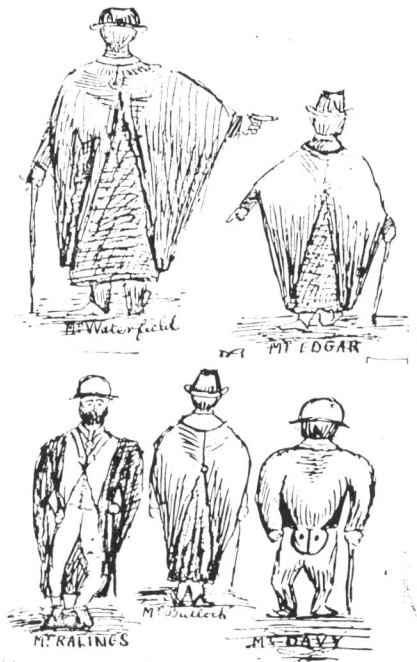

Mr Waterfield and his staff as depicted by one of the Grenfell brothers (1869)

fellows. The idea, I suppose, was that we should feel our disgrace acutely. My recollection is that we felt rather heroic, and fragments of birch picked up on the floor after Waterfield had left the room were valuable souvenirs. As we entered the dining-room we were expected to hold up fingers to show how many strokes we had received. He was indeed a proud boy who could hold up all fingers of both hands'.

However severe himself, Waterfield would never tolerate bullying in the school. 'Two or three times' (Benson recalled) 'I remember his assembling the school, and saying a few words of admonition and reproof about some unpleasant incident that had occurred. He did this admirably, with great emotion and even with tears. His splendid presence, his clear, penetrating voice, his admirable little pauses and gestures, gave me a deep thrill. He used no threats; he expressed his astonishment and disgust at the occurrence, and added that he expected every boy to do his best to prevent the repetition of such a thing. "Mind, I mean *every* boy, the smallest and weakest as well as the biggest and strongest. We are all at one in hating this kind of thing, and in preventing it as far as we can; and if we all hate it, it *is* prevented." At that moment, under the thrill of his persuasive presence, I do not think there was a single boy in the school who would not have done his best to intervene'.

This was Waterfield at his best as a setter of moral standards which boys could understand and which they would wish to emulate. But he himself had a temper he found difficult to control and outbursts of fury detracted from his aura of Olympian majesty. James Pitman gives a clear-cut example.

'I happened to be in the big schoolroom when the door was thrown open; the Head burst in with a face like a turkey cock; he took the school list off its peg and put his pen through all the red crosses in the list, hung it up again, and banged out of the room. Two deep lines through each cross! It transpired afterwards that he had been round the grounds and discovered that some bamboos of which he was very proud had been cut down and made into assegais. On looking back I must say I think he was justified in withdrawing the privilege, but he should have done so calmly, and I am glad that he did not then adopt a plan which I remember he once adopted when he discovered some iniquity, viz. he summoned the whole school and asked each boy in turn whether he had done the deed. It was asking too much of a boy to expect him to confess in front of the school to an iniquity of which he was probably ashamed, and we all lost half a holiday as a punishment, but without complaint as we disapproved of the method adopted to discover the miscreant'.

Lionel Byrne, in a letter, emphasises an attractive side of Waterfield's character: 'He was a genuine enthusiast. He had spent his holidays in Brittany. He came back filled with interest in Ogam inscriptions, which he had found – or thought he had found – cut on some of the circle stones. He began talking to us about them just at the end of morning school, the time when everyone rushes out and shrieks for a quarter of an hour before dinner. While he was talking to us the crowd surged past our window. I glanced at them, said to myself: "You don't in the least want to be out there, you would much rather be here"; and I remember nothing more beyond the fact that my interest in casual history was permanently stimulated. But the scene up to that point is indelible'.

The range of Waterfield's scholarship is well illustrated by an account of an animated discussion on 13th century French stained glass between the headmaster and the future Archbishop of Canterbury, listened to with awe by the latter's son, Arthur Benson.

The two least favourable assessments of Waterfield as a headmaster come from Fred Benson and Norman Macleod and it may be no coincidence that they were contemporaries whose days at the school came at the end of his reign. Over twenty years in a position of unquestionable authority is a severe test of any man's (or woman's) character and it may well be that his natural authoritarianism had hardened into a certain rigidity of outlook and method, that some of the earlier enthusiasm had evaporated and that the undeniable strains of a demanding job had taken their toll of his health (although still only in his mid forties). Benson (*Our Family Affairs*) begins by declaring, 'I believe him to have been about the best private schoolmaster who ever lived, for he ruled by love and fear combined in a manner that while it inspired small boys with hellish terror, yet rewarded them with the fruits of hero-worship'. But he continues, 'He exacted blind obedience, under peril of really infamous torture with a thick ruler with which he savagely caned offending hands, but he managed at the same time to make us appreciate his approbation. The ruler was kept in a convenient drawer of the knee-hole table in his study, and was a perfectly brutal instrument, but the approach of the ruler, like a depression over the Atlantic, was always heralded by storm-cones. The first of these was the taking of keys from his trousers-pocket, and then you had time to pull yourself together to retract an equivocation, to confess a fault, or try to remember something you had been repeatedly told. The second storm-cone was the insertion of the key into the drawer where the ruler was kept. You had to be of very strong nerve when that second storm-cone was hoisted, and divert your mind from the possible future to the supine which you could not recollect, for when the key was once inserted there might at any moment be a sudden startling explosion of wrath, and out flew the ruler. Then came a short agonising scene, and the blubbering victim after six smart blows had the handle of the door turned for him by somebody else, because his hands were useless through pain. The ruler was quite rare, and probably well-deserved; anyhow it was the counter-balance to the hero-worship born of Mr Waterfield's approval.'

Benson's judgment of Waterfield as a teacher falls well below that of earlier witnesses. 'The top form had certain specified lessons every week taken by Waterfield, and he did not teach regularly in other forms. But he was liable to make meteoric appearances soon after the beginning of a lesson in the big schoolroom where the next three forms were at work, and take any lesson himself. A hush fell as he strode in, and we all cowered like partridges below a kite, while he glared round, selecting the covey on to which he pounced. This was a subtle plan, for you could never be sure that it would not be he who would hear any particular lesson, and the chance of that made it most unwise to neglect any preparation altogether'.

Norman Macleod is a still more severe critic: 'His personality was of that kind which made one hold one's breath and shake in one's shoes. He seemed to us small boys well over six feet, and wore a full black beard. I can still remember him as he stalked through the great school with everyone on the tremble, lest he should take a fancy to ordering one of the forms to his study for their lesson. He may have had the reputation of a great schoolmaster, and undoubtedly Temple Grove was one of the four famous preparatory schools round about London, but in 1877 we were not so far removed from the days of Dickens and a school, if it were run nowadays (1935) on the lines on which Temple Grove was run, would find its classrooms empty. Waterfield took no interest in his boys, and never attempted to make contact with them. They brought him his livelihood, and it was his duty to instil into them as much knowledge as they were capable of imbibing, according to the notions of those days, with a view to their securing, if possible, scholarships at the public schools and so enhancing his prestige'.

A few minor matters need to be referred to before a final summing up can be attempted. Macleod's accusation that Waterfield took no interest in his pupils has to be set against the recollection of Arthur Benson (by then an Eton housemaster) that when Waterfield had occasion to visit him many years later, 'He remembered, I found, with extraordinary exactness, some of the small incidents of my boyhood'. Meston Batchelor (*Cradle of Empire*) says that Waterfield was covertly known as the 'Cow' (from his initials transposed) but curiously the only direct evidence of this interesting fact is supplied by James Hannay (*Pleasant Places*). Perhaps, even after his death, it was considered unwise to commit such sacrilege to print. Batchelor also avers that Waterfield, after inflicting punishment, would say to a boy, "Well, that's all over, my boy", and then kiss his victim. The only reference to this practice is made by Fred Benson (*Our Family Affairs*) who claims that that particular form of forgiveness was administered 'when he had frightened the life out of you by terrible harangues'.

Arthur Benson's assessment of Waterfield as a Headmaster seems the fairest: 'Altogether I feel that Waterfield was really a great educator . . . What was lacking was just the further touch of imagination, which alone could have made him able to revise the old tradition of hardy and severe barrack-discipline. If he had seen a little more of the boys, and expanded more into his delightful talk; if he could have kept more in control the natural irritability of a highly-strung, imperious man; if he could have introduced a little more amenity into the life of the place, without sacrificing its simplicity and liberty, it would have been at the head of all private schools'.

Whatever their feelings about the Headmaster the boys were united in their adoration of Mrs Waterfield. Arthur Benson describes her as 'a handsome impulsive woman' and even when her husband spoke somewhat abruptly to her 'it never for a moment ruffled her good-natured and kindly composure'. As an instance of her thoughtfulness, Benson cites the occasion when it was decided that he and his brother, suffering from whooping-cough, should spend an hour being fumigated at the neighbouring gas works, "Mrs Waterfield came to our rescue, gave us an excellent luncheon in the private dining-room, drove us over to the works, paced about with us on the filthy, smouldering rubbish, where we coughed and spluttered, and told us amusing anecdotes till the process was completed'. Her angelic status had been established a few years earlier, probably in 1870, in an episode which really does deserve the epithet 'legendary'. The most reliable eye-witness accounts appear to be those of William Young and Edward Graham.

'The episode which made a heroine of Mrs Waterfield in the eyes of the boys occurred at our early dinner [i.e. lunch] in the school dining-room, where the boys were seated at long tables round the room, Mr and Mrs Waterfield and the masters lunching at a large central table. In a corner of the room was another table, called 'The Pig Table', where the servants deposited the dirty dishes and where any boy, guilty of some trivial offence, was ordered to stand for the rest of the meal. A boy, detected in concocting a most nauseating mixture ['beer, mustard and salt', according to Graham who was sitting 'close to the culprit when he mixed his cocktail'], was ordered by Mr Waterfield to take his tumbler to the Pig Table and drink the horrible mixture. The end of dinner came but the culprit had not complied. Mr Waterfield then announced that all should sit there till the mixture was drunk. Twenty minutes of horrible strain ensued. At last Mrs Waterfield, who had evidently been protesting to her husband, dashed from her place, seized and drank the noxious mixture, and bolted from the room.

'Dear and gallant lady, she had not gone far before she was terribly sick but she had extracted her husband from an impossible position and saved the boy'. Later versions have Mrs Waterfield agreeing to drink half if the boy would do the same but one feels that the evidence of two eye-witnesses is more compelling.

As far as one can tell Mrs Waterfield played almost no part in the actual running of the school. The arrival of her eight children would, in any case, have kept her fully employed. In the 1871 census are listed Charles (b.1862), Margaret (1863) and Frank (1865), Henry, the fourth child (1866) having already died. Family records supply the remaining details: Arthur Hugh (1870), Alice Katharine (1871), May Constance (1872) and Aubrey (1874). Curiously one can't be sure which of the boys were actually pupils at Temple Grove. None is entered in the *Name Book*, although three other Waterfields *are* listed: the brothers, Harry (later Headmaster of Temple Grove himself) and Reginald, who were cousins and a nephew, Bertram. From the 1905 Register it appears that Charles entered Charterhouse from Temple Grove in 1875, and that Frank went on to Marlborough in 1878 after four years at his father's school. Nothing, however, seems to be recorded about the education of Hugh or Aubrey, or indeed of the three girls. And, although five of Ottiwell's children married, only Mary and Aubrey themselves are known to have had families. The Register is both defective and, in places, inaccurate so the two youngest boys *may* have been pupils in Edgar's time, though neither is listed in his record of pupils.

Mrs Waterfield outlived her husband by sixteen years and although obituaries then were more glowing than those of today there is no reason to doubt the sincerity of the tributes paid to her. 'To all who came within the family circle, to the poor of her district [in Kent] and to the inmates of the Union, she gave the full measure of rich friendship and love . . . To her fell that golden capacity for expressing sympathy and interest . . . Her laugh to the end of her days was musical and exhilarating . . . Life had brought its burden of losses and sorrows, but they left her unembittered . . . She was religious because she was good, not good because she was religious'.

2   The Running of the School

Evidence as to the financial aspects of running a school like Temple Grove is scanty. The best one can hope to do is to present an assessment which is coherent and not obviously false, in other words one in which the sums add up. The underlying assumption is that prices and wages did not change much in the second half of the 19th century. It is not a hard-and-fast rule. Some prices came down as competition increased and production costs fell; some wages rose slightly with unemployment remaining at a low level. But if one's calculations assume stable prices and wages the error will not be serious.

Fees at Temple Grove were £35 per term, to which must be added no more than £10 for extras, including one guinea yearly to the Doctor for ordinary attendance, 15/- to the Church and varying sums for school books (compulsory items), and three guineas for instrumental music, two guineas for German and Drawing and one guinea for Carpentry (optional extras). For most of Waterfield's time there were around 120 pupils, giving an annual income of somewhere around £15,000.

These figures compare closely with those of St Andrews, Eastbourne (kindly supplied by David Mockler) at the same period. There the fees were also £35 a term, but there was a wider range of extras viz. drawing (£2.3.6), gymnastics and boxing (a guinea), medical attendance (a guinea), games and library subscriptions (each a guinea), dancing (two guineas), vocal music (half a guinea), and instrumental music (two – four guineas). Together with a host of minor items they produced bills that averaged £45 to £50 a term, a figure that had scarcely altered when the present writer went to boarding school in 1938. For comparison at Lancing in 1878 the fees were from £60 to £100 per annum; at Hurstpierpoint £30 – £35 and at Ardingly 15 guineas p.a.

What were Waterfield's principal outgoings? First there were the Articles of Agreement between Waterfield and his second-in-command, and eventual successor, J.H. Edgar, which will need to be looked at. Then come ten members of the teaching staff, including the drill sergeant; an indoor domestic staff of about twenty, headed by a lady superintendent, housekeeper and butler, and including four matrons; and finally six men, the estate carpenter, three gardeners, the swimming instructor and the coachman (although the latter may have been self-employed).

Supplies consumed week by week for three-quarters of the year – and some items throughout the year – would include food, heating, books and other school equipment, cleaning materials and general household needs together with medical stores (most of the laundry was done on the school premises). And finally we have to consider the cost of leasing the entire estate from the Pinckney family, together with the leases of *Clarence House, Temple Grove Cottage* and *Wedderlie*, maintenance costs and taxes (which included the tax on all servants, not repealed until 1937).

The Articles of Agreement provided that Joseph Edgar was to receive a capitation fee of £3 per boy for the first year, £4 for the second, £5 for the third to the twelfth and £10 for the thirteenth to the seventeenth. In addition he was to receive handsome remuneration for the boys (at least 20) who lodged at his house. For much of the time, therefore, Edgar was receiving upwards of £1000 p.a. The comment of Mr Kekewich, the solicitor, that 'the Agreement appears to be in many respects a very favourable one for Mr Edgar' seems fully justified. We have no means of knowing whether Waterfield ever came to regret either his choice of partner – or his own apparent generosity. The only other reference is a letter from W.P. Pinckney (November 1st, 1875): 'My Dear Mr Waterfield – I have ascertained that there is nothing in the Lease to forbid our giving consent to your underletting Temple Grove, and my brothers, as well as myself, will be ready to give the needful formal consent when you shall have found such a person as yourself and we may consider likely to prove a satisfactory occupant of the premises'.

What were the other masters paid? Taking into account the comments of Eric Parker and also the salaries paid by Hawtrey at Aldin House, my feeling is that the six form masters at Temple Grove probably earned £100-150 p.a. but that the other teaching staff received a good deal less. All would have received full board and lodging without charge. To Edgar's £1000, must be added at least another £1000 for the remainder of the teaching staff. With £600 covering the entire domestic staff and a further £200 the outdoor men, the total bill for wages and salaries comes to around £3000 p.a.

In calculating food and other costs one must remember that the boys were away from school for a quarter of the year. In term time rather more than 160 mouths had to be fed each day – perhaps £80 a week for 40 weeks, with an addition for domestic staff present in the holidays, making a total of perhaps £3500 p.a. A further £1000 would almost certainly have covered all other material costs, with coal at around 50/- a ton as the biggest single item. If one allowed a fairly generous £3500 p.a. for the cost of the lease of the estate and the other three buildings, wear and tear and taxes, one arrives at outgoings of somewhere around £10,000 p.a., giving Waterfield a fairly handsome profit on which to bring up his family and indulge his taste for overseas travel.

The prospectus consists of nothing more than a single printed sheet, giving purely factual information, with none of the references to 'friendly atmosphere' or 'encouraging the children to achieve their full potential' favoured by the elaborate brochures of today. 'Boys are admitted at the age of eight provided they can read fluently and write fairly . . . No one can remain in the school after his fifteenth birthday'. Prospective parents are informed that 'The year is divided into three terms, with about five weeks' holiday at Christmas, a fortnight at Easter, and six weeks in the latter part of the Summer'. After an explanation of the arrangements for exeats there is a stern admonition that 'no leave can be granted when a boy has behaved ill, or has not worked properly during the preceding month'. Since parents were to receive 'reports made of each boy's progress, work and conduct' twice a term, cancellation of leave should not have come as a surprise. Health concerns dominated the admission form. It was 'particularly requested that no boy may be sent back to school who has been exposed to infectious illness, without due notice and a proper quarantine approved by the School Doctor'. And the form, printed on the back of the prospectus, apart from seeking the obvious family details and asking, 'Can he read fluently and write legibly?', concentrates on medical interrogation. 'When was he vaccinated?' 'Is his general health good?' 'Is there any peculiarity of constitution which requires to be considered?' And finally, 'Has he had smallpox, chickenpox, whooping cough, measles, scarlatina and – if so – when?'

3   The Teaching Staff

In the 1871 census appear the names of six members of Waterfield's staff living at *Clarence House*. They were: Charles Rawlings (26), Charles Browne (25), William Bullock (24), all with Oxford or Cambridge degrees, Charles Geoghegan (34) (Trinity College, Dublin), Frederick Ranscher (25), described as 'Professor of German', and Henry Prescott (20). All were unmarried. It is almost impossible to ascertain how long each remained on the staff; certainly other names appear in the caricature drawn by two of the Grenfell brothers in 1869 and fresh faces are in evidence in the first school photograph (1880). Mere names and ages tell us very little; we have otherwise to rely on brief descriptions occurring in the reminiscences of former pupils. Of these by far the greater number refer to Charles Rawlings, Charles Geoghegan and Mr (no ascertainable Christian name) Prior. By good fortune those three men personify three types of schoolmaster to be found in almost every prep school: the fine scholar who teaches well but is somewhat remote from the boys; the man of marked idiosyncrasies, genuinely popular with his pupils; and the hopeless disciplinarian for whom the boys, much as they rag him, have a real affection.

Charles Rawlings was a Cornishman who had taken his degree at Cambridge in the '60s and may well have come straight to Temple Grove from university. He took the First Class and the verdicts on his teaching are unanimous. Lionel Byrne, who himself taught at Eton for many years, describes him as 'really good and extremely conscientious' and Monty James attributed a good deal of his own scholarly grounding to Rawlings. Even Norman Macleod (a stern critic of most things Temple Grovian) calls him 'a sound teacher [though] he made no contact with the boys he taught'. Arthur Benson gives a fuller picture: 'Rawlings, a burly bonhomme, who took the First Class, was liked and reverenced mainly for the reason that in speaking to the Headmaster he called him "Waterfield" and showed him no particular respect. He was an excellent scholar, and a man who might have made a place for himself in the world; I daresay he was unhappy and felt a failure; but one never thought of one's elders in those days as anything but quite complacent and self-satisfied'. Fred Benson is more sparing in praise, saying 'he used habitually to read *The Sporting Times* in school with his feet up on the desk until the time came for him to hear us construe.' Bulmer La Terrière recalls 'a very clever man whom I met in future days on racecourses as a bookmaker', an unexpected attribute confirmed by Maurice Headlam who called him 'a keen racing man' and added 'We always used to look anxiously at him as he arrived on Saturday morning. If he was carrying *The Sporting Times* under his arm we knew we were sure to end school about an hour earlier than the scheduled time'. The final word on Rawlings comes from Ernest Mackintosh: 'He was a wonderful teacher, and the ten or so scholarships which T.G. gained annually were entirely due to him. To make an extra penny or two, he donned a bowler and satchel on Saturday afternoons and was a bookie in the ring at suburban racecourses. Something must have gone amiss for (in 1893) he took his own life'.

Of Charles Geoghegan all his pupils speak with affection. He arrived at Temple Grove in the mid '60s, Edward Graham remarking, 'Not long before I left we had a new master, a tall Irishman, who had only one arm. He is said – I believe with truth – to have lost his other arm in battle'. Two other versions of the loss exist: one has him 'saving his brother from drowning in a well' and the other, more dramatically, has him diving into a mill stream to rescue his brother and 'being struck by the flail of the water-wheel'. One fancies that the Irishman may have helped to encourage the rumours. At all events the loss of an arm caused him, when walking, to swing himself with a slight lilt from side to side and to have the empty sleeve pinned across his chest.

'Everyone liked Giege and Giege liked all of us. He called me "Cuddy" because I was Scotch' (Tom Pitman). 'Geoghegan – we called him Gairgan or Geege – was a hard-featured, bearded, one-armed man, with a quick temper and incisive of speech, but really fond of boys and much beloved by them' (Arthur Benson). 'There was a one-armed Irish master who was uniformly kind and sympathetic to generations of boys, and we all loved him' (William Palmer). 'The popular one-armed Giege stands out in my memory – he seemed to have grown on the place and to have become part and parcel of it. After all these years I can see him sitting at his desk, hunched up writing, the paper being steadied with his right shoulder whilst, with his left hand, he produced that wonderfully clear handscript always in purple ink' (R.J. Yarde-Buller). Only Fred Benson strikes a less complimentary note: '[He] used always to have some favourite in his class, who sat on his knee in school time and was an important personage, for he could, if you were friends with him, always persuade Geege not to report misconduct to Waterfield. One such boy . . . I well remember: he pulled Geege's beard, and altered the marks in his register, and ruled him with a rod of iron. Geege was otherwise an effective disciplinarian, and had an unpleasant habit, if he thought you were not attending, of spearing the back of your hand with the nib of his pen, dipped in purple ink'. This was the man who gave yeoman service to the school, and to four Headmasters, for a total of 38 years, a record unlikely to be broken. He died, aged 66, in November, 1904.

Of Mr Prior we have no personal details and no photograph but both Arthur Benson and Monty James have left unforgettable portraits of him. Benson introduces him as 'the writing-master, a good-humoured man, with large side-whiskers, of incredible and yet lovable absurdity, for whom we wrote specimens'. And later he says: 'Mr Prior was really the joy of the place; he had almost menial duties to perform, such as serving out ink and paper, but he threw into the whole process such mystery and dignity, his threats of punishment were so far-reaching, the status of his family and connections was so exalted, by his own account, that we never tired of discussing him. I shall never forget how in one of the sudden silences that occur without apparent cause in large gatherings, the rich voice of Mr Prior . . . was heard all over the room saying, "My uncle is a man of large property in the North". He was visibly and unashamedly terrified of Waterfield, and would do much to avoid an official interview . . . He was entirely incapable of preserving any semblance of discipline in his classes'. When things got completely out of hand he would march the chief offender off in the direction of Waterfield's study but stop well short, give the culprit a final warning and send him back to the classroom, returning himself to announce, "I thought I might give him one more chance." Inevitably the moment came when master and boy were intercepted by Waterfield. "Mr Prior, what's the meaning of this?" Murmuring something about wishing to consult Matron, the unfortunate man bore his 'victim' back to the classroom where he told the boys that he had "successfully interceded for the offender". '

M.R. James, his mind already delighting in the mythical and the occult, relished school walks with Mr Prior on a winter's afternoon. 'Fixed in my mind is a fragment of one of his sagas, told with great solemnity, of adventures which befell him when staying at the castle of his wealthy uncle . . . He was returning from a long day's shooting over the wide estate when in a narrow lane he encountered, to his great surprise, a large and formidable dragon. No doubt the combat was graphically described, but I only retain the concluding phrase: "Providentially I was enabled to slay the terrible creature". '

Many years later, when Provost of Kings, Monty James became famous as a writer of ghost stories and, inevitably, *A School Story* was set at Temple Grove. Although the school is not named it was near London and 'was established in a large and fairly old house – a great white building with very fine grounds about it'. A new master called Sampson, 'tallish, stoutish, pale, black-bearded [who] had travelled a good deal, and had stories which amused us on our school walks', became inexplicably agitated when one of his pupils chose to write 'Memento putei inter quattuor taxos' (Remember the well among the four yews) and even more so when the words, 'Si tu non veneris ad me, ego veniam ad te', appeared, in red ink (which no boy was allowed to use) in an exercise book. The story must not be spoilt by a hurried re-telling but one feels that James may well have borrowed characteristics of two or three of his prep school mentors to portray Mr Sampson.

The remainder of the staff, if they are referred to at all, receive only the briefest of mentions: Bullock was 'horrid but efficient' (Byrne); 'Daubeny (we called him 'Dubs') was tall and lanky with a long nose that sniffed' (Tom Pitman) and was said to have 'no thought but for the encouragement of a small moustache' (Fred Benson); Davey 'once got angry with my brother; seized him and attempted to throw him out of the schoolroom but my brother's head came into contact with the side of the door and was badly cut. The matter came before Waterfield and it rested with my mother finally whether Davey got the sack or not: he was allowed to stay' (R.J. Yarde-Buller).

Foreigners, throughout the 19th century, were almost invariably mocked – or pitied – by true Englishmen and it was an accepted canon of faith that their function in the classroom was to be ragged unmercifully. According to Fred Benson, 'one French master, whose name really was M. Voltaire, conducted a dancing class as well as teaching French and being, I think, slightly immoral'. Foreigners were also required to accept whatever spelling of their outlandish names one chose to give them. 'The French master was Chezey or Chassée or something like that. The boys used to rag him and he was succeeded by a man with the good old French name of Bourke. There was no ragging him, or Neumann (German master) either' (S. Hare). The latter, and another French master, M. De Costa (both appear in the 1880 photograph) shared a classroom of sorts, partitioned down the middle, on opposite sides of which they endeavoured to keep order amongst their not-too-well behaved pupils. It was said that these two men had been, respectively, outside and inside Paris during the Siege. 'Neumann was fat and looked well-fed but De Costa from his appearance might easily for a time have subsisted on cats and rats, but we liked them both' (Tom Pitman). To English school boys for whom the Franco-Prussian War was the culmination of modern warfare, the daily sight of a German and a Frenchman on opposite sides of a partition must have provided a source of endless speculation.

Julius Neumann, known as Poucher or Pooch, remained at Temple Grove for fifteen years. 'He was a fat man with a beard and wore spectacles with lenses about half-an-inch thick which magnified his eyes fearsomely. He always wore very loud check suitings. One had to be continually passing through his classroom on the way to another part of the building and was almost invariably greeted with an order to "Peck up zat orwange peel", with which, for some reason, the floor of his class always seemed to be littered'.

These, then were the men who, in their varied ways, laboured to serve Waterfield. Since they received no pension from the school or from the State most of them would have needed to keep working into their seventies. Perhaps it would be better not to attempt to peer into their futures but to turn our attention to what, and how, they taught at Temple Grove.

4   The Curriculum

It is inevitable that there should be inconsistencies in the accounts written, and the figures quoted, by Old Boys concerning their schooldays fifty years earlier. Almost the only records of Temple Grove which were written down at the time and which have survived are the annual reports of the Classical Examiner (from 1865 onwards), letters written to his parents by Martin Benson (1870-74) and the *Name Book*. From the latter we can form a reasonably accurate estimate of numbers. Waterfield took over a school with 64 pupils. In the next twenty years he took in 860 new pupils. Each boy stayed (on average) two and a half to three years, so that from around 1863 the school nearly always had between 120 and 130 pupils.

The most precise and most informative of those who later wrote about the school is Lionel Byrne. *Changing Eton* devotes the first ten pages to a detailed description of the life and work of the school in the 1870s. Each day contained six hours of lessons: 7.30-8.30 (i.e. before breakfast), 9.00-12.30 (with a break for drill), 4.00-6.00, and an hour for prep (7.00-8.00). Afternoon lessons were not done on Wednesdays and Saturdays, and occasionally the evening was 'given', so that in most weeks 38 hours of work were done by the boys, compared with around 38 lessons (of 35 or 40 minutes each) and four hours of preparation in a typical prep school today. Terms were slightly longer and there were almost no breaks in them, so that approximately 25% more time was spent each year in the classroom by the pupil of a century ago.

'The subjects taught to all boys were Latin, Greek, divinity, mathematics, history, geography and French. The lower classes, those few which did no Greek, learnt writing under a master of inferior status. This writing was done in copy-books . . . and a favourite punishment was the setting of so many copies. The bulk of the serious work consisted of Latin and Greek, both well taught throughout, and, towards the top of the school, with such accuracy as to leave an ineffaceable impression through life. In the middle of the school Latin verses were begun, and the chief Latin author was Cornelius Nepos. Caesar was taken in the second class, Virgil in the first. Greek began with the Eton *Sertum*, followed by Xenophon's *Anabasis*, Homer's *Odyssey* not being reached till the second class. A shortened Greek play was often not beyond the capacity of the first class. Divinity was taught by means of Maclear's Old and New Testament Histories. Mathematics (chiefly arithmetic, varied by a little Euclid and towards the end some simple algebra) did not reach a very high standard. History was extracted from a text-book; geography was inculcated by means of a series of maps done with a J pen and paint smeared on with the fingers. French was very sketchy, though some was evidently learnt by the naturally proficient' (Byrne). German is not mentioned, perhaps because it was always an optional extra. In *Dreadnought* Robert Massie observes that, in the famous Liberal Cabinet (1906-16), 'none of the ministers even spoke French, except Churchill who spoke it with a grandly atrocious accent'. Richard Haldane, having attended the University of Gottingen, was fluent in German. The most obvious omissions are English and Science, together with Art and Music. Of the three main branches of English, grammar, composition and literature, the first would be supplied by the intensive analysis to which all passages of prose were subjected before translation into Latin or Greek; the second by the exact shade of meaning and its rendering into the appropriate English equivalent required in translation *from* the classical languages; the third by general encouragement (school library etc.) outside the classroom. For Science, as we have already seen at most Public Schools, there was no time, no inclination and no spur. The question of the arts will be left to later consideration.

From the above it is apparent that almost every member of staff was required to teach Latin, and most Greek as well, and it is likely that most of the four hours' work up till dinner time was devoted to the classics. 'Of the value of the classical teaching which was given at this school there is no doubt. It was clear, systematic, unhurried, and each individual realised that what he had learnt he was expected to know, and know accurately. Of course the masters varied in efficiency, but all were kept up to the mark by the knowledge that at any moment their class might be summoned to the Headmaster's study to have its lesson heard. A grammar in Latin which bore the name of *Principia Latina*, written by the Headmaster himself, was learnt by heart and applied in every sort of way, so that its fundamental principles became to the learner part of his being . . . The Headmaster always set and corrected the weekly Latin prose, making the class learn and say to him his own fair copy, a fearful joy, but a joy none the less, for boys appreciate good teaching however stern'. Byrne's stamp of approval for the system is, inevitably, challenged by Norman Macleod. 'A lesson with Waterfield was truly a penance, but we endeavoured to avoid trouble by preparing the appointed passage from some Greek or Latin author in the most meticulous fashion. How on earth can boys enjoy the beauties of Cicero's *De Amicitia* when taught in that way, while we knew that the heavy ruler was at hand to come down upon our unfortunate hands as the penalty for any lapse from grace?' And later he says, 'I have always thought very poorly of the system in vogue in those days. We were very thoroughly grounded in the mysteries of the Latin and Greek Grammars, but we learnt very little about the beauties of their respective literatures . . . Latin verse was my abomination, I could never see the object of turning good English poetry into bad schoolboy Latin elegiacs'. And he ends with a diatribe against the whole system of education then practised in prep and public schools:

'I don't know how things are managed now [1940], but it has always struck me as curious that so little attention used to be paid in schools towards giving boys instruction in their own language. How many hours we used to have to devote to mastering the mysteries of the Latin gerundive and the Greek particles, while English grammar and composition were neglected. History and Geography were not "taught". We had an elementary *History of England*, very little superior to *Little Arthur's History*, and all we could learn from that was a lot of dates, the names of the Kings and Queens, and a few salient but not very important facts. Geography consisted of learning boundaries, the names of rivers with their tributaries, and the towns on their banks, and every month we had to make a tracing of the contents of a page of our atlas, filling in all rivers, mountains and names of towns with a particular kind of tracing pen . . . English literature was entirely neglected, and even at Wellington when we had a chance of getting acquainted with some of the English classics, the method of introduction was such that we learnt nothing by it'.

Each year in July (at least from 1865 onwards) Waterfield invited a University don, or a Public School Classics master, to come to Temple Grove for a short visit and to examine the Upper School on their competence in Latin, Greek and Divinity. His report was, in due course, printed and circulated to parents but there is no reason to believe that the examiner's comments were unduly flattering: indeed the testing seems mostly to have been pretty rigorous. Six classes were normally examined, Class I being the top boys of the school. Normally Class II and Class III were divided into alpha and beta ('one half preparing while the other was being heard') but numbers often required that there should also be a gamma. In none of these classes were there ever more than a dozen boys; eight seems to have been the average. Class I varied between two and six.

With such small numbers and intensive teaching of a relatively restricted field of knowledge, unusually good results were to be expected. Even so the level of attainment was extremely high and the reports themselves breathe the very spirit of Victorian scholarship and indeed of Victorian academic ideals. And since the teaching of the classics was the essence of preparatory school education – some would say its primum mobile – the assessments of the different examiners seem to justify quotation at some length.

'The Boys who have come under my notice appear to me to be well taught and thoroughly grounded in their Greek and Latin Grammar; even in the cases in which the passages set before them were not so well construed as I could have wished, I almost invariably received accurate answers to questions in Parsing' (C.J. Evans, 1865).

'The behaviour of the boys was quiet, attentive and gentlemanly, and gave me the impression of their being happy in their life and work' (H.J. Roby, 1866).

'On the whole my Divinity questions were fairly answered; and if no answers were VERY good, few were bad; and I observed with satisfaction that nearly every boy answered correctly the only question that touched upon the Creed – a proof that dogmatic teaching is effectively carried out' (E. Walford, 1867).

'Class III beta. In this Class also the Greek Grammar and Translation were very good; the Nepos was fairly done; but a large number broke down completely in the Ovid' (J. Sharp, 1873).

'Class I. All four boys did well. Benson ma. [Martin, the future Archbishop's eldest son] is, I should say, the best and on this occasion *knew his work* better than the others. In the unseen work and composition there was very little difference between Benson and Boyle . . . A very little more care would have rendered some of the pieces irreproachable' (A.W. Verrall, 1874).

'Class I. This class consisted of seven boys. In composition they were all good . . . The Latin Verse was fairly done; the Latin Prose . . . was wonderfully good. The Translations of Kirkpatrick and Morton were the best. In Latin, *Viva Voce*, Kirkpatrick and Capper were first and second. In Greek, Scudamore was first' (E.W. Bowling, 1875).

'Class III gamma. The work done in school was well known, and a piece of unseen Greek was fairly translated by all – best by Goodenough, Rawlinson and Beckwith. The Latin Verse seemed to be a weak point, but in the case of boys so young that is nothing to be surprised at' (Winfield Bonser, 1877).

Perhaps the most comprehensive and most revealing of these reports was the one rendered on July 1st, 1876, by H.W. Paul (scholar of Corpus Christi, Oxford).

'Class I. These boys knew the work which they had already done in school so well, that I found it desirable to let them translate a good deal of Greek and Latin which they had not read before. The result was highly satisfactory, especially in the case of Thucydides, an author who often presents insuperable difficulties to much older boys. The only weak point which I observed in these boys while I was examining them viva voce was their inability to read Horace without perpetrating false quantities . . . The Latin Prose of this Class was correct but not polished, the boys appearing to think that they had done enough when they had avoided grammatical blunders. James's verses (a translation of a Sonnet of Spenser's) were remarkably good, and those of the others very fair, but not sufficient attention was paid to the exact meaning of the English, and epithets were used without much discrimination. James translated best into English, Davidson wrote the best Prose'.

In the outcome the future Provost of Eton topped the future Foreign Secretary (Grey) by 32 marks. Equally impressive were Mr Paul's comments on some of the lower forms:

'II beta and II gamma. These divisions had only recently begun to read Homer, and I was therefore agreeably surprised to find that they translated him generally with ease and fluency, a circumstance which I cannot but regard as highly creditable to their teacher . . . In II beta Brinton showed great ability in translating passages of Virgil which he had not previously seen . . . In II gamma Byrne displayed considerable powers of mind, and his success in the examination would have been far greater, if he had taken a little more pains with the books which he read in school. III alpha and beta. The only criticism which I have to make upon their translation is that the boys showed acuteness in grasping the general meaning of a passage rather than that verbal accuracy or nicety in which technical scholarship principally consists.

'The characteristic of this school, or at least of those portions of it examined by me, which struck me most forcibly was that every boy had received the individual attention which was his due. Clever or stupid, they all bore marks of having had their minds skilfully and diligently cultivated'. The marks obtained by all fifty boys are then given.

It would be remarkable if teaching of such high standard and intensity had not yielded a heavy crop of scholarships. Twelve in 1874 appears to have been the highest total, out of 41 boys leaving that year. Eight a year seems to be the average but there are a number of instances of individual boys gaining more than one scholarship.

No copy of *Principia Latina* is known to have survived but a copy of *Elementa Latina* ('compiled and privately printed for the use of Temple Grove School', 1897) has recently come into the author's possession. Its 50 pages give a comprehensive analysis of the grammar and syntax of the Latin language. Explanations are given of Tertiary Predicates, of Model or Indeterminate Verbs and of the division of a conditional sentence into Protasis and Apodosis. Each usage is accompanied by an appropriate example. Do the first three quoted here give an insight into the pedagogic mind? The fourth would surely have evoked heartfelt schoolboy responses.

Mori nemo sapiens miserum ducit.	No wise man thinks death a misfortune.
Mens alitur discendo.	The mind is nourished by learning.
Dum trahitur periit.	He died while being dragged along.
Edamus ut vivamus.	Let us eat to live.

There are some interesting sidelights on the actual taking of scholarship examinations. Waterfield always conducted the little group to the school where the test was to take place, 'and saw that we were well fed at the Hotel. He did not go in for last minute coaching, but he used to try to find out how we had prospered, and when I had to show him a rough copy of a composition which I had been foolish enough to keep, I was told I hadn't a ghost of a chance of success' (Norman Macleod). In spite of this encouragement, when the Wellington scholarship list for 1880 was published, the first three names were: N.C. Macleod, H.E. Stockdale and J.W. Cave. It was Stockdale who was chosen to go round the school, 'to announce the half-holiday given in honour of the occasion'. One cannot help feeling that the post mortem conducted by Waterfield was considerably more alarming than anything the examiners themselves could devise. Little imagination is required to picture the scene (at Eton) when it transpired that Monty James had actually used an ablative absolute as the *subject* of a sentence. As Waterfield remarked in a letter to Monty's father, 'The examiners are the Provosts of Eton and King's. This sort of thing will make their hair stand on end with horror'. And indeed a second attempt the following summer became necessary. On that occasion Edgar accompanied the candidates and was suitably encouraging. Monty's name appeared second on the list. A similar experience fell to Arthur Benson: 'I was taken by Waterfield himself with three other boys to try for an Eton scholarship. He was in his liveliest mood, bought us books and papers at the station, made jokes, propounded unanswerable dilemmas . . . Our first paper was at 7a.m. Waterfield came to see that we were getting up in good time, comforted us with coffee and rolls and . . . when we returned . . . had a great substantial breakfast ready for us. But then he spoilt the whole affair by asking us what we had written; and I remember his casting down a fork on the table with a dramatic gesture of disgust at something I had put down'.

Almost nothing has been said about the teaching of mathematics. How much time was devoted to it, what standard did it reach etc.? The very silence, in all the wide-ranging accounts of the school's life and work, is indicative and it is not even clear whether candidates at that date were required to take a maths paper in order to enter a public school. A revealing comment was made by Matthew Hill (descendant of one of the famous Hill brothers), who was awarded an Eton scholarshp in 1885, when he wrote – on reaching Eton – 'Never had I been taught geometry before. All I had done was to learn Euclid's prepositions by heart; a rider was Hebrew to me'. Was Temple Grove an extreme example of the worship of classics and the neglect of mathematics? That other schools *did* aim higher is indicated by the level of questions set for would-be Westminster scholars in the 1880s:

(a) Four merchants, A, B, C and D trade together. A's stock of £200 was in trade 8 months; B's stock of £396 for 12 months; C's of £375 for 8 months and D's of £395 for 6 months; how much ought each to receive of the profits amounting to £2169?

(b) Define the tangent and secant of any angle. Prove (1) $Sin^2A + Cos^2A = 1$ (2) $Cos(180\text{-}A) = -CosA$.

(c) Explain the method of determining the position of a point by means of Cartesian Coordinates. Prove that the equation $Ax + By + C = O$ represents a straight line.

By the standards of today Temple Grove's contribution to the Arts must be reckoned abysmal. That it was no better, and probably no worse, than that of other prep schools at that time is simply a recognition of the worship of the classics at the High Altar while relegating Art and Drama, Music and Literature to the crypt. Although a number of future authors *were* educated at the shool, the records indicate only two distinguished musicians, Sydney Nicholson and H.B. Gardiner, and one great man of the theatre (Tyrone Guthrie) in fifty years. There must, one feels, have been artistic talent present among the Waterfields or the Candys (manifesting itself in the work of Maragaret and Aubrey) but a conventional Victorian preparatory school had little truck with such activities. A piano stood in the entrance hall 'where a frail widow lady called Mrs Russell gave music lessons'. Her own preference was for 'sugared melodies' but she did allow Fred Benson to 'entrap my awkward fingers in Bach for which I owe her an undying debt of gratitude'.

It used to be confidently stated that Arthur Sullivan, as a young man, had been the music master at Temple Grove but his biographer shows conclusively that this was not so: 'It is from this time, April, 1862 (Sullivan being then just on 20) that Sir Arthur dates his public career as a composer . . . Whatever doubts and fears he may have entertained up to that time, he then definitely decided to avoid teaching and to rely on composition' (Arthur Lawrence). Sullivan, however, *was* a personal friend of the Waterfields and 'one day came to lunch and spent the afternoon with them. Now for some reason I had been roped in to the school choir, though I expect my voice was more fitted to scare the crows than sing the *Te Deum*. However the choir was invited to go into the Drawing Room and hear Mr Sullivan sing and play to us. There was a large grand piano, the top opened, and on the wires he laid a sheet of music and when he played it was exactlly as if he was playing a banjo. He sang *Camptown Races* and other nigger songs, but from that day the banjo was the musical instrument for me' (Edward Graham).

Other references to the choir appear in Old Boy recollections. 'The efforts of the choir were confined to practising some few part-songs on a Saturday evening, after which they had cake and a curious kind of pineapple jam' (Monty James). The inducement of food is confirmed: 'We sang part songs and glees but . . . I remember best the supper afterwards . . . I have seldom met "Choir Cake" since; it had a special flavour of its own, and there were biscuits with gooseberry jam in between' (Tom Pitman). The meal was apparently served at the Headmaster's dining-table and on occasion orange wine appeared on the menu. 'Beneath the table crawled unmusical friends of the singers, begging for scraps. To pass cake down in the dark was easy enough and, though the orange wine was apt to lead to detection, it was hard for a privileged boy to resist the blandishments of his best friend . . . stroking his knees in silent supplication' (Lionel Byrne).

Drama seems to have been confined to fairly elaborate charades. 'I remember during my last Christmas half we had some theatricals, which Waterfield took a great interest in, and coached us for. Sleep Walking was what we acted as a Shakespearian Charade. Sleep was Queen Catherine's death scene from *Henry VIII* . . . Wall was the scene from Pyramus and Thisbe in *Midsummer Night's Dream*. King was the scene from *Henry V*, the King's soliloquy before the battle . . . Sleep Walking itself was, of course, Lady Macbeth's scene. I think Rowe was Lady Macbeth and I was a varlet. We were well coached, and I remember the thing was so well done that there were favourable notices in the papers about it' (Bulmer La Terrière).

If the Artistic and Literary world received only a slender contribution from Temple Grove, the Church of England was kept well supplied: several bishops (notably George Bell) and deans, numerous canons and others in holy orders. Surprisingly little information is obtainable, however, about the part played by religion in school life, and even less about the attitude of the boys towards it. Every morning 'brief prayers' in the big schoolroom preceded early school, and prayers were again said at 8p.m. after preparation had ended. On Sundays breakfast (at 8.30) was followed by an hour's divinity. During this hour the boys were required to recite various passages. Martin Benson, in an early letter home, writes: 'Before every Sunday we have to learn part of the Catechism, part of the fifth chapter of St John, and the Collect and say it on Sunday morning'. A divinity lesson taken by the Headmaster in person, on a Sunday afternoon after letter-writing, is compellingly described in *David Blaize* and, although the school is called 'Helmsworth', the description of Mr Acland leaves one in no doubt that it is really Temple Grove. On this occasion Catechism and Commandments were safely negotiated but the interrogation concerning St Paul's second voyage revealed horrifying gaps in the boys' knowledge. A quantity of additional learning was therefore set and the lesson ended with the ominous words, "Next Sunday, then, I shall hold the class in my study".

In Waterfield's early years the boys attended Mattins at St Mary's, the parish church in Mortlake, a walk of some twenty minutes, but the growth of East Sheen led to the building of Christ Church (p.113), consecrated in January, 1864, and for the rest of Waterfield's headship the school walked to morning service there every Sunday, occupying most of the south aisle. There was no question of the school choir adopting that role at the parish church. Indeed the local choirboys took strong exception to the silk hats worn by privileged boys from wealthy homes and 'delighted in giving chase to them and removing the offending headgear'. On the services themselves only two comments have been preserved: 'On Sundays we went to an extraordinarily ugly and tasteless modern church, where the services were hideously performed' (Arthur Benson); 'I hope I got some good from the services but the only thing that comes to my remembrance is the vigorous and dramatic force with which the preacher quoted the words of Balaam to the ass: "I would there a sword in my hand for now would I kill thee" ' (Yarde-Buller).

It seems fair to conclude that Sunday mornings were not enjoyed and that the lengthy sermons, intended for adult parishioners, were quite unsuitable for small boys. But, if Christ Church failed to leave its mark on many of the Temple Grove pupils, the opposite is certainly not true. Meston Batchelor, in 1976, copied down some 150 names of pupils carved on eleven different side pews in the church.

Christ Church, East Sheen

The Orchard (previously Wedderlie), a view from the S.E.

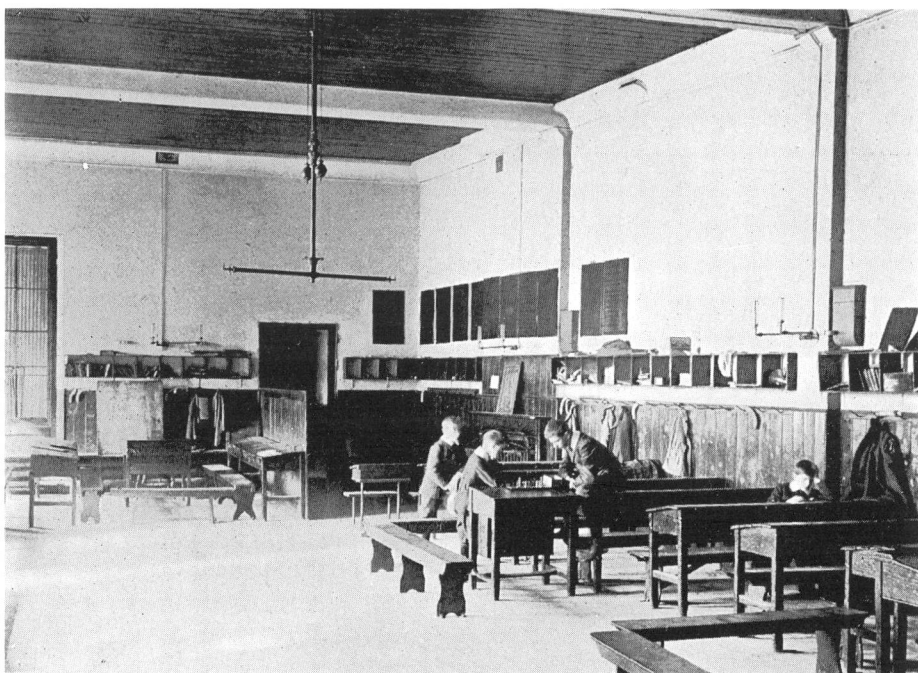

Most of the classes were taught in this large schoolroom

Each dormitory was divided up into cubicles

5   Domestic Arrangements

One approaches the subject of the school's domestic arrangements with considerable caution. They impinged on the boys' lives in two important departments: the dormitories and the dining-room. But whereas the matrons were key figures who feature in most Old Boy recollections the cooks and other domestic staff remain shadowy figures, mostly unknown personally to the boys and certainly unremembered afterwards, their efforts largely unappreciated and only their failings commented on. Their names and ages can be retrieved from the census returns, though many would have come and gone between those decadal events, and after 1851 only one further census (that of 1881) was taken in term-time so that the information is far from complete. More significantly no account books have survived from those early days. To get some idea of what domestic staff were paid in Waterfield's day one must turn to that interesting and informative guide, *The Rise and Fall of the Victorian Servant*, by Pamela Horn. The figures which Mrs Horn quotes for wages paid by a variety of establishments are consistent enough to serve as a guideline. A butler might expect to receive £60 p.a., a coachman slightly less and a footman £30. A housekeeper would receive £50, cooks £40-50, housemaids varied from £14-24, kitchen maids slightly less and scullery maids no more than £10. Of course all board and lodging was supplied and also uniforms together with a good deal of cast-off clothing (often of excellent quality) and wages were paid during the fortnight's annual holiday. Matrons are not specifically listed in Mrs Horn's book but about £25-30, depending on length of service, would be the probable figure.

By the time that Waterfield had built up the number of pupils in the school to over a hundred the domestic staff appears to have numbered about twenty, comprising: a lady superintendent, a housekeeper, four matrons, a nurse (at *Temple Grove Cottage*), a butler, a footman, two cooks, five or six house-maids, a kitchen maid, a scullery maid and a laundryman and laundress. Apart from the senior women none of the domestic staff – unsurprisingly – appear to have stayed very long at the school. They were nearly all young people and would be trying constantly to better their lot. In 1871 ten junior domestic staff have an average age of 21; in 1881 it has gone up to 26 (with a lady's maid of 50); in 1891 it is 22 (with Julia Chapelin – if the census is correct – a kitchen maid aged eleven).

Over this constantly changing entourage presided for some twenty years Anna Abbott, always referred to as Dame Abbott. She must have been engaged very soon after the 1871 census in which she appears, aged 29 and a widow, as a 'Visitor'. Her position as Lady Superintendent gave her many of the duties that later would be performed by a headmaster's wife, and one feels that she probably had the responsibility for engaging at least the junior domestic staff. Her room on the ground floor was strategically placed at the very hub of the school's daily life. James Pitman recalls: 'Our ailments were looked after by

Dame Abbott who kept our money, dosed us, tied up our wounds, examined our heads once a week and gave out our journey money'. The remedies she used as doses appear to have been few and simple but 'devastatingly effective for every malady from pneumonia to a broken leg. The first, licorice powder, was pretty awful but at least had a taste that was not objectionable to all; the second, castor oil, was only a spoonful but invited immediate vomiting; the third and final deterrent was Gregory Powder, dreadful in its almost tastelessness, utterly and completely loathsome, but calculated to restore all but the dying to immediate health' (C. Storrs).

The bathroom (which was situated on a half-landing in the Centre Wing) was reached by a few steps down 'into a sort of well room. Half-a-dozen large coppers, each in the shape of an egg, were let into a wooden platform two or three feet above the ground. One reached one's particular vacant egg by mounting a short flight of steps and walking along the back of this L-shaped platform. The smaller boys were bathed by a matron, and as the baths were raised to a convenient level, she did not have to bend over' (T.V. Scudamore). La Terrière remembered 'the big copper baths, all kept spotlessly bright and heated by a big central stove. A jolly fine row we made; a dozen little naked beggars all shouting and playing the fool at once'. James Pitman adds, 'We ran down here every morning for a cold bath, and once a week for a hot bath at night. After this latter entertainment, matron produced a pot of yellow-coloured train oil which she called pomatum which was rubbed into our heads. The following morning we had to parade outside Dame Abbott's room for what was called "Showing up". The hair was combed through and a thorough search made for any disease of the scalp'. 'She was a dear old lady, and we all liked her very much'.

In prep school mythology almost the largest part is played by the dormitory. The rules were strict, the punishments for breaking them severe but offset by the better-than-even chance of not being caught. And presiding over the dormitory world were the matron and her assistants. Matron might, at some schools, be a dragon, feared and heartily disliked. Much more often she was a substitute mother and (the headmaster's wife being often a shadowy figure) one of the few sympathetic listeners to whom a boy could turn in a harsh world. At Temple Grove, by 1870 anyway, there were four matrons, one responsible for the West Wing (one dormitory with 25 cubicles), two for the Centre Wing (five dormitories – one on the top floor – with 38 cubicles) and one for the East Wing (two dormitories with 27 cubicles). And so a total of some 90 boys slept in the main building. La Terrière describes the dormitories as being 'cut up into rows of horsebox-like bunks' and of course it was absolutely forbidden to go into another boy's cubicle between 'lights out' and the dressing bell. Fred Benson relished 'going on tiptoe into the next dormitory and, after waking up my special friend, sitting on his bed . . . and talking in whispers till there came the sound of the dressing bell which portended the entrance of the matron. Or else it would be I who was awakened by the soft-stepping night-shirted figure . . . and there would be plans to be made and then we would take the stag-beetles (called 'The Monarch of the Glen' and 'Queen') out of my washing basin . . . and refresh them with a breakfast of elm leaves and perhaps the half of a strawberry. They had to be put back into two match-boxes which were their travelling carriages before Jane came round, for she had said that if ever she found stag-beetles in basins again she would throw them out of the window'.

Jane Deeming (then aged about 23) came to Temple Grove in 1863 and Mary Underwood (seven years older) shortly afterwards. It is safe to say that no two other women have given such long and devoted service to the school. To a boy arriving in 1879 Jane already seemed old, 'her grey hair was brushed smooth back over her forehead', and Underwood, too, looked 'distinctly matronly' (Tom Pitman). By then, of course, they had learned all the tricks of the trade. After the morning wash, 'we filed past a hatchway . . . from which Jane handed us each a halfpenny bun which was nearly always a day old but still had that lovely shiny varnished look that buns had in those days. The hatchway was 3 ½ feet above the floor level and it was possible for a greedy boy to duck down after he had passed, crawl back and try for a second bun, but I never knew Jane to be taken in' (John Brereton).

Matrons, of course, were expected to enforce discipline in their dormitories. Anyone misbehaving or caught talking after lights out had his name put on a report. 'Just inside the door of the library was a table where [these reports] were placed for the Head to pick up as he came from his study into prayers in the morning. Opposite the library door was the staircase leading up to the dormitories. It was not difficult to open this door, nip in, gather up the reports and nip off to Prayers. A sin, yes, and one with great risk of the direst punishment if one were caught, but a sin with a redeeming quality if it was to save a pal' (Reginald Yarde-Buller). Norman Frank describes Jane and Underwood as 'kind and capable women, but their motto was "No nonsense", and they both got me swished at different times for something or other, but there was no ill-feeling about it and both were absolutely reliable in cases of illness'.

By the end of the century Jane and Underwood had become living legends to generations of Temple Grove boys. Not all the stories related about them could possibly be true but some of the accounts have an authentic ring about them. Christopher Storrs recalled 'standing on my bed in my pyjamas, bending over and uncovering my behind and saying in a commanding voice, "My bus, my ba, my bum," and then hearing Jane (who had just looked in) say, "Very pretty, Master Storrs, I shall report you in the morning".' Underwood who had the senior dormitory, was 'much too kind-hearted for the job and got ragged unmercifully' (that would be in her later years, one feels). Her dormitory retired to rest chanting some ditty, the best-remembered being:

> 'Old Mother Underwood jumped out of bed,
> And out of the window she popped her head,
> "Now, you boys, you're making too much noise,
> So you'll arl be reparted in the marning".'

In 1902 an appeal was sent out to "all Temple Grovians on behalf of two old Matrons, Jane and Underwood, who, after fulfilling their duties faithfully for 38 years, during last Summer Term terminated their long connection with the school. They are now in poor circumstances, Jane especially, who has supported her family for many years *out of her earnings*' (my italics). No record of the response has survived: one hopes, and believes, it would have been a generous one.

As the school grew bigger Temple Grove did not possess enough sleeping accommodation and, in 1871, Waterfield took the opportunity of leasing a house built in 1856 by the wine merchant, Charles Ellis, and originally called *The Orchard* but renamed *Wedderlie* (after an estate in Scotland belonging to Edgar's forbears). The Rev J.H. Edgar and his family moved in and took charge of some twenty to thirty boarders, of whom the majority appear to have been new boys. According to Dermot Freyer, however, *Wedderlie* had been acquired 'with a view to providing special accommodation for boys who were delicate or might require extra care and attention for one reason or another, but with the rapid growth of the school . . . it had come to serve in addition as overflow quarters for the latest entrants. In it were housed around two dozen boys in conditions more approximating to the snug comforts of home'.

Delicate or not the Wedderlie brigade were subject to assaults from the lads of the neighbourhood as they walked back from the main school in the evening. Harry Waterfield who came as a day boy for his first two terms (his father was then living at *Lime Grove Villas* – plot 169 on the map) remembered that 'it was a recognised thing that any T.G. boy found alone on the road was always fallen upon and attacked by the boy bandits of the village, and so I made a point of going out of my way past *Wedderlie* in order to have the escort of Wedderlie boys on at least part of my journey. But even that escort did not always save us from attack and many is the time that I arrived at home with a bleeding nose and the proud satisfaction of having made someone else's nose bleed'. On the other hand a walk across the school grounds (known to be haunted) carried its own perils, and it is those perils that pervade Dermot Freyer's chilling little story, *Bogey Baxter*.

One small consolation for these assorted hazards was the large biscuit, called a 'morning' which was issued to Wedderlie boys before they set out for school and was considered greatly superior to the standard school issue so that useful swaps could be achieved with it.

When numbers dropped in the later '90s the lease of *Wedderlie* was given up.

The care of the sick, in the days before antiseptic precautions were properly understood, when knowledge of hygiene was in its infancy and when mothers, over-protective or over-concerned about their children's health, were almost unknown, did not occupy an important place in the school's arrangements. Vermin (i.e. head lice) must be guarded against at all costs – for social as much as for hygienic reasons, one suspects – but ordinary ailments received only cursory attention. However 'a comfortable sick-room at the top of the house with a fire and a shelf of books' *was* provided though, as it doubled as the birching room, the occupants might find themselves suddenly ejected and 'on return [might] find the floor littered with tiny morsels of birch twig. These scraps were highly prized, for they were said to taste of brine and vinegar' (Byrne).

Temple Grove Cottage, in the Upper Richmond Road, having once served as the lodging for unmarried staff became (after the leasing of *Clarence House* in 1865) the school sanatorium. To Arthur Benson's delight it turned out to be presided over by Miss Louisa Cox, 'a sturdy motherly woman [she would have been 40 in 1872] who had once been a maid in our own household, who petted and made much of me, let me help her in the kitchen, and in whose presence alone I was conscious of some irrational human affection. The happiest days I spent at Temple Grove were in that little house'. And when the time came for Benson to leave the school, 'On the way to the station we stopped for a moment at the little sanatorium, I flew in . . . and was clasped for a moment tearfully to Louisa's ample bosom'.

A careful search reveals only two boys (for certain) who died at Temple Grove in Waterfield's day. The *Name Book* is understandably reticent, merely recording 'May 1873' opposite the name of James Alexander McNeile, who had arrived at the school with his brother, John, only a few weeks earlier. And similarly 'July 1875' appears opposite the name Brien Ibrican Cokayne, who came in September 1873. The information, but not the cause of death, is given in the 1905 register. There may have been others since, on the right hand page where the public school and awards are entered, there are a number of instances where no more than a date is given, but again parental dissatisfaction would *not* have been recorded. If the care of illness fell a long way below modern standards, the authorities (remembering Eagle House's unfortunate experience) were certainly not foolhardy, and the story of a football team with bandages round their throats because of mumps must be accounted apocryphal.

For much of Waterfield's time the housekeeper was Jane Lowes, again always referred to as Dame Lowes. Her responsibility was almost entirely confined to the catering department, never the easiest of jobs and an unrewarding one in those days. It is as unsafe to make generalisations about school food as it is about other aspects of life. On the whole it may be assumed that Temple Grove pupils were as voracious as any other schoolboys and that, coming from well-to-do homes, they were accustomed to good food in the holidays. These factors need to be borne in mind when considering the often conflicting verdicts of the boys. A reasonably unbiased account is that given by James Pitman: 'For breakfast and tea, tea, milk and bread and butter were provided, supplemented by anything the boys brought from home, such as jam, sardines and potted meat. These were given out to each boy in turn, who had to distribute his contribution to his table. Sometimes we did not give up all the grub that we took back with us, but kept some for a private feast. I remember sitting under some bushes with John McNeile eating sardines neat when the Head came round with a dog which yapped at us, but we were saved from discovery by giving the dog a sardine to keep quiet. The meal of the day was, of course, mid-day dinner, and there were never complaints as to the quantity – rather the opposite. The quality was not good, and the cooking dreadful, but we all got beer to wash it down with. There was a rule that we had to eat everything we were given. This led to great competition to sit near one of the three or four rat holes which were in the dining-room floor. Failing a rat hole, unwelcome food was surreptitiously shovelled into a tin or an envelope and taken away in our pockets. On one occasion I left the room with a helping of pudding (suet and raisins) wrapped in paper in my pocket, and threw it onto the roof of an outhouse [presumably the w.c's]. Unfortunately, I threw it too far, the parcel landed in the road and opened out in front of a master who had been on duty and was going home. I think he must have had some of the pudding himself and sympathised with the perpetrator, as he contented himself with returning and threatening pains and penalties to all and sundry'.

Norman Macleod, on the other hand, claims that 'it was not good for young boys to be quite so hungry as we were. Very often there was nothing for breakfast but bread and butter, or what passed for such, [though] sometimes we were allowed one sardine each. After tea was finished in the evening we used to hang about while the servants were clearing the tables in the hope of being able to get the end of a loaf from the masters' table'.

Dame Lowes was credited with fiendish ingenuity in her efforts to economise. 'The tapioca pudding was made (so we absolutely believed) from fishes' eyes and frogspawn. Someone had seen Dame Lowes making it herself through the peephole into the kitchen. Even those boys who did not believe the story could not be seen eating it, and so the bulk of the pudding was put secretly into envelopes and either poured down the rat holes or carried out to the back of the cloisters' (Tom Pitman). In the dining-room Dame Lowes mounted guard over a high table on which stood 'galvanised urns with a spout half way down. The current belief was that the taps were placed high on purpose so that she could collect all the dregs (of the cocoa) at the end of the week to make into chocolate pudding for Sunday'. She is also credited with describing the mutton as venison if ever it was 'high'. Another witness recalls how 'the chocolate shape that should have arrived cold and shapely came hot and shapeless, lakes of brown stuff in the dishes' and how Dame Lowes put her head through the hatch and called out, "O Chef, jamais, jamais!" (Reginald Yarde-Buller). (Two young Swiss chefs were employed at that time).

Along the north wall of the dining-room stood the table on which the dirty plates were stacked. This also did duty as the 'Pig Table', to which 'boys whose behaviour was too outrageous were sentenced for a week by the Headmaster [and] where they had to stand and scramble for what food could be got amid the jostling of the serving maids' (Byrne).

It was not only the tapioca which was subject to the boys' taboo. 'It was said that discarded gloves were chopped up and introduced into mincemeat pies'. Potatoes too, were the subject of unworthy rumours and the butler had been observed to water the beer. 'Most poisonous of all were supposed to be the sausages which we had for breakfast now and then: it was a point of honour not to eat a single mouthful of this garbage', and above all 'it was the rule to consume one's meals with an air of intense and reluctant disgust' (Arthur and Fred Benson). But schoolboy fashions – and taboos – quickly change and, at a later date, 'we gobbled up our sausages, asked for more, and got them'.

About the rats, however, no legends were necessary. Whether they actually gnawed the holes in the dining-room floor or whether they simply availed theselves of the boys' offerings through existing apertures that gradually widened is not clear, but several witnesses testify to the contribution they made to waste disposal, and one learns of rats 'waiting in hordes for the food eagerly passed down to them'. It was as well, however, not to presume too trustingly on an alliance of mutual interest. William Palmer remembered 'in my very first term a large and confidential rat coming out and allowing me to pick it up by the tail. Well, the rat of course at once turned and fastened on my thumb'.

When one remembers how little boys saw of their parents in term-time, and their craving for food, one can readily sympathise with Tom Pitman: 'What I remember best was the joy of a hamper from home, the delicacies contained in which were shared with one's table. Crosse and Blackwell's (those two gentlemen must have come to England with William the Conqueror) potted chicken and ham or turkey and tongue, and the pots of jam, what a godsend they were. When the jam was to all intents finished someone would call out, "Bags I first scrapings", then "Second scrapings", but it was an understood rule that whoever secured third scrapings was allowed to pour milk into the jampot and that was the end of it'.

All in all one may conclude that the food was of much the same quality and quantity as in most other prep schools and that the cooking arrangements and general hygiene were probably little worse. But if the school's attitude to food ('eat up every scrap on your plate') is typical and comprehensible, the prohibition on drinking, except at meals, seems totally unreasonable. James Pitman describes it as 'hard to believe but absolutely true. The powers that be had ordained that drinking between meals was bad for small boys and, however thirsty one was, one could not get a drink of water. There was, however, a long lavatory [we would say, washroom] with some half dozen basins on each side. The water was turned off but there would always be a little left in the pipes. The only way to get it was to shut all the intermediate pipes, get a friend to blow at one end and oneself suck at the other end, with the result that a few drops of warm liquid could be obtained'. 'But sometimes, a few minutes before 4 o'clock, a small window would be opened in the passage leading from the classrooms to the library and a servant would hand out cups of water. As soon as some passing boy discovered this, the cry went down the school, "Water, no squash" (meaning, "Don't all rush") and the resulting scramble to get a drink before lessons was terrific' (Macleod). Arthur Benson even records the discovery of 'an old well in one of the fields, closed by a stone disc. We used to let down a blue fluted lotion-bottle into the darkness by a string. The water came up full of black living creatures, leeches and beetles. The fluid was carefully strained, the beetles cast away, and the rest eagerly drunk'. Such remedies bear out his assertion that 'the agonies of thirst on really hot days were almost insupportable'.

6   Games and Leisure

The playing field nearest the school was reserved for cricket. Good pitches were prepared and matches were watched by enthusiastic spectators. The latter did not, however, include Monty James who was observed – on at least one occasion – to be reading his copy of Aristophanes and chuckling over it, oblivious of the match. But even by 1870 the only opponent appears to have been Cheam. 'We just played cricket and football for exercise and fun. Our Eleven played Cheam School once a year, but I do not think we produced anyone very great in the cricket line except, perhaps, Wickham who afterwards played for Oxford' (La Terrière). The first match of which a precise record exists was the one played at Temple Grove on July 25th, 1859, (Waterfield's first summer term). The Temple Grove XI, which included six pupils and five adults, won by six wickets. 'The play of Gyll, only twelve years of age, was much admired: he played in thoroughly cricket-like style, and did not give a chance. He was rewarded by a prize bat after the match, presented by J.E. Vernon Esq.' Gyll's score of 20 in the school's second innings was the highest of the match, byes and wides being otherwise the most notable contributors, though 'Messrs Peatfield and Ommanney each scored 6 by drives out of the ground' (Bell's *Life in London*). No other score has been preserved until the match played on June 2nd, 1875, against Slough (which one takes to mean Hawtrey's). Slough made 106 (Law ma. taking six wickets) and winning by an innings and 37 runs, Law ma., with 15 in the first innings, also being Temple Grove's top scorer. A year later Temple Grove played Charterhouse (presumably an under 15 team), a match in which Edward Grey took two wickets and scored two runs.

In the early days the Cheam match was the highlight of the season. 'Our great opponent at cricket was Cheam, and how keen we were to win', wrote one Old Boy while another recalled, 'one of those perfect summer days, with that lovely hollow sound as bat drives ball to the fence. The school is playing a match against Cheam and all along the cloister wall sit a row of boys cheering on the home team'. But gradually the fixture list increased so that by 1879 Temple Grove was playing 'home and away matches with Eagle House at Wimbledon, Tabor's at Cheam and Hawtrey's at Slough', and sometimes 'Old Boys at Eton brought down a scratch team and, on such occasions, the professional was allowed to help the school'. Equipment was of the most primitive kind and Fred Benson was probably not exaggerating when he credited David Blaize with possessing 'one right-hand batting glove, which he

Temple Grove v. Slough (Hawtreys) 2nd June, 1875

Charterhouse v. Temple Grove, 22nd June, 1876. E. Grey is the future Foreign Secretary

The First Club

The only surviving photograph of a game of domestic cricket (The Second Club c. 1890)

had bought second-hand . . . Stone's glove being the only other one in the school'. 'For matches the First Eleven had a complete set of white flannels with a green elastic belt round the waist; the others merely took off superfluous garments, retaining only what was indispensable' (Byrne).

The other playing fields seem to have had minimal attention paid to them. Evidence for this comes from a number of sources but is best illustrated by the ability of Monty James, apparently on numerous occasions, to lose his spectacles in mid-pitch, 'whereupon the game was suspended and a couple of dozen small boys went down on all-fours to search. The feel of those spectacles triumphantly extracted from the coarse dry grass lives with me still' (Byrne). Macleod caustically speaks of 'the rest of the school being left to fend for themselves in a small field which was used by cattle. The pitches were of the roughest, and there was no on in charge to see that the games were properly played' . . . And even at a later date (around 1898), although the field looks well mown, there is no sign of a master and the fielders' demeanour scarcely radiates enthusiasm (picture p.164). 'On rare occasions a football was produced and the whole school turned out to kick it about, but no one had the foggiest idea about any sort of rules' (Macleod). This verdict is not wholly supported by James Hannay: 'Football was quite informal. We played according to rules of our own which I have never seen used anywhere else. You were allowed to pick up the ball and run with it, as in Rugby, but only if you caught it full pitch or after a single bounce. Otherwise you kicked it as in Association'. Unsurprisingly Monty James, even thirty years later, was unclear as to which code of football was played in his day. It must be remembered that the laws of Association Football were not codified until 1863, nor those of Rugby until 1879, and that the majority of the older public schools retained their own versions of the game for longer.

The map shows an enclosed area of perhaps half an acre with a five's court on one side and the gymnasium in the southern corner. It was bounded on the road side by a high wall which was further extended by netting. This playground had a hard but uneven surface and 'in rainy weather a little pool would form in the centre', making it very tempting to hurl half a brick into the puddle, 'especially if there was someone near enough to get a ducking from the splash'. Here 'we played rounders and hoppy, a game in which the players had to hop when they left their base to attack that of their opponents. They then charged any hopping opponent, and the one who lost his balance had to retire, and so on until one side had lost all its men'. Otherwise the main use of the playground was, by an Orwellian distortion of its name, for 'Extra Drill', 'a gloomy but not altogether ineffective form of punishment which consisted in marching in fours for an hour round and round the playground', supervised by a 'much-scarred drill-sergeant'. The sergeant was also in charge of the gymnasium, described as being 'small and rather dark, floored with deep and dusty cocoanut fibre and fitted with a rope, parallel bars, a horizontal bar and a horse'. The sergeant 'was supposed to be there to instruct us in gymnastics and also to teach us boxing and singlestick, but there was no system and few persevered at these things'. Steuart Hare, however, remembered Sergeant Bell as 'a decent old boy. He was always particularly nice to me as my father was a soldier'. If fives was played no record of it has survived. There was, however, an annual athletic competition, 'but our notions as to training were somewhat crude though very simple. All we did was to plod round and round the playground in our spare time just to get into training. The longest race was the half-mile and it was the only event I won. As a prize I got a set of three gold studs which I wore for many years' (James Pitman).

Close to the playground stood a covered shed and although it was 'already cumbered with playboxes, a sort of hockey was played there with walking sticks on wet days'. It was also the venue for Highcockalorum in which 'sides of eight or more were picked and innings tossed for. The losing side then turned itself into what we will call a human centipede, the first boy supporting himself on an old desk and the others taking support from each other. Then the side that won the toss had to get all its members on to the back of the centipede; no foot might touch the ground and if it did the side was out and then became the centipede. You will understand it was no mean feat for the first boy to jump to the head of the centipede, and I can remember my admiration for one boy who used to take a flying leap, nimbly using his hands on the vertebrae of the centipede as he reached its head, leaving the necessary room behind for the rest of his team' (Yarde-Buller).

'Close to the end of the covered shed there was a large chestnut tree and in the autumn there was much competition in looking for ripe chestnuts; occasionally they were used for missiles in pitched battles fought in the shed, but they were mostly used for the chestnut game. A hole was neatly bored down the centre of the nut and a string or bootlace inserted. Then you would sing out, ''First six, three a stringer'', and in this manner challenge all comers. If you could not break your adversary in six shots then you in turn must hold up your champion while the enemy battered six times at him and so on until one of the two broke into fragments. Should the two strings wind round each other then the opponent had three extra shots. Since the victor took to himself all the scores of defeated chestnuts, he might attain the rank of a 'thousander'. When the conker season was over a few hardened survivors were hidden away for next year'. The keeping of stag beetles (by Fred Benson) has already been mentioned. Tom Pitman enlarges on this by explaining that 'In Richmond Park, in summer, we collected the beetles from oak trees; some of them were enormous creatures and fought like tigers when matched against a champion from another boy's locker'. The only other kind of 'pet' was the silkworm, fed on 'stray lettuces and mulberry leaves from a tree growing near the first form classroom'. Mechanical toys of anything greater than basic simplicity were still in their infancy, but Fred Benson and a friend, occupying the two bottom places in their form, were fascinated by steam engines. 'We got illustrated catalogues from the makers of models, and copied and recopied diagrams of slide-valves, waste pipes and eccentrics with a zeal and accuracy which, if devoted to lessons, must speedily have pulled us out of the humble positions we so contentedly occupied'. 'One of the masters was interested in Electricity. He had a contraption for generating it. This seemed to consist mostly of a large glass wheel and with this he used to charge his Leyden jar. Some of us then formed an incomplete circle with one boy holding the jar. When the boy at the other end touched the knob of the jar we all received a mild shock'. That was the nearest approach to a Science lesson at Temple Grove.

Martin Benson, in an early letter home, refers to another 'Craze', long since forgotten: 'There has been a great Panorama here which consisted in a kind of small windowpane set in wood with two rollers at each side on which several pictures were wound and then one roller was turned and the pictures went on and on'. And, not long afterwards, 'The Panorama mania has broken out again and is very fierce now. A fellow called Hobday who is the best drawer in the school does nothing all his play hours hardly except draw, draw, draw for them'. What strikes one about these and many other 'manias' referred to by former pupils is that they all involved *doing* something – however infantile – rather than sitting back and waiting to be entertained. And if the school authorities may be thought to have lacked the impetus to encourage boys to seek out fresh pursuits, they did at least provide time and opportunity. Of course it was only an exceptional boy like Monty James who would plunge into a study of archaeological, epigraphic and apocryphal matters. One of Arthur Benson's earliest memories of Monty at Temple Grove was of being shown 'an immense list of Roman saints, with mysterious symbols appended'. And the library was sufficiently well stocked for the boy to discover a copy of Curzon's *Visits to Monasteries in the Levant*, which he afterwards believed had 'first inspired me with a curiosity about manuscripts'.

Like the Roman Catholic Church, Victorian headmasters had a very clear idea of what was proper for their pupils to read and what was not. A notice sent out to parents in 1867 makes this very clear: 'Mr Waterfield particularly wishes that inferior Novels should not be introduced into the School. As it is very difficult for him to exercise an effective censorship, he would prefer that no reading books should be sent with boys, except such as are intended for Sunday or devotional use. The School Library, open to all, includes several hundred volumes, among which are many of such works of fiction as are suitable for boys' use'. Censorship, however, was effectively used in the case of *Jane Eyre*. 'When I went to my private school in January, 1862, I bought a copy, bound in yellow cloth at Southampton railway bookstall, and the purchase emptied my slender pocket. It was discovered the next morning by my bed, and handed over to the Headmaster, who sent for me and told me that, as it was not a proper book for a boy to read, I should not be given it again. This piece of Victorian tyranny rankles in my mind to this day' (Prothero). Waterfield, however, *was* justified in extolling the merits of the school library. Its use was a privilege appreciated by hundreds of boys at various times, as Benson's nostalgic memory of it (p.115) shows clearly.

The school is fortunate to have preserved two handsomely bound volumes which record all withdrawals from the library between January, 1869, and

October, 1873. New additions are listed, and also books which boys had failed to return. Interestingly, separate pages are allocated to books taken out by members of staff. From all this we learn that, on average, around ten books a day were borrowed. Fennimore Cooper's *Last of the Mohicans* and Captain Marryat's *Masterman Ready* were apparently the two most popular books, but R.M. Ballantyne, with a new boy's adventure story ready for Christmas each year, was probably the most widely-read author. Books bought for, or presented to, the library in those years include: Southey's *Life of Nelson*, Dean Farrar's *World of School*, Hughes's *Tom Brown's Schooldays*, Clintock's *The Fate of Sir John Franklin*, Layard's *Nineveh and Babylon*, Samuel Smiles's *Character* and Henty's *Out on the Pampas*. There was a strong emphasis on brave deeds, especially those achieved by Britons in remote corners of the globe. But there was also a special compartment of books for Sunday reading where titles like *Tales from Church History, Sacred Allegories* and *The Companions of St Paul* were to be found. The register also records the acquisition of eight volumes of *Good Words for the Young* and nine of *Sunday at Home* in November, 1870. The *Boys Own Paper* (published by the Religious Tract Society) made its first appearance in the very last year of Waterfield's reign. He might, one feels, have approved. Among its early contributors were Jules Verne, Conan Doyle, R.M. Ballantyne, G.A. Henty and Talbot Baines Reed. The passage quoted here seems not untypical.

' "I fear", said Black McTavish quietly, "we are surrounded. Meanwhile, do you, MacIntosh, hand me MacNab's knife. I fear his leg is no longer of service".

It was the work of a moment to amputate MacNab's leg. A substitute was briskly carved from the nearest njama tree by Sandy. MacNab at once leapt up and expressed himself ready to march all day. As he spoke a shower of poisoned spears rattled through the bush and laid three of the little party prostrate.

MacTavish gnawed his lip. '

As regards the staff one has the unworthy feeling that withdrawals were made in the knowlede that Waterfield would undoubtedly scan those pages from time to time and, mentally, award marks of approval. *Pupils of St John* (Mr Bullock), Schiller's *Thirty Years' War* (Mr Ranscher) and Prescott's *Critical and Historical Essays* (Mr Rawlings) would certainly have received a nod of approval, while Mr Browne's choice of *La Grèce Contemporaine* could only have enhanced his prospects, even more so if he had managed to drop a quotation from it – casually – into the conversation. Perhaps one is being cynical. There is at any rate, no reason to question the sincerity of Dame Abbott, an avid reader of *Good Words*. But it gives added point to the little scene with which *David Blaize* opens where Mr Dutton, while ostensibly supervising letter-writing, is actually engrossed in 'a yellow-backed volume of stories by Guy de Maupassant'. It is, inevitably, discovered – and immediately destroyed – by the Headmaster.

For the senior boys almost the best hour of the week came on Sunday evening when Waterfield read aloud to them in the drawing-room. 'He would tell us all to make ourselves comfortable, and for the rest of the hour entranced us with *The Pilgrim's Progress*. His delightful voice melodiously rose and fell; he asked us no inconveneint question to probe the measure of our attention; his object, in which he strikingly succeeded, was to let us hear magnificent English magnificently read, and to leave us to gather our own honey' (Fred Benson). For Monty James the pleasure was enhanced when the episode of Doubting Castle was reached. '[We] were agog to see whether he would give us the author's words and say, "That lock went damnable hard". When he came to the point, he checked slightly, and said "desperate hard".'

The most envied privilege attainable at Temple Grove was to be 'on the snore'. 'At the end of the big schoolroom hung a board containing a roll of the boys' names, divided into classes, and placed in order of merit according to the last examination. Under the first three (some authorities say two) names in each class was a thick red line. A red star at the foot of the roll indicated that this meant "Honoris Causa" [to the boys, "on the snore"] and carried with it much honour' (Tom Pitman). Not only could one go into those parts of the grounds forbidden to the common herd, which included 'the Dell', but one could also go outside the grounds with a friend and wander over Sheen Common or into Richmond Park. 'As long as one was back by the proper time, one could go where one liked. I and others used to go away for miles over Wimbledon Common and hunt for bugs of all sorts, and we made a really good collection, the battered remains of which I have even now. In Richmond Park our walks frequently took us to the big ponds, where we caught fish and also collected birds' nests to our hearts' content when the keepers were not looking'. On one such expedition his companion was Charles Hooker, one of the four sons of Sir Joseph Hooker, a close associate of Darwin, who had recently become Director of Kew Gardens. 'I remember meeting him [Sir Joseph] and being taken to visit Professor Owen who lived at a house on the edge of the Park' (La Terrière).

But of course the greatest delight was to go to Dame Abbott and draw sixpence of one's pocket money. 'At the end of my first month my name appeared at the top of my class with a red X opposite it, and in due course I found myself in Richmond on the Saturday. There I met Edward Grey who stopped and congratulated me on having got "on the snore", and he proceeded to double my ready cash by tipping me sixpence, he being in a position to do so as the first class were allowed to keep their own pocket money' (James Pitman). 'The way [to Richmond] led past the Church [Mortlake] and through a large area of market gardens . . . to the Richmond Road and so to the High Street . . . the usual limit being the Maid of Honour Shop. The problem was to appease one's own hunger with that sum, and also bring home sufficient to appease the hungry crowd which awaited one's return' (Macleod). However, 'It was wonderful what one could get for 6d. There was always a large new seed cake waiting on a Saturday afternoon at a baker's shop on the outskirts of Richmond – Baldwin's, I think. One penny for a slice of cake and another for a bottle of stone ginger beer to wash it down with. "Black Jack" (sticks of toffee rolled up in paper) was a favourite sweetmeat which had the advantage of being cheap. There was also a shop in East Sheen, called Gaunts, which would take stamps in payment but I think it was only when in quarantine for scarlet fever in the sick house opposite Gaunt's that I made that discovery' (James Pitman). Tom Pitman, and others, favoured 'Voysey's, the fruit shop, where one could get full value when bulk was the chief consideration – dates, brazil nuts and raisins' being specially favoured. 'Then there was Black Jack and Privatum, the latter perhaps the most popular of all because, under the circumstances in which it was disposed of, it could hardly be given away. A pennyworth would last one nearly a week; the pink and white stodgy cream was placed in a folded sheet of paper and squashed flat. You could then open the paper and start to lick and, like the 'Widow's Cruse', it never seemd to get any smaller. If we could not get as far as Richmond there was a shop on the London – Richmond road near Mortlake called Easton's. Here we bought Callard and Bowser's butterscotch and I forget what else'. La Terrière also testifies to the lure of Black Jack and another favourite, known as 'Baton de Dame' but he also, on occasion, purchased 'shrimps and prawns and all sorts of sweets and cakes which we sold to our less fortunate scholfellows at, I fear, double the price they cost'.

In the days when nearly all transport was horse-drawn and a railway excursion was still a considerable 'event', school outings were almost unknown. Slough (to play cricket against Hawtrey's), some fifteen miles away, represented a considerable journey involving an early start. But 'the great event of the summer term was Waterfield's birthday . . . [which was] kept as a whole holiday [and] was celebrated by everybody being taken [in a flotilla of horse-drawn brakes] to the Crystal Palace for the day and furnished with half-a-crown [say £5 today] to spend as he pleased . . . With what rapture I beheld that amazing edifice glittering in the sun, and went through its Palm Court and its Egyptian Court and its Assyrian Court, and beheld all that the Prince Consort had done to educate the love of beauty in these barbarous islanders. All day I wandered enchanted, and laid out most of the half-crown in a glass paper-weight with a picture of the Crystal Palace below, and the remainder in a small nickel ornament in the shape of an ewer, undoubtedly made in Germany'. Rowland Prothero, taken there by the German master a good many years ealier, was less impressed. 'The visit, which promised so much, proved a disappointment. Herr Fliedner's main interest was machinery, and to me machines made no appeal. We hurried rapidly through the Picture Gallery . . . [and] spent hours in examining machine after machine with the microscopic thoroughness of a mechanical enthusiast'.

La Terrière records seeing his first Boat Race while a boy at Temple Grove 'from a place Waterfield took for us, but although actual attendance was later discontinued (possibly because the crowds around the finish at Mortlake became too great), the school was divided into Oxford and Cambridge, few remaining neutral'.

There is no mention anywhere of a firework display at the school but Fred Benson was fortunate enough to be invited 'On November 5th each year to a children's party given by Princess Mary, Duchess of Teck, at the White Lodge, Richmond Park, and there was an immense tea followed by fireworks in the garden. There we were given squibs and told to be sure to throw them away as soon as they burned low before the explosion came at the end. On one of these occasions the Duke of Teck, wanting a light for his cigar, told me to give him my squib for he had no matches. I told him it was already burning low, but he said "Wass?" rather alarmingly, and so I handed it to him. He had just applied the burning end of it to his cigar when the explosion came, and his face and hair were covered with sparks and he danced about, and said sonorous things in German, and I gathered that he was vexed'.

Fred's invitation is accounted for by the fact that the Duke's two elder sons, Prince Adolphus and Prince Francis, entered Temple Grove (as day boys) in September, 1879. They 'used to ride down every day from the White Lodge'.

The boys were granted one exeat each term. 'We were allowed away for one weekend to our 'people' at home. There was no fixed date . . . it could be taken when it suited our parents. Our exeats were always to London where my father had rooms at Fenton's Hotel. There was always a theatre on the Saturday night and it was on one of those occasions that I had my first introduction to Gilbert and Sullivan in their earliest effort. On the Sunday afternoon we invariably went to the Zoo and I made my first acquaintance with some of the animals which I was to see more of on the plains of Africa and India. I remember my young brother writing of his first visit, "The elephants made a noise like a thousand Papas blowing their noses". We knew so well the sound of the paternal nasal organ that the idea was most expressive' (Tom Pitman).

In addition to the exeat there might be one other occasion when one's people came down to take one out. 'My father and old Mr Pascoe Grenfell used to come down together – they were old business friends – and it was on the latter's recommendation that we had all in turn been sent to Temple Grove as had the Grenfell family. I think that at this time there had already been six Grenfells at the school, and I was the sixth representative of our family, but there were more to follow from 'both sources'. These termly 'outings' consisted of a visit to the Maid of Honour shop, where a supply was given us to take back to school, followed by an expedition to the Star and Garter. I shall never forget those dinners: whitebait, devilled whitebait, "more devilled whitebait", as the waiter ran round the table with his silver salver, shovelling more on to your plate each time (and that was only the first course). I remember on one of these occasions a telegram arriving during dinner, announcing the birth of the Grenfell twins, Riversdale and Francis' (Tom Pitman).

A good many boys lived too far away for such visits to be possible. Bulmer La Terrière usually spent his exeats with the Candys (Mrs Waterfield's family). 'I saw a lot of May Candy on these occasions and at about the age of twelve I was madly in love with her; and she, though of course I must not say so, led me on all she knew, as is the way of girls, and encouraged me in all sorts of pranks . . . Waterfield was always a man of his word and once, when the Candys kept me on a day after my leave expired, his butler was sent (to Chipstead) to say that if I did not return at once with him I could stay away altogether. I was led back ignominiously, leaving the young lady in tears!'

7 The Ethical and Moral Code

The time has come to attempt an assessment of the ethical code that existed at Temple Grove in Waterfield's day and, no doubt, in the majority of preparatory schools. From the little evidence available (and its scantiness is itself a pointer) it would appear that organised religion (attendance at morning and evening prayers, the Sunday morning service and the learning of Creed and Catechism) was not an important factor in moulding the characters of the boys or influencing their moral judgment. Indeed lengthy and boring sermons and the weekly learning of the Catechism were probably what would now be described as counter-productive. David Blaize, on a visit to Marchester (Marlborough) to sit for a scholarship, is reproved when he remarks to a boy already at the school, "I suppose chapel's pretty good rot", a remark which is at least indicative of underlying assumptions.

Conformity was all-important. No boy wished to stand out from the crowd for the wrong reasons. It was acceptable to make runs and take wickets in a cricket match, especially against Cheam, but one must guard against showing 'side'. A boy who was caned or 'swished' might, temporarily, be a hero, especially if he took it well. But the offence itself needed to gain the approval of his fellows: failure to prepare one's Latin, talking after 'lights out' or, better still, ragging 'Old Dubs', or removing the list from the Library (p157), earned enviable fame. Cribbing, particularly to save oneself from a swishing, was perfectly acceptable, but 'chousing', which implied swindling or taking unfair advantage of, someone else, was not. And although one might lie (as we have all done) in an attempt to avoid drastic punishment, one did not do so to one's fellows. Fred Benson had got into the way of magnifying several minor episodes into dramatic events (a rough Channel crossing became a near-shipwreck, a heavy snowfall gave rise to a hazardous coach journey along icy roads) which he related in the dormitory. But 'great was my horror when an implacable enemy handed me one morning a scrap of paper . . . headed 'BENSON'S LIES', and there below, neatly summarised, were all those stories which I thought had been listened to with such respectful envy'. The genuine story-teller, of course, was welcomed. Monty James fondly remembered 'stories which certain boys used to tell when we were out for walks or marching round and round the playground at extra drill. They went on for days, and were of thrilling interest'.

Waterfield may not have been quite as severe with a detected liar as Dr Arnold had been, but it was always a heinous offence; equally he could be amazingly forgiving of a crime if the offender owned up. The little episode in *David Blaize* where the Head refrains from chastising the boy, thanks him for telling the truth and ends the interview by saying, "I see I have made you late for breakfast so come and have breakfast with me", suggests a personal experience recalled. And Waterfield himself respected the schoolboy code concerning sneaking. 'While we were preparing our lessons in the large schoolroom', recalled William Young, 'a boy hit me in the eye with an apple core. Wiping that eye with my handkerchief I was spotted and called up before the master. On my refusing to give the name of the boy who threw it I was ordered to stand upon a form until I did. After a considerable lapse of time, I was again asked to name the offender, and on another refusal I was asked why. To which I replied that my mother did not approve of 'informers'. The master was nonplussed, and the incident closed'.

In normal circumstances one avoided mentioning either of one's parents in term-time. One might say, "My people are coming down for the exeat" and, if necessary, refer briefly to one's 'Pater' and 'Mater' ('Papa' and 'Mama' when at home) but details of home life must be eschewed at all costs. 'When my father once suggested that we might wish to have some boys to stay with us in the holidays, I regarded the suggestion with incredulous horror. To admit boys from school to the sanctities of home, where they would see that one had *sisters*, hear one called by one's Christian name, carry back stories of our family ways, was a thought of deep tragedy, like casting what was holy to dogs' (Arthur Benson). William and Patrick Young, arriving together at Temple Grove in 1868, utterly unaware of the schoolboy code, 'fools that we were, we let them know the names of our sisters . . . [and] all through that term we were mockingly asked how Annie was, or the colour of Mya's or Grace's eyes'. Their crime was compounded by the fact that Patrick was attired in 'a brown velveteen suit, of which he was so proud'. James Pitman, too, 'appeared in a somewhat babyish suit of clothes on my first Sunday' and came in for unmerciful derision which was stopped by the arrival of Edward Grey, who chanced to have accompanied the new boy on his train journey from the north. Walter Yarde-Buller was another victim. 'He used to get horribly bullied and teased because he had some knickerbockers made of corduroy; and the boys would have it that his father was a working-man and couldn't afford anything else!' (La Terrière).

In *David Blaize* E. F. Benson not only makes David's father a laughing stock by appearing in full clerical gear ('shovel-hat and odd, black, wrinkled gaiters') but piles on the agony by causing him to bowl – hopelessly wide – in the nets, preach an excessively long sermon in chapel, address his son as "David" and not only, at lunch, consume with relish his portion of 'resurrection-bolly' (firmly believed to be composed of all the scraps left on the plates during the last week) but ask for a second helping. Mercifully he redeemed himself by requesting an extra half-holiday for the school.

With reference to Christian names, a pupil of Allen's time remembered that he had made friends with another new boy called Harland. 'It was a terrible disgrace if anyone at school found out your Christian name. But it was the custom for great friends to reveal to each other their Christian names under an oath of secrecy'. Brothers avoided the obvious trap by referring to 'My major' or 'My minor' as required. New boys arrived at the school a day before the rest, and so enjoyed 24 hours of comparative calm. 'The next day our troubles began, and it was a really miserable day. There were over a hundred boys in the school and they came pouring in all day. We new boys sat on forms feeling utterly dejected while a string of boys passed us, each one asking, "What's your name?" "What is your father?" (meaning, "What sort of position does he hold?").' A refinement of the ordeal is recorded by a boy who arrived one November day (and from Dublin); 'To arrive in the middle of term, when even the most timorously fresh are no longer capable of making a single one of those foolish and often costly blunders, all too easily achieved, in the day's curriculum; to be the solitary new 'snob', meeting through each successive break, the same inconsequent but relentless fusillade of queries as to name, age, domicile, place of birth and other haphazard details; alone to receive at every unsuspecting turn the taunts and barbs, the slings and arrows of ridicule at some fresh and wholly innocent breach of hallowed code or custom [intensified by the pleasurable baiting of his Dublin brogue] – these, and the like, in their sum, made the days at first seem long and ever full of some new lurking danger' (Dermot Freyer).

Some boys blush more easily than others. The weakness is quickly apparent and full advantage taken of it. 'My tormentors discovered that it was only necessary to call attention to me and say, "Don't smoke", to make me blush and they would sometimes pretend to toast a bit of bread at my cheeks' (James Pitman). Harmless enough, one may feel, and even a necessary introduction to the harsh, unfeeling realities of the outside world. But teasing quickly leads on to more unpleasant forms of bullying, and to the kind of victimisation, terrifying for its target, recounted by William Young: 'A Dublin boy, Robert Harrison, in his first term at the school had been placed under my care. One evening when we boys were congregated round the big fireplace at the end of the room, some mischievous spirits found out that . . . the footman had forgotten to padlock the huge steel construction round the fireplace to the wall. Seizing my protegé with cries of "We'll roast you", they thrust the unfortunate boy between this construction and the fire. I don't for one moment believe that they had any intention of more than frightening him, but his shrieks were so appalling that I had to hurl myself single-handed on the crowd and rescue him'. And Claude Palmer testified: 'The bullying was very bad and the elder boys half frightened the new boys out of their wits with constant tormenting; I remember a boy named Sharpe who was a perfect fiend'.

If teasing and even tormenting were acceptable, outright bullying certainly was not and Waterfield seems to have been more successful than a good many prep school heads in countering it. One of his general admonitions to the school has already been described (p.135). Its sequel is related by Arthur

Benson: 'The boys ranged themselves down the big schoolroom and the culprit had to run the gauntlet. I can see his ugly, tear-stained face coming slowly along among a shower of blows. I joined in with a will, I remember, though I hardly knew what he had done. He came out a truly pitiable object'. Afterwards Waterfield summoned the senior boys to the study: 'I hear you have given T– a school licking. I think he deserved it; but I won't have that kind of thing done again . . . He has had his punishment and is not to be molested or teased about it or reminded of it in any way. I count upon you all to help him to do better, and I shall be very proud of the school if this is carried out'. James Pitman adds, 'A school licking . . . was a very cruel way of meting out schoolboy justice [but] they were seldom resorted to, only one, or possibly two, in my four years at Temple Grove'. He also remembers being 'let in for' a school fight, his opponent being Frank Waterfield, the Head's son. 'There was never any ill feeling between the two of us, before or after. However, Providence came to my rescue, for we had hardly begun when my opponent slipped and fell against the steel cage that protected the fireplace and cut his cheek rather badly. This brought the fight to a close. Frank had to confess to his father how he came to get hurt, but was not asked, and did not say, with whom he had been fighting'. James Hannay is more explicit over the question of fights between individual boys: 'They were of two kinds: The 'face fight', a kind of duel à l'outrance in which you were allowed to damage your opponent as best you could by hitting him on the nose, eyes, mouth or anywhere else you chose; and a milder kind of duel in which you were allowed only to hit on the body, though of course never below the belt. These fights took place in the big school before prayers and were the only things which ever got us out of bed before the last possible moment. On the morning of a fight we were all dressed and downstairs half an hour before prayer time'.

Whether Waterfield was wise not to appoint prefects must be a matter of opinion. Boys know better than staff what is really going on in a school and the inculcation of a sense of responsibility at the age of 12 or 13 is not a bad thing. But power can go to a boy's head and a prefect's judgment as to what deserves punishment (and how much) and what is better left unpunished is necessarily unreliable.

Ethics are not too difficult to assess: the moral tone of a school, if such a thing can ever be said to exist, is a very different matter. Two judges may arrive at conflicting verdicts on much the same evidence about a school at a particular period, especially when a good deal of the evidence is necessarily obscure and the witnesses are no longer there to be cross-examined. In the vocabulary of many Victorian headmasters lying, swearing and disobedience were sins of such extreme culpability that really serious offences were beyond description and could, in any case, only be hinted at, never openly named and certainly never rationally discussed. When Edward Benson, then headmaster of Wellington, wrote in a letter to his eldest son, Martin, 'Be sure you let me know if there is the *least* budding out again of the Abominable', *we* would assume that he was referring to unnatural sexual activity between males, but that assumption would presuppose that Martin had referred to some such practice in an earlier letter, which seems highly unlikely. Martin could well have been shocked by bad language or possibly by a lavatorial joke. His brother, Arthur, declares, 'I never heard a single word of improper or indecent talk, nor saw an indecent action of any kind, and I left school after two years as entirely innocent and ignorant as I had entered it'. William Palmer confirms this when he writes, 'The teaching was good and the school was clean morally' and the absence of the topic of sex from most of the published and unpublished reminiscences of the pupils of Waterfield's day is, if only in a negative way, an indication that, as David Newsome puts it, 'The tone of the school seems to have been good and wholesome'. The one episode on record where the evidence is reasonably clear occurred in 1878 and is described in *Our Family Affairs*:

'There occurred a scene which I still look back on as among the most awful I have ever witnessed. Two boys, one high in the school, a merry handsome creature, the other quite a small boy, suddenly disappeared. They were in their places at breakfast, but . . . were sent for by Waterfield and at school that morning their places were empty . . . Curious whisperings were about of which I could not grasp the import. Next morning there came a sudden order that all the school should be assembled, and we crowded into the big schoolroom. Presently Waterfield entered with his cap and gown on, followed by the two missing boys. He took his place at his desk, and motioned them to stand out in the middle of the room. There was a long silence. Then Waterfield began to speak in a low voice that grew gradually louder. He told us all to look at them, which we did. He then told us that they had brought utter ruin and disgrace on themselves, that no public school would receive them, and that they had broken their parents' hearts. They were not going to stop an hour longer amongst us, for their presence was filthy and contaminating. They were publicly expelled and would now go back to the homes on which they had brought disgrace'.

The sequel is almost equally significant. Benson himself was summoned to the study. 'Waterfield was sitting at his table and he was crying. He indicated to me that I was to sit down, which I did. Then he blew his nose with an awful explosion of sound, and came with his rocking walk across to the chimney-piece.

"I want to ask you a question", he said, "Do you understand why those two boys were sent away?"

"No, Sir", said I.

His voice choked for a moment.

"I am very glad to hear it," he said, "I thank God for that. You may go".'

Young Benson departed, totally baffled. Later he recalled having been invited not long before the expulsion by the elder boy concerned for a stroll round the wooded grounds and being asked, or rather half-asked, questions like, "I say, do you ever – ?" And then Waterfield had appeared, took them into the greenhouse and gave them some grapes. The school, of course, was forbidden to talk about the subject at all, 'which gave an additional zest to discussion. Some knew a great deal, some knew a little, some knew nothing'. And he ends by asking, 'In heaven's name, why could we not all have been given clear lessons in natural history?' (by which, one takes it, he meant human biology). But, in Victorian times, that was to ask for the impossible. Boys, as well as girls, of the upper classes were to be kept in ignorance of all sexual matters for as long as possible. No one took the responsibility for making it all clear.

In *David Blaize* Benson transposed the above interview to David's last day at Helmsworth. ' "David," he said, "You are on the point of going out into the bigger life of a public school", and he then recounted several of the boy's minor offences which he fondly imagined had gone undetected. "But you will find that there are worse things than smoking, worse even than stealing and that many quite good chaps, as you would say, don't think there is any harm in them. Do you know what I mean?"

David looked up in quite genuine bewilderment.

"No Sir," he said.

"Thank God for it, then", said the Head. "You don't understand me now, but you will. And when you do understand, try to remember, for my sake . . . or for your own sake . . . or for God's sake . . . that there are worse things than stealing. Things that damn the soul, David. And now, forget all I have said till the time comes for you to remember it". '

And that, no doubt, or something like it, is how most prep school boys were equipped with the facts of life to face the world of adolescence and the public school.

Impressions of schooldays are as varied as the characters of those who experience them. Arthur Benson regarded Temple Grove, from first to last, with aversion: 'I used to awake on summer mornings . . . [and] count the days before the holidays . . . and then dreamily reflect over the tiresome formal day before me with its dull work, its frozen intercourse, its unappetising meals . . . I think I should have responded eagerly to the influence of anyone who could have shown that he was personally interested in me. But I cannot remember that any piece of work that I ever did was praised and no touch of human emotion was ever exhibited to me by any person in the place, except by the beloved Louisa (Cox)'. Bulmer La Terrière, on the other hand, entered zestfully on all the activities, legal and otherwise, which the school had to offer and thoroughly enjoyed most of his time there. Members of large families, too, like the Pitmans and the Grenfells, generally looked back on their schooldays with affection, all except the eldest having had the way paved for them and thus being able to avoid frightful gaffes and the inevitable teasing of bewildered new boys. Norman Macleod came, alone, to Temple Grove from his home in the north of Scotland and viewed the school, and especially its teaching methods, through anything but rose-coloured spectacles. Monty James, easy-going, well-liked and very clever, adapted without difficulty to the system. Fred Benson, too, 'plunged headlong into this riot of school life' and although, after the first year, his academic progress gave both his father and the Headmaster cause for grave concern, he himself concentrated on getting as much enjoyment out of school life as it had to offer with a minimum of exertion.

But, 'much as we enjoyed (and pretended to hate) our time at school, we shared the greatest of all pleasures and that was, looking forward to the holidays. As the time grew near we scratched off the days on a specially prepared chart and that musical spirit, a sure sign of joy, entered into us as we sang, "This time next week, where shall we be? Not in this Acadamee! No more Latin, no more Greek, no more cane to make me squeak, No more spiders in my tea, making googly eyes at me".' If the original author of that ditty had received his copyright dues he would have died a *very* rich man. Then there were the endless schoolboy jokes and riddles: "Last week but one, take it all in fun"; "Last week, don't squeak", entitled one to thump or pinch one's neighbour and avoid retaliation. "Why is a tie like a telescope?" "Because it pulls out", as one gave it a sharp tug. Tradition, rather than wit, was the guiding principle, and showed how eagerly end-of-term was awaited. 'We Scotch boys', wrote Tom Pitman, 'stole a march on our English companions for we were allowed to decamp the night before the great day. About seven o'clock in the evening a specially chartered Great Northern Railway Omnibus [horse-drawn obviously] drew up outside the gate and old Montague the butler tucked us all inside, with the luggage on the top and Montague himself on the box. Shortly after starting we stopped at Eastons in Mortlake to buy food and a supply of peas for our pea-shooters. There were some eight or nine of us and we sang throughout the journey to King's Cross, *John Brown's Body, Phairson swore a feud, Green Grow the Rushes, O*, and when we had come to the end of our repertoire we started again on *Omne Bene*. We were too fully occupied to follow the route we were taking but the glare of lights as we passed down the Marylebone and Euston Roads is still fresh in my memory'.

Little is known for certain concerning the origins of *Omne Bene*, though clearly it has a long history. Dr Orme considers it 'quite similar in principle' to late medieval school songs he has come across, but he doesn't *feel* that it is pre-Reformation, more probably 17th century. This would accord with the theory that a link exists with the Winchester College song, *Domum*. James Sabben-Clare, who has investigated the origins of *Domum*, surmises that it may have been written when boys from Winchester were lodged temporarily on a farm at Crawley (Hants.) to escape the plague then (1665) raging, but the actual tune was composed (by John Ridding) later, not till after 1681. It seems at least possible that adaptations of *Domum* (accompanied by an earlier and simpler tune) reached schools in contact with Winchester and that increasingly degenerate versions were passed on to other schools. Iona Opie points out that the song which begins 'Amo, amas, I love a lass' (in *The Agreeable Surprise* by the Irish playwright, John O'Keeffe – 1781) has for its chorus:

Home for the Christmas holidays by R. Seymour (1836)

TEMPLE GROVE SONG

Here's to the new boy of bashful eight years, Here's to the old one of forty
Here's to the boy who has always done well, And here's to the one who is naughty

CHORUS:
For whether they're bad or whether they're good,
or whether they shouldn't or whether they should
We'll hope for the best and we find as a rule,
They some day do honour to Temple Grove School.

Here's to the boy who has always a smile, Here's to the one who is glum, Sir.
Here's to the Pickle whose tongue's never still, And here's to the one who is *mum*, Sir
Chorus
Here's to the boy who a soldier would be, To follow Her Majesty's will, Sir,
He can't start too soon to prepare for the line, And that's why he's always on drill, Sir.
Chorus
Here's to the boy who at work, or at play Will struggle with main and with might, Sir
To follow what's manly and never give in On the side which he knows is the Right, Sir.
Chorus
Here's to the boys who at football last week Were licked but they'll never give in, Sir
They must bear with misfortune as well as success They'll play them again and they'll
win, Sir *Chorus*
Here's to the Scholars at Eton last week Who distinguished themselves at Exams, Sir
'Twas good honest work that accomplished the trick, We don't owe success to a cram, Sir.
Chorus
Here's to the Master, whose excellent sway Has made us the boys you have seen, Sir
For many long years the King of his day He has won our regard at East Sheen, Sir.
Chorus
Here's to the others, his Colleagues at work Of such the world has but a few, Sir.
Then, Temple Grove boys! make a jolly good noise, Three Cheers for our friends, the True
Blues, Sir. Hip-Hip-Hurrah!

Rorum, corum sunt Divorum!
Harum, scarum, Divo!
Tag rag, merry derry, periwig and hatband!
Hic hoc horum Genitive!

At all events the first verse of *Omne Bene* appears at the start of the chapter entitled 'The Stage Coach' in Washington Irving's *Sketch Book* (published in 1820) and there described as an 'Old Holiday School Song'. The chapter portrays 'three fine, rosy-cheeked schoolboys . . . returning home for the holidays in high glee . . . it was delightful to hear the gigantic plans of the little rogues . . . during their six weeks' emancipation from the abhorred thraldom of book, birch and pedagogue.' One would dearly love to know where Irving first heard the words.

By Waterfield's day that version of *Omne Bene* had become the authorised one at Temple Grove and has been sung – or roared – at the end of almost every term since. The generally accepted rendering, and a valiant attempt by Reginald Waterfield to 'translate' it, are given opposite.

It would be absurd to ponder too deeply over a schoolboy jingle which expresses his delight in being free of the tyranny of Latin for a few weeks. Readers who have studied Latin will, however, remember the simple pleasure of chanting various nouns, adjectives and pronouns and will recall that 'Hic, haec, hoc (meaning 'this') is practically a nonsense rhyme in itself. Although the Latin for 'home' is 'domus', the accusative case, 'domum', without a preposition signifies 'motion towards', just as we say, "I'm going home". Thus 'dulce domum' may be taken as short for 'dulce est ire domum' (It's wonderful to be going home) and 'domum felix' as short for 'ire domum felix sum' (I'm happy to be going home').

The Temple Grove song could well have been the work of Jack Hawtrey (p. 206), in which case the Master in verse 7 would be H.B. Allen. Football matches were more frequent by then but it is not possible to pinpoint a date. The anthem was almost certainly not composed until after the opening of the school chapel in 1891. It may even have been written by Edgar himself with help from his daughters.

Footnote: DAVID BLAIZE:
Several references have been made in the foregoing chapter to *David Blaize*. The book was written by E.F. Benson in 1916 and, as Brian Masters comments, 'It was an entirely new departure for Fred, and an (almost) final break with those talking robots who prattle away to no effect in so many of his early novels'. Although the reader is likely to remember it chiefly for the chapters on 'Marchester', three-eighths of the book are in fact given up to an account of David's last year at 'Helmsworth'. There was never much secret about the identity of the two schools but it was not till near the end of his life that Fred stated unequivocally, 'The scene and the general mode are those of my own private and public schools, Temple Grove and Marlborough, with a certain dash of Winchester thrown in with regard to certain members of the staff'. And there can be no doubt that in Mr Acland, he has drawn an accurate portrait of O.C. Waterfield:
   'He was a tall, grizzle-bearded man, lean and wiry, with cold grey eyes. To the boys he appeared of gigantic height, and was tragedy and fate personified. Terror encompassed him: he scintillated with it, as radium scintillates and is unconsumed. But he scintillated also with splendour: he was probably omniscient, and of his omnipotence there was no doubt whatever . . . Though there was not a boy in the school who did not dread his displeasure above everything else in the world, there was none who would not have taken an infinity of trouble to be sure of winning his approbation'.
   The book, appearing in the middle of the First World War, was an immediate success and one officer on the Western Front wrote describing it as 'the best school story yet written bar none' and said that a copy was being passed round among the soldiers in the trenches. As well as giving an immediately recognisable picture of life at prep and public school, Fred unashamedly weaves a story of the romantic friendship between two adolescent boys in and out of the plot, a daring theme for those days, although – as Brian Masters says – 'it contains nothing overtly erotic'. Arthur Benson, who was allowed to read the manuscript before publication, was distinctly unhappy about several passages and advised Fred to leave them out. Fortunately Fred took no notice of his brother's advice, and it is a tribute to the book's realism and its sincerity of purpose that Fred later confessed he had had more letters about *David Blaize* than about any other book he had written.

OMNE BENE

Omne bene, sine poena	All, all is well! No penal bell!
Tempus est ludendi	It is the time for play
Venit hora absque mora	Now comes the day, without delay
Libros deponendi	For putting lesson books away

Quomodo vales, mi sodales	How goes with you, my comrades true?
Visne edere pomum?	Would you an apple like to chew?
Si nonvis, mirabilis	If not, you *are* a 'Wunderbar'!
Dulce adire domum	Its jolly having home in view.

Domum, domum, dulce domum	Home, home, sweet home
Domum, domum felix	Home, happy home
Domum, domum, dulce domum	Home, home, sweet home
Domum, domum felix	Home, happy home

Horum scorum sancti morum	These silly old rules are sacred – pooh!
Harum scarum ibo (or 'scribo')	These stinking fish I offer you
Rag tag, willy willy wag	Rag Tag, willy willy wag
Hic haec hoc redibo (or 'et ibo')	Hic haec hoc, but I'll be back.

Jolly good song and jolly well sung,
Jolly companions everyone
Holla, boys, holla, boys, this is the day
Holla, boys, holla, boys, hip hip hooray.

7 *Edgar and Allen 1880-1902*

'In the early part of 1879 Waterfield fell ill, and had to go abroad to convalesce, so that during the summer term Edgar was in charge', Norman Macleod recalls. There is no indication, here or anywhere else, of the nature of the illness. Lengthy recuperations were to be expected in those days and it seems highly unlikely that his wife and family would have accompanied him. Whether such a serious illness may have hastened Waterfield's intention to retire – he was still only 48 – or whether he had already marked 1880 as the year to do so is unclear. The 17 year length of his agreement with Edgar, signed in 1863, makes the latter at least likely.

The school photograph, in all probability, was taken in the final month of Waterfield's 21 year reign. The benevolent old man has Aubrey sitting on his knees and Mrs Waterfield standing behind him. An unidentified woman has Maisie sitting on her lap and Alice on the ground in front of her. The faces are nearly all remarkably clear but it has been possible to put names to no more than two dozen of the boys. Tom Pitman (sitting between Mr Starkie's knees) wondered, many years later, what had become of them all. Relatively few of *that* generation were killed in the First War since even the youngest would have been in their 40s when it broke out, but several reached high command. 'With Tom Brand, who was then captain of the Eleven, I shared a dug-out on the edge of Bourlon Wood at the battle of Cambrai . . . it is rather a strange coincidence that two very small boys sitting at Mr Waterfield's feet side by side are Mullens minor and myself. We commanded the First and Second Cavalry Divisions respectively and so fought side by side throughout the whole war in France'. Elsewhere he says, 'but the war had a lighter side on occasion. One day in 1918 the Division was resting in billets and a horse show had been got up for the amusement of all. So dense was the crowd of soldiers that it was difficult to see. Some of the men, seeing the red bands on our hats, brought out a form [bench] for us to stand on. Two other generals got up, one on each side of me. "Haven't done this for a long time", said one of them, "It used to be a punishment when I was at school". "Where was that?" said I. There was a similarity about the two replies for they both said, "Temple Grove". I looked at them and recognition came at once: Shute and Wellesley. I could almost hear a faraway voice saying, "Stand on the form, you idle boy". '

Waterfield's later career does not properly concern us and will be only briefly sketched in. The family after living for a time in Guernsey moved into the country near Canterbury and bought Nackington House. He became a director of the Imperial Ottoman Bank, (on one occasion he was entrusted by the bank 'to undertake a delicate enterprise in the Near East because he was the

The oldest surviving photograph of the whole school is the one taken in O.C. Waterfield's last term (1880)

A photograph of O.C. Waterfield at about the age of 50

School Photograph 1880 *(opposite page)*

O.C. Waterfield sits in the centre with Aubrey on his knee. To the right of him is Margaret Waterfield, half-facing Dame Abbott, with Maisie Waterfield on the knee of an unidentified woman. Directly behind Maisie is 'Poucher' Neumann and on the far right (wearing a boater, tilted) is a figure identified as H.A.P., who *could* be Mr Prior. Mr Daubeny (in a bowler) stands in front of an open window and a little to the left of him is Charles Geoghegan (bare-headed). Mrs Waterfield stands behind her husband. To the left of O.C.W. are Mr Starkie, Mr Morgan and two unidentified masters (in bowlers), with Mr de Costa standing between two of them.

About thirty of the boys have been identified with some certainty. Of those mentioned in the text, four are sitting on the ground in the middle: Leese N. (pale suit and boater), Leese W., Pitman T. and Mullens R. (in front of O.C.W.). In front of Geoghegan is Mordaunt G., with Richardson W. immediately to the right of him. Well over on the left, in a pale suit, stands Hare S.

sort of man to whom no door-keeper would dare deny admittance') and of several railway companies and chairman of the Nizam of Hyderabad's State Railway. His death occurred on August 24th 1898. 'He had come down to Windsor on business and was going away by train when he saw an old soldier on the platform and, sitting on one of the station benches, offered him a cigar. A moment later he leaned back and died almost instantaneously from a sudden heart attack . . . He had nothing on him that could bring about identification . . . till the mention by a police inspector of the initials on his cigar-case caused him to be recognised' (Arthur Benson). Mrs Waterfield lived on at Nackington until her own death sixteen years later.

Joseph Haythorne Edgar, Waterfield's successor, is a difficult character to assess. As a young man he had served in the army before becoming ordained, after which he appears to have combined an undemanding curacy in Putney with a variety of other posts, lecturing at King's College, teaching at the School of Mines and coaching pupils for Army exams. The circumstances in which he first met Waterfield are not known but in 1863 he came to Temple Grove, giving up all his other appointments except that at the School of Mines. He later took charge of Wedderlie where – as the school got bigger – a number of the younger boys were to board and, as we have seen, he entered into an agreement with Waterfield on extremely favourable terms to himself, terms which virtually made certain that he would eventually succeed as Headmaster. Whether the letter of Nov., 1875, from W.P. Pinckney already quoted (p.140) suggests that Waterfield was having second thoughts or whether the Pinckney brothers guessed that the succession had already been decided on is unclear. In the event of contrary evidence we must conclude that Waterfield and Edgar worked harmoniously together. What is *not* clear is exactly what duties Edgar performed, other than the supervision of Wedderlie. We know that he had a study beyond the big schoolroom, and that he taught a good deal of Scripture and that he had authority to beat boys, but that is all. References to him in the accounts written in Waterfield's day are scanty. 'Of Edgar, who later succeeded Waterfield, I remember almost nothing', is a typical one. Waterfield would certainly not have appreciated any apparent diminution of his own authority in the school and Edgar may have been content – on a handsome salary – to bide his time.

At all events, when he did take over, one of his first actions was to publish a small booklet headed *General Rules for the Conduct of Temple Grove School*. From this we need not infer that he intended to alter the statutes laid down by Waterfield or, except in minor detail, to change the style of discipline of the school. From the fact that more of the rules are directed at the masters than at the boys one gets the feeling that, although himself a member of staff for seventeen years, he is now letting his former colleagues know that *he* is in charge. The Rules begin: 'Every Master is expected to be in school and at his desk whenever his division is at work . . . Punctuality is of the utmost importance; especially as, where several classes have to be taught in the same room, each class is so dependent upon the order and quiet of its neighbours

. . . No class should be left until the relieving Master has come.' Then come regulations about punishment: 'School punishments are to be, as far as possible, avoided . . . the Headmaster objects strongly to punishments of written lines, much preferring the repetition of grammar rules or . . . of Latin and Greek poetry . . . No Master is on any account permitted to strike or to wilfully hurt a boy . . . Boys are not to be kept standing longer than is necessary'. But cases of grave misconduct should be referred to the Headmaster, such complaints to be 'made in writing, though they may be accompanied by verbal explanations'.

Outside the classroom further vigilance by 'the Masters on duty for the week' is invoked. 'All damage to, or climbing on, trees, shrubs, railings, wall or gates; all cutting of desks or forms; all throwing stones out of doors, or balls etc. indoors, is forbidden; all buying and selling, bartering or other dealing among the boys; and all wilful mischief whatever is strictly prohibited . . . Every case of fighting or bullying is to be reported to the Headmaster . . . Playing or dawdling about in the lavatory [wash room], yard, closets, or in their neighbourhood, is strictly forbidden . . . No boy is at liberty to enter the servants' passage without permission'. At meals, 'The Senior Master of the day is more especially responsible for the observance of order in the dining room at breakfast; the Junior at dinner and at tea . . . The boys are to be called by tables for going into the dining room and dismissed in the same way . . . Books may be read at breakfast and tea'. On Sundays, 'On the road to and from Church, the line is to be strictly kept; and any cases of disorderly conduct, or any misbehaviour in Church, are to be reported to the Headmaster on Monday morning . . . It is expected that the boys will show a due observance of the day by quiet and orderly conduct'. The booklet concludes with further instructions to the Assistant Masters: 'It is expected that all Masters . . . should be present . . . at morning and evening prayers. The Headmaster considers this attendance of very great importance . . . as boys are taught religious duties and reverence by example even more than by precept'.

Edgar had kept his own *Name Book* (a less impressive one than Waterfield's) since at least 1875. In September 1880 the page is headed 'First Term in Temple Grove'. He does not entrust the writing of a pupil's name to the boy himself – which is a distinct loss – but he does give the boy's age on arrival in years and months. Twenty-six new boys started at the school that term, four 8 yr. olds, three 9s, eight 10s, seven 11s, three 12s and the eldest 13.2. Throughout the next 13 years the average stay of a boy at Temple Grove continued to be just under three years and numbers remained between 120 and 130. Halfway through his time, Edgar added in the class in which each boy was placed on starting and, from September 1888 for four terms, he recorded each boy's height and weight (in lbs.). The last entry in Edgar's reign is that of Dermot Freyer, whose three sons followed their father in the 1920s.

Very few assessments of Edgar by his pupils appear to have survived, two of them coming from his last year at the school. In *Our Family Affairs* there is a good description of the man: 'The new Headmaster was distinguished for a very long clerical coat, two most amiable daughters and a gold-rimmed eye glass which he used to clean by inserting it into his mouth and then wiping it on his handkerchief [together with] the most remarkable hat ever seen. The nucleus of it, that is to say the part he wore on his head, was of hard black felt, like the ordinary bowler, but it was, geometrically, quite round, so that he could put any part of it anywhere . . Outside that circular nucleus came an extremely broad black felt rim, turning upwards on all sides in what I can only describe as a saucy curve. His aimiability was unbounded, his driving power that of a wad of cotton-wool. Indeed he was so pleasant that for his sake it became the fashion to fall in love with either of his two daughters'.

Matthew Hill on the other hand confesses to having 'very few pleasant memories of the school as it was under J.H. Edgar, and many distasteful ones. I do not think Edgar took any pleasure in beating the boys; but he had a 'tariff' of punishments quite irrespective of any difference in one boy from another which was applied to all and sundry . . . His attitude towards us was always that of a strict drill sergeant. He had some insight into character but he never tried to get on friendly terms with boys. None of us would have dreamed of going to him for advice in any difficulty . . . He was a dull teacher and unpopular'. Harry Waterfield, however, praises his Scripture teaching, remarking, 'I can honestly say that under his teaching I learnt my Bible more thoroughly than at any other period of my school days'.

Mrs Edgar is described as 'a warm-hearted motherly woman for whom no on ever had a bad word. One evening she went into a dormitory to say "Good night" to a sorrowful new boy whom on leaving she kissed. Hearing a titter or two from some boys who had witnessed the operation, she went back into the room and kissed them all' (Hill) though elsewhere he also says, 'I think she was always afraid of her husband'.

Their two daughters, Mabel and Edith, did not mix much with the boys. 'But they were kind souls. The latter had many pet animals and birds, kept in a summerhouse. This she gave me the run of as I was then, and have always been, a devoted pet keeper' (Hill). Neville Leese bears out Edith's fondness for pets: 'She used always to have white rats appearing out of her dress. Good Lord, how she used to make us jump.'

Ernest Mackintosh also leaves a fairly unfavourable impression of Edgar. 'I sat for the Eton scholarship . . . but came 12th on the list, bottom but one. Back at East Sheen, I at once got mumps and was segregated in the sanatorium. One day the door opened and Mr Edgar stood there with a paper in his hand and said, "Mackintosh, this is the Eton scholarship list, but your name is so low down on it that I don't think you'll even get in." And he shut the door . . . I'm afraid none of us liked him or saw much of him, except for corporal punishment during the term or receiving a prize at the end of it'. Next term Mackintosh found himself Captain of the School. 'Both Edgar and I were leaving at Christmas, so I had to sign innumerable begging letters to parents for a parting present to Edgar, and later at a vast parental gathering declaim the speech (in Latin) of giving (and forgiving?) a mine of silver tea-things'.

A new boy arriving that term retained a happier memory of Edgar. 'The Headmaster who welcomed us was Mr Edgar, an elderly, benevolent and grey-haired clergyman. We had a good tea and then played games and when I went to bed I felt a bit more cheerful. Mr Edgar had two daughters who tried to make us feel at home on that first night' (Cyril Barclay). One cannot avoid the feeling that Joseph Edgar's thirty years at Temple Grove had brought light and joy to remarkably few of his pupils.

The impression one receives is that the standard of the teaching staff showed a gradual decline in the later years of the 19th century. Matthew Hill (himself a fine science teacher at Eton for many years) was in no doubt: 'The masters were on the whole a very ill chosen set of men. Rawlings, who taught the 1st class, was a good classical coach, no more and no less, and a human being also. This could not have been said of most of the staff . . . Teaching was on stereotyped lines, but very little attention was given to mathematics, the standard of which was ludicrously low. Though I was successfully crammed for an Eton scholarship, I can recall nothing in the work I did at Temple Grove which gave me the least interest, except a course of lectures on Electricity which somebody gave us once a week of an evening during one Christmas term'. In *Eton and Elsewhere* Hill exposes the poor teaching methods in greater detail: 'My first term I was put into the class of an incredibly bad teacher. He was never known to smile, he spoke so indistinctly . . . that it was almost impossible to hear him except when . . . he bellowed like the bull of Bashan at our misdeeds. He seemed to know little or nothing of any subject and to be too apathetic to impart even that . . . The master of my next form took an active dislike to me, and made it as unpleasant for me as he could. He was again a very poor teacher, cold, cynical and conceited. How such men can ever take to teaching is a mystery. He clearly hated his work and his pupils and took little pains to hide it'. However for one year a quite exceptional young man was on the staff at Temple Grove. Robert Morant (1863-1920) had been educated at Winchester and New College, Oxford, where he took a First in Theology in 1885. Of his year at Temple Grove, sadly, we know nothing. He was later described as 'a giant with a large, pale, melancholy face, the eyes glowing out of deep black pits'. At the end of 1886 he went out to Siam to tutor the King's nephews and, within a few years, had laid the foundations of a system of public education in that country. His greatest achievement, however, was to construct 'an orderly and comprehensive system . . . out of incoherent and antagonistic elements' in England itself. The passing of the great Education Act of 1902 'was largely due to his vision, courage and ingenuity'.

That there was no falling off in the number of scholarships gained is a tribute to the solid teaching of one or two men like Geoghegan in the middle forms and the consummate skill of Charles Rawlings at the top. 130 scholarships were won from 1881 to 1893 inclusive, with 18 in 1889 as the highest (that included three boys who each won two scholarships).

The earliest reports that have survived date from 1887 and '89. The earlier prospectus stated that reports would be sent to parents twice a term but these appear to be monthly ones. For each subject a grading of Good, Fair, Moderate or Poor was indicated by a line drawn across that square (or between squares) though occasionally a comment was added. Thus Benyon, in his first month at Temple Grove, was placed sixth out of nine in the class. The only comment occurs in Divinity, 'Sent up an untidy exercise'. Conduct in School and Conduct out of School were both assessed 'Good' and Edgar comments: 'Rather slack and inattentive at first, as was most natural amidst the new surroundings, but now is settling to work and improving properly. I think him a capable little man'. A month later he was first in the class, with only a 'Moderate' in Divinity counting against him. Edgar wrote: 'I am very much pleased with this little man's position and think it most creditable to him', but clearly there had been difficulties with some of the other boys as he adds, 'I expect next term R. will have a better time, and not be foolish enough to provoke teasing by calling boys childish names'. Basil Kemball-Cook, aged 11, was in IIα, coming sixth out of six. 'Basil is the ablest boy in this class, but one of the most difficult. He requires constant watching and keeping up to the mark and his carelessness is at times marvellous e.g. 'pulchrus' for 'pulcher'; 'afferit' for 'affert' in a recent piece of composition shown up to me. I notice that he is more disposed to make friends with the troublesome boys than with those who are good and industrious and the alliance does not help him'. A few words of encouragement follow but Victorian headmasters were not the men to spare feelings.

The major event of Edgar's headship was the building of the school chapel. This was sited on the N.E. perimeter at the far end of the cloisters so that there would be no problem in wet weather. It was designed by Edgar himself and the expenses of construction 'were met partially by subscription among Old Boys but . . . I believe the greater part of the cost was defrayed by himself' (Harry Waterfield). It is described as 'a neat building, 52 feet long, seating about 170 people. It is of stock brick, with a green slate roof, and the latticework windows are from Powell's'. The foundation stone was laid on All Saints Day, 1890, and only seven months later, on June 3rd, 1891, a service of dedication was held at which the sermon was preached by Alfred Earle. The first entry in the Chapel Register reads: 'Alfred Marlborough Bishop Suffragan of West London dedicated the chapel to the honour of God and the service of the School – was a scholar of this School in 1838'. He chose as his text the 15th Psalm, 'Lord, who shall dwell in thy tabernacle?' Apparently he dubbed it 'the Gentleman's Psalm' and – inevitably – the boys were required to learn it by heart afterwards. To refresh their memories the text was painted on the wall. The dedication service was also marked by the presentation of a silver trowel 'to Miss Edgar with the love and respect of the Household of Temple Grove'. Nineteen names are inscribed on the back, including those of Jane Deeming and Mary Underwood.

For just over a century visiting preachers have inscribed their names in the Register and, until 1946, the text or the theme is always given. There was

Temple Grove.

O LORD, Who from Thy throne above
  Dost heed our simplest prayer,
And with a Father's tender love
  For all our wants dost care;
To Thee within this holy place
  Once more Thy children kneel:
O turn Thou not away Thy face,
  Give ear to their appeal!

For some there are whose sojourn here
  Ends with the morrow's sun;
Thou call'st them, in another sphere
  Their onward course to run:
Dear LORD, if Thou thro' childhood's hour
  Didst shield them from all ill,
Withdraw not Thy protecting pow'r,
  Be Thou their Guardian still!

And, LORD, to all Thy children give
  The strength, where'er they be,
A pure and holy life to live,
  And make them worthy Thee:
So those who go and those who stay,
  When earth's last term is o'er,
May gather at the Final Day,
  And part again no more!

The Rev. Joseph Edgar wearing the famous hat

Built in 1891 the Chapel continued in use as the Edgar Memorial Hall until its destruction by a German bomb in 1944

Interior of the chapel. The organ is still in use at Temple Grove, Herons Ghyll

surely a stirring sermon from the Rev G.H. Perry (22.11.91): 'Yet now be strong, O Zerubbabel, saith the Lord' (Haggai 2.4). Bishop Montgomery (father of the Field Marshal) (16.12.06) chose the famous charge from Isaiah (40.3), 'Prepare ye the way of the Lord'. Percy Dearmer, compiler of *The English Hymnal*, whose two sons were then pupils at the school, took 'Sons of God' as his theme (12.2.05). And on July 14th, 1957, the present writer was privileged to hear George Bell, Bishop of Chichester, preach at Evensong.

Two further innovations followed. The chapel was equipped with an organ and Frank Figg (who was organist and choirmaster at Christ Church from 1888 to 1912), the visiting music master at the school for some years, may well have played at services. Organs in those days needed organ-blowers. Though the boy changed from time to time the blowing mechanism followed the organ – and the school – to new homes and was not superseded until about 1970. The anthem, as already noted (p.180), may well have been composed by Edgar. The last lines are, 'So those who go and those who stay, When earth's last term is o'er, May gather at the Final Day, And part again no more'. It is easy to scoff at Victorian sentimentality, but the occasional tear has been surreptitiously wiped away by unsentimental children of the present era.

Sadly the chapel retained its original use for only sixteen years and it has to be said it seems to have aroused no affectionate memories in the pupils of that time. 'I do not remember much about the services' is one comment while a future bishop's only recollection is that 'a boy had changed round most of the organ pipes, so that when the first hymn was played on a certain Sunday, the results were horrific'. When the move to Eastbourne took place and Temple Grove was demolished, only the chapel survived. It appears on the 1933 O.S. Map and was known as the Edgar Memorial Hall. By 1940 when Meston Batchelor made a pilgrimage to it, the building was in a sorry state but with the help of the vicar he was able to rescue the school bell, the silver trowel previously mentioned, and a wealth of other relics. In 1944 a German bomb scored a direct hit on the building.

Another innovation introduced by Edgar was swimming. The bath must have been constructed quite early in his headship as it is mentioned by Matthew Hill as being in existence when he arrived (1883). 'An ex-sailor, Davidson, kept watch on the boys in play time and in the summer term presided over the swimming bath'. The charge for its use and for instruction in swimming was one guinea for the term. The bath being totally unheated one assumes the sport had nothing like the popularity it enjoys today. And one wonders how effective a supervision 'Daddy' Davidson exercised when one reads Christopher Storrs' account: 'I can only remember my first bathe. I had been told that the best way to learn to swim was to jump in at the deep end: which I did, with my specs on, felt that I was drowning, was pulled out and became rather a water-funk thereafter'.

Edgar was a keen cricketer and an all-round sportsman. It is fair to assume that, during the '80s and '90s, the standard of games coaching slowly improved. The fixture list (at least as regards cricket) was extended to include

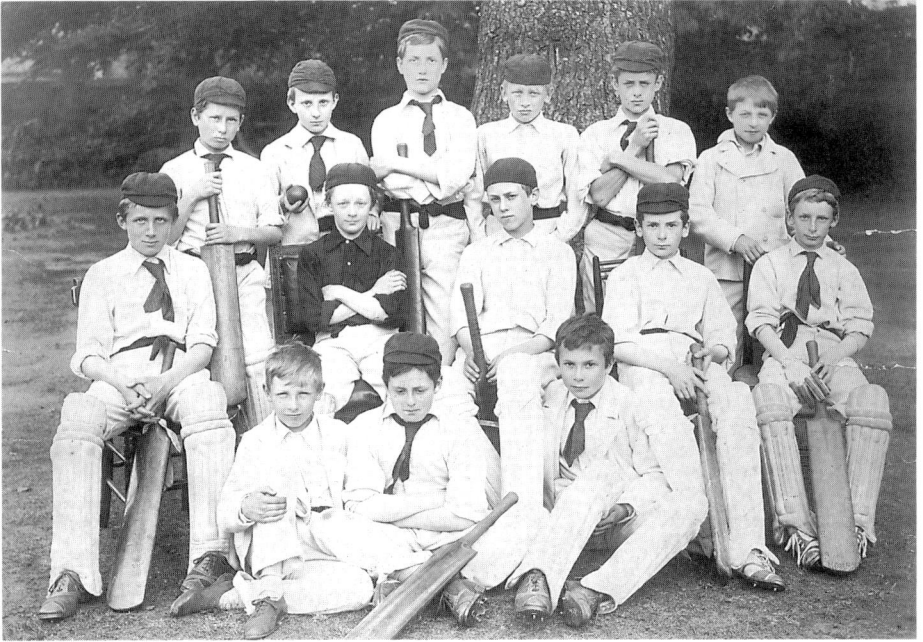

Left to right, back row: O.West, R. Waterfield, W. Patten, H. Salt, A. Smith, A. Fox; middle row: W. Richardson, W. Leese, T. Brand, W. Smith, V. Leese; front row: L. Harrison, A. Cowie, S. Hare

Left to right, back row: Hodgson, Gardener, Stewart; middle row: ?, Fordyce, Mordaunt, Rayner, Scott Elliot; front row: Ware, Macnamara, Sherring

The swimming bath was installed in c. 1881

The winners of scholarships and exhibitions in 1889: F.B. Sherring, H.R. Ward, H.S. Hearn, E.W. Grubb, B.A. Kemball-Cook, E.M. Loughborough, J.F. Crowdy, H.S. Barwell, W.C. Eaton, M.F. Clarke, R.C. Moore, R.S. McClintock, T.C. Musgrave and A.G. James. Unfortunately only James (one from the right, centre row) can be positively identified.

Elstree, Colet Court, Sandroyd, Stanmore Park and others. The earliest team photographs date from this period. The cricket team of 1881 shows fourteen players, perhaps indicating that there was no fixed 1st XI. Providentially the names of all its members have been preserved. The boy in the green and black shirt (worn apparently, to identify the wicket-keeper) is William Leese and on his right is William Richardson who carefully preserved this precious relic. The equipment in use seems very little different from that of a hundred years later. By then caps were beginning to have peaks – none is visible in the 1880 school photograph. The football team of 1886 numbers the conventional eleven and again all but one have been identified. The captain is Gerald Mordaunt, a fine games player. Robert Fordyce (on his right) and Frank Sherring (on the ground, right) later gained Cambridge blues for soccer. The kit is interesting with only the face, hands and part of the neck actually visible. One wonders how many layers there were underneath those jerseys. The trousers appear to be of plus-four variety with the tops of the thick woolly stockings firmly fastened inside them. The boots have no studs but seem capable of doing extensive damage to opponents' shins and to the somewhat flimsy football. From the following year comes the first (surviving) report of a football match, taken from *Elstree Notes*: 'Elstree v. East Sheen – this, the first of a series of brilliant victories, was played at East Sheen. The result might have been still more disastrous for our opponents but for the excellent play of their goal-keeper'. Four years later (1891) a more detailed account is given:

'Elstree v. Temple Grove – Played at East Sheen on Nov. 21st. The Elstree eleven met in this match a very quick, clever team of opponents, and had for half the time the worst of the game. Two goals were scored by the enemy in the first ten minutes and at half-time they led by three to two. That Elstree won eventually by six to three was greatly due to the excellent play and leadership of Wilson . . . Weston, of East Sheen, captain and centre-forward, played a fine game and scored all three goals for his side'.

As far as everyday games were concerned, 'We played Soccer nearly every afternoon and . . . we did not do much changing. We were each provided with two jerseys, one red and the other blue. The sides were always divided into Reds and Blues, so if you were a red you simply put your red jersey over the blue one, and vice-versa. I don't think we changed shirts, knickerbockers or stockings'. In the summer term 'we were encouraged to play golf in the really spacious grounds, a pleasant change from compulsory cricket' (de Lotbinière). An indication of the adoption of muscular Christianity at Temple Grove is the growing involvement of the masters in the boys' games. Around the turn of the century Mr Bellamy had the reputation of being a very fast bowler who had broken a boy's leg during the staff match (he also played the organ in chapel on Sunday). One further aspect of the growing cult of games is the proliferation of caps. By 1880 a considerable variety is apparent, and either by then or soon afterwards each class had its own distinctive cap: 1st class – light blue with thin white vertical stripes; 2nd class – light blue with wider crimson vertical stripes; down to 6th class with brown and pink rings.

The straw boaters were also fitted with appropriate hat-bands and of course teams and team colours had their own caps. This system, with all its expense and wastage, continued up to 1902 when Harry Waterfield became Headmaster. A group of senior boys 'said it was such a pity we had not a school cap' and Mrs Waterfield at once said, "Why not?" and, without delay, a decision was reached to have a green cap with black piping on which was superimposed the Waterfield family crest and motto in silver. Team caps, however, continued in use much longer and when 'colours' were awarded were highly prized.

Apart from these innovations one feels that Edgar altered very few of the arrangements which Waterfield had made for the running of the school. Opinions about the food varied as much as they had done earlier. According to John Brereton, 'the meals were ample though not appetising, the usual watery mince and cabbage (with the occasional caterpillar) for lunch, and musty-tasting rice to follow, or sometimes suet pudding with treacle'. Maurice Headlam thought that 'the food was beastly, if not actively bad. And we were compelled to eat it. We used to hold envelopes under the edge of the table and shovel the custard and marmalade pudding in, thence to be transferred to what we called 'The Boggs'. I was discovered doing this when captain of the school and condemned to stand on the form and eat a fresh helping'. Another boy goes so far as to say, 'I think we were fed very well', though he qualifies his praise by adding, 'It was the custom in those days to rely on parents to provide additional luxuries', which would include fruit. 'On various occasions parents would send down cases of apples or oranges for the whole school. These were handed out after lunch', though another version has it that 'when any boy had a crate of oranges sent him, next day we always had marmalade pudding'. Hungry boys could not always blame the school. 'It was not 'the thing' to eat the food provided, though it was quite good, and I had not the moral courage to go against the fashion. Hence I was undernourished for three years' (Hill). On the other hand they may have had some justification for this attitude if one is to believe Christopher Storrs: 'In the middle of the dining-room was the Master's Table, which seemed to our famished eyes to be loaded with every possible appeal to the stomach – joints, hams, cheeses, puddings, fruit – while we rebelliously swallowed our watery mince and 'fust' (bread and margarine)'.

At all events pity on starving youngsters was taken by Alice Bassett, one of the matrons, described as being 'young and dark with rosy cheeks and – some said – gypsy blood in her veins'. 'Every Sunday night her supper used to be sent up to her room, consisting of cold beef and pickles. She never finished her supper but used to make up the most delicious sandwiches, cut thick, and after lights out would give them in turn to selected boys . . . One Sunday night I was just dozing off when I felt a warm hand feel for mine in the dark and a large beef sandwich was pressed into my palm and a faraway voice said, "Good night, Scudie dear". However cold beef and pickles are not the best things for ten-year-old stomachs and I awoke later with a dreadful nightmare

R.L. Morant who taught for a year (c.1886) at Temple Grove and was later to be the architect of the 1902 Education Act.

Temple Grove School c.1884 *(opposite page)*

Most of those who can be identified are standing in the middle row i.e. the one containing the headmaster. *From the left*: two boys, a master, J. Spottiswoode, a boy (half concealed), W.A. Collins, three more boys, T.S. Furniss (light jacket), a boy, Mrs Mackinnon, Mrs Edgar, Rev. J.H. Edgar, Miss Edith Edgar with a boy half-hidden behind her, Gardiner, five more boys, H.A.A. Cruso (light jacket, dark waistcoat), W.E.L. Stewart, one of the Grenfells, two more boys, T.A. Headlam(?), four more boys. It *may* be Jane Deeming at the window immediately above the door.

and screamed the house down. Sandwich suppers had to cease for a while and I became the most unpopular boy in the dormitory. Nevertheless kind words were rare to me and from that moment I was more in love with Alice than ever' (T.V. Scudamore).

Nightmares of a different kind were not unknown. On a certain night in 1881 Mother Shipton had foretold Armageddon. 'We all lay in bed in our dormitory waiting for the end of the world. A most terrific thunderstorm broke over us. Lord, how frightened we were. The other and more awful scare was when we wanted to go to the lavatory at night. We always thought Charles Peace (hanged in 1879) was going to nab us' (Neville Leese).

Crazes, or manias, changed with the usual frequency though nib battles, once introduced, had a long period of popularity (the present writer enjoyed them in the early 1940s). The fountain pen was unknown and the school kept a large supply of nibs for the boys' pen-holders. If one acquired a dozen or more one could array one's army on any flat surface and challenge an opponent. The trick was to push or flick one's nib so that the point penetrated below the enemy's unprotected stern. A quick flip and if the foe landed on its back it became the attacker's property. Did the advertisement hoardings of those days, proclaiming 'They come as a boon and a blessing to men, the Pickwick, the Owl and the Waverley pen', have Temple Grove specifically in mind? (Strictly speaking a nib *was* a pen and it fitted into a pen-holder). The risk of carrying a pen-holder in one's teeth was made clear when 'a boy named Bullock tripped up while running across the room with one in his mouth and fell on his face forcing it right down his throat. The doctor had to be called to pull it out' (Brereton).

'On wet half-holiday afternoons we used to spread out our lead soldiers on the dining-room tables and stage battles. A box of seven, with a mounted officer, cost a shilling. A few of us saved up our pocket money and bought a box of Mountain Battery soldiers with field guns which fired match sticks by means of a spring breech' (Brereton). 'Another amusement was 'Shooting Galleries' set up in one's desk. Some of them were real works of art. One shot at targets by means of paper clips fired from elastic bands. Prizes were given but naturally one had to pay to shoot' (de Lotbinière). 'In our lockers we kept, as well as books, such things as silkworms, stag beetles and even slow worms. We bought the silkworm eggs for 6d. a hundred from Gamages, or through advertisements in *Exchange and Mart*, and hatched them out with the aid of the sun and a magnifying glass and we carried out the whole life cycle from egg to moth, winding the golden silk on to spools' (de Lotbinière). Large pets, such as guinea pigs, white mice and even rats, were kept in hutches in the cloisters.

Macramé, called netting by the boys, enjoyed a vogue. 'When I arrived I was already a skilled netter, so I was much in demand by big boys to do some of their netting. We used a nice soft yellow string which was called Macramé thread. Everyone was making a hammock but I never remember one being finished or used to lie on'. An unusual accomplishment was credited to

Payne-Gallwey who 'had a thin copper stencil with which he used to print the outlines of animals on sensitised photopaper in a small printing frame, which we all thought wonderful! R.M. Worsley's skill was of quite a different order: 'The gym had a vaulting horse, parallel bars and a thick rope hanging from the roof timbers. Swallows used to build up in the rafters and he used to climb up the rope and bring down the eggs in his mouth to avoid breaking them' (Brereton). Nor was his skill restricted to the gymnasium. 'I and another boy named Worsley held the record for climbing every climbable tree in the grounds. Worsley was actually a bit better than I was, but I was not bad and had a very good head for heights' (Gerald Burrard).

Meston Batchelor seems to feel that standards slipped during the reigns of Edgar and Allen from what they had been in Waterfield's day, but after a hundred years it is very difficult to be objective with the benefit of only scanty evidence. Reading the *General Rules* one might feel that this was a well-run school where both masters and pupils were kept up to the mark by a vigilant head, but theory and practice do not always coincide. Maurice Headlam, for example, said, 'I don't remember having been bullied – except a stray kick or two', but another boy recalled that 'Outside the little room from which the sweets (3d worth) were distributed there was always a row of big boys holding out their hands and saying "Give me, I'm stopped" (meaning that they had been denied their weekly sweets as a punishment). By the time the little boy had run the gauntlet there was very little left for him'. If that occurred regularly it certainly argues weak supervision. John Brereton remembered having his tuck removed by more forceful means: 'One of the big boys was the chief bully. His method was to sit on your lap and squeeze your stomach in by pushing his feet against the form in front. By this means he forced me to give him my weekly tuck allowance for four weeks in exchange for a story book I did not want'. To give him his due, Edgar was severe on any bullies who were caught. 'I only got it [bullying] badly once. Three big boys caught me, two held my hands behind my back and the third punched me in the wind; as I began to recover he punched me again and again'. But the episode was reported to Edgar by one of the footmen. 'The result was that the boy who did the punching was expelled at once by Mr Edgar. The two who held me were both flogged later. One of them had come to me weeping and offering me sweets, begging me to tell Edgar that he had never hit me. But I was not even sent for and questioned' (Cyril Barclay).

Matthew Hill is probably Edgar's severest all-round critic. As we have seen he had a poor opinion of the teaching capacity of most of the staff and he described the standard of mathematics as 'ludicrously low'. Outside the classroom he comments, 'In spite of there being a gymnasium we were never taught gymnastics; fives, tennis and racquets were unknown . . . No boy was ever encouraged to have a garden of his own; no form of nature study was included in the curriculum, let alone any sex instruction. The value of handicraft was entirely ignored, the school had no workshop'. Elsewhere he remarks, 'The morality of the boys was not high. Although there was little actual bullying, cribbing was rampant and I remember one evil-faced lad, who managed to open a certain desk wherein the marks awarded in examination papers were kept. These he manipulated so that when they were added up he became entitled to a prize. I can see him now walking up to the dais to receive his book with a horrid leer on his face'. He also remarks that 'There was a good deal of Rabelasian talk, but I do not think that bad practices were indulged in except by one small clique'. It sounds a good deal less serious than the situation at many other schools. In *Eton and Elsewhere* Matthew Hill relates a far more ingenious method of cribbing at Temple Grove. It sounds almost too good to be true but it is also too good to omit.

'Another lad, J., was an expert cracksman. Sometime before we were due to write an examination paper in Euclid [Geometry], Edgar used to send a note to one of the masters giving the numbers of the propositions we should be required to write out. The letter was put into the master's desk under lock and key; when the master's back was turned and his attention engaged by confederates, J. would pick the lock, open the desk, and steam open the note by means of a small kettle secreted in his locker. He then took down the numbers of the propositions, fastened down the envelope and put it back in the desk . . . He and his friends then withdrew and learnt their Euclid hard'. A nasty suspicion lurks that the earlier story has been embellished for publication, but the verdict must be 'Not Proven'.

In October 1893, Edgar (who earlier in the year, had acted as the first chairman of the Association of Headmasters of Preparatory Schools) sent out a notice to all the parents informing them of his decision 'to retire at Christmas . . . leaving the care of the School to my colleague, The Rev H.B. Allen', and giving a brief resumé of Allen's scholastic career: 'Mr Allen [born 1857] was a Scholar of Winchester, and of B.N.C., Oxford. In 1880, shortly after leaving Oxford, he joined me at Temple Grove as second-in-command, and remained until he became Headmaster of Gore Court School in 1883. In 1889, leaving Gore Court to his partner, he rejoined me at Temple Grove with the view of taking my place whenever I should retire. He and Mrs Allen have had charge of my second house, Wedderlie, with the pupils residing in it'. The notice was accompanied by one from Mr Allen, modestly commending himself to the parents and ending, 'I hope therefore that you will have confidence in trusting your boy to our care'. Allen, who became ordained priest in 1882 and had married a Miss Gossett in 1887, was 36 when he succeeded to the headship, a post he held for only nine years. Thereafter he farmed in the Cotswolds for ten years before becoming vicar of Didbrook. He died, aged 93, in 1950.

Assessments of his character vary considerably. One has of course met headmasters who have become almost human after relinquishing office but in Allen's case the variety of opinions concerning him is so great that it is sometimes hard to believe they refer to the same person. A good deal hangs on events that occurred early in his headship. It is only fair to say at once that, in Meston Batchelor's opinion, the episode took place in Edgar's last year. One witness, however, merely refers to 'the Headmaster' while the other states categorically that the Head concerned was Allen. There are also important differences concerning the origins of the affair. According to Ronald Storrs (who arrived at Temple Grove in May, 1892, aged 10½), 'One of the boys had cheeked the music-master who, swept with fury, put him across his knee and gave him six. As in other schools, none but the Headmaster had the right to flog. Although this was no flogging the boy, who was popular, complained to his friends and they summoned a secret meeting of the school at which it was resolved that they should register their indignation (comically enough) by maintaining a silence of icy disapproval in the Headmaster's presence throughout luncheon; further that, if this conduct was called in question, there should be no ringleaders. I knew nothing of these events having at that time been absorbed in a game of chess with the German master. As I walked into the dining room with the others I was told to hold my tongue, but neither explanation nor warning was added. The moment the meal was over we were summoned to the big school room where the Headmaster, a highly strung scholar, asked in broken accents the names of those responsible for the dastardly plan of humiliating him before all his pupils. Fully expecting to be joined at least by the boy who had instructed me, and surely by some of his companions, I stood up, to find myself the only erect figure in the room. There was a brief and terrible pause. The Headmaster then looked mournfully upon me and said with disgust that he had had enough of such behaviour, that I was a bad influence and that I must pack up and leave the school that very afternoon. In a word, I was expelled. I walked out before my comrades in a dreadful silence and the meeting broke up in mingled discomfort and awe'.

The second witness, Cyril Barclay, (whose first term was Edgar's last) gives the following evidence: 'The first time I was beaten by the Headmaster, Mr Allen, was a day I shall never forget. I was nine years old and I had done a bad Latin prose, so Mr Crowe sent me into Allen. I had done many worse proses than this one and nothing had been done, so I arrived in the study feeling quite cheerful. But Allen said, "You are a lazy boy and I am going to flog you". He did not use a cane which stings but does not injure. He used a long black solid ruler and made me bend over in mid air without the support of a chair or anything. He then gave me ten blows. After the first five it hurt so terribly that I put my hand in the way. The next blow fell on my wrist and broke all the skin, and blood poured down my hand. Allen remarked, "Stupid boy. Don't put your hand in the way" and gave me four more.

'There had been a lot of violent flogging and the boys were getting tired of it, but did not know what to do. Then one big boy had a brainwave. He announced one day, "No one speaks a word at dinner tomorrow". It was the perfect plan. There was always a great noise at dinner, over 120 boys and all the masters. On this occasion Allen said Grace and sat down. Instead of the usual clamour there was not a sound. Allen looked surprised but said nothing. In a few minutes he appeared to think that there had been a misunderstanding, for he said, "You can talk, boys". But not a word was spoken. Again he said, "You can talk, boys". And then it began to dawn on him that this was a silent protest against himself. He was perfectly livid. It was so clever. You could punish a boy for talking too much and making a row, but you could not punish a boy for not talking at all. Allen had the whole school into the big schoolroom after lunch, and said, "Who was responsible for your behaviour at dinner?" Then, to everyone's surprise, Ronald Storrs stood up. Allen looked at him and said, "You, Storrs. You leave the School tomorrow". I can remember Storrs' face, and how the tears came and he sat down and buried his head in the desk'.

In Storrs' own account the school sergeant 'who had seen me playing chess . . . put the matter to the German master and between them they persuaded the Head of my innocence. This he might perhaps have established more handsomely than by sending for me and announcing in public that I would be pardoned this time'. In Barclay's account, 'when Allen had cooled down and come to his senses, three of the top boys went and explained this [Storrs' innocence] to him, so of course Storrs did not leave, but the flogging became much less frequent, so the plan worked'.

Coincidentally one of his later victims was the youngest of the four Storrs brothers, Christopher, and again the sadistic streak is all too apparent. 'He had heard me calling the servants "Sluts". This was a horrible but quite universal term, and I had used it with no idea of insult. Anyhow Allen gave me a deserved licking using his customary ritual. He scattered paper widely over his study floor and told me to pick it up; then while I did so, with what speed I could, he whacked me with righteous wrath'. In spite of that Storrs describes him as 'cultured, a good scholar and teacher, kindly, I think, but a weak chief and serenely unconscious of what went on in his school'. T.V. Scudamore remembers him 'only as a vague form, always in a hurry, appearing and disappearing through the door leading from his private rooms into the class rooms'. John Brereton gives a fuller description: 'The Headmaster, H.B. Allen, was a short, loosely-built, kindly man with a dark drooping walrus moustache, and he gave the impression of a conjuror, with his long black coat-tails from which he might produce a rabbit at any moment, low-church white collar and tie. His wife . . . was a very kind, quiet person who seemed to be somewhat stitched into her coat and skirt and always to have the same round flower-trimmed hat perched precariously on the front of her head'. It is at least reassuring to read that B.C. Watson, one of Allen's later pupils, reckoned that 'Allen did very little beating. I escaped entirely'.

However much opinions of H.B. Allen differed, those of his dog did not. 'We were greeted by Sturk, the white bull terrier with a pink nose, belonging to the Headmaster. I was fond of dogs, as we had three at home, but having been brought up with the idea that if a bull-terrier did bite, it never let go, I kept well away' (Brereton). Mr Allen was 'devoted to his unpleasant dog, Sturk' (C. Storrs) and Dorrit Waterfield, arriving with her father, who became headmaster in 1902, remembers that 'the plumbing [of the boys' bathroom] was such that the day we arrived *all* the water went straight through to the kitchen; a few hours later all the gas and water supplies were cut off because H.B.A. had not paid up; and in the dressing-room it at once became apparent

Let to right, back row: Neumann, Purves, Cheese, Whitelaw; seated: Hawtrey, Westoby, H.B. Allen, Crowe, Samuel. Charles Geoghegan was still on the staff but, for some reason, not in the photograph

First Class Summer Term 1898. Left to right, back row: K.E. Bittleston, J.R. Trench, G.G. Knox, A.C.O. Morgan, A.C. Aldous, J.V.S. Wilkinson, A.J. Whittall; middle row: B.C. Reade, A.J. Carter, C.W. Orde, Albert Cheese, W. Duranty, J.P. Gibbs; front row: M.G. Anderson, W.H. Ricardo, G.W. Anderson

that Sturk had lived in the room without benefit of water-closet'.

Having given a generally unflattering picture of Allen as a headmaster one must in fairness add that he earned golden opinions during his many years, first as curate and latterly as vicar, at Didbrook. He features as one of two 'Cotswold Characters' in *Remember and Be Glad* by Cynthia Asquith, who writes, 'He had become a famous coach whom [many of his pupils] remember with grateful affection. A wonderfully inspiring teacher, he was much given to praise, often as unexpected as it was stimulating'. Even after being appointed vicar in 1940 (at the age of 83) he continued to take in pupils and to the end of his days he remained a distinctly unconventional cleric. 'Because of [his] unclerical appearance; his unholy glee in gossip; his habit of gabbling through the service; his chronic impecuniosity; there were some who considered him unfitted to be a vicar. He may not have been a very earnest theologian . . . yet some of the most moving sermons I have heard came from his lips'.

Crowe is described by John Brereton as 'a fat, dark, greasy man. He took a delight in beating little boys' bottoms with a short round black wooden ruler about ten inches long. His technique was to grab the unfortunate, and often terrified, pupil by the back of his collar and force his head down between his (Crowe's) knees, and administer several juicy strokes with a deft flick of the wrist'. Cyril Barclay describes him as 'an elderly and jovial person' but confirms that he used 'a black ruler' though transferring its application to 'heads and knuckles'. T.V. Scudamore also mentions 'good hard raps from a round black ebony ruler', adding that 'his favourite form of punishment was to take his bunch of keys by the longest key and hit one hard over the knuckles with the remainder'. Evidently not a pleasant man and probably a poor teacher though with a talent for writing lyrics. According to Adam Fox he later committed suicide. 'Bowler' Bellamy and Albert Cheese were to remain on the staff for many years. By all accounts both were fine teachers, Cheese – in charge of the First Form – largely responsible for the long line of scholarships the school continued to accumulate. Christopher Storrs describes him as 'a hard man; I'm sure he must have smiled occasionally, but I never saw it. At the top of the central blackboard was his motto for the term (the only words never rubbed out): "Oderint dum metuant" (Let them hate, so long as they also fear). This invocation was reinforced by 'Blackie', a long formidable ruler with which he swiped our knuckles or heads impartially'. Charles Hobart reckoned he 'learned more from Albert Cheese in two years than from anyone else in my life'. De Lotbinière describes Cheese as 'a most interesting man. He and his brother were great travellers during the holidays – and also keen photographers. The result was lantern shows which we thoroughly enjoyed'. It is appropriate then that the earliest First Form photograph (taken in 1898) has Albert Cheese seated amidst his scholars. S.O. Purves was obviously well liked: 'a tall pleasant man with a good voice' and 'a *grand* man, very tall and well built' are two surviving descriptions. Most interesting of the new intake of masters, however, was Jack Hawtrey, a cousin of the well-known actor,

Charles Hawtrey. Theatrical talent clearly ran in the family, though like most of the staff he maintained discipline when necessary by strict methods. Indeed Cyril Barclay describes him as 'ferocious and rather frightening'. Christopher Storrs modifies this judgment by saying that 'he really held the place together. He taught with a stick in his hand, but he didn't use it much, and he was, at least, smart, disciplined, efficient and kindly'. On one occasion when acting as master on duty he 'chose to use a different word in dismissing each table from the dining-room: "First table, go; Second table, bunk; Third table, vamoose" etc. and ending: "Inside, hook it; Outside, skedaddle". We looked on this as the highest form of humour'. But it was his theatrical talent that implanted itself in the memory of Temple Grove boys. 'His star turn was as the Pied Piper of Hamlyn when he slithered about the stage in scarlet tights while rows of little boys scuttled along behind boards which revealed to the audience only their rat-head masks'. He certainly produced the school play at the end of each Christmas term and, in all probability, he wrote the script as well. 'The two that I remember were *The Knave of Hearts*, in which I was the Knave, and *The Clockwork King*. The audience, largely of parents, were of course extravagantly appreciative of these plays' (C. Storrs).

In 1898 he was invited to present a performance of *Bombastes Furioso* (an 1810 burlesque by W.B. Rhodes) on the stage at White Lodge by the Duke of Teck. In the oldest surviving play photograph four boys, Orde, Williamson, Morgan and Wilkinson, appear to be dancing a gavotte in front of a beautifully painted backdrop.

For the following year Jack Hawtrey had written *The Beadle and the Burgomaster*, a new and original Fairy Drama in 3 Acts, with lyrics by Mr Crowe and the music arranged by Mr Westoby. Most of the action takes place in the market place of Sourkriontenberg and the characters include Herr von Stumm (the Burgomaster), Herr Plunk (the Beadle) and Lurline (Queen of the Fairies), together with choruses of citizens and citizenesses, guards, fairies and children. There may be echoes, in some of the lyrics, of W.S. Gilbert but they clearly had a real swing to them. In Act 2, the 3rd citizen sings:

> 'Now here's a pretty state of things
> Our burgomaster's missing, O;
> His robe and office off he flings,
> His duties all dismissing, O'.

And in Act 3 there's a fetching duet between Naughty Bobby and Gentle Jane:

> '*Bobby*   Oh! I am a naughty young scamp of a lad
> *Jane*   And I am a girl that is good;
>   A sweet little creature that couldn't do wrong.
> *Bobby*   And I'm a young chappie that would'.

A few more events in the last years of the century seem deserving of mention. On June 23rd, 1894, the future Edward VIII was born at White Lodge. The following letter, written by ten year old John Benyon, records the event:

'Darling Mother

There is something to put in my letter this week although I am sure you know it. The new King of England was born last night at White Lodge son of Princess May in Richmond Park at 10 A.M. [sic] last night. They are both quite well. We went today for a walk up to White Lodge and saw Dr Williams her dꭓter doctor. The term is nearly finished now and we will all be home soon. Give my love to all at home. Come and see me after Ascot week and bring some grub. All the royalty are at White Lodge. Some half-holiday either Wednesday or Saterday I think it is next Saterday the whole school are comeing to Windsor. Will you write and ask Mr Allen wether I can come for an exeat next Saterday don't tell him that I asked you to. It is a lovely day. I enjoy the batheing so much. I am,

your loving,

John

Other Temple Grove boys took a keen, indeed personal, interest in the event. Spencer Weston reported: 'On Sundays and other days too wet for games we were taken for a walk in Richmond Park. Once in the Park we were allowed to scatter. In June, 1894, the steps of some of us gravitated towards White Lodge. Outside there was a tent and a table and a book. In the book distinguished visitors had inscribed their names. Greatly daring, some of us added our signatures and then fled. Unless I am mistaken, the two 'Pongos', the nickname of the twin Grenfell boys, were of the party and set the example'.

Evidence of the virulence of school epidemics is provided by a note (dated Feb. 15th, no year given) sent out by Allen to all parents:

'One case of measles was brought to us from home and yesterday, exactly a fortnight from the day of that case being declared, I sent a notice to 50 parents that their boys showed every symptom of having contracted the infection . . . We have now over 50 boys down with measles and we have isolated them and are nursing them in this House. So far the complaint is running its normal course and there is no reason in any instance to apprehend any complication. You will receive a further bulletin on Monday and succeeding days and, if at all necessary, a wire will be sent at once . . .'

A crafty innovation in Allen's day, when the food provided day by day was probably no better than before, was 'the last supper of the term when the authorities put on a really excellent tea, called by us the 'Gut'. For [there] every iced cake and cream bun ever invented was spread before us, so that we went home with our bellies distended and criticisms hushed'.

In April 1897, a school magazine makes its appearance for the first time. It was edited entirely by the boys and, since it was a ten-page, double-column job and was cloth-bound one assumes that a very limited number of copies was produced. A photograph of the editorial staff, R.P. Keigwin (editor), C. Anderson (General Manager), C.W. Orde, G.H. Staveley and F.E. Storrs, is given prominence and a somewhat pompous Introduction takes the place of an editorial. In it we are told that *The Temple Grovian* 'contains all those points essential to a school journal. It has a sports page, a general topic column, a poem column and many original jests'. The prize offered for the best original anecdote (not exceeding 100 words) was to be 'sixpenny worth of succulent confectionery'. An 'Ode to the Athlete' contains the verse:

'O would I were an athlete
But then I'd have to train
And fast at every blooming meal
And sprint with might and main'.

There is an interview with a Mr Joseph Williams on the subject of photography extending over one and a half pages in which he recommends a Pocket Kodak camera, Eastmans ABC developing outfit and the Ilford P.O.P. for printing. 'Tête à Tête' with the editor opens confidently with: 'Editor hic sidit quo non operosior alter, et calamum trepida nunc tenet ipse manu', with no fears that his readers will be deterred or defeated. Much of the page of 'Sporting Notes' is devoted to Old Boys who have rowed in the Boat Race, but reference is also made to the approaching cricket season at both Universities, and there is a brief mention of Sports Day: 'We hope that everyone is fit and in good training which is probably questionable'. Finally come a Chess Problem set by F.E. Storrs, a weak anecdote about W.G. Grace and full details of the School fives competition, won by G.H. Staveley and R.P. Keigwin. There is, however, no mention of a craze which was probably not unique to Temple Grove and which was described by Adam Fox:

'There was a great deal of hypnotism at Temple Grove in the years 1896 and 1897. Once or twice an attempt was made to 'send me off' but I never felt like it at all, and I never tried it on anyone else. Some boys certainly were in a trance and were thought to do things by suggestion. E.A. Bell told me recently that there was sometimes a difficulty about getting a boy round again and that the only thing to do was to hit him in the wind. He also told me how the masters discovered the matter. One of them, Millington, had a nephew who used to be hypnotised by Cobb mi. Cobb used to draw two circles on the blackboard and tell the victim that they were his (Cobb's) eyes. This was enough to send him off. One day Cobb forgot to rub the circles out. In the afternoon Millington's nephew came into school, saw the circles and went off'.

In the article on Temple Grove in *Country Life* (Aug 18th, 1900) there is a great deal of inaccurate information but presumably the statement that Mr H.B. Allen . . . who after some years as Headmaster has finally purchased the whole estate' from the descendants of Dr Pinckney is correct. Omission of any description of the building itself is defended on the grounds that 'it is hardly necessary to say very much in the face of common knowledge and of the pictures which we show' and even the grounds receive scant treatment, the writer drifting off into further – mainly inaccurate – historical surmise, and ending with the deferential assertion that 'at Temple Grove little boys of the best class are well taught in mind and body'.

It seems possible that Allen, who was not an astute business man, had got himself into difficulties as a result of the purchase of the estate freehold. It is certainly true that the number of pupils in the school had been slowly dropping from over 120 to just under 100. A big leave in July, 1902, was not replaced and, with only 78 pupils, Allen would have been running the school at a loss. At all events he sold Temple Grove to Harry Waterfield (a pupil in the '70s) who brought some extra pupils with him from Arnold House at Hastings, and Allen retired to the Cotswolds.

Just four years later Waterfield made the momentous decision to move the school. In the notice sent out to parents (Aug. 1st, 1906) he refers to the school 'becoming surrounded by London suburbs', and to the purchase of two large tracts of land nearby for the Hammersmith and Fulham Cemeteries. But a boy of that era (Wilfrid Jervois) remembered the 'splendid view across market gardens to the Richmond Road. There was not a house to be seen but in the distance, rising among the trees, was the Kew Gardens pagoda'. Waterfield's notice continued, 'I feel sure that the time has come to move the school further away from London'. He announces that he has taken, 'from next Easter', a large building at Eastbourne. This was the former New College, a minor public school which had closed in 1903, and – naturally – he extols the merits of a move to the salubrious Sussex coast. 'We shall, of course, retain the name and traditions of Temple Grove'.

One may, perhaps, be forgiven for ending this account of a Victorian preparatory school with verses written by the mother of Archibald and Colin Gordon:

"COELUM NON ANIMUM MUTANT"
Peace and great calm on Sheen's fair hill!
    No more the merry football rout,
    No sports, no games, no joyous shout,
The fields of Temple Grove are still.

Calm and deep peace in those old halls,
    Where echoes reigned of honest work, –
    And, year by year, of lazy shirk,
Of swish of cane, – now silence falls.

But far away from smoke of towns,
    And din of jerry-builders' strife,
    There shall arise a freer life,
New Temple Grove on Sussex Downs.

Yet not all new, for on the wall
    The roll of honour still survives,
    And echoes from a thousand lives
Still from the past to duty call.

And still, as 'Omne Bene' thrills,
    Each heart the old tradition hears,
    The heritage of a hundred years, –
Old Temple Grove on Sussex hills!

A performance of Bombastes Furioso was performed for the Duke and Duchess of Teck at White Lodge in 1898. From the left: C.W. Orde, Williamson, A.C.O. Morgan, J.V.S. Wilkinson

Appendix I

Prince **Adolphus of Teck** was the eldest of Queen Mary's three brothers. After Wellington he entered the Army, serving in the 17th Lancers, transferring later to the Life Guards. He took part in the first year's fighting in the Boer War but was recalled in 1900 on the death of his father, whom he succeeded as Duke of Teck. On his brother-in-law's accession to the throne, he was made Marquess of Cambridge and later took Cambridge as his surname. He died in 1927.

Anderson, Charles. No information.

Barclay, Cyril C. After Harrow, he entered Trinity College, Cambridge, and was ordained priest in 1909. For twenty years he served in Australia, returning in 1932 to become chaplain of Bloxham School. During the war he was vicar of Helmsley and from 1946 for four years, vicar of Buxted. For some thirty years he acted as Clerical Organising Secretary for Accra Diocesan Association. He retired to Bognor where he died in 1967.

Beckwith, Sidney. He won a scholarship to Eton and went on to Corpus Christi, Oxford. He was later branch manager of an insurance company.

Bell, George K.A. George was the eldest of three sons of a clergyman. He won scholarships to Westminster and to Christ Church, Oxford, and trained at Wells Theological College. His work in Leeds brought him into contact with working people and had a profound effect on him. He later became chaplain to the Archibishop of Canterbury and married Henrietta Livingstone, sister of a Temple Grove contemporary. Dean of Canterbury in 1924, where he staged religious drama in the Cathedral itself, he was appointed Bishop of Chichester in 1929. In four separate fields he proved an outstanding bishop. His first care was for the clergy of his diocese, showing himself a true pastor by his involvement in their problems and their welfare. His concern for the poor, and especially for the unemployed, took the form of setting up occupational centres and community halls. His contacts with clergy overseas gave him a clearer insight than most into the perils posed by the rise of the Nazi party. And he is remembered, too, for his encouragement of the Arts in support of religion. In the war he pursued two aims over and above his diocesan commitments: his secret contacts with the persecuted clergy of Nazi Germany, and his condemnation of the indiscriminate bombing of German cities. This latter may well have counted against him when the time came to choose a successor to Archbishop Temple. But perhaps in any case Bell was best fitted to serve the Anglican Church as a bishop. He retired from Chichester in 1958 after 29 years as bishop, and died not many months later.

Bell, Edward A. The second son in the Bell family, Edward followed George with scholarships to Westminster and Christ Church and entered the teaching profession, first at Giggleswick, then at Eton, becoming headmaster of St Bees in 1926.

Benson, Martin. Martin was the eldest of the four sons of E.W. Benson, head-master of Wellington, bishop of Truro and Archbishop of Canterbury. His short life is illuminatingly told by David Newsome in section III of *Godliness and Good Learning*. At the age of 14 he gained the top scholarship to Winchester where he gave promise of a brilliant career but in his 18th year he was struck down by a fatal illness. The account of his last days, written by his father, is a most moving document.

Benson, Arthur C. Arthur, the second of the Archbishop's sons, was a man to whom scholastic and literary success came easily. Scholarships to Eton and Kings were crowned by first-class honours in the classical tripos. He then returned to Eton for eighteen years, becoming a highly regarded housemaster. He found that the demands of schoolmastering were hampering his literary output and he moved back to Cambridge where he became first a Fellow, and then Master, of Magdalene. He published over a hundred books – essays, literary criticism and novels – mostly popular but not profound. He also wrote the words for *Land of Hope and Glory*. His diaries were edited by David Newsome and published as *On the Edge of Paradise* and reveal him as a much more complex character than the outside world ever realised. He died in 1925.

Benson, Edward F. Always known as Fred, E.F. was the third of the four brothers and the one who wrote most illuminatingly about his own family. He was less scholastic and more athletic than his brothers, captaining the 1st XV at Marlborough and going on to Kings, Cambridge, from there. At first he toyed with a career in archaeology, but the success of his first novel, *Dodo*, caused him to abandon all else in pursuit of a literary career. His acute social observation gave him a 'witty and malicious delight' in mocking pretentious women and social climbers. The scene of his *Lucia* novels, Tilling, is actually Rye where he lived for many years and of which he was four times Mayor. His novel, *David Blaize*, is considered one of the very best school stories ever written. He died at Lamb House in Rye in 1940.

Benyon, G. John B. From Temple Grove, John won a naval cadetship, gaining rapid promotion and being appointed to HMS *Good Hope* as Lieutenant-Commander in 1913. Admiral Cradock's squadron of four ships encountered five powerful German cruisers off Coronel (on the coast of Chile) on Oct. 31st, 1914. Salvos from *Scharnhorst* and *Gneisenau* put *Good Hope's* guns out of action and at 8pm she blew up with the loss of over a thousand lives.

Benyon, Reginald K. He went from Eton to Pembroke, Oxford, (with scholar-ships at both) and travelled out to South Africa just as the Boer War began, serving throughout it in the Johannesburg Mounted Rifles. Afterwards he was employed by the Mines Department in Johannesburg.

Boyle, William H.D. William gained scholarships to Eton and King's and later became a clerk in the Treasury. He became a great authority on ecclesiastical music and compiled a chant book used at King's.

Brand, Thomas W. After Eton and Trinity, Cambridge, Thomas entered the

Army and served with distinction in the Boer War as brigade-major in the Second Cavalry Brigade, gaining the Queen's medal with six clasps. On suceeding his father as the third Viscount Hampden in 1904 he retired from the army but rejoined in 1914, and served throughout the 1st World War, commanding a brigade from 1915 onwards. After the war he acted as a Lord-in-Waiting to King George V and as Lord Lieutenant of Hertfordshire.

Brereton, John L. John went on to Clifton and then to Trinity, Cambridge. He was a Land Agent by profession and died in December, 1973.

Brinton, Hubert. He won the top scholarship to Marlborough and afterwards took First Class honours in Mods and Greats at New College, Oxford, before going to teach at Eton where he later became a house master, remaining on the staff for 37 years. He died in 1941.

Bullock, Robert S. His father had served in India and Robert joined the Indian Army in 1908 after five years at Malvern (where he had won a scholarship). He was a fine linguist, acting as an interpreter in Persian and Pushtu. In the early part of the 1st World War he served in Egypt and in France but in 1916, in the rank of captain, he was transferred to the Kut Relief Force in Mesopotamia. He was killed leading his men on April 17th, 1916.

Burrard, Gerald. Gerald was born in India and after Cheltenham and Woolwich he joined the Royal Field Artillery in 1909, serving in India till the outbreak of the 1st World War, when he went to France with the Indian Expeditionary Force. He was in action at Ypres in 1914, at Loos (where he gained a DSO) in 1915 and at the Somme, where he was severely wounded, in 1916. Invalided out of the Army, he became Gun Expert to *The Field*, the editor of *Game and Gun* and the author of a number of books on field sports. He inherited a baronetcy in 1943.

Byrne, Lionel S.R. From Eton he went on to Trinity, Oxford, and rowed in the Oxford boat in 1886. Three years later he began a long teaching career at Eton and in 1937, published (with E.L. Churchill) *Changing Eton*. He died in 1948.

Capper, John E. From Wellington, where he gained a scholarship, John went on to Woolwich and then embarked on one of the most unusual and varied careers in the British Army. After early service in India and Burma and in the Boer War (where he was responsible for some spectacular bridging feats) his great interest in the possibilities of military ballooning caused him to be posted to Aldershot as O.C. Balloon Sections. Every one of the early attempts at flight, kites, balloons, gliders, airships and finally aeroplanes, he investigated personally to assess their potentialities. He was next appointed Commandant of the Balloon School at Farnborough and took every opportunity to impress on senior officers the part that war in the air might play in future conflicts, resulting eventually in the formation of an Air Battalion. In 1911 be became Commandant of the School of Military Engineering but when war broke he was given command in the field, leading the 24th Division through some of the most severe fighting on the Western Front. In 1917 he took over as Director-General of the newly-formed Tank Corps, raising its fighting capacity to play an important part in Haig's final offensive in 1918. He died in 1955.

Cave, John W. John gained a scholarship to Wellington where he finished as Head of School. At Cambridge (Trinity College) he played rugger for the University in 1887 and '88 and afterwards gained an England cap. He later returned to teach at Wellington. He died in 1950.

Champion de Crespigny, Claude. The career of Sir Claude Champion de Crespigny (Bart.) belongs to the pages of the *Boys Own Paper*. Born in 1847 he entered the Navy as a cadet aboard the training ship HMS *Britannia* in 1860, and later served in HMS *Warrior*, the Navy's first ironclad. In 1865 he transferred to the Army and it was while he was serving in Ireland that he took up steeple-chasing which became the passion of his life. One famous race he won on *Maid of the Mist*, a 100 – 1 outsider. After service in India, however, he retired from the Army and most of the rest of his life was devoted to sport, particularly dangerous sport. Hunting, shooting and steeple-chasing occupied every winter, but the monotony was varied when he rescued a man from drowning or when he made an early crossing to Belgium in a balloon (in 1883) or even when circumstances compelled him to act as hangman at an execution in Colchester (Sir Claude was High Sheriff and the official hangman was blind drunk). Expeditions abroad included a hazardous journey through the Florida swamps, an attempt to join the Nile expedition of 1889, during which he became the first man ever to swim the First Cataract, and service as an observer in the Boer War. Sadly he found himself too old to be considered for active service in the 1st World War but he continued to play a role in county affairs almost up to his death in 1935 at the age of 88.

Cobb, Reginald. He won a scholarship to Malvern and later became an electrical engineer.

Davidson, Walter L. He went on to Marlborough and entered the Army (R.A.) in 1883. He served in Burma in 1889 winning a medal with clasp but died while on service in the Punjab on July 7th, 1901.

Deane, Henry B.F. After Winchester and Balliol, Henry became a barrister of Inner Temple in 1870, and was appointed Recorder of Margate in 1885, a post he held for twenty years after which he was knighted and made a judge of the High Court (Probate, Divorce and Admiralty Division).

De Lotbinière, Henry A. Joly. After Wellington he entered the Army as Lieutenant R.E. During the 1st World War he served on the N.W. Frontier, with the Mesopotamia Expeditionary Force and in Egypt. In the 2nd World War he was, for four years, Commandant of the R.E. Transportation Training Centre. He retired in the rank of Brigadier and died in 1974.

Fordyce, Robert D. At Charterhouse Robert captained the football XI in his last year (1893). He then went on to Trinity, Cambridge, before being commissioned into the Scots Greys in 1898. He served in the Boer War gaining the Queen's medal and the D.S.O. and being severely wounded at the relief of Kimberley. He continued to serve throughout the 1st World

War, retiring in 1919 in the rank of major. He died at the family home, Brucklay Castle, in 1935.

Fox, Adam. He won a scholarship to Winchester and an exhibition to University College, Oxford, and then taught at Lancing for twelve years, during which he took Holy Orders. From 1918 to 1924 he was Warden of Radley, afterwards going out to South Africa to teach at Diocesan College, Rondebosch. On his return he became a Fellow and Dean of Divinity at Magdalen College, Oxford. He had great gifts as a preacher, as a poet and as a writer, becoming Professor of Poetry at Oxford in 1938 and being awarded the James Tait Black Memorial Prize for his biography of Dean Inge. He came to Westminster in 1942, serving as Treasurer, then Archdeacon and finally sub-Dean of the Abbey up to the age of 80. He died in 1977, aged 93.

Prince **Francis of Teck** was the second of Queen Mary's brothers. He followed Prince Adolphus to Wellington and into the Army. He served with Kitchener's expedition to the Sudan and in the Boer War, being awarded the D.S.O. But 'from his mother he inherited an endearing extravagance and an optimistic temperament'. Finding himself £1000 in debt, he managed to borrow £10,000 which he placed on 'Bellevin' at odds of 10 – 1 on. The horse lost and it fell to the future George V to rescue his brother-in-law with a loan (never repaid) of £11,000. When his mother died he secured her emeralds and 'bestowed them on his aging paramour' but Queen Mary retrieved them when he died in 1910 after a minor operation.

Freyer, Dermot. From Wellington he went on to Cambridge (Trinity College) where he read medicine, continuing his studies at St Thomas's Hospital and at Edinburgh University. By then he had had a number of pieces published and he abandoned the medical profession in favour of journalism. During the 1st World War he served in the London Irish Rifles, for part of the time on the staff of Brig.-Gen. H.M. Grenfell. After the war he returned to journalism and also wrote short stories and light verse. He was active in local politics, serving on the Cambridge Borough Council and he also stood (unsuccessfully) as a Labour candidate in several general elections. He died at Cambridge in 1970.

Gardiner, Henry B. From an early age he displayed a notable talent for music and after Charterhouse and Oxford (New Coll.) he went to Germany to study under Ivan Knorr. Fellow students included Frederick Delius, Roger Quilter and Percy Grainger, the latter describing Gardiner as 'a magnificent pianist, a resourceful conductor and one of the most inspired composers of his generation'. From then until the war he devoted himself to composition, his output including a symphony, quartets and quintets for strings, and ballads for chorus and orchestra, though his most popular piece, *Shepherd Fennel's Dance*, comes from an opera that was never completed. For, on returning from service in the 1st World War, he found that the romanticism of pre-war music had given way to 'a more austere and intellectual approach', and after *Philomena* (1923) he abandoned composition, devoting the remainder of his life to country pursuits. He died at Salisbury in 1950.

Goodenough, William E. He joined the Navy immediately on leaving Temple Grove and by 1905 had attained the rank of Captain. At the outbreak of war he was in Command of the First Light Cruiser Squadron which he led with great dash in the actions of Heligoland and the Dogger Bank. At Jutland, flying his flag in *Southampton*, he commanded the Second L.C. Squadron which reported to the Grand Fleet the approach of the main German battle fleet. In the second phase of the battle, which took place at night, his squadron became heavily engaged and *Southampton* sustained very severe damage but sank the *Frauenlob*. He was immediately promoted Rear Admiral, then Vice-Admiral in 1920 and Admiral, and C in C, Nore, in 1925. After retirement in 1930 he was actively involved in the affairs of the Royal Geographical Society, the British Sailors' Society and the Port of London Authority. He died in 1945.

Gordon, Colin. From Winchester he went on to Wye Agricultural College and gained the Beadle prize for Agriculture. On the outbreak of war he joined the London Regiment and from January, 1915, he was constantly in action in France, being wounded in March, 1915, and, in 1916, being promoted captain and mentioned in dispatches. He was killed in action on Aug. 16th, 1917.

Gordon, Archibald. Colin's younger brother, Archibald, won a scholarship to Westminster. In 1916 he was granted a commission in the Royal Naval Air Service and early in 1918 he was sent to Italy as an Observer Sub-Lieutenant. While his aircraft was patrolling the coast, the pilot got into difficulties and the plane plunged into the sea. Both men were drowned.

Graham, Edward R.C. From Eton Edward joined the 22nd Regt. of Foot, transferring later to the Cheshire Regiment. He served initially on the staff in the Boer War and later in command of the Second Battalion being awarded both the Queen's Medal and the King's Medal. Promoted Major-General in 1912, he served through the 1st World War, being mentioned in dispatches eight times, and being appointed K.C.B. and K.C.M.G. He retired in 1920, and died in 1952.

Grenfell, Francis Octavius and Riversdale Nonus. The twins, Francis and Riversdale were the eighth and ninth sons of Pascoe du Pre Grenfell. All nine attended Temple Grove and eight went on to Eton (the other into the Navy). At Eton cricket and beagling were their chief preoccupations. After the deaths of both parents their uncle became their guardian. He gave them sound advice: 'Read your Bibles, and shoot well ahead of the cock pheasants'. Both had set their hearts on military careers but family finances would support only one and for the remainder of their lives Rivy loyally supported Francis's Army career while going into business himself. During the reign of Edward VII they became two of the best known young men in England. They were outstanding polo players and excelled in other sports, too. But Francis also took a keen interest in military developments, being present at both German and French army manoeuvres. When war came Rivy, a Territorial officer, soon got himself attached to Francis's regiment, the 9th Lancers. In an heroic action on the Belgian frontier Francis led his squadron in a charge on the advancing Germans and later, although badly wounded, organised the successful withdrawal of the guns of an RFA battery. For these actions he was awarded the first V.C. of the war. Rivy had meanwhile been in almost equal danger, acting as galloper for the General, 'riding for hours under heavy fire on a tired horse on missions of vital importance'. A month later he was killed in action near the Chemin des Dames. Francis survived him by eight months, losing his life on May 24th in the terrible Second Battle of Ypres.

Grey, Edward. Edward's father died when he was still at Temple Grove and from an early age he made himself responsible for his mother's welfare and that of his six brothers and sisters. At Winchester, although a fine cricketer, he developed a passion for fishing and for the peace of the countryside. At Oxford (Balliol) he was actually sent down for idleness but was allowed to return for the examinations in which he took a poor third in Jurisprudence. But with his marriage to Dorothy Widdrington and his election as Liberal M.P. for North Northumberland his life became altogether more purposeful, without ever losing his love of Nature and of country pursuits. In 1892 he was appointed Under Secretary of the Foreign Office but from 1895 to 1905 the Conservatives were in office and Grey was able to enjoy the happiest ten years of his life. When the Liberals returned to power he was appointed Foreign Secretary, a post he was to hold for eleven momentous years. Tragically his wife died as the result of an accident early in 1906 and Grey, from a combination of personal loneliness and a strong sense of duty, devoted himself to public affairs. He never mastered French (or any other foreign langauge) and he almost never went abroad but from then until August, 1914, he worked unceasingly in the cause of peace and of better understanding among the Great Powers of Europe. When Germany declared war on France and her army invaded Belgium it was Grey's speech to the House of Commons which convinced the waverers that England could not remain neutral. He remained Foreign Secretary, in spite of failing eyesight, until Asquith's resignation in December, 1916. After the war he remarried and lived quietly on his estate at Fallodon, seeing friends, writing and enjoying the companionship of his ducks. He died in September 1933.

Guthrie, Tyrone. Wellington, which Tyrone entered soon after the outbreak of the 1st World War, bred in him 'a disregard for conventional comforts that endured all his life', but it was not until he reached Oxford (St Johns) that his talent for acting and direction was given full scope. During the 1920s he gained experience of theatrical work in various parts of Britain, in Ireland (for which he had a special affection) and in Canada, work that included both radio and theatre productions. But his great opportunity came, soon after his marriage, when he was appointed producer at the Old Vic, directing a string of classic plays in which a new breed of actors and actresses first made their reputations. During the war he toured Britain with companies that kept opera, ballet and drama alive and his wartime experiences emboldened him to become more bravely experimental when the war was over. At the Edinburgh Festival of 1948 he first made use of the open stage which was to revolutionise theatrical production, and when a Canadian businessman persuaded Guthrie to launch a Shakespeare Festival in Stratford, Ontario, he was able to give full rein to his new ideas. By then a world-famous director, he worked in Dublin, in New York, in Tel-Aviv and elsewhere, his productions ranging from *HMS Pinafore* to *Oedipus Rex* and including *Traviata* at the Metropolitan Opera House. There followed work in America where, in Minneapolis, he directed Aeschylus' *Orestia* trilogy; and in Australia with *Oedipus Rex* in Sydney and *All's Well that ends Well* in Melbourne. He died at his home in Ireland in May, 1971.

Gyll, Brooke. No information.

Hailstone, Arthur. He went on to Rugby and spent much of his life as a solicitor in Bradford. Not much is known of his three brothers: Walter joined the navy as a cadet in HMS *Britannia*; Herbert won a scholarship to Eton and from there went on to Peterhouse, Cambridge, where he afterwards taught for several years, but committed suicide in 1896; Samuel followed Arthur to Rugby, went on to Cambridge (Caius) and was ordained in about 1878.

Hannay, James O. James came from the most staunchly Protestant part of Belfast where his father was a clergyman. After Haileybury he went on to Trinity College, Dublin, and not long after he was ordained he married Adelaide Wynne. In 1892, they moved to Westport, a small town in the poorest part of Ireland, where James was rector. There his four children enjoyed an idyllic upbringing but his stipend was so modest that he took to writing, for which he had a natural gift, to supplement the family income. His first novel, *The Seething Pot*, appeared in 1905, under the pseudonym, George A. Birmingham, and some thirty others followed, many of them with an Irish setting. They are characterised by improbable plots and lively dialogue and were immensely popular in their day. In the 1st World War James served for a time as a chaplain to the forces and when the war was over he helped to restart the English church in Budapest during which he conducted the funeral service of a miner (who had had an English mother) in a remote village at the foot of the Carpathians. On returning to England he was offered the living at Mells in Somerset where he spent ten happy years. After his wife's death he moved to London as rector of Holy Trinity in Prince Consort Road. He continued to write novels up to the time of his death in February, 1950.

Hare, Steuart W. He went on to Sandhurst after Eton and served in a series of expeditionary forces in the 1890s. At the outbreak of the 1st World War he commanded the 86th Fusilier Brigade and was wounded during the Gallipoli landing. He later saw service at Salonika and, under Allenby, in Egypt and Palestine. He was promoted Major-General in 1916 and appointed KCMG three years later. He died in 1955.

Harland, Reginald W. Seven of the nine Harland brothers attended Temple Grove, of whom Reginald was the last. After Wellington and Sandhurst he joined the Hampshire Regiment in 1903, reaching the rank of Captain in 1911. He was killed in action on Oct. 30th, 1914.

Harrison, Robert F. He won a scholarship to Wellington and was later Senior Moderator at Trinity College, Dublin. He embarked on a legal career becoming a Scholar in Constitutional and International Law at Middle Temple and subsequently a barrister at King's Inns, Dublin. He was made Q.C. in 1899.

Headlam, Hugh R. He won a scholarship to Wellington where he played in the cricket XI, went on to Sandhurst and joined the York and Lancaster Regiment. He served in the Boer War, gaining both the Queen's medal and the King's medal and throughout the 1st World War, in the latter stages as (Brigade) Commander of the 64th Infantry Brigade. He was awarded the D.S.O. He retired (1930) in the rank of Brigadier-General and died in 1955.

Headlam, Maurice F. He won a scholarship to Eton and went on to Corpus Christi College, Oxford. He acted as Private Secretary to Austen Chamberlain, who was then Chancellor of the Exchequer, for two years, before pursuing a career in the Civil Service which culminated in his appointment as Comptroller-General and Secretary of the National Debt Office. He died in 1955.

Hill, Matthew D. After Eton and Oxford (New College) where he read Natural Science and obtained a First, he returned to Eton to teach biology. Etonians still regarded any form of science as a joke, or as some sort of conjuring trick and it proved a long struggle to raise the status of science teaching at the College. In 1905 he became a house master, taking over the last of the Dames' houses from the famous Miss Evans. His autobiography, *Eton and Elsewhere*, is full of delightful anecdotes, many of them centring on the varied menagerie that he maintained at Eton. 'Two lemurs . . . emitting blood-curdling shrieks' or ' "There's a live python somewhere in the room", I warned' or the roar of laughter when a queen bee stung him on the lip, make one realise that Hill's lessons must have been a popular, as well as an educative, part of the syllabus. On retiring from Eton in 1927 he went to live in Shropshire where he farmed 300 acres.

Hobart, R. Charles A.S. At Charterhouse (to which he won a scholarship) he played in the football XI of 1899 before going on to Trinity, Oxford. He joined the Indian Civil Service as an assistant commissioner in United Provinces. In 1937 he was made a Companion of the Order of the Indian Empire. He died in 1955.

Hobart, Percy C.S. A keen rugger player he captained the 1st XV at Clifton (where he had won a scholarship) and the 2nd XV at Woolwich. Joining the Royal Engineers he served in India for ten years before the outbreak of the 1st World War and went to France with the Indian Expeditionary Force. He won an M.C. at Neuve Chappelle, a DSO later in Mesopotamia, and was awarded the OBE when the war ended. In 1923 he joined the Royal Tank Corps, predicting that the outcome of the next war would be decided by the tank. When, in 1934, he was appointed to command the 1st Tank Brigade, he 'evolved new tactical methods based on mobility, flexibility and speed', and he incorporated those principles into the training of the 7th Armoured Division in Egypt but his methods, and his outspokenness, did not endear him to the War Office and by 1941, incredibly, he was a mere corporal in the Home Guard. Rescued by Churchill, he was promoted Major-General and commanded, first, the 11th Armoured Division and then the 79th with which he prepared for the Normandy landings. He developed a wide range of tanks adapted for assault duties, collectively known as 'The Funnies' and in June, 1944, they played a vital role during and after the D-Day landings. Made K.B.E. in 1943, he retired soon after the war and for five years acted as Governor of Chelsea Hospital. He died in Feb. 1957.

Hobday, Edmund A.P. He joined the Army (Royal Artillery) and served with the Malakand Field Force and other expeditions during the 1890s. He retired as full colonel in 1909 but was recalled to the Army in the 1st World War. He died in 1932.

Hooker, Charles P. After Marlborough he undertook medical training at St Bartholomew's Hospital and was a surgeon in Edinburgh for many years. He died in 1933.

James, Montague Rhodes. From an early age he developed an interest in esoteric subjects to the detriment of those officially approved and it required a second attempt to secure the Eton scholarship that his father was counting on. At Eton a brilliant scholastic career (his nickname was 'The Learned Boy') was crowned by the Newcastle Prize and the top scholarship to King's, Cambridge. There in 1887 he became a Fellow and, in 1905, Provost. He had a deep love of France and undertook walking and bicycling tours almost every year. As a relief from his more serious researches into palaeography, ecclesiastical architecture and Aprocryphal literature he took to writing ghost stories, the first of which were published in 1904 and which remain among the very best of the genre. When the 1st World War ended he was appointed Provost of Eton, showing a keen interest in the activities of the College, ranging from the Shakespeare Society to the Scout Troop, but reserving his deepest affection for the Chapel, both its furnishings and its services. He was appointed to the Order of Merit in 1930 and died in June, 1936.

Keigwin, Richard P. He was probably the most versatile all-round games player to have emerged from Temple Grove. He won a scholarship to Clifton where he played in the 1st XV and captained the cricket team. At Cambridge (Peterhouse) he won blues for cricket, football, hockey and racquets. He later went to teach at R.N.C., Osborne.

Kemball-Cook, Basil A. After scholarships to Eton and King's, he passed into the first class of the Civil Service and entered the transport department of the Admiralty. In the 1st World War he held the vital post of director of naval sea transport at the Ministry of Shipping. His harmonious relations with Britain's allies, notably France and the U.S.A., led to his appointment as a delegate on the Reparation Commission, at the end of which he was made K.C.M.G. He then retired from the Civil Service to become managing director of the British Tanker Company and later a director of British Guiana Goldfields. 63 when the 2nd World War began he nevertheless served as an air-raid warden and from 1942 to '48 as divisional food officer for London. He died in November, 1949.

Kirkpatrick, Frederick A. He gained a scholarship to Wellington and an exhibition to Trinity College, Cambridge. He afterwards became University Lecturer in Spanish and an acknowledged expert on the Spanish Empire in Central and South America.

La Terrière, F. Bulmer de Sales. After Eton and Oxford (Magdalen) he entered the army and saw active service with the 18th and 19th Hussars in Egypt and the Sudan. After retiring from the Army he was appointed exon of the King's Body Guard of the Yeoman of the Guard. His entertaining autobiography, *Days that are gone*, gives a good picture of military and social life in late Victorian England.

Law, Charles W.A. He went on to Marlborough and, in 1882, joined the 4th Dragoon Guards. He served in the Sudan and was killed at Abu Klea on the Gordon Relief Expedition in January, 1885.

Leese, Neville. After Winchester he became a mining engineer, working in New Zealand and later in British Columbia, where he gained the Royal Humane Society's medal for saving life. In the Boer War he served in Strathcona's Horse and was afterwards manager of the Waterson Gold Mine in Mexico. He served throughout the 1st World War, being awarded the D.S.O. and O.B.E. He died in 1949.

Leese, William H. He was the eldest of the four Leese brothers and after Winchester and Cambridge (Trinity Hall) he was called to the Bar and practised as a solicitor. He succeeded to the baronetcy in 1914 and afterwards served as a J.P. His son, Oliver, followed Bernard Montgomery in command of the 8th Army.

Mackintosh, Ernest E.B. After Eton (where he won a scholarship) he entered Woolwich and became 2nd Lieutenant, R.E. He was seconded to the Egyptian army for nine years, acting as ADC to the Sirdar (1908-11). During the 1st World War he served in France (1915-18) and for the last 18 months he was Assistant Engineer-in-Chief GHQ, being awarded the D.S.O. On retirement from the Army in the rank of colonel in 1933 he became Director of the Science Museum. In the 2nd World War he was recalled to service and was appointed Commandant of the School of Military Engineering at Chatham, returning later to the Science Museum from which he finally retired in 1945.

Macleod, Norman C. He won a scholarship to Wellington where he was head of the school before going on to New College, Oxford. In 1890 he became a barrister of Inner Temple but almost the whole of his legal career was spent in India where he became, in turn, Official Assignee (Insolvent Debtors' Court) Bombay, a member of the Legislative Council of India, a judge of the Bombay High Court and finally (1919-26) Chief Justice of the Bombay High Court.

McNeile, John. He went on to Eton and Trinity College, Cambridge, and joined the Coldstream Guards in 1885. In the Boer War he was awarded the Queen's medal with five clasps and the King's medal with two clasps. Promoted major in 1903, he retired from the Army two years later.

Mordaunt, Gerald J. He won a scholarship to Wellington where he was head of school, captain of the cricket XI and a member of the racquets pair. At Oxford (University College) he again captained the cricket team, appearing for the Gentlemen against the Players at Lords in 1896 and '97. He played for Kent for four seasons before devoting his attention, full time, to the Stock Exchange.

Morgan, Arthur C.O. At Winchester he won the King's Gold Medal for a Latin essay and at Cambridge (Trinity) the Chancellor's Gold Medal for English verse. He was also President of the Union. He was called to the Bar in 1909 but joined up soon after war broke out, serving in the RFA. In October, 1915, in an attack on a German redoubt, 'recognising the great importance of taking the trench, he dashed in front . . . all the infantry officers having been shot down. He was killed in the act of cheering others on who were still behind'.

Morton, William R. From Marlborough he joined the Royal Engineers, reaching the rank of lieutenant-colonel in 1908. He was fluent in Arabic and Turkish, acting as an interpreter. In the Burmese expedition of 1885 he gained the Queen's medal and he later served as Executive Engineer in the Public Works Dept. of the Punjab.

Mullens, Richard L. He went on to Sandhurst after Eton, becoming 2nd Lieutenant, 16th Lancers, in 1890. He served in India and Egypt and in the Boer War he was employed on special service, gaining both the Queen's medal and the King's medal and being severely wounded towards the end of the war. In the 1st World War he commanded the 1st Cavalry Division in France from 1915 to 1918, being promoted Major-General in the latter year.

Nicholson, Sydney H. He went to Rugby and Oxford (New College) before entering the Royal College of Music. In 1903 he became organist of Lower Chapel at Eton and later of Carlisle Cathedral before accepting the organistship at Manchester Cathedral (in preference to Canterbury) in 1908. He brought the cathedral services to a very high standard and, in 1918, he succeeded Sir Frederick Bridge as organist of Westminster Abbey. After ten years he left the Abbey to found the English (later Royal) School of Church Music. By the time of his death over 2000 church choirs were affiliated to it. He organised and conducted great festivals of church music, both at the Albert Hall and at Crystal Palace in the 30s. He published *Quires and Places where they sing* and in 1938 he was knighted for his services to church music. He died in May, 1947.

Ommanney, Frederick G. After Winchester, he went into the brewery business and became a partner in the firm of Thompson & Son of Walmer. He devoted much of his time to Kentish affairs and died at his home, Sheen House, Upper Walmer, in June, 1889.

Orde, Charles W. He won a scholarship to Eton and afterwards to King's. He entered the Foreign Office in 1909, eventually becoming a senior clerk, 1st Division.

Palmer, Claude B. From Cheltenham he went into the Army, rising to the rank of Lt. Colonel. On retirement he was appointed J.P. (Co. Durham) in 1895 and as a Knight of Grace of the order of St John of Jerusalem in 1907.

Palmer, William W. At Winchester, where he won the English silver medal, he had Edward Grey as his fag. He gained first class honours for modern history at Oxford (University College) and then went into politics, being elected Liberal M.P. for E.Hants in 1885, but split with Gladstone over Home Rule and thereafter sat as a Liberal Unionist until succeeding his father (Roundell) as Earl of Selborne in 1895. He had earlier married Beatrix Cecil, daughter of the 3rd Marquess of Salisbury (three times Prime Minister). He was appointed Under Secretary for the Colonies under Joseph Chamberlain, his five year tenure covering the Jameson Reid, the Commonwealth of Australia Act and, of course, the outbreak of the Boer War. But it was during his next five years as First Lord of the Admiralty that he played his most crucial role in British history. He made Jackie Fisher 2nd Sea Lord in 1902 and Fisher began to undertake the momentous changes (continuing when he became 1st Sea Lord in 1904) that culminated in the launching of HMS *Dreadnought*, changes that fitted the Royal Navy to face the German challenge on the High Seas in the 1st World War. The introduction of submarines, of steam turbines and of the Naval Colleges of Osborne and Dartmouth, date from this period. The next five years were spent in South Africa as High Commissioner and as Governor of the recently defeated provinces of Transvaal and the Orange Free State. He proved 'remarkably successful in winning the confidence of the Boers', and the setting up of the Union of South Africa in 1910 was a fitting culmination to his efforts. He was awarded the Knighthood of the Garter. During the 1st World War he was for a time President of the Board of Agriculture and Fisheries, and a committee he set up on home-grown timber led to the formation of the Forestry Commission. His rooted distrust of Lloyd-George, however, caused him to refuse further appointments, including that of Viceroy of India. After the war his varied concerns were reflected in such offices as Warden of Winchester College, an elder brother of Trinity House and chairman of the House of Laity in the Church Assembly. He died in February, 1942.

Payne-Gallwey, Reginald F. He went on to Lancing and to Mercer's Hall, Oxford, and afterwards took a post in industry in Birmingham. He succeeded to the family baronetcy in 1955.

Pinckney, William J. He took his degree at Cambridge (Trinity College) and was ordained in 1870. From 1884 to 1901 he was vicar of Quatford, near Bridgnorth. He died in 1927.

Pitman, James C. He was one of eight brothers, all of whom attended Temple Grove. From there he went on to Eton and to New College, Oxford. He was called to the Scottish Bar, holding the post of Standing Counsel to the Commissioners of Woods and Forests, to the Board of Trade and to the Postmaster-General for seventeen years. He was then Sheriff of Caithness, Orkney and Zetland for a further eight years before (in 1928) becoming one of the Senators of the College of Justice in Scotland. He died in 1941.

Pitman, Thomas T. Like all his brothers, Thomas went on to Eton where he rowed in the VIII and then entered the Army as 2nd Lieutenant, 11th Hussars, in 1889. He served on the N.W. Frontier and throughout the Boer War; and for much of the 1st World War he commanded the 4th Cavalry Brigade, being promoted Major-General in command of the Second Cavalry Division in 1918. He retired in 1930 and died in 1941.

Prothero, Rowland E. At Marlborough he played cricket in the 1st XI and went on to Oxford (Balliol) where he took first class honours in History and Law. A deterioration in his eyesight caused him to abandon a legal career and he spent the greater part of two years touring France on foot, studying with equal relish French farming methods and the places associated with the great figures of French medieval history. Back in England he became a regular contributor to the learned journals of the day and he published several books, eventually becoming editor of *The Quarterly Review*. An unplanned change of career took place when, in 1898, he was invited by the Duke of Bedford to become Agent-in-Chief of his vast estates, work that revived his dormant interest in farming and to which he responded with vigour and enthusiasm. An uncontested election, just before war broke out, enabled him to become M.P. for Oxford University. In December, 1916, he succeeded Lord Selborne as President of the Board of Agriculture and instituted vigorous measures to combat the food shortage resulting from the all-out U boat campaign. War Agricultural Committees were set up and a Corn Production Bill was passed by Parliament 'to make wheat-growing profitable and to improve farm workers' wages and conditions of employment'. Even more valuable was the creation of the Women's Land Army, which brought 16,000 badly needed women to work on the farms as 'tractor drivers and thatchers, cowherds and threshers, and even as plough girls'. After the war, ennobled as Lord Ernle, he continued to give public service and in 1924 was elected President of M.C.C. He died in July, 1937.

Rawlinson, Henry S. He went from Eton to Sandhurst, becoming 2nd Lieutenant, King's Royal Rifle Corps, in 1884. He served with the Burmese Expedition of 1886 and with Kitchener's army in the Sudan in 1898 (having meanwhile transferred to the Coldstream Guards), at the battles of Atbara and Omdourman. In the Boer War he was present throughout the siege of Ladysmith. Later in the war he commanded a mobile column, was mentioned in dispatches five times and promoted colonel. Soon after the outbreak of the 1st World War, in the rank of major-general, he was given command of the 7th division with the task of covering the right flank of the Belgian Army. Throughout 1915 he commanded the 4th Corps, leading the corps through the grinding battles of Neuve Chapelle, Festubert and Loos. The following year he was promoted Lieutenant-General in command of the Fourth Army, and as such the responsibility for a crucial part of the Battle of the Somme fell on him. A detailed study of his generalship was published in 1992 under the title, *Command on the Western Front*. In 1918 he had to take command of the remnants of the Fifth Army, shattered by Ludendorff's offensive, and by heroic efforts the Germans were prevented from capturing Amiens. Rawlinson's counter-attack, on Aug.8th, was described by Ludendorff as 'the black day in the history of the German Army'. At the end of September, reinforced by American troops, Rawlinson broke through the Hindenburg Line. Six weeks later the Germans asked for an armistice. In 1919 he had to supervise the evacuation of British forces from Archangel and the following year, now a Baron, he was appointed Commander-in-Chief, India. He died at Delhi in March, 1925.

Richardson, William P. He went to Winchester, of which his father was Second Master, and later became a solicitor in the City. Much of his life thereafter was lived in Hampshire for which he served as Under-Sheriff, and also as Mayor of Winchester. He died in 1950.

Rowe, William H.P. He won top scholarship to Harrow in 1870 and finished as Head of School. Going up to Oxford he took a First in Classical Mods and rowed in the Balliol boat which went head of the river. Next year he became a student at Lincoln's Inn but died suddenly in April, 1880.

Scudamore, Charles P. He won a scholarship to Wellington, went on to Sandhurst, where he was chosen Queen's cadet and joined the Royal Scots Fusiliers. He served in the Burma campaign of 1886-8, winning the D.S.O., served with the Hazara Field Force and in the Tirah campaign and, although officially Inspector of Army Signalling in the Punjab, he volunteered for service in the Boer War, in which his company was the first to reach Mafeking. He was later appointed Commandant of the Army School of Signalling. Recalled for service in the 1st World War he was awarded the D.S.O. and three times mentioned in dispatches. He reached the rank of Brigadier-General and died in December, 1929.

Scudamore, T.V. After Wellington, he joined the Army but was wounded and taken prisoner at Ypres. After the war he emigrated to Canada where he became a Notary Public and estate agent in Vancouver. He returned to England before war broke out and in 1944 commanded a Pioneer Corps Group in Italy, afterwards serving with the Allied Military Government in Germany.

Sharpe, Henry C. He won a scholarship to Marlborough and afterwards went on to Trinity College, Cambridge. He became a partner in Townsend and Sharpe solicitors, in the City.

Sherring, Frank B. He won a scholarship to King's Canterbury, but transferred a year later to Westminster, also with a scholarship. From there he went on to Cambridge (Trinity) where he gained a blue for soccer and 1st class honours for Law. He entered the Indian Civil Service in 1897, first as an Assistant Magistrate and eventually as District and Sessions Judge at Lucknow. He died in 1939.

Shute, Cameron D. He went on to Marlborough and Sandhurst, joining the Welsh Regiment in 1885. He served in Kitchener's expedition to the Sudan and was present at the battle of Omdourman. At the outbreak of the 1st World War he was a colonel commanding the 2nd battalion, the Rifle Brigade. Subsequently he commanded the 59th Brigade, the Royal Naval Division, the 32nd Division and finally the 5th Corps (3rd Army), all these commands being on the Western Front. In 1919 he was awarded the K.C.B. and K.C.M.G. His final command was G.O.C.-in-C. Northern Command and on its termination he retired with the rank of full General.

Staveley, George H. After Charterhouse he went on to Sandhurst and then joined the Yorkshire Light Infantry. He was a member of the winning team for the Army Cup at Bisley in 1904. He had reached the rank of major when he was killed in action in April, 1917.

Stockdale, Herbert E. He won a scholarship to Wellington and afterwards went on to Woolwich for which he played cricket and football. He was commissioned into the Royal Artillery and served in the Boer War, gaining the Queen's medal with six clasps. In the 1st World War he rose to the rank of Brigadier-General. He retired in 1919, and served as a J.P. in Northamptonshire.

Storrs, Christopher E. He gained a scholarship to Malvern, and won first class honours in classics at Cambridge (Pembroke College) before being ordained in 1912 and returning to Malvern to teach. His sixteen years there (for much of the time as housemaster) was interrupted by three years' service as Chaplain to the Forces. In 1930 he became the first Sub-Warden of St George's College in the University of W. Australia and in 1946 he was appointed Bishop of Grafton (NSW). On his return to England he became Warden of St John's College, Morpeth, and later Rector of Hazelbury Bryan in Dorset. He retired in 1964 and went to live in Chichester where he died in 1977.

Storrs, Francis E. He won scholarships to Radley and to Jesus College, Cambridge. In the 1st World War he served with the RNVR, at one stage, working with Compton Mackenzie for British Intelligence. Towards the end of the war he was employed in the War Cabinet Office but contracted Spanish 'flu and died on Nov.10, 1918.

Storrs, Ronald H.A. He won a scholarship to Charterhouse, followed by an exhibition to Pembroke Colege, Cambridge. He was accepted for entry to the Egyptian Civil Service and returned to Cambridge for a fourth year to study Arabic. He arrived in Egypt in 1904 and immediately began to immerse himself in the languages, customs and culture of the Near East. Within five years he had become oriental secretary to the British Agent (Sir

Eldon Gorst) who, at that time, was effectively the ruler of Egypt. When Kitchener replaced Gorst, Storrs worked closely with him and together they visited almost every inhabited part of the country. When war broke out, the threat to Egypt from Turkey was to be countered by fomenting rebellion among the Arabs who were then Turkish subjects. Storrs during the course of 1916 made three journeys into the Hejaz to gain the confidence of the Arab chieftains and on the last one he took with him a young archaeologist, T.E. Lawrence. Throughout the Arab revolt Storrs maintained contact with Lawrence and when Jerusalem had been liberated from the Turks, Storrs was appointed military governor of the city. For the next seven years he worked tirelessly to restore the city to its former prosperity and beauty. The first priority was to save the inhabitants from starvation. Later he could attempt the task of bringing harmony to people of different races and religions. He founded Pro-Jerusalem, a Society for the preservation of the City's antiquities and on its behalf he became 'a convincing and successful Schnorrer' (Yiddish for beggar). From Jerusalem he was transferred to Cyprus, as Governor of the Colony. Here the main problems he faced were those of poverty and Enosis (the movement for political union with Greece). He set to work to revitalise the Island's economy, working with a tiny budget, and after six years was beginning to see some reward for his efforts but shortly before his planned departure a riot broke out, Government House was set ablaze and Storrs lost his lifetime's collection of books and works of art. His active career ended with a two-year spell as Governor of Northern Rhodesia. He enjoyed a long and busy retirement and died in London in November, 1955.

Twining, Daniel S. His father was Edward Twining, a surgeon, and his grandfather the Rev. David Twining, a younger brother of Richard Twining, whose son, also Richard, is mentioned on p.92. Edward, who may have been a naval surgeon (Daniel was born in Hull) was frequently away from home and young Daniel appears to have been brought up by maiden aunts. It is not known where he went on to school in 1859, but he died, aged 21, at Walthamstow in 1865.

Waterfield, Bertram C. He entered Marlborough and went on to Sandhurst but nothing seems known about his career. He died in 1929.

Waterfield, Charles R. He was O.C.Waterfield's eldest son and went on to Charterhouse and King's, Cambridge. He became Private Secretary to Sir Arthur Gordon, Governor of Fiji, but died there of dysentery in 1886, aged 24.

Waterfield, O. Frank. O.C.Waterfield's second son, he went on to Marlborough and to Trinity College, Cambridge. He was a cotton merchant in Manchester for many years and died in 1941.

Waterfield, Harry W. He won a scholarship to Westminster where he was captain of the school and of the football XI, and an exhibition to Trinity College, Cambridge. He taught for eight years at Bradfield, before becoming Headmaster of a prep school at Hastings, Arnold House. He returned to

Temple Grove as Second Master in 1900, and lived at Observatory House. He was ordained priest in 1901 and the following year when H.B. Allen retired, he bought the school from him. Four years later he moved Temple Grove to Eastbourne and remained there as headmaster until 1935. His three sons and two daughters were all pupils in the school. When Meston Batchelor became headmaster and moved the school to Herons Ghyll, Waterfield continued to teach and acted as chaplain till 1944, driving over from his home at Lamberhurst every day. He died in March, 1951.

Waterfield, Reginald. He was Harry's younger brother and went on, with scholarships, to Winchester and New College, Oxford. For a time he acted as tutor to Prince Arthur of Connaught and then taught at Rugby for six years before becoming principal of Cheltenham College in 1899, a post he held for twenty years. On leaving Cheltenham he became Dean of Hereford, retiring in 1947. He died in 1966, aged 99.

Watson, Bertram, C. On leaving Temple Grove he went straight to HMS *Britannia*. His first sea-going appointment was as midshipman on the flagship HMS *Majestic*. Throughout the 1st World War he served with the Harwich Force, winning the DSO, and later commanded, in turn, the 4th Destroyer Flotilla, the Tactical School at Portsmouth, HMS *Curlew* and HMS *Valiant*. In 1934, in the rank of captain, he was appointed Director of the R.N. Staff College, Greenwich. He retired in the rank of Vice-Admiral.

Wellesley, Richard A.C. He won a scholarship to Charterhouse and went on to Woolwich. During the 1st World War he commanded 126 Battery RFA (1914), 64th Brigade RFA (1915), was CRA 21st Division (1916) and finished as GOC RA 13th Corps (1917-18). He retired in 1920 in the rank of Brigadier-General, devoting himself to organisations for ex-service men. He died in June, 1939.

Weston, Spencer V.P. He joined the Army after Wellington and served throughout the First World War, being awarded the DSO and two bars and the M.C. and rising to the rank of Brigadier-General. He left the Army to become a member of the Stock Exchange but rejoined in 1940 and, as Inspector of Transport, spent more time at sea than many naval officers, being on one occasion torpedoed, swimming all night and then spending several days in an open boat. At the time of his death in 1974 at the age of 90 he is believed to have been the last surviving holder of the rank of Brigadier-General.

Wickham, Archdale P. Seven Wickhams attended Temple Grove in the 19th century, but Archdale is the only one whose dates would qualify him for the reference on p.162. He won a scholarship to Marlborough and to New College, Oxford, and was ordained in 1880. He served as vicar of Martock (Som.) for twenty-two years (and Rural Dean of Ilchester); and for a further twenty-four years he was vicar of East Brent and a Prebendary of Wells Cathedral. He died at E. Brent in 1935.

Wilkinson, James V.S. A great all-rounder, he won scholarships to Rugby and Oxford (University Coll.), played in Rugby's 1st XV and won a rugger blue in

1904. A member of the Indian Civil Service for 13 years he worked mainly on the N.W. Frontier, but in 1924 he joined the staff of the British Museum. He already possessed a sound grasp of several Indian languages and was not only a fine Persian scholar but possessed a deep understanding of Persian and Moghul art. He was largely responsible for the Persian Art Exhibition at Burlington House in 1930 and the one devoted to India and Pakistan in 1947. In 1950 he undertook the installation of the great Chester Beatty Library in Dublin which he continued to oversee till his death in 1957.

Williamson, G. After Temple Grove he was educated in America, at St Pauls, Concord, and then at Harvard, but returned to take a degree at Oxford (New College) in 1905.

Worsley, R.M.M. He won a scholarship to Malvern in 1901. After Oxford (Balliol) he joined the Ceylon Civil Service and became Assistant Government Agent at Hamlantota, being awarded the C.M.G. During the 1st World War he served in the R.G.A. and was awarded the M.C. He died in 1939.

Yarde-Buller, Reginald J. He went on to Radley and to Christ Church, Oxford, and was ordained in 1887. He served, in turn, as vicar of Haresfield, Warden of Navy House, Chatham, rector of Peasemore and vicar of Newquay. He died in 1950.

Yarde-Buller, Walter. From Eton he joined the Army in 1878. He served in the British South African Police for fifteen years (till 1911) and with the 1st Mounted Division in the 1st World War.

Young, Patrick S. He went on to Harrow, and all that appears to be known of him thereafter is that he worked as an Exchange Broker in Chile.

Young, William R. After Harrow he devoted himself to public affairs in Antrim, being sworn of the Privy Council for Ireland in 1921. He was the author of *Fighters of Derry*. He died in 1940.

To draw conclusions, however tentative, from such a tiny sampling (one hundred and ten individuals out of around 2,000 pupils who attended Temple Grove between 1859 and 1906) may seem both rash and unjustified. Yet the very fact that it has been a random sampling with boys selected on the basis of proficiency in classics or at games, on dislike of school food, on frequent receipt of punishment or other escapades, or simply through appearing on a particular list, emboldens me to make the attempt. With those whose writings about the school have appeared in print one must be more careful, remembering that birching, bullying, bad teaching and worse food, make more interesting topics for the writer than kindly schoolmasters and attentive pupils. Yet here, too, the verdicts are sufficiently varied to make one feel that such men are not untypical of their fellows.

Eleven of the names appear twice so that the final total comes to 121. Almost a third (34) were soldiers (excluding those who served only in the 1st World War). Eleven were clergymen and eleven were dons or schoolmasters (two of whom were ordained), nine were scholars or writers, nine were members of the legal profession and eleven were in business or commerce; five were civil servants or in local government; five were in the Royal Navy, five in the Colonial or Indian Civil Service and three in politics (all three held ministerial office). Of the remainder four died young, one was a lifelong sportsman, three were engineers (other than Royal Engineers), one went into medicine, one into farming, three followed careers in music and one in the theatre. That leaves four boys of whose later careers very little seems known.

The most obvious fact that stands out is that they almost all worked. In Victorian and Edwardian days, when the country gentleman was part of the social fabric, the typical product of Temple Grove was a member of one of the professions. To the extent that any civilised country will always have need of such men, how did the education (in its widest sense) that they received at Temple Grove and at public school prepare them for later life? Would they have been better at their jobs (taking into account the times in which they lived) if, say, the curriculum had included literature, music, drama, art, science and conversational modern languages? Would they have followed different careers if they had attended day, or co-educational, schools or ones at which there was no corporal punishment (such schools *did* exist, even in 19th century England), and would the country have been poorer as a result? If the findings for Temple Grove are not dissimilar to those of other prep schools of the time, does it mean that, by and large, the fields of scientific research, of engineering, of the Arts, of land management, of medicine and of local government were occupied by men who had had a very different educational grounding? And, if that was so, did it matter? Obviously these are questions to which no definite, and certainly no quantitative, answers can be given, but that does not mean that they should not be asked. At a time when educational aims and methods are under intense scrutiny, it behoves us to consider the strengths and weaknesses of earlier systems.

The world of the Victorian prep school, and of the men educated in them, was a world of certainties. Few people of that class were in much doubt as to what was 'right' and what was 'wrong'. Cruelty was wrong and so was untruthfulness. One respected one's elders and betters; one obeyed orders, one did not question them. And one certainly did not question the assumptions that lay behind the educational system: the supreme value of the classics; the importance of accuracy in all one's work; the necessity for corporal punishment when standards of behaviour and of study were not achieved. Had their names ever been mentioned in the schools, Charles Darwin, Charles Bradlaugh or Josephine Butler would surely have been execrated.

The underlying belief was that a gentleman served his country in peace and war. Far-reaching decisions would be made by a handful of men and, if they happened to be Liberals, those decisions might well evoke howls of protest but even Liberals, thank God, at least were mainly gentlemen. For the remainder the good governance of the realm and its vast overseas possessions was both the privilege and the duty of the English gentleman. 'Governance', of course, would include the spiritual and moral welfare of its people. Let others, whose education had fitted them for such tasks, build roads and railways, design buildings and ships, invent machines for every conceivable purpose, improve the nation's health and – if absolutely necessary – paint pictures, compose music or even act on the stage. But although there might well be changes in the mechanism of life and work and a gradual rise in overall prosperity there could, there should, be little or no change in the basic fabric of society. A gentleman and his family could expect to enjoy the full-time services of cooks and chambermaids, gardeners and governesses, if they themselves were to fulfil their duties and enjoy their deserved privileges.

It is impossible to assess what part religion played in the lives of those families. Even if one knows that they attend church Sunday after Sunday and may even discuss the sermon afterwards; even if they deplore the spread of atheism and socialism (the two heads of the hydra); even if they denounce immorality and divorce (and I can find no single instance among nearly a hundred former pupils of a divorce case), that does not constitute proof of genuinely Christian standards of behaviour and morality. What is certain is that the Church of England was a vital element in society, buttressing society's stability by its own certainties. A clergyman's private life must be above suspicion and his conduct of divine service must be free from innovations, avoiding both the 'bells and smells' of the Roman Catholics and the over-enthusiastic preaching and hymn singing of the nonconformist churches.

And so we approach the year 1914 and the coming of a war that was to shake all those certainties to their very foundations and indeed to shatter many of them. All but a handful of the Old Boys listed above were still alive and most of them, in one way or another, played an active role in that terrible drama. To the gallantry and self-sacrifice of so many of them, long lists of decorations and – even more – the Roll of Honour, pay tribute. To the loss of so many brave lives and the shattering of so many certainties Britain's subsequent decline has been attributed by numerous writers. The theme is too vast to be more than touched on here. But certain ideas may, very tentatively, be put forward.

The ethos of schools like Temple Grove, in which funking, lying, cheating and bullying are scorned and in which 'taking one's punishment like a man', involvement in risky escapades and stoical acceptance of hardship are part of the approved standard of behaviour, breeds the fearless infantry officer; the officer who leads his men against the enemy trenches into a storm of fire and, when the fighting is over, whose concern is first and foremost for his men's welfare. Does the method of teaching, and the content of what is taught, also breed rigidity of mind, suspicion of new ideas and the lack both of inventiveness and empathy that characterised the tactical and strategic outlook of so many senior officers on the Western Front during those prolonged battles? Even more fundamentally was the ignorance of – and, perhaps, contempt for – painting and sculpture, music and drama, literature and true scholarship, which the narrowness of much Victorian education tended to produce, partly responsible for the war itself? Too rigid an adherence to codes of conduct, too great an unwillingness to face the problems of the world from the viewpoint of other classes, creeds, cultures and colours were very far from being purely British failings. But perhaps Britain, with its immense Empire, with its commercial supremacy and with a form of government then superior to all others, had a greater responsibility to preserve peace than any other country. No amount of special pleading or specious argument can disguise the fact that Germany, and Germany alone, was responsible for the outbreak of the First World war, nor that – in the end – Britain had any choice but to become involved. But the future of every country lies in the hands of its ablest sons and daughters. Their education, or lack or limitation of it, is crucial to its welfare and its success. The education of Britain's sons in Victorian England had many strengths – and virtues. Did it also have weaknesses and limitations for which the country was to pay an incalculable price?

Appendix II

Early in 1870 the Royal Society for the encouragement of Arts, Manufactures and Commerce published a most interesting document, some 15,000 words in length, entitled *Inquiry into the Existing State of Education in Richmond, Twickenham and Mortlake and Neighbourhood*. Its author, T. Paynter Allen (previously Master of the Russell School at Petersham), seems to have been fair-minded and was certainly most conscientious in his investigations. Of the area (some 15 sq.miles) under survey he writes, 'The district comprising these semi-rural villages is occupied in the centre by Richmond . . . Waterside industries are common [but the district] is still studded with the seats of a great number of families of social eminence'. Apart from widespread adult illiteracy, 'my examination has revealed a mass of ignorance and of juvenile preparation for vice and crime which has surprised me'.

His investigations revealed that barely half the children of school age (3-14) were in regular attendance at any school, 'one third are absolutely uninstructed' and the remainder 'alternate attendance at school with fluctuating labour, injuriously to themselves and to the school'. And of the children receiving regular schooling, half were under 5. Mr Allen goes on to give details of the children who were not attending any school. Some, he acknowledged, were in some form of 'productive labour'. The majority were to be found 'loitering about their doors, in the streets or in the fields without acknowledged occupation' and in addition there was a class of 'lawless youths . . . who respect no authority but that of the police' and he underlined 'the mischievous influence these lads exercise over the younger ones'.

Was inability to pay even the few pence weekly that schools then charged a major factor, Mr Allen wondered. But with 180 public-houses or beer houses (one to every 176 persons) and a weekly expenditure on beer per family of 4/2 ¼ d, as opposed to that on education of 2 ¼ d, he concluded the answer was 'No'. He went deeply into the question of irregular attendance and, while recognising many valid reasons, he concluded that 'mere occasional schooling operates far more prejudicially against a child's chances of success in life, than entire absence where that is due to regular occupation'. He even quotes 'a justice of the peace and a liberal patron of education' who was beginning to doubt the value of education among the poor 'as being of questionable benefit to them'.

In a most interesting section of his report, Mr Allen examined the teaching – and its effects – at the North Surrey School, 'to which the orphan and the destitute are sent', describing the 'low type of physiognomy, immobility of feature . . . and stunted growth [which] prove them to belong to the most unfortunate and degraded classes'. His investigation led him to believe that 'teaching and training power, properly applied, could produce better moral as well as intellectual results'.

His comments on the system commonly known as 'Payment by Results' are also enlightening. Out of almost 1600 children of the necessary age, only 60% could be presented to the inspectors, the remainder having 'not even a remote chance of success'. Of those who were presented rather less than half passed the tests in Reading, Writing and Arithmetic. 'It is notorious that instruction has become more mechanical and that mental training is . . . quite neglected'.

An important section of the Inquiry is devoted to the views of the teachers themselves. At the sixteen schools he visited, with 3,474 children on their rolls, 32 Masters and Mistresses, 10 Assistants, 21 Pupil Teachers and 28 Monitors and Monitresses were employed (one to every 38 children), nearly a third being children themselves. Many of the teachers were critical of the training colleges where they 'learned to acquire knowledge but not to communicate it'. Not only were the teachers poorly paid but they were severely handicapped 'through the want of apparatus and fittings [many of the children being] without books of any kind to read in'. The total expenditure of those sixteen schools amounted to £4,000 p.a. so that each child cost about 24/- a year to 'educate'. Mr Allen explained how the money was raised (17% coming from Government grant) and described the thankless task undertaken by schoolmasters of going round the great houses to ask for annual subscriptions and having to submit to 'indignities from the servants'.

Some of the blame lay with the local clergy, especially in National Schools where the Vicar was invariably chairman of the managers. Not only were they often poor teachers of Scripture themselves but a (typical) requirement that all children at a particular school should be instructed in the principles of the Established Church and 'attend divine service in the parish church every Sunday morning and afternoon' caused dissenting families to send their children to more distant schools.

One section of the report is devoted to the aspirations of 'the commercial and trading classes', the choice for them lying between technical and classical schools. The latter offered 'superior facilities to persons of social position and means . . . One of these, a private adventure school, has the repute of being surpassed by few of its class in England'. This must surely refer to Temple Grove. Mr Allen quotes a tradesman applying to one such school but, concerned that his son would have to learn 'Lat'n', pleading anxiously 'Don't give it 'im too strong'.

The Inquiry ends with recommendations for a complete reorganisation of the system of inspection, so that art, science and physical training could be taken into account and for a wholesale improvement in the standards of teacher training colleges. If such reforms were carried out, the children 'of the wage-class would receive a complete primary instruction and those among them of better attainments would enjoy the benefit of an advanced secondary education' so that gradually, 'hereditary mendicancy, pauperism and delinquency would be extirpated'.

Index

The principles behind the compilation of the index have been 1) to include the names of all persons, places and schools of significance mentioned in the text and 2) to group together references to certain themes which recur at intervals, such as bullying, the teaching of Science, School food etc., but not, for example, boarding or the teaching of the Classics, references to which occur throughout the book. Names of individual pupils at Temple Grove are omitted in those instances where they merely supply a brief reference to school activity, and titles which they may have acquired or inherited later in life are not normally given in the index. Appendix I is indexed only to enable the reader to link up a pupil mentioned in the text with his later career. Appendix II is not indexed.

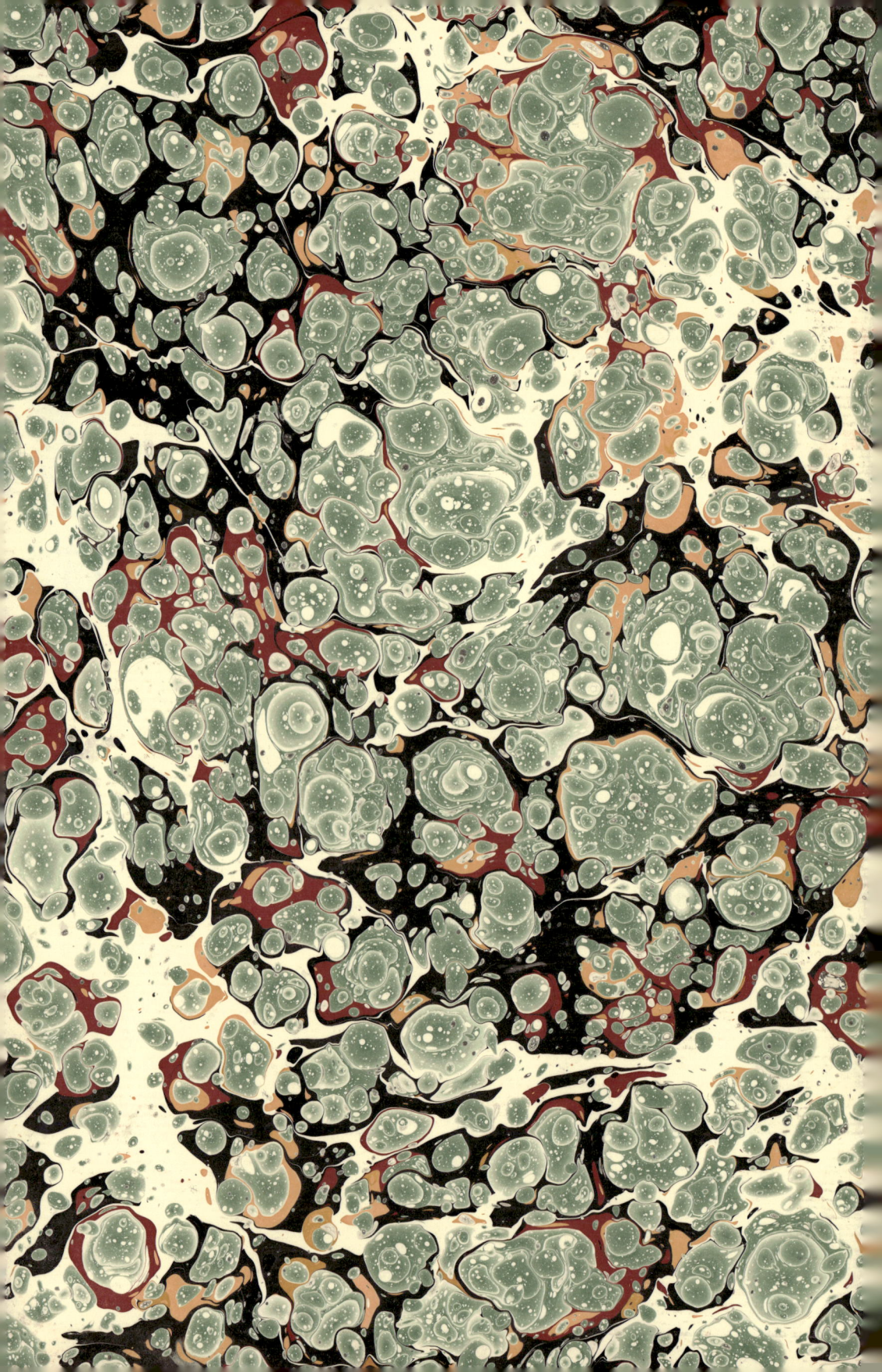